D0023190

Profiling Fabrics

Language of Fashion Series

Profiling Fabrics:

Properties, Performance & Construction Techniques

Debbie Ann Gioello
Adjunct Associate Professor: Fashion Design Department
Fashion Institute of Technology

Fairchild Publications
New York

Standard Book Number: 87005-259-4

Library of Congress Catalog Card Number: 80-68747

Printed in the United States of America

Contents

Preface

The cycle of fashion embraces the new, rediscovers the old and often combines making certain fabrics more prevalent and more in demand at different times. The advances and changes in the textile industry have resulted in the production of a variety of fabrics with different and complex characteristics and properties.

The purpose of *Profiling Fabrics: Properties, Performance & Construction Techniques* is to help clarify these differences and complexities and identify existing and newly emerging fabrics.

This book is third in a series entitled *Language of Fashion.* It is intended to be used by the fashion designer, textile designer, stylist, production room technician, textile manufacturer and converter; in the textile showroom by management and sales personnel; by the retail merchandiser of textiles and apparel, educator, student, as well as the consumer and layperson.

Profiling Fabrics: Properties, Performances & Construction Techniques is necessary for all who use, sell or buy fabrics.

At one time, fabrics were made of a single natural or man-made fiber and often presented practical or aesthetic limitations. Recognizing and understanding their limitations necessitated the need for change and/or the acceptance of compromise. Today's fabrics are not only made of a single natural or man-made fiber, but may also utilize combinations of two or more fibers, all natural, all man-made, variations or combinations of both. Regardless of the fiber content, generically named fabrics may be simulated. The characteristics of the generic fabric are maintained and allow for recognition and identification of the fabric or fabric-type. Through usage in the trade, the terms cotton, linen, silk and wool have become associated with particular fabric characteristics without regard to fiber content.

This text is presented as a systematic and comprehensive study of fabrics, their technology, properties, performance expectations, care factors and construction techniques. Fabrics included in this book are grouped according to:

1. Generic or classical name with regard to the original natural fiber classification;
2. Structure.

Due to the inavailability of fabrics and lack of space, some judgment regarding the selection of fabrics to be included was required.

Similarities of generically classified fabrics include:

- Yarn type and construction
- Structure
- Appearance
- Texture and surface interest
- Hand and feel
- Drapability qualities
- Weight
- Opacity
- Some finishes

Differences of generically classified fabrics include:

- Fiber content
- Performance expectations
- Some finishes
- Hand
- Construction methods of garments
- End use
- Care
- Durability or wear factors

Properties and characteristics of fabrics depend on:

- Inherent fiber properties
- Interrelationship of fiber and yarn
- Size, shape and construction of yarn

- Arrangement of yarn and thread count
- Type and method of fabric structure
- Texture and surface interest
- Type and method of finishes
- Type and method of color or surface design application

Properties and characteristics of fabrics relate to:

- Performance expectations and comfort
- Hand
- Weight
- Texture or appearance
- Drapability qualities
- Care

Profiling Fabrics: Properties, Performance & Construction Techniques is visually presented so that the reader can see how the drape or drapability qualities of the fabric interrelates to the fabric's properties, weight, hand and texture. Knowledge and understanding of the interrelated factors that influence fabric performance will serve as a guide for the selection of the proper fabrics for a particular end use.

Accuracy has been of great importance, but new works are rarely free of errors. The author hopes that the reader will call attention to errors of commission or omission.

Other books in *Language of Fashion* series include *Fashion Production Terms* and *Figure Types and Size Ranges*. Books to follow will cover:

Fabrics including definitions and photographs of all the fabric specifications, performance expectations, drapability qualities listed in this book
Silhouettes including parts of the garment and design details
Trimmings including embroidery, decorative stitches and buttons
Fitting including fitting problems, figure types and styles for different figures
Machinery including attachments, parts and accessories

1981

Debbie Ann Gioello
New York

Acknowledgments

A pictorial encyclopedia as inclusive as *Profiling Fabrics: Properties, Performance & Construction Techniques* required the assistance of many people. Without their continued cooperation, generosity and encouragement, the monumental task of compiling all the fabrics to be photographed and information would have been impossible.

The author wishes to acknowledge the following people and fabric companies for information, aid and their generous contribution of fabrics used throughout this book: Milton Adelman, Kabat Textile Corp.; Jay T. Badgley, Burlington Men's Wear; Harry Banks, Hamilton Adams Imports Ltd.; Herb Berman, Ameritex Division, United Merchants and Manufacturers, Inc.; Michael Block, Earl Glo/Erlanger Blumgart and Co. Inc.; Helen Boyd and Laurinda Finelli, Crompton-Richmond Inc.; Bill Brandt, American Silk Mills Corp.; Jennifer Butler and Peggy Wing, Springs Mills Inc.; Fred Burkhardt, Milan Textile Machines Inc.; Bob Calabro and Joan Carron, J.P. Stevens & Co. Inc.; Bill Caldwell and Mark Jackson, Cinderella Knit Mills Division of Reeves Brothers Inc.; Teresa Cox, Applause Fabrics Ltd.; Herb Cron, Printsiples and Company B; R. Dash, Lafitte Inc.; Vince DiGeorge, Stēvcoknit, Inc.; James Donahue, Klopman Mills; Arnold Entine and Joan Gehrlein, Loomskill/Gallery Screen Print; Murray Fishman, Cameo Fabrics Inc.; Julie Frohman and Miss Lumming, Jack Larsen Incorporated; Gregory Gehring, Gehring Textiles; Nancy Goldsmith and Susan Laskey, Russell Corporation; Teresa Gramuglia, Successful Creations/Shartex Inc.; Murray Grobstein and Ellen Halpern, Bloomsburg Mills; James Hahn, Drexler Associates Inc.; Jack Halpern, M. Lowenstein & Sons, Inc.; Larry Heller, Carleton Woolen Mills Inc.; Carol Hillman, Burlington Industries Inc.; Mark Jackson, Loom Tex Corp.; Mr. Jackson, Moiré Corporation; Art Kelman, Marcus Brothers Textile Inc.; Mrs Keogh, J.B. Martin Co. Inc.; Nathan Kleiderman, Associated Lace Corp.; Mr. Kohn, Ambassador Lace and Embroidery Co. Inc.; Stefan Koref, Kortex Associates Inc.; Gill Kotch, Pervel; Barbara Langeland, Milton Sherlip Inc.; Mr. Lawrence, Lawrence Textile Co. Inc.; Irving Lederich, Truemark Discount Fabrics Inc.; Archie Leppo, North American Lace Co. Inc.; Bob Levin, Gloversville Mills Inc.; Sam Levin and Steward Epstein, Twintella Fabrics; Bernard Liebes, Lace of France; Janet Liverance and Jan Landau, Embroidery Council of America; Ed Newman, Dan River Inc.; Jim Noon, Whelan Lace; Edward Orlando, Deluxe Velvet Co.; Richard Poses, Alba-Waldensian Knits; Al Ranaudo, Sequin International Corp.; Barry M. Richards, Walden Textiles/Barry M. Richards Inc.; Sheldon Ritter, Universal Knitting Machines Corp.; Chris Sadowski and Gloria Veeder, Cone Mills Corporation; Mr. Seigel, Anglo Fabrics Co. Inc.; Mr. Shapiro, Hargro Fabrics Inc.; Boris Schlomm and George Bass, Amical Fabrics; Mario Sherwood, Native Textiles; Marjory Shor, Continental Felt Division, Daniel H. Price Inc.; Mr. Schwartz, Novick & Co. Inc.; Mr. Sormani and Valerie McLean, Sormani Co. Inc./Taroni; David T. Starky, Brunswick Associates Inc.; Jerry Suffrin and Barbara Winfield, Furtex; Everett T. Turk, Auburn Fabrics Inc.; Iva Udis, Liberty Fabrics of New York, Inc.; Joan Weinstein and Sam Schreiber, Concord Fabrics Inc.; Fred Werber, Private Collections Fabrics Ltd.; Christopher Zaharian and Ed Lynch, Reeves Brothers Inc.; Arthur Zeiler Woolens Inc.

The following people and companies supplied technical information regarding fiber and yarn specifications which related to specific fabrics: James Adshead Jr., E.I. du Pont de Nemours & Company; James P. Allen, Celanese Corporation; Walter J. Bartlett, American Cyanamid; Walter C. Caudle, Allied Chemical Corporation; Libby Clark, Cotton Inc.; Donald R. Clark and Harold W. Young, American Enka Co.; Wanda H. Coffen, Dow Badische Company; R.D. Colman, Eastman Chemicals Products Inc.; Elizabeth Dagget, Avtex Fibers Inc.; James Donovan, American Textile Manufacturers Institute (ATMI); F.C. Flint, Monsanto Textile Co.; Dee Graper, Hoechst Fibers Industries; Thomas C. Haas, The Wool Bureau, Inc.; Hercules Inc.; International Silk Association; Francesca Joelion, Mohair Council of America; Frank McNeirney, International Nonwoven & Disposables Association (INDA); Linda Muller, U.S. Department of Agriculture; National Knitted Outerwear Association; Julie Rlymes, Man-made Fiber Producers Association, Inc.; Lynn Sanders, Belgian Linen Association; Maria Sciandra, Courtaulds North America Inc.

The following designers from garment manufacturing firms have supplied fabrics and reference sources: Jonan Forman, Jonan Enterprises; Richard Gold, Bill Ditport Inc.; Sherri Mannuzzi, College Town; Angel Martinez, free lance designer.

Special thanks and appreciation are extended to the following members of the faculty and staff of the Fashion Institute of Technology for their invaluable assistance: Lita Konde, Hedda Gold, Marge Citkovick and Gail Strauss, Fashion Design Department; Clara Branch, Lotis Mallard and Charles Turner, Fabric Room Fashion Design Department; Susan Pavlos, Fashion Design Department Fabric Resource Room; Arthur Price (chairman) and George Tay, Textile Science Department; Robert Riley (director), Laura Sinderbrand, Don Petrillo and staff, Dorothy Tricarico (Curator of Textiles) and Lynn Felsher (Docent), Design Laboratory Textile Department; Marjorie Miller and Lorrain Weber, Library Media Service.

My personal gratitude to my husband Dr. David T. Novick, my daughter Donna Gioello, my friends Herman Novick and Dorothy Burricelli and my editor at Fairchild Publications, Olga Kontzias.

How to Evaluate Photographs

The photographs in *Profiling Fabrics: Properties, Performance & Construction Techniques* illustrate:

1. A close-up view of the fabric;
2. The fabric on the straight grain with gathered waistline draped on the model form;
3. The fabric on the bias grain draped on the model form.

The width of the fabrics range from 36 to 70 inches (91.4 to 177.8 cm). The width is given for each fabric.

Fabrics with border designs, finished edges, horizontal and one way motifs may be and have been gathered on the crosswise grain. The amount of fabric take up and fullness handled for the crosswise drape is 45 inches (114.3 cm).

Amount of fullness and quality of fullness may be judged by how the fabric gathers with regard to width. From this comparison one may determine if more or less fullness is desired or required and plan accordingly when selecting fabric for a design.

Bias draping was done on the 45° angle of the fabric. Bias draping of less than 45° may be desired. Evaluating the size, amount and position of flares, cones and ripples on a bias drape will assist the designer in the selection of a fabric for a particular design. Some fabrics cannot be used on the bias without consideration to the edge finishes, weaves, patterns, nap or light-reflecting qualities.

Stretch fabric has been pulled to eliminate fullness and molded to the contours of the body to show elongation and potential of the fabric. Fabric may be stretched more or less with regard to a specific garment design, type and end use of fabric.

In some cases, only a close-up view of the fabric is given due to the size of the sample available. When the information for these fabrics is the same as the preceding term, text is not included.

1 ~ Cotton/Cotton-type Fabrics

Airplane Cloth / Byrd Cloth®
Bark Cloth
  Bark Cotton
Batiste
  Cotton Batiste
  Cotton Batiste with Cord
Beach Cloth
Bedford Cord
  Baby / Junior Cotton Bedford Cord
  Sueded / Brushed Cotton Bedford Cord
Bengaline
  Cotton Bengaline
Broadcloth
  Cotton Broadcloth
Brocade
  Cotton Brocade (medium weight)
  Cotton Brocade (heavyweight)
Calcutta / Bangladesh Cloth
Cambric
  Cotton Cambric
Canvas
  Cotton Canvas
Challis
  Cotton Challis
Chambray
  Chambray
  Corded Chambray
Chameleon / Iridescent Cotton
Chino / Chino Cloth
Chintz / Glazed Cotton
Clokay / Embossed Cotton / French Damask
Covert
  Cotton Covert
Crash
  Cotton Crash
Crepe
  Cotton Crepe
Creponne
  Cotton Creponne
Cretonne
Damask
  Cotton Damask
Denim
  Denim
  Brushed-faced / Brushed Denim
Dimity
  Dimity
  Dimity with Cord / Satin-striped Dimity
Dotted Swiss
  Dotted Swiss
  Dotted Swiss-type / Flocked Swiss
Drill Cloth
Duck
End and End Cloth
  End and End Cloth (lightweight "topweight")
  End and End Cloth (medium weight "bottomweight")
Eyelet
Flannel / Flannelette
  Cotton Flannel/ Flannelette
  Outing / Dommet Flannel
Gabardine
  Cotton Gabardine
Gauze Cloth
  Gauze Cloth/Bunting/Cheesecloth
  Gauze Cloth/Cotton Net
Gingham
Homespun
  Cotton Homespun
Hopsacking
  Cotton Hopsacking
Jean Cloth / Middy Twill
Khaki Cloth / Dyed Drill Cloth
Lawn
  Lawn
  Glazed-faced Lawn
Madras
Monk's Cloth / Druid's Cloth
Muslin / Muslin Sheeting
Organdy
Osnaburg
Oxford
  Oxford Shirting
  Oxford Stripe
Percale
  Percale
  Percale with Yarn Embroidery
Plissé
Polished Cotton
Pongee
  Cotton Pongee / Silk Noil Cotton
Poplin
  Cotton Poplin (soft finish)
  Cotton Poplin (crisp finish)
  Cotton Poplin (heavyweight)
Printcloth
  Printcloth
  Calico Printcloth
Ratiné
  Cotton Ratiné
Sailcloth
Sateen
Seersucker
  Cotton Seersucker (even stripe)
  Cotton Seersucker (novelty stripe)
  Cotton Seersucker (checkered effect)

- Sheeting / Muslin Sheeting
- Suede Cloth
  - Cotton Suede Cloth
- Tattersall
- Ticking
- Tricotine
  - Cotton Tricotine
- Voile
  - Cotton Voile
  - Crinkled Cotton Voile
  - Voile with Yarn Embroidery
- Whipcord
  - Cotton Whipcord
- Novelty Cottons / Cotton Novelty Weaves
  - Bouclé Yarn Cotton
  - Dobby Weave Cotton
  - Double Weave Cotton / Double Cloth
  - Jacquard Weave
    - Jacquard Weave Cotton
  - Leno Weave
    - Leno Weave Cotton (lightweight)
    - Leno Weave Cotton (heavyweight)
  - Skip Denting
    - Skip Denting—Open Effect Cotton
    - Skip Denting—Corded Effect Cotton
  - Spot Weave
    - Spot / Unclipped Spot / Unclipped Dot Cotton (lightweight)
    - Spot / Unclipped Spot / Unclipped Dot Cotton (heavyweight)
    - Spot / Clip Spot / Clip Dot / "Eyelash" Cotton (close-up only)
    - Spot / Clip Spot / Clip Dot Cotton
  - Twill Weave
    - Reverse Twill Weave Cotton
  - White-on-White Cotton
    - White-on-White Cotton Variation (close up only)
- Piqué
  - Bull's Eye Piqué
  - Corded Piqué / Piqué Cord
  - Embossed Piqué
  - Jacquard Piqué
  - Waffle Piqué
  - Piqué with Yarn Embroidery
- Piqué Variations (close-up only)
  - Bird's Eye Piqué
  - Crow's Foot Piqué
  - Diamond Piqué
  - Ladder Piqué
  - Square Piqué

Cotton and cotton-type fabrics, as listed in this unit, refer to fabrics generically named and classified as cotton fabrics which were originally made of one-hundred percent natural cotton fibers. This unit includes cotton and cotton-type fabrics which may be made of:

- Natural cotton fibers
- Man-made fibers simulating cotton
- Man-made fiber blends
- Natural and man-made fiber blends
- Mixed or combined yarns of different fiber origins

Regardless of the fiber content, cotton and cotton-type fabrics maintain or simulate the same outward characteristics of the original generically named fabric. The following characteristics remain the same:

- Appearance
- Surface texture and interest
- Hand or feel
- Yarn type and construction
- Fabric structure
- Drapability qualities
- Weight (usually)

Although the original outward characteristics of the fabric are maintained, the fiber content used *changes* the performance, thread selection, pressing and care factors of the fabric. Each fiber has its own particular characteristics and properties.

There are variations within the basic generically named fabric. Variations may be obtained by changing, altering, modifying or combining:

- Yarn type
- Yarn construction

- Yarn count and size
- Yarns in warp and filling
- Usual weave structure
- Finishing processes

By combining or varying the components the textile designer can create new fabrics. When one component is changed, the fabric is changed. The newly designed or produced fabrics receive a new trade name. The new fabrics or newly named fabrics presented each season are usually placed into categories which may refer to their hand or feel, texture, surface appearance, structure or weight. Different textile companies have their own trade names or trademarks for the fabrics they produce.

Airplane Cloth/Byrd Cloth®

Fiber Content 80% polyester/20% cotton
Yarn Type combed staple; filament fiber
Yarn Construction conventional; fine; 2-ply; hard twist
Fabric Structure closely woven and balanced plain weave
Finishing Processes treated with dope; mercerizing; water repellent
Color Application piece dyed-union dyed
Width 44-45 inches (111.8-114.3 cm)
Weight medium-light
Hand compact; firm; rigid
Texture boardy; papery
Performance Expectations subject to edge abrasion; dimensional stability; durable; heat-set properties; moth resistant; nonpilling; nonsnagging; resilient; high tensile strength; water and stain resistant
Drapability Qualities falls into stiff wide flares; maintains crisp folds, creases, pleats; retains shape or silhouette of garment

Recommended Construction Aids & Techniques

Hand Needles betweens, 3-5; cotton darners, 5-10; embroidery, 5-8; milliners, 6-8; sharps, 6-8
Machine Needles round/set point, medium-fine 9-11
Threads poly-core waxed, 50 all purpose; cotton-polyester blend, all purpose; cotton-covered polyester core, heavy duty; spun polyester, all purpose
Hems book; double-fold or single-fold bias binding; bound/Hong Kong finish; machine-stitched; overedged; seam binding
Seams flat-felled; lapped; plain; safety-stitched; welt; double welt
Seam Finishes book; single-ply bound; Hong Kong bound; edge-stitched; single-ply overcast; pinked; pinked and stitched; serging/single-ply overedged; untreated/plain; single-ply zigzag
Pressing steam; safe temperature limit 325°F (164.1°C)
Care dry clean
Fabric Resource **Reeves Brothers Inc.**

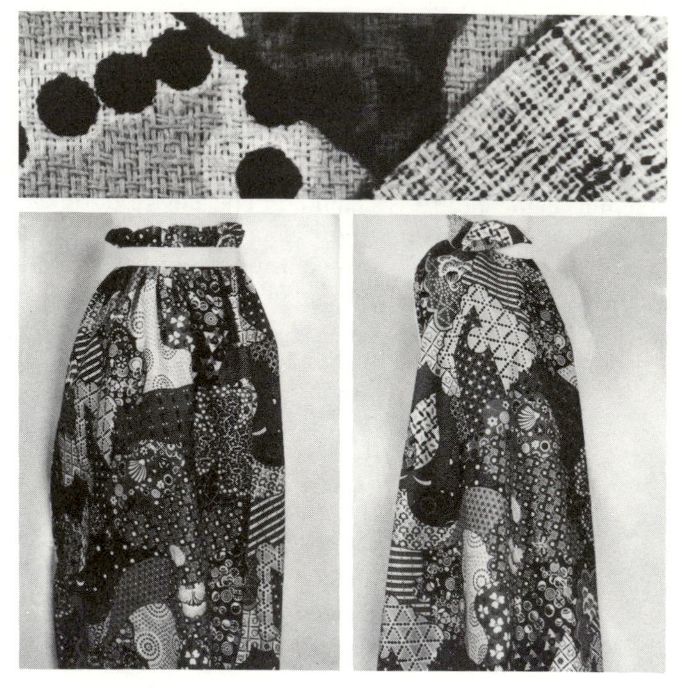

Bark Cotton

Fiber Content 100% cotton
Yarn Type carded staple
Yarn Construction conventional; loose twist
Fabric Structure plain weave variation
Finishing Processes beetling; stabilizing; water and stain repellent
Color Application piece dyed; roller printing
Width 54 inches (137.2 cm)
Weight heavy
Hand coarse; stiff
Texture rough surface; woven to simulate tree bark
Performance Expectations absorbent; antistatic; durable; flexible; resilient; sun resistant; subject to abrasion and mildew; high tensile strength; water and stain resistant
Drapability Qualities falls into stiff wide cones; fullness maintains bouffant effect; better utilized if fitted by seaming and eliminating excess fabric

Recommended Construction Aids & Techniques

Hand Needles betweens, 3-5; chenilles, 18-22; embroidery, 1-7; milliners, 3-7; sharps, 1-5
Machine Needles round/set point, medium 14
Threads mercerized cotton, 40; six-cord cotton, 40; cotton- or poly-core waxed, 40; cotton-covered polyester core, heavy duty
Hems double-fold or single-fold bias binding; bound/ Hong Kong finish; stitched and pinked flat; overedged; seam binding
Seams flat-felled; lapped; plain; safety-stitched; welt; double welt
Seam Finishes single-ply bound; Hong Kong bound; single-ply overcast; pinked; pinked and stitched; serging/single-ply overedged; untreated/plain; single-ply zigzag
Pressing steam; safe temperature limit 400°F (206.1°C)
Care dry clean
Fabric Resource **M. Lowenstein & Sons, Inc.**

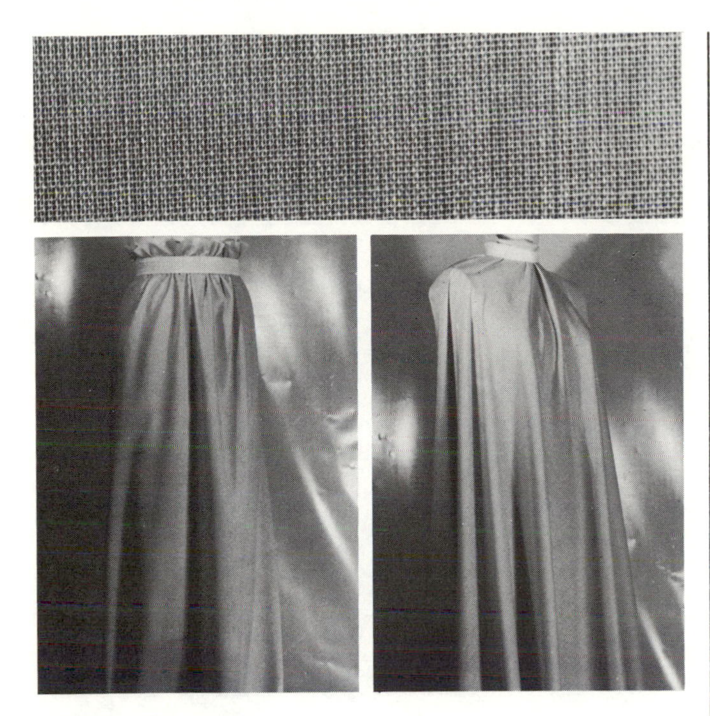

Cotton Batiste

Fiber Content 65% polyester/35% cotton
Yarn Type combed staple; filament staple
Yarn Construction conventional; fine; tight twist
Fabric Structure plain weave
Finishing Processes bleaching; calendering; mercerizing
Color Application piece dyed–union dyed
Width 44–45 inches (111.8–114.3 cm)
Weight[1] light
Hand[1] delicate; limp; soft
Texture[1] fine, smooth to coarse with nub depending on yarn; transparent
Performance Expectations abrasion resistant; absorbent; air permeable; dimensional stability; resilient; wash-and-wear properties
Drapability Qualities falls into soft flares; accommodates fullness by gathering, elasticized shirring, smocking; fullness retains soft graceful fall

Recommended Construction Aids & Techniques

Hand Needles betweens, 6–7; cotton darners, 9–10; embroidery, 8–10; milliners, 8–10; sharps, 11–12
Machine Needles round/set point, fine 9
Threads cotton- or poly-core waxed, 60 fine; cotton-polyester blend, all purpose; cotton-covered polyester core, all purpose; spun polyester, all purpose
Hems double fold; edge-stitched; double edge-stitched; machine-stitched; hand- or machine-rolled; wired
Seams flat-felled; French; mock French; hairline; plain; safety-stitched; tissue-stitched
Seam Finishes self bound; double-ply bound; double-stitched and overcast; double-stitched and trimmed; single-ply overcast; serging/single-ply overedged; untreated/plain
Pressing steam; safe temperature limit 325°F (164.1°C)
Care launder
Fabric Resource **Hargro Fabrics Inc.**

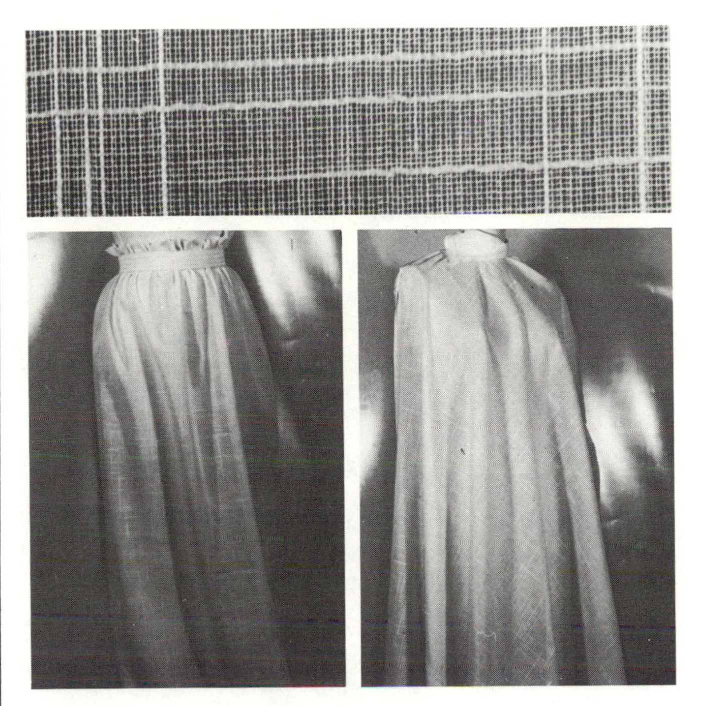

Cotton Batiste with Cord

Fiber Content 65% polyester/35% cotton
Yarn Type combed staple; filament staple
Yarn Construction conventional; fine; tight twist
Fabric Structure plain weave
Finishing Processes bleaching; calendering; mercerizing
Color Application piece dyed–union dyed
Width 44–45 inches (111.8–114.3 cm)
Weight[1] light
Hand[1] delicate; limp; soft
Texture[1] fine, smooth to coarse with nub depending on yarn; transparent
Performance Expectations abrasion resistant; absorbent; air permeable; dimensional stability; resilient; wash-and-wear properties
Drapability Qualities falls into soft flares; accommodates fullness by gathering, elasticized shirring, smocking; fullness retains soft graceful fall

Recommended Construction Aids & Techniques

Hand Needles betweens, 6–7; cotton darners, 9–10; embroidery, 8–10; milliners, 8–10; sharps, 11–12
Machine Needles round/set point, fine 9
Threads cotton- or poly-core waxed, 60 fine; cotton-polyester blend, all purpose; cotton-covered polyester core, all purpose; spun polyester, all purpose
Hems double fold; edge-stitched; double edge-stitched; machine-stitched; hand- or machine-rolled; wired
Seams flat-felled; French; mock French; hairline; plain; safety-stitched; tissue-stitched
Seam Finishes self bound; double-ply bound; double-stitched and overcast; double-stitched and trimmed; single-ply overcast; serging/single-ply overedged; untreated/plain
Pressing steam; safe temperature limit 325°F (164.1°C)
Care launder
Fabric Resources **Printsiples and Company B; Milton Sherlip Inc.**

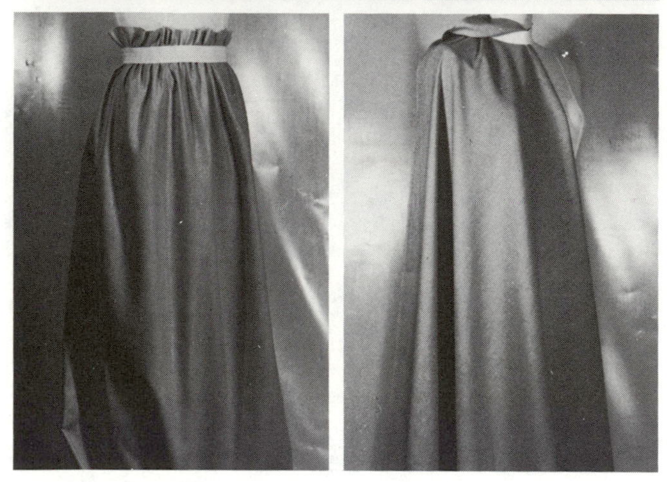

Beach Cloth

Fiber Content 50% cotton/50% polyester
Yarn Type carded staple; filament staple
Yarn Construction conventional; thick
Fabric Structure plain weave; low thread count
Finishing Processes beetling; mercerizing; softening
Color Application piece dyed-union dyed
Width 44-45 inches (111.8-114.3 cm)
Weight medium-heavy
Hand linenlike; thick
Texture coarse; grainy; opaque
Performance Expectations high covering power; durable; moth resistant; subject to seam slippage
Drapability Qualities falls into firm wide flares; fullness maintains lofty effect; retains shape of garment

Recommended Construction Aids & Techniques

Hand Needles betweens, 3-5; cotton darners, 1-7; embroidery, 1-7; milliners, 3-7; sharps, 1-5
Machine Needles round/set point, medium-coarse 14-16
Threads mercerized cotton, 50; six-cord cotton, 50; cotton- or poly-core waxed, all purpose; cotton-covered polyester core, all purpose
Hems double-fold or single-fold bias binding; bound/ Hong Kong finish; machine-stitched; overedged; seam binding
Seams flat-felled; lapped; plain; safety-stitched; welt; double welt
Seam Finishes single-ply bound; Hong Kong bound; single-ply overcast; pinked; pinked and stitched; serging/single-ply overedged; untreated/plain; single-ply zigzag
Pressing steam; safe temperature limit 325°F (164.1°C)
Care launder; dry clean
Fabric Resource Loom Tex Corp.

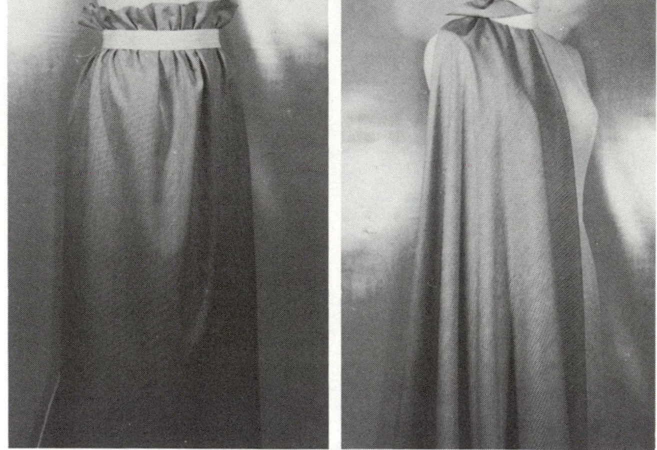

Baby/Junior Cotton Bedford Cord

Fiber Content 65% polyester/35% cotton
Yarn Type combed staple; filament staple
Yarn Construction textured; tightly twisted fine yarns and heavier lightly twisted yarns
Fabric Structure rib weave; raised lengthwise rib
Finishing Processes mercerizing; shrinkage control
Color Application piece dyed-cross dyed; yarn dyed
Width 44-45 inches (111.8-114.3 cm)
Weight medium-light
Hand compact; crisp; nonelastic; firm
Texture raised lengthwise rib
Performance Expectations abrasion resistant; dimensional stability; elongation; heat-set properties; nonpilling; high tensile strength; wrinkle resistant
Drapability Qualities falls into firm stiff flares; accommodates fullness by pleating, gathering; retains shape of garment

Recommended Construction Aids & Techniques

Hand Needles betweens, 5-7; cotton darners, 8-10; embroidery, 7-10; milliners, 8-10; sharps, 9-10
Machine Needles round/set point, medium-fine 9-11
Threads cotton- or poly-core waxed, 60 fine; cotton-polyester blend, all purpose; cotton-covered polyester core, all purpose; spun polyester, all purpose
Hems book; bound/Hong Kong finish; edge-stitched; machine-stitched; overedged; seam binding
Seams flat-felled; lapped; plain; safety-stitched; welt; double welt
Seam Finishes book; single-ply bound; Hong Kong bound; single-ply overcast; pinked; pinked and stitched; serging/single-ply overedged; untreated/plain; single-ply zigzag
Pressing steam; safe temperature limit 325°F (164.1°C)
Care launder
Fabric Resources Dan River Inc.; Russell Corporation

Sueded/Brushed Cotton Bedford Cord

Fiber Content 65% polyester/35% cotton
Yarn Type combed staple; filament fiber
Yarn Construction bulked; conventional; 2-ply
Fabric Structure rib weave; raised lengthwise rib
Finishing Processes brushed or sueded face; stabilizing
Color Application piece dyed–union dyed; yarn dyed
Width 47–48 inches (119.4–121.9 cm)
Weight medium-heavy
Hand heavy; soft; solid; thick
Texture napped face; raised lengthwise rib; opaque
Performance Expectations dimensional stability; durable; elongation; heat-set properties; subject to abrasion and yarn slippage; wrinkle recovery
Drapability Qualities falls into wide cones; accommodates fullness by pleating, gathering; fullness maintains lofty effect

Recommended Construction Aids & Techniques

Hand Needles betweens, 3–5; cotton darners, 5–10; embroidery, 5–8; milliners, 6–8; sharps, 6–8
Machine Needles round/set point, medium-coarse 14–16
Threads cotton- or poly-core waxed, 50 all purpose; cotton-polyester blend, all purpose; cotton-covered polyester core, heavy duty; spun polyester, all purpose
Hems double-fold or single-fold bias binding; bound/ Hong Kong finish; machine-stitched; overedged; seam binding
Seams flat-felled; lapped; plain; safety-stitched; welt; double welt
Seam Finishes single-ply bound; Hong Kong bound; single-ply overcast; pinked; pinked and stitched; serging/single-ply overedged; untreated/plain; single-ply zigzag
Pressing steam; safe temperature limit 325°F (164.1°C)
Care launder; dry clean
Fabric Resources Concord Fabrics Inc.; Reeves Brothers Inc.

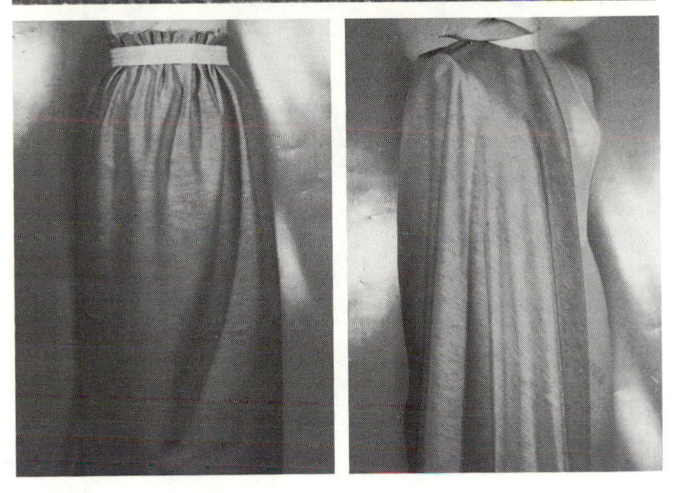

Cotton Bengaline

Fiber Content cotton filling; polyester warp
Yarn Type combed staple; filament fiber
Yarn Construction conventional; coarse filling yarn; fine warp yarn
Fabric Structure firmly woven plain weave; thick filling yarns form rib in crosswise direction
Finishing Processes beetling; calendering; mercerizing
Color Application piece dyed–union dyed; yarn dyed
Width 45 inches (114.3 cm)
Weight medium-heavy
Hand crisp; firm; rigid
Texture raised crosswise rib
Performance Expectations dimensional stability; subject to flex abrasion and yarn slippage
Drapability Qualities falls into firm wide cones; fullness maintains crisp effect; retains shape of garment

Recommended Construction Aids & Techniques

Hand Needles betweens, 3–5; cotton darners, 5–10; embroidery, 5–8; milliners, 6–8; sharps, 6–8
Machine Needles round/set point, medium-coarse 14–16
Threads mercerized cotton, 50; six-cord cotton, 50; cotton- or poly-core waxed, fine; cotton-covered polyester core, all purpose
Hems double-fold or single-fold bias binding; bound/ Hong Kong finish; machine-stitched; overedged; seam binding
Seams flat-felled; lapped; plain; safety-stitched; welt; double welt
Seam Finishes single-ply bound; Hong Kong bound; single-ply overcast; pinked; pinked and stitched; serging/single-ply overedged; untreated/plain; single-ply zigzag
Pressing steam; safe temperature limit 325°F (164.1°C)
Care launder; dry clean
Fabric Resource Stevcoknits, Inc.

Cotton Broadcloth

Fiber Content cotton/polyester blend
Yarn Type combed staple; filament fiber
Yarn Construction conventional; fine; high twist
Fabric Structure tightly woven plain weave; high thread count
Finishing Processes calendering; mercerizing; Sanforized®
Color Application piece dyed-union dyed
Width 44-45 inches (111.8-114.3 cm)
Weight light
Hand firm; soft
Texture flat; soft luster; smooth
Performance Expectations absorbent; abrasion resistant; dimensional stability; nonpilling; nonsnagging; high tensile strength
Drapability Qualities falls into soft flares; accommodates fullness by gathering, elasticized shirring, smocking; fullness retains soft fall

Recommended Construction Aids & Techniques

Hand Needles betweens, 5-7; cotton darners, 8-10; embroidery, 7-10; milliners, 8-10; sharps, 9-10
Machine Needles round/set point, medium 9-11
Threads mercerized cotton, 50; six-cord cotton, 60-80; cotton-covered polyester core, extra fine
Hems bound/Hong Kong finish; double fold; edge-stitched; machine-stitched
Seams flat-felled; French; lapped; plain; safety-stitched; welt; double welt
Seam Finishes single-ply bound; Hong Kong bound; edge-stitched; pinked; pinked and stitched; single-ply overcast; serging/single-ply overedged; untreated/plain; single-ply zigzag
Pressing steam; safe temperature limit 325°F (164. 1°C)
Care launder
Fabric Resources **Concord Fabrics Inc.; Marcus Brothers Textile Inc.**

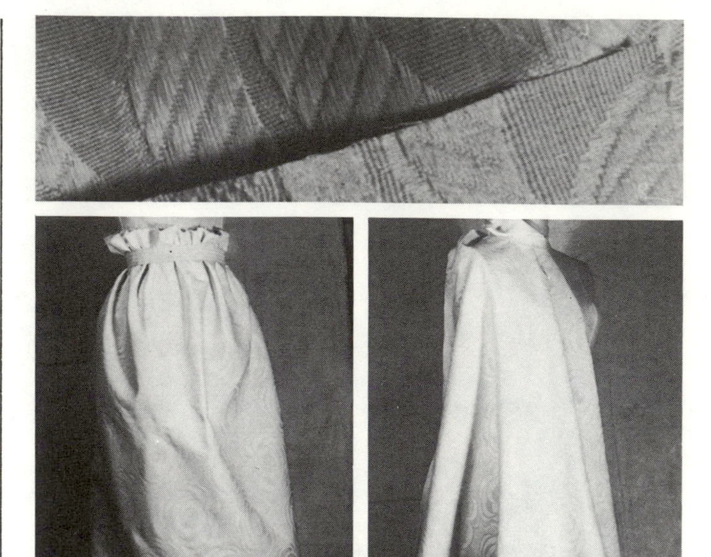

Cotton Brocade (medium weight)

Fiber Content 65% Dacron® polyester/35% cotton
Yarn Type combed staple; filament fiber
Yarn Construction conventional; textured
Fabric Structure Jacquard weave
Finishing Processes beetling; bleaching; mercerizing; shrinkage control
Color Application yarn dyed
Width 45 inches (114.3 cm)
Weight medium-light to medium
Hand crisp; firm
Texture opaque; raised pattern (high and low areas); sculptured appearance
Performance Expectations dimensional stability; durable; elongation; subject to snagging due to weave; wash-and-wear properties
Drapability Qualities falls into wide cones; fullness maintains crisp effect; retains shape of garment

Recommended Construction Aids & Techniques

Hand Needles betweens, 3-5; cotton darners, 5-10; embroidery, 5-8; milliners, 6-8; sharps, 6-8
Machine Needles round/set point, medium 14
Threads mercerized cotton, 50; six-cord cotton, 50; cotton- or poly-core waxed, all purpose; cotton-covered polyester core, all purpose
Hems single-fold or double-fold bias binding; bound/Hong Kong finish; machine-stitched; overedged; seam binding
Seams flat-felled; lapped; plain; safety-stitched; welt; double welt
Seam Finishes single-ply bound; Hong Kong bound; single-ply overcast; pinked; pinked and stitched; serging/single-ply overedged; untreated/plain; single-ply zigzag
Pressing steam; safe temperature limit 325°F (164.1°C)
Care launder; dry clean
Fabric Resource **White Rose Fabrics (sample courtesy of College Town)**

Cotton Brocade (heavyweight)

Fiber Content 100% cotton
Yarn Type combed staple
Yarn Construction conventional; thick-and-thin
Fabric Structure Jacquard weave
Finishing Processes mercerizing; stabilizing; water and stain repellent
Color Application yarn dyed
Width 54–58 inches (137.2–147.3 cm)
Weight heavy
Hand hard; rough; springy; thick
Texture opaque; raised pattern (high and low areas); sculptured appearance
Performance Expectations absorbent; antistatic; elongation; flexible; moth resistant; subject to shrinkage, snagging, mildew
Drapability Qualities falls into wide cones; retains shape or silhouette of garment; better utilized if fitted by seaming and eliminating excess fabric

Recommended Construction Aids & Techniques

Hand Needles betweens, 3–5; cotton darners, 1–7; embroidery, 1–7; milliners, 3–7; sharps, 1–5
Machine Needles round/set point, medium-coarse, 14–16
Threads mercerized cotton, 40; six-cord cotton; 40; cotton- or poly-core waxed, heavy-duty; cotton-covered polyester core, all purpose
Hems double-fold or single-fold bias binding; bound/Hong Kong finish; faced; stitched and pinked flat; overedged; seam binding
Seams plain; safety-stitched; welt; double welt
Seam Finishes single-ply bound; Hong Kong bound; single-ply overcast; pinked; pinked and stitched; serging/single-ply overedged; untreated/plain; single-ply zigzag
Pressing steam; safe temperature limit 400°F (206.1°C)
Care dry clean
Fabric Resource **Demarco Fabrics (sample courtesy of Fabric Room, FIT)**

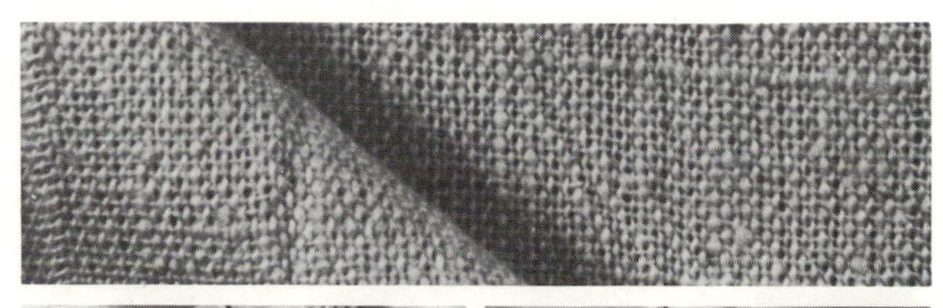

Calcutta/Bangladesh Cloth

Fiber Content 100% cotton
Yarn Type carded staple
Yarn Construction conventional; irregular; thick; low twist
Fabric Structure plain weave
Finishing Processes mercerizing; slack mercerization (stretching); crinkled process
Color Application piece dyed
Width 44–45 inches (111.8–114.3 cm)
Weight heavy
Hand dry; hard; loose
Texture coarse; rough; crinkled face and back
Performance Expectations absorbent; air permeable; antistatic; high covering power; durable; subject to mildew and snagging; moth resistant; nonpilling; high tensile strength
Drapability Qualities falls into firm flares; fullness maintains crisp effect; retains shape of garment

Recommended Construction Aids & Techniques

Hand Needles betweens, 3–5; cotton darners, 1–7; embroidery, 1–7; milliners, 3–7; sharps, 1–5
Machine Needles round/set point, medium 11–14
Threads mercerized cotton, 40; six-cord cotton, 40; cotton- or poly-core waxed, heavy duty; cotton-covered polyester core, heavy duty
Hems double-fold or single-fold bias binding; bound/Hong Kong finish; faced; stitched and pinked flat; overedged; seam binding
Seams plain; safety-stitched; welt; double welt
Seam Finishes single-ply bound; Hong Kong bound; single-ply overcast; pinked; pinked and stitched; serging/single-ply overedged; untreated/plain; single-ply zigzag
Pressing steam; safe temperature limit 400°F (206.1°C)
Care launder
Fabric Resource **Iselip Jefferson Co. Inc. (sample courtesy of Jonan Enterprises)**

Cotton Cambric

Fiber Content 100% cotton
Yarn Type combed staple
Yarn Construction conventional; tight twist
Fabric Structure closely woven plain weave
Finishing Processes calendering; glazing; mercerizing; softening
Color Application piece dyed; yarn dyed
Width 44–45 inches (111.8–114.3 cm)
Weight light
Hand delicate; fine; soft to crisp depending on finish
Texture compact; smooth; slightly glossed or glazed face
Performance Expectations abrasion resistant; absorbent; antistatic; subject to mildew; moth resistant; nonpilling; nonsnagging; high tensile strength
Drapability Qualities falls into soft flares; accommodates fullness by gathering, elasticized shirring, pleating; fullness retains soft graceful fall

Recommended Construction Aids & Techniques

Hand Needles betweens, 5–7; cotton darners, 8–10; embroidery, 7–10; milliners, 8–10; sharps, 9–10
Machine Needles round/set point, medium-fine 9–11
Threads mercerized cotton, 50; six-cord cotton, 80–100; cotton-covered polyester core, extra fine
Hems book; double fold; edge-stitched; machine-stitched
Seams flat-felled; French; lapped; plain; safety-stitched; welt; double welt
Seam Finishes self bound; double-ply bound; double-stitched and overcast; edge-stitched; single-ply or double-ply overcast; serging/single-ply overedged; untreated/plain
Pressing steam; safe temperature limit 400°F (206.1°C)
Care launder
Fabric Resource **Earl Glo/Erlanger Blumgart and Co. Inc.**

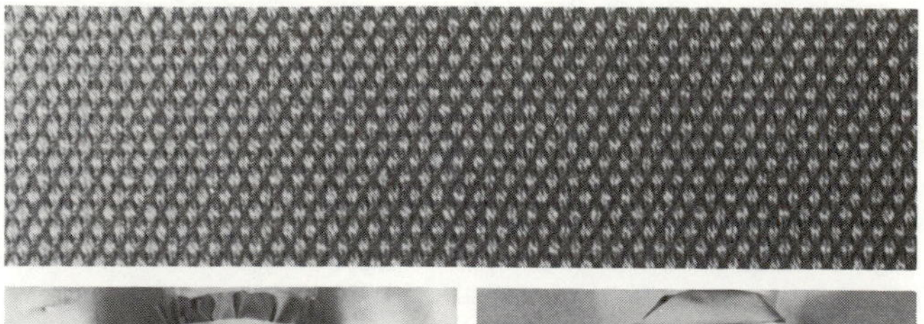

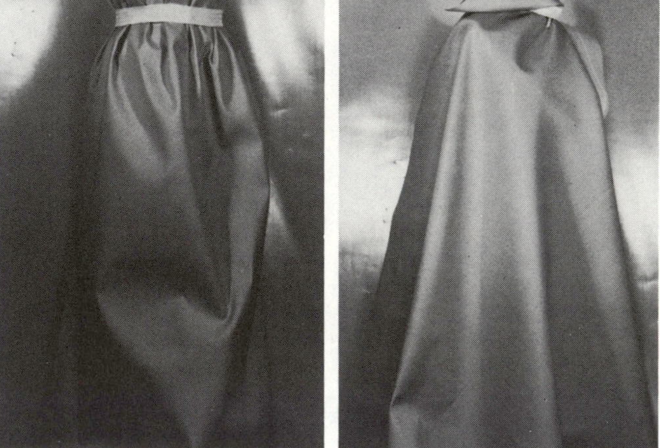

Cotton Canvas

Fiber Content 65% polyester/35% cotton
Yarn Type carded staple; filament staple
Yarn Construction conventional; textured; thick; low twist
Fabric Structure plain weave
Finishing Processes calendering; mercerizing; Zepel® water/stain repellent; may be softly finished or heavily sized to produce variation in hand
Color Application piece dyed-union dyed
Width 44–45 inches (111.8–114.3 cm)
Weight heavy
Hand compact; firm; heavy; rigid; stiff
Texture coarse; grainy
Performance Expectations high covering power; crease resistant; durable; high tensile strength; water and stain resistant
Drapability Qualities falls into stiff cones; retains shape or silhouette of garment; better utilized if fitted by seaming and eliminating excess fabric

Recommended Construction Aids & Techniques

Hand Needles betweens, 3–5; chenilles, 18–22; cotton darners, 1–7; embroidery, 1–7; milliners, 3–7; sharps, 1–5; sailmakers; tapestry, 18–22
Machine Needles round/set point, coarse 16–18
Threads cotton- or poly-core waxed, 50 all purpose; cotton-polyester blend, all purpose; cotton-covered polyester core, heavy duty; spun polyester, all purpose
Hems double-fold or single-fold bias binding; bound/Hong Kong finish; faced; stitched and pinked flat; overedged; seam binding
Seams plain; safety-stitched; welt; double welt
Seam Finishes single-ply bound; Hong Kong bound; single-ply overcast; pinked; pinked and stitched; serging/single-ply overedged; untreated/plain; single-ply zigzag
Pressing steam; safe temperature limit 325°F (164.1°C)
Care launder; dry clean
Fabric Resource **Milton Sherlip Inc.**

Cotton Challis

Fiber Content 50% Fortrel® polyester/50% rayon
Yarn Type filament fiber; filament staple
Yarn Construction conventional; tight twist
Fabric Structure plain weave; may also be twill weave
Finishing Processes calendering; mercerizing; softening
Color Application piece dyed; printing: direct, roller, screen methods
Width 44–45 inches (111.8–114.3 cm)
Weight light
Hand fine; soft; supple
Texture flat; lustrous; smooth; may be slightly napped
Performance Expectations abrasion resistant; dimensional stability; mildew and moth resistant; nonsnagging
Drapability Qualities falls into soft flares; accommodates fullness by gathering, elasticized shirring, pleating; fullness retains soft graceful fall

Recommended Construction Aids & Techniques

Hand Needles betweens, 5–7; cotton darners, 8–10; embroidery, 7–10; milliners, 8–10; sharps, 9–10
Machine Needles round/set point, medium-fine 9–11
Threads mercerized cotton, 50; six-cord cotton, 80–100; cotton-covered polyester core, extra fine; spun polyester, all purpose
Hems double fold; edge-stitched; double edge-stitched; machine stitched
Seams flat-felled; French; lapped; plain; safety-stitched; welt
Seam Finishes single-ply bound; Hong Kong bound; edge-stitched; single-ply overcast; pinked; pinked and stitched; serging/single-ply overedged; untreated/plain; single-ply zigzag
Pressing steam; safe temperature limit 325°F (164.1°C)
Care launder
Fabric Resource **Printsiples and Company B**

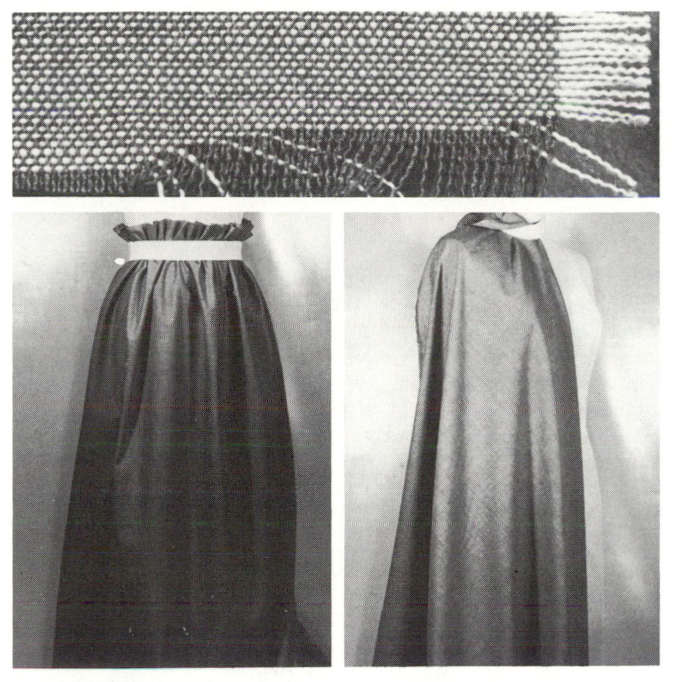

Chambray

Fiber Content 50% cotton/50% polyester
Yarn Type combed staple; filament fiber
Yarn Construction conventional; textured
Fabric Structure plain weave; almost square count (80x76)
Finishing Processes Russpress® permanent press precured; Sanforized®
Color Application piece dyed–cross dyed; yarn dyed (colored warp/white filling)
Width 60 inches (152.4 cm)
Weight medium-light
Hand dry; firm; harsh; soft
Texture flat; smooth; two-colored appearance produced by colored warp and white filling yarns
Performance Expectations durable; nonpilling; nonsnagging; 1% residual shrinkage; high tensile strength; wash-and-wear properties; wrinkle resistant
Drapability Qualities falls into moderate flares; accommodates fullness by gathering, elasticized shirring, pleating; fullness retains soft fall

Recommended Construction Aids & Techniques

Hand Needles betweens, 5–7; cotton darners, 8–10; embroidery, 7–10; milliners, 8–10; sharps, 9–10
Machine Needles round/set point, medium-fine 9–11
Threads mercerized cotton, 50; six-cord cotton, 60–80; cotton-covered polyester core, extra fine; spun polyester, all purpose
Hems bound/Hong Kong finish; edge-stitched; double edge-stitched; machine-stitched
Seams flat-felled; French; lapped; plain; safety-stitched; welt
Seam Finishes single-ply bound; Hong Kong bound; edge-stitched; single-ply overcast; pinked; pinked and stitched; serging/single-ply overedged; untreated/plain; single-ply zigzag
Pressing steam; safe temperature limit 325°F (164.1°C)
Care launder
Fabric Resources **Dan River Inc.; Russell Corporation**

Corded Chambray

Fiber Content 50% cotton/50% polyester
Yarn Type combed staple; filament fiber
Yarn Construction conventional; textured
Fabric Structure plain weave; almost square count (80x76)
Finishing Processes Russpress® permanent press precured; Sanforized®
Color Application piece dyed–cross dyed; yarn dyed (colored warp/white filling)
Width 60 inches (152.4 cm)
Weight medium-light
Hand dry; firm; harsh; soft
Texture flat; smooth; two-colored appearance produced by colored warp and white filling yarns
Performance Expectations durable; nonpilling; nonsnagging; 1% residual shrinkage; high tensile strength; wash-and-wear properties; wrinkle resistant
Drapability Qualities falls into moderate flares; accommodates fullness by gathering, elasticized shirring, pleating; fullness retains soft fall

Recommended Construction Aids & Techniques

Hand Needles betweens, 5–7; cotton darners, 8–10; embroidery, 7–10; milliners, 8–10; sharps, 9–10
Machine Needles round/set point, medium-fine 9–11
Threads mercerized cotton, 50; six-cord cotton, 60–80; cotton-covered polyester core, extra fine; spun polyester, all purpose
Hems bound/Hong Kong finish; edge-stitched; double edge-stitched; machine-stitched
Seams flat-felled; French; lapped; plain; safety-stitched; welt
Seam Finishes single-ply bound; Hong Kong bound; edge-stitched; single-ply overcast; pinked; pinked and stitched; serging/single-ply overedged; untreated/plain; single-ply zigzag
Pressing steam; safe temperature limit 325°F (164.1°C)
Care launder
Fabric Resource **Russell Corporation**

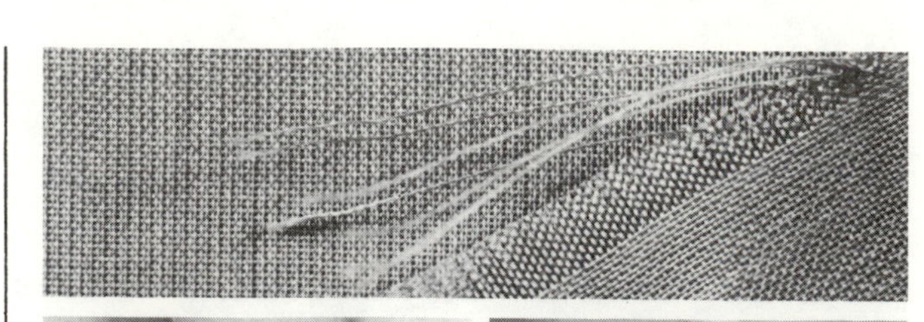

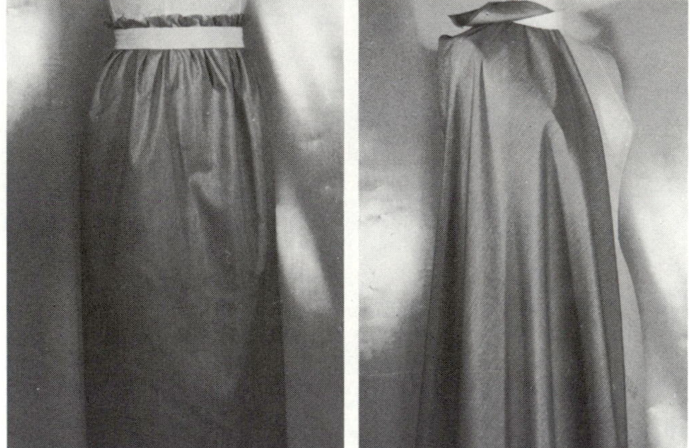

Chameleon/Iridescent Cotton

Fiber Content 65% Kodel® polyester/35% cotton
Yarn Type combed staple; filament fiber
Yarn Construction conventional; tight twist
Fabric Structure plain weave
Finishing Processes calendering; Russpress® permanent press precured
Color Application piece dyed–cross dyed; yarn dyed (warp is one color/two filling yarns are two different colors)
Width 60 inches (152.4 cm)
Weight medium-light
Hand firm; thin; soft to crisp depending on finish
Texture flat; smooth; multicolor appearance produced by different colored warp and filling yarns
Performance Expectations abrasion resistant; durable; nonsnagging; high tensile strength; wash-and-wear properties; wrinkle resistant
Drapability Qualities falls into moderate flares; accommodates fullness by gathering, elasticized shirring, pleating; fullness retains soft fall

Recommended Construction Aids & Techniques

Hand Needles betweens, 5–7; cotton darners, 8–10; embroidery, 7–10; milliners, 8–10; sharps, 9–10
Machine Needles round/set point, medium-fine 9–11
Threads cotton- or poly-core waxed, 60 fine; cotton-polyester blend, all purpose; cotton-covered polyester core, all purpose; spun polyester, all purpose
Hems bound/Hong Kong finish; double fold; edge-stitched; machine-stitched
Seams flat-felled; French; lapped; plain; safety-stitched; welt; double welt
Seam Finishes single-ply bound; Hong Kong bound; edge-stitched; single-ply overcast; pinked; pinked and stitched; serging/single-ply overedged; untreated/plain; single-ply zigzag
Pressing steam; safe temperature limit 325°F (164.1°C)
Care launder
Fabric Resources **Marcus Brothers Textile Inc.; Russell Corporation**

Chino/Chino Cloth

Fiber Content 50% polyester/50% cotton
Yarn Type combed staple; filament fiber
Yarn Construction conventional; 2-ply; textured
Fabric Structure twill weave
Finishing Processes mercerizing; preshrunk; schreiner calendering
Color Application piece dyed–union dyed
Width 44–45 inches (111.8–114.3 cm)
Weight heavy
Hand compact; nonelastic; rigid; thick
Texture coarse; grainy; opaque; noticeable diagonal twill line
Performance Expectations abrasion resistant; dimensional stability; durable; nonsnagging; high tensile strength
Drapability Qualities falls into firm wide flares; fullness maintains lofty effect; retains shape or silhouette of garment

Recommended Construction Aids & Techniques

Hand Needles betweens, 3–5; cotton darners, 1–7; embroidery, 1–7; milliners, 3–7; sharps, 1–5
Machine Needles round/set point, medium-coarse 14–16
Threads cotton- or poly-core waxed, 50 all purpose; cotton-polyester blend, all purpose; cotton-covered polyester core, heavy duty; spun polyester, all purpose
Hems double-fold or single-fold bias binding; bound/Hong Kong finish; machine-stitched; overedged; seam binding
Seams flat-felled; lapped; plain; safety-stitched; welt; double welt
Seam Finishes single-ply bound; Hong Kong bound; single-ply overcast; pinked; pinked and stitched; serging/single-ply overedged; untreated/plain; single-ply zigzag
Pressing steam; safe temperature limit 325°F (164.1°C)
Care launder; dry clean
Fabric Resources **Loom Tex Corp.; Reeves Brothers Inc.; Dan River Inc.**

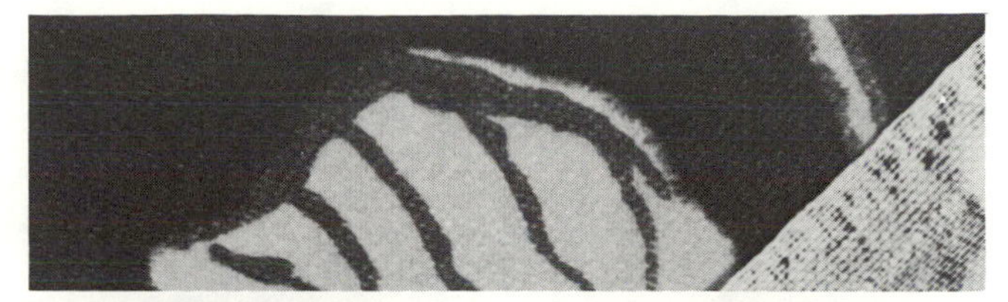

Chintz/Glazed Cotton

Fiber Content 100% cotton
Yarn Type combed staple
Yarn Construction conventional; light twist
Fabric Structure closely woven plain weave; high thread count
Finishing Processes calendering; glazing; mercerizing
Color Application printing: direct, roller, screen methods
Width 44–45 inches (111.8–114.3 cm)
Weight medium
Hand crisp; firm; inflexible
Texture highly glazed face; lustrous; smooth
Performance Expectations abrasion resistant; antistatic; glaze tends to crock and dull with wear; nonpilling; nonsnagging; water resistant
Drapability Qualities falls into stiff flares; fullness maintains crisp bouffant effect; maintains crisp folds, pleats

Recommended Construction Aids & Techniques

Hand Needles betweens, 5–7; cotton darners, 8–10; embroidery, 7–10; milliners, 8–10; sharps, 9–10
Machine Needles round/set point, medium-fine 11
Threads mercerized cotton, 50; six-cord cotton, 50; cotton- or poly-core waxed, fine; cotton-covered polyester core, all purpose
Hems bound/Hong Kong finish; double fold; edge-stitched; machine-stitched; overedged
Seams flat-felled; lapped; plain; safety-stitched; welt; double welt
Seam Finishes single-ply bound; Hong Kong bound; edge-stitched; single-ply overcast; pinked; pinked and stitched; serging/single-ply overedged; untreated/plain; single-ply zigzag
Pressing steam; safe temperature limit 400°F (206.1°C)
Care launder
Fabric Resource **White Rose Fabrics (sample courtesy of Fabric Room, FIT)**

Clokay/Embossed Cotton/French Damask

Fiber Content 100% cotton
Yarn Type carded or combed staple
Yarn Construction conventional; tight twist
Fabric Structure closely woven plain weave
Finishing Processes embossing; mercerizing; stabilizing
Color Application piece dyed
Width 45 inches (114.3 cm)
Weight light
Hand compact; soft; supple
Texture raised pattern (high and low areas); sculptured appearance
Performance Expectations absorbent; air permeable; antistatic; nonpilling; nonsnagging; embossed design tends to flatten with wear
Drapability Qualities falls into moderately soft flares; accommodates fullness by gathering, elasticized shirring; fullness retains soft fall

Recommended Construction Aids & Techniques

Hand Needles betweens, 5–7; cotton darners, 8–10; embroidery, 7–10; milliners, 8–10; sharps, 9–10
Machine Needles round/set point, medium-fine 9–11
Threads mercerized cotton, 50; six-cord cotton, 60–80; cotton-covered polyester core, extra fine
Hems bound/Hong Kong finish; double fold; edge-stitched; machine-stitched; overedged
Seams flat-felled; lapped; plain; safety-stitched; welt; double welt
Seam Finishes single-ply bound; Hong Kong bound; edge-stitched; single-ply overcast; pinked; serging/single-ply overedged; untreated/plain; single-ply zigzag
Pressing steam; safe temperature limit 400°F (206.1°C)
Care launder
Fabric Resource **Ameritex Division, United Merchants and Manufacturers, Inc.**

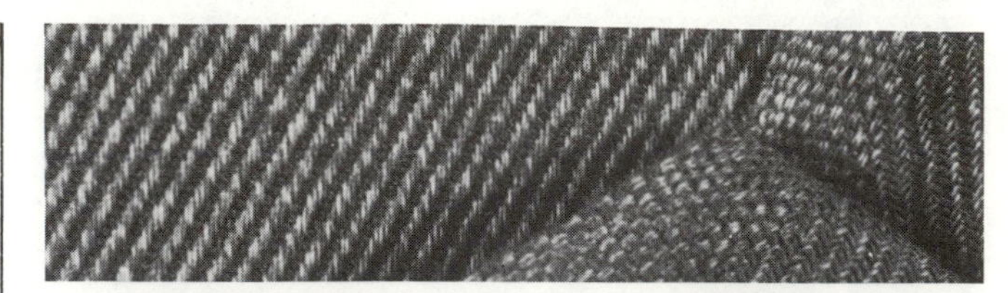

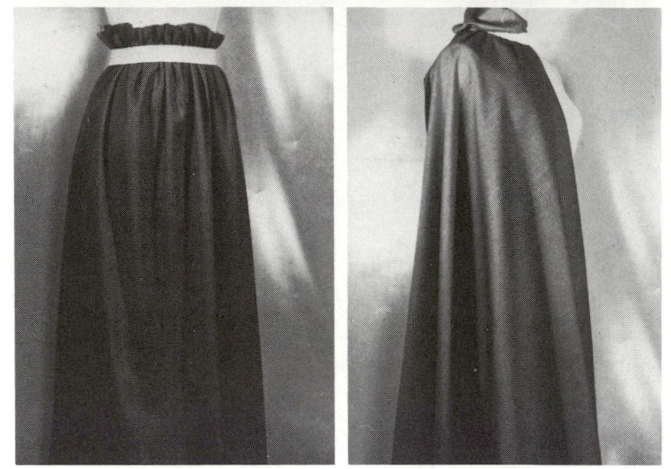

Cotton Covert

Fiber Content 65% Dacron® polyester/35% rayon
Yarn Type filament fiber; filament staple
Yarn Construction 2-ply; textured; slack twist
Fabric Structure 2x1 warp-faced left-hand twill weave
Finishing Processes calendering; mercerizing; stabilizing
Color Application piece dyed–cross dyed; yarn dyed (one colored/one white)
Width 44–45 inches (111.8–114.3 cm)
Weight medium to medium-heavy
Hand compact; nonelastic; hard; heavy; thick
Texture raised diagonal twill line; grainy; mottled appearance produced by colored and white yarns; opaque
Performance Expectations durable; mildew and moth resistant; subject to edge and flex abrasion, pilling, static buildup; wrinkle resistant
Drapability Qualities falls into firm wide flares; fullness maintains lofty effect; maintains crisp folds, pleats

Recommended Construction Aids & Techniques

Hand Needles betweens, 3–5; cotton darners, 5–10; embroidery, 5–8; milliners, 6–8; sharps, 6–8
Machine Needles round/set point, medium-coarse 14–16
Threads cotton- or poly-core waxed, 50 all purpose; cotton-polyester blend, all purpose; cotton-covered polyester core, heavy duty; spun polyester, all purpose
Hems double-fold or single-fold bias binding; bound/Hong Kong finish; machine-stitched; overedged; seam binding
Seams flat-felled; lapped; plain; safety-stitched; welt; double welt
Seam Finishes single-ply bound; Hong Kong bound; single-ply overcast; pinked; pinked and stitched; serging/single-ply overedged; untreated/plain; single-ply zigzag
Pressing steam; safe temperature limit 325°F (164.1°C)
Care launder; dry clean
Fabric Resource **J. P. Stevens & Co. Inc.**

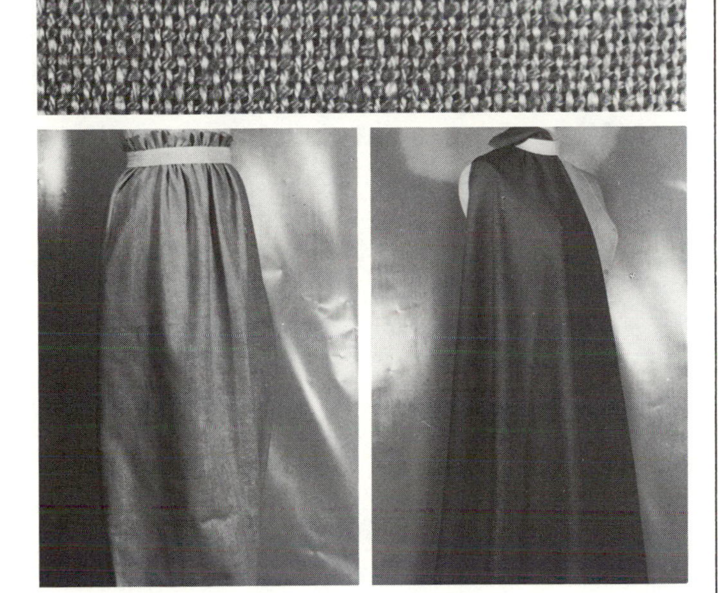

Cotton Crash

Fiber Content 65% cotton/35% polyester
Yarn Type carded staple; filament fiber
Yarn Construction conventional; coarse; uneven
Fabric Structure loosely woven plain weave; may also be twill weave
Finishing Processes beetling; mercerizing; shrinkage control
Color Application piece dyed-union dyed
Width 44-45 inches (111.8-114.3 cm)
Weight medium-light to medium-heavy
Hand hard; harsh; linenlike
Texture grainy; rough; uneven surface
Performance Expectations poor crease retention; subject to abrasion, bagging, seam slippage
Drapability Qualities falls into wide flares; fullness maintains lofty effect; maintains soft folds, unpressed pleats

Recommended Construction Aids & Techniques

Hand Needles betweens, 3-5; cotton darners, 5-10; embroidery, 5-8; milliners, 6-8; sharps, 6-8
Machine Needles round/set point, medium-coarse 14-16
Threads mercerized cotton, 50; six-cord cotton, 50; cotton waxed, all purpose; cotton-covered polyester core, all purpose
Hems double-fold or single-fold bias binding; bound/ Hong Kong finish; machine-stitched; overedged; seam binding
Seams flat-felled; lapped; plain; safety-stitched; welt; double welt
Seam Finishes single-ply bound; Hong Kong bound; single-ply overcast; pinked; pinked and stitched; serging/single-ply overedged; untreated/plain; single-ply zigzag
Pressing steam; safe temperature limit 325°F (164.1°C)
Care launder; dry clean
Fabric Resource Concord Fabrics Inc.

Cotton Crepe

Fiber Content 50% cotton/50% polyester
Yarn Type combed staple, filament fiber
Yarn Construction textured; crepe twist
Fabric Structure plain weave
Finishing Processes crepe-set; mercerizing; stabilizing
Color Application piece dyed-union dyed
Width 44-45 inches (111.8-114.3 cm)
Weight medium-light
Hand pliable; soft; spongy; supple
Texture crepey; pebbly; rough
Performance Expectations abrasion resistant; durable; nonpilling; subject to snagging; high tensile strength; wrinkle resistant
Drapability Qualities falls into soft flares and ripples; accommodates fullness by gathering, elasticized shirring, smocking; fullness retains soft graceful fall

Recommended Construction Aids & Techniques

Hand Needles betweens, 5-7; cotton darners, 8-10; embroidery, 7-10; milliners, 8-10; sharps, 6-8
Machine Needles round/set point, medium-fine 9-11
Threads mercerized cotton, 50; six-cord cotton, 50; cotton- or poly-core waxed, fine; cotton-covered polyester core, all purpose
Hems bound/Hong Kong finish; stitched and pinked flat; machine-stitched; seam binding
Seams flat-felled; lapped; plain; safety-stitched; welt; double welt
Seam Finishes single-ply bound; Hong Kong bound; edge-stitched; single-ply overcast; pinked; pinked and stitched; serging/single-ply overedged; untreated/plain; single-ply zigzag
Pressing steam; safe temperature limit 325°F (164.1°C)
Care launder; dry clean
Fabric Resource Concord Fabrics Inc.

Cotton Creponne

Fiber Content 100% cotton
Yarn Type combed staple
Yarn Construction conventional; fine; high twist
Fabric Structure plain weave
Finishing Processes crinkle pleating; mercerizing
Color Application piece dyed; yarn dyed
Width 45 inches (114.3 cm)
Weight light
Hand fine; loose; stretchy
Texture crinkles create high and low ridges
Performance Expectations abrasion resistant; absorbent; antistatic; subject to mildew; nonpilling; nonsnagging
Drapability Qualities falls into stiff flares; accommodates fullness by gathering; fullness maintains crisp effect

Recommended Construction Aids & Techniques

Hand Needles betweens, 5–7; cotton darners, 8–10; embroidery, 7–10; milliners, 8–10; sharps, 9–10
Machine Needles round/set point, medium-fine 9–11
Threads mercerized cotton, 50; six-cord cotton, 80–100; cotton-covered polyester core, extra fine
Hems bound/Hong Kong finish; edge-stitched; machine-stitched; overedged
Seams flat-felled; French; lapped; plain; safety-stitched; welt; double welt
Seam Finishes single-ply bound; Hong Kong bound; edge-stitched; single-ply overcast; pinked; pinked and stitched; serging/single-ply overedged; untreated/plain; single-ply zigzag
Pressing steam; safe temperature limit 400°F (206.1°C)
Care launder; dry clean
Fabric Resource American Silk Mills Corp.

Cretonne

Fiber Content 100% cotton
Yarn Type carded staple
Yarn Construction conventional; thick
Fabric Structure plain weave; may be twill or satin weave
Finishing Processes beetling; mercerizing; schreiner calendering
Color Application printing: direct, roller, screen methods
Width 56–58 inches (142.2–147.3 cm)
Weight medium to medium-heavy
Hand compact; hard; harsh; thick
Texture coarse; grainy; characteristic large design
Performance Expectations absorbent; antistatic; durable; nonpilling; nonsnagging; subject to seam slippage; high tensile strength
Drapability Qualities falls into crisp wide cones; retains shape or silhouette of garment; better utilized if fitted by seaming and eliminating excess fabric

Recommended Construction Aids & Techniques

Hand Needles betweens, 3–5; cotton darners, 1–7; embroidery, 1–7; milliners, 3–7; sharps, 1–5
Machine Needles round/set point, medium-coarse 14–16
Threads mercerized cotton, 40; six-cord cotton, 40; waxed, 40; cotton-covered polyester core, heavy duty
Hems double-fold or single-fold bias binding; bound/Hong Kong finish; machine-stitched; overedged; seam binding
Seams plain; safety-stitched; welt; double welt
Seam Finishes single-ply bound; Hong Kong bound; single-ply overcast; pinked; pinked and stitched; serging/single-ply overedged; untreated/plain; single-ply zigzag
Pressing steam; safe temperature limit 400°F (206.1°C)
Care dry clean
Fabric Resource M. Lowenstein & Sons, Inc.

Cotton Damask

Fiber Content 65% polyester/35% cotton
Yarn Type combed staple; filament fiber
Yarn Construction conventional; textured; tight twist
Fabric Structure tightly woven figure weave on Jacquard loom
Finishing Processes mercerizing; schreiner calendering
Color Application piece dyed-cross dyed; yarn dyed
Width 44-45 inches (111.8-114.3 cm)
Weight light
Hand fine; soft to firm depending on finish
Texture smooth; shiny and dull contrasting surface on face and back; pattern is visible due to light-reflecting glossy yarns and weave; reversible
Performance Expectations abrasion resistance; dimensional stability; durable; high tensile strength
Drapability Qualities falls into soft flares; accommodates fullness by gathering, elasticized shirring; fullness retains soft graceful fall

Recommended Construction Aids & Techniques

Hand Needles betweens, 5-7; cotton darners, 8-10; embroidery, 7-10; milliners, 8-10; sharps, 9-10
Machine Needles round/set point, medium-fine 9-11
Threads cotton- or poly-core waxed, 60 fine; cotton-polyester blend, all purpose; cotton-covered polyester core, all purpose; spun polyester, all purpose
Hems bound/Hong Kong finish; double fold; edge-stitched; machine-stitched; overedged
Seams flat-felled; lapped; plain; safety-stitched; welt; double welt
Seam Finishes single-ply bound; Hong Kong bound; edge-stitched; single-ply overcast; pinked; pinked and stitched; serging/single-ply overedged; untreated/plain; single-ply zigzag
Pressing steam; safe temperature limit 325°F (164.1°C)
Care launder
Fabric Resource **Concord Fabrics Inc.**

Denim

Fiber Content 50% cotton/50% polyester
Yarn Type carded staple; filament fiber
Yarn Construction conventional; textured; hard twist
Fabric Structure closely woven warp-faced twill weave; heavy warp yarns
Finishing Processes calendering; decating; mercerizing
Color Application piece dyed-cross dyed; yarn dyed (colored warp/white or natural filling)
Width 44-45 inches (111.8-114.3 cm)
Weight medium-heavy to heavy
Hand nonelastic; firm; hard;
Texture compact; coarse; grainy
Performance Expectations subject to edge and flex abrasion; durable; nonsnagging; subject to shrinkage unless treated; high tensile strength
Drapability Qualities falls into stiff wide cones; fullness maintains crisp effect; retains shape or silhouette of garment

Recommended Construction Aids & Techniques

Hand Needles betweens, 3-5; chenilles, 18-22; cotton darners, 1-7; embroidery, 1-7; milliners, 3-7; sharps, 1-5; sailmakers; tapestry, 18-22
Machine Needles round/set point, coarse 16-18
Threads poly-core waxed, 50 all purpose; cotton-polyester blend, all purpose; cotton-covered polyester core, heavy duty; spun polyester, all purpose
Hems double-fold or single-fold bias binding; bound/Hong Kong finish; faced; stitched and pinked flat; overedged; seam binding
Seams flat-felled; plain; safety-stitched; welt; double welt
Seam Finishes single-ply bound; Hong Kong bound; single-ply overcast; pinked; pinked and stitched; serging/single-ply overedged; untreated/plain; single-ply zigzag
Pressing steam; safe temperature limit 325°F (164.1°C)
Care launder
Fabric Resource **Dan River Inc.**

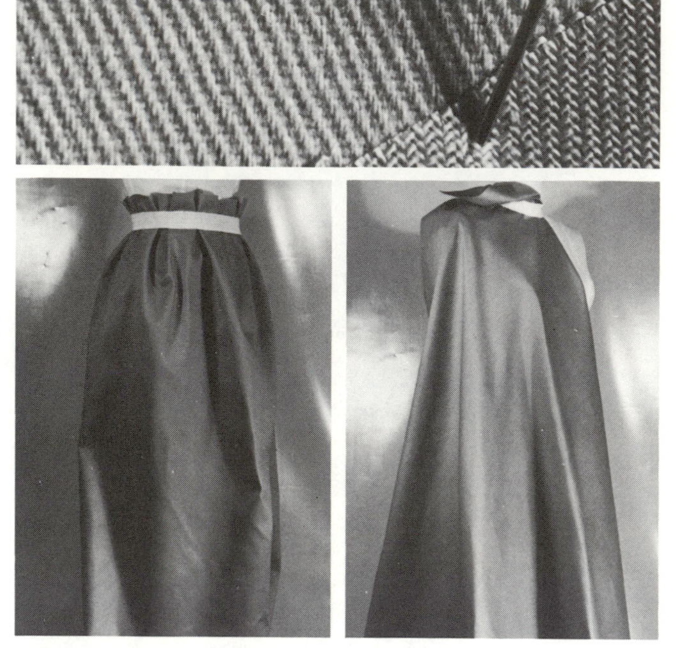

Brushed-faced/Brushed Denim

Fiber Content 50% cotton/50% polyester
Yarn Type combed staple; filament fiber
Yarn Construction conventional; textured
Fabric Structure closely woven warp-faced twill weave; heavy warp yarns
Finishing Processes brushing; mercerizing; napping; sanding/sueding
Color Application piece dyed–cross dyed; yarn dyed
Width 47–48 inches (119.4–121.9 cm)
Weight heavy
Hand firm; soft; thick
Texture compact; napped face
Performance Expectations subject to edge and flex abrasion; durable; nonsnagging; subject to shrinkage unless treated; high tensile strength
Drapability Qualities falls into stiff wide cones; fullness maintains crisp effect; retains shape or silhouette of garment

Recommended Construction Aids & Techniques

Hand Needles betweens, 3–5; chenilles, 18–22; cotton darners, 1–7; embroidery, 1–7; milliners, 3–7; sharps, 1–5; sailmakers; tapestry, 18–22
Machine Needles round/set point, coarse 16–18
Threads poly-core waxed, 50 all purpose; cotton-polyester blend, all purpose; cotton-covered polyester core, heavy duty; spun polyester, all purpose
Hems double-fold or single-fold bias binding; bound/Hong Kong finish; faced; stitched and pinked flat; overedged; seam binding
Seams flat-felled; plain; safety-stitched; welt; double welt
Seam Finishes single-ply bound; Hong Kong bound; single-ply overcast; pinked; pinked and stitched; serging/single-ply overedged; untreated/plain; single-ply zigzag
Pressing steam; safe temperature limit 325°F (164.1°C)
Care launder; wash-and-wear
Fabric Resources Cone Mills Corporation; Reeves Brothers Inc.

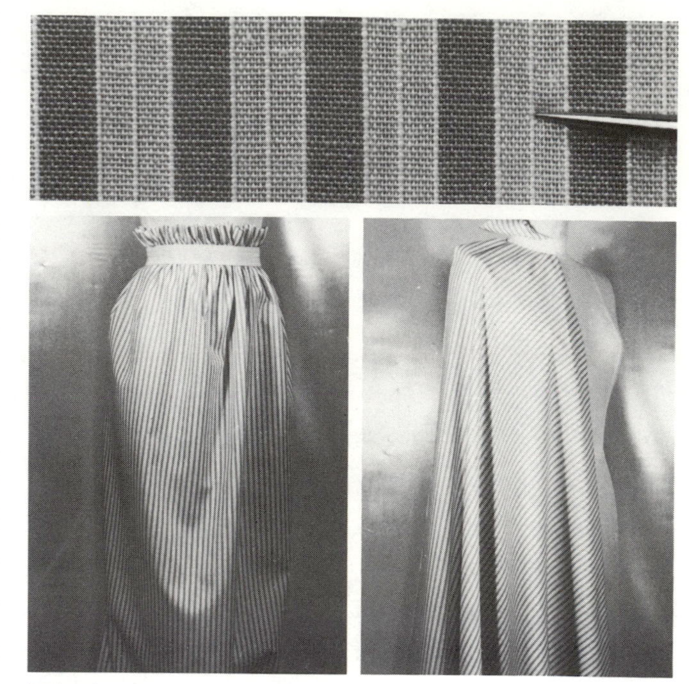

Dimity

Fiber Content 65% Kodel® polyester/35% cotton
Yarn Type combed staple; filament fiber
Yarn Construction conventional; high twist
Fabric Structure rib weave variation with vertical ribs
Finishing Processes mercerizing; Russpress® permanent press precured; Sanforized®
Color Application yarn dyed
Width 44–45 inches (111.8–114.3 cm)
Weight light
Hand crisp; fine; firm
Texture fine cord creates uneven surface
Performance Expectations abrasion resistant; dimensional stability; durable; flexible; nonpilling; 1% residual shrinkage; high tensile strength; wash-and-wear properties
Drapability Qualities falls into moderately soft flares; accommodates fullness by pleating, gathering, elasticized shirring; fullness retains crisp effect

Recommended Construction Aids & Techniques

Hand Needles betweens, 5–7; cotton darners, 8–10; embroidery, 7–10; milliners, 8–10; sharps, 9–10
Machine Needles round/set point, medium-fine 9-11
Threads mercerized cotton, 50; six-cord cotton, 50; cotton-covered polyester core, extra fine
Hems double fold; edge-stitched; machine-stitched; overedged
Seams flat-felled; French; lapped; plain; safety-stitched; welt; double welt
Seam Finishes double-stitched; double-stitched and overcast; double-stitched and trimmed; edge-stitched; pinked; pinked and stitched; serging/single-ply overedged; untreated/plain; single-ply or double-ply zigzag
Pressing steam; safe temperature limit 325°F (164.1°C)
Care launder; drip dry; permanent press cycle of dryer
Fabric Resource Russell Corporation

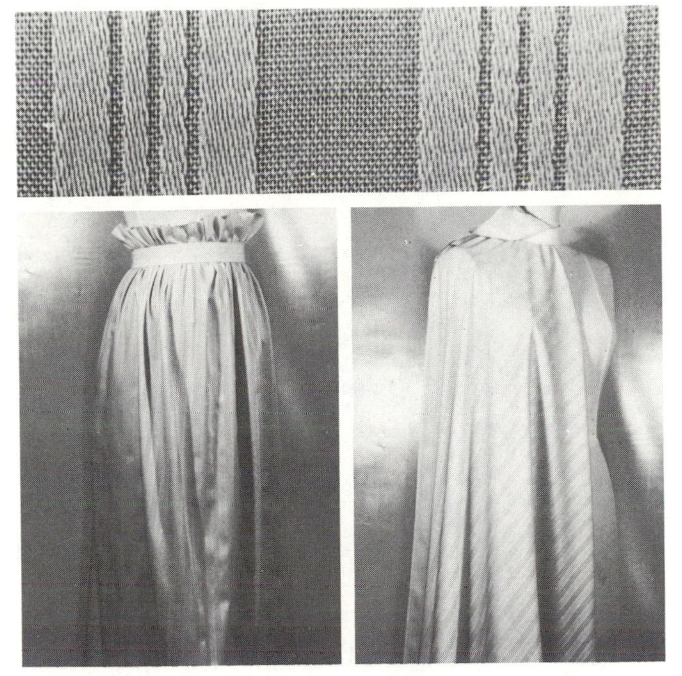

Dimity with Cord/Satin-striped Dimity

Fiber Content 65% Kodel® polyester/35% cotton
Yarn Type combed staple; filament fiber
Yarn Construction conventional; high twist
Fabric Structure plain weave variation with satin weave stripe
Finishing Processes mercerizing; Russpress® permanent press precured; Sanforized®
Color Application piece dyed–union dyed; yarn dyed
Width 48 inches (121.9 cm)
Weight light
Hand firm; soft to crisp depending on finish
Texture stripe creates slightly raised uneven surface
Performance Expectations abrasion resistant; dimensional stability; durable; nonpilling; resilient; high tensile strength; wash-and-wear properties
Drapability Qualities falls into moderately soft flares; accommodates fullness by pleating, gathering; fullness maintains crisp effect

Recommended Construction Aids & Techniques

Hand Needles betweens, 5–7; cotton darners, 8–10; embroidery, 7–10; milliners, 8–10; sharps, 9–10
Machine Needles round/set point, medium-fine 9–11
Threads mercerized cotton, 50; six-cord cotton, 50; cotton-covered polyester core, extra fine
Hems double fold; edge-stitched; machine-stitched; overedged
Seams flat-felled; French; lapped; plain; safety-stitched; welt; double welt
Seam Finishes double-stitched; double-stitched and trimmed; edge-stitched; pinked; pinked and stitched; serging/single-ply overedged; untreated/plain; single-ply zigzag
Pressing steam; safe temperature limit 325°F (164.1°C)
Care launder; drip dry; permanent press cycle of dryer
Fabric Resource **Russell Corporation**

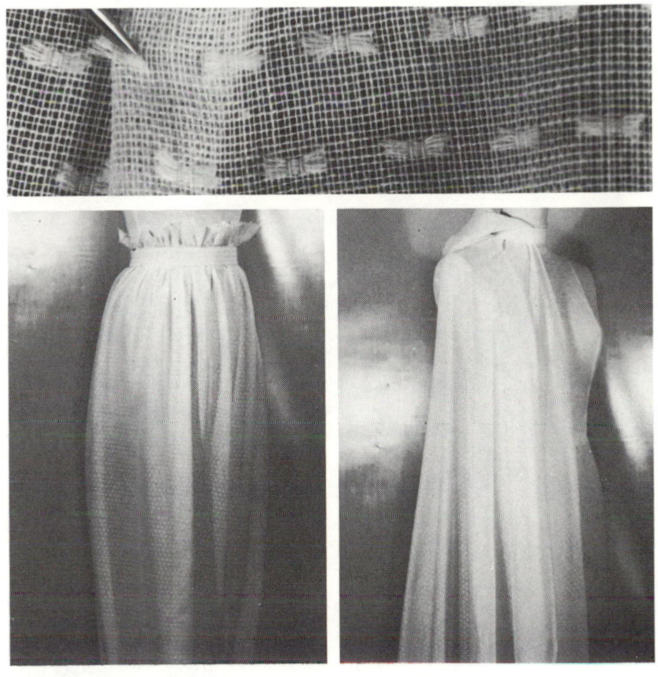

Dotted Swiss

Fiber Content 65% polyester/35% cotton
Yarn Type combed staple; filament staple
Yarn Construction conventional; high twist
Fabric Structure figure weave; swivel weave
Finishing Processes clear finish; mercerizing; sizing
Color Application piece dyed–union dyed; yarn dyed
Width 44–45 inches (111.8–114.3 cm)
Weight light
Hand delicate; thin; soft to crisp depending on finish
Texture raised fuzzy dots create uneven surface; sheer
Performance Expectations air permeable; dimensional stability; flexible; nonpilling; resilient; wash-and-wear properties
Drapability Qualities falls into soft flares; accommodates fullness by gathering, elasticized shirring, smocking; fullness retains soft fall

Recommended Construction Aids & Techniques

Hand Needles betweens, 5–7; cotton darners, 8–10; embroidery, 7–10; milliners, 8–10; sharps, 9–10
Machine Needles round/set point, medium-fine 9–11
Threads mercerized cotton, 50; six-cord cotton, 50; cotton-covered polyester core, extra fine
Hems double fold; edge-stitched; double edge-stitched; machine-stitched; overedged
Seams flat-felled; French; lapped; plain; safety-stitched; welt; double welt
Seam Finishes double-stitched; double-stitched and trimmed; edge-stitched; pinked; pinked and stitched; serging/single-ply overedged; untreated/plain; single-ply zigzag
Pressing steam; safe temperature limit 325°F (164.1°C)
Care launder; drip dry; permanent press cycle of dryer
Fabric Resource **Earl Glo/Erlanger Blumgart and Co. Inc.**

Dotted Swiss-type/Flocked Swiss

Fiber Content cotton/polyester/rayon flock
Yarn Type filament fiber; filament staple; carded staple
Yarn Construction conventional
Fabric Structure compound structure: plain weave fabric with flocked design
Finishing Processes electrocoating; mercerizing; stabilizing
Color Application flock printing; piece dyed
Width 44-45 inches (111.8-114.3 cm)
Weight medium-light
Hand dry; firm; soft
Texture raised fuzzy dots create rough uneven surface
Performance Expectations flocking tends to abrade, wear off, wash off
Drapability Qualities falls into moderately soft flares; accommodates fullness by pleating, gathering, elasticized shirring; fullness maintains lofty effect

Recommended Construction Aids & Techniques

Hand Needles betweens, 5-7; cotton darners, 8-10; embroidery, 7-10; milliners, 8-10; sharps, 9-10
Machine Needles round/set point, medium-fine 9-11
Threads mercerized cotton, 50; six-cord cotton, 50; cotton-covered polyester core, all purpose
Hems double fold; edge-stitched; machine-stitched; overedged
Seams flat-felled; French; lapped; plain; safety-stitched; welt; double welt
Seam Finishes double-stitched; double-stitched and trimmed; edge-stitched; pinked; pinked and stitched; serging/single-ply overedged; untreated/plain; single-ply zigzag
Pressing steam; safe temperature limit 325°F (164.1°C)
Care launder
Fabric Resource Printsiples and Company B

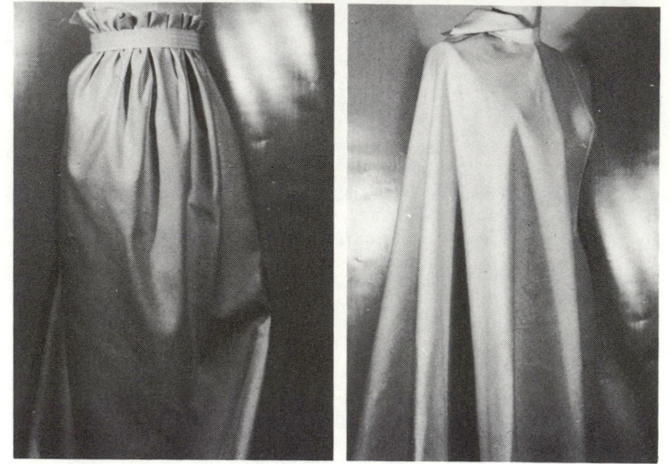

Drill Cloth

Fiber Content 65% cotton/35% polyester
Yarn Type carded staple; filament fiber
Yarn Construction conventional; ply; textured
Fabric Structure 2x1 warp-faced left-hand twill weave; count includes 60x36, 80x48
Finishing Processes calendering; mercerizing; sized and pressed for compact surface or brushed for raised nap surface
Color Application piece dyed-union dyed
Width 46-47 inches (116.8-119.4 cm)
Weight medium to heavy
Hand compact; nonelastic; hard
Texture coarse; grainy; rough
Performance Expectations subject to edge and flex abrasion; durable; nonsnagging; subject to shrinkage unless treated; high tensile strength
Drapability Qualities falls into stiff wide cones; fullness maintains crisp effect; retains shape or silhouette of garment

Recommended Construction Aids & Techniques

Hand Needles betweens, 3-5; chenilles, 18-22; cotton darners, 1-7; embroidery, 1-7; milliners, 3-7; sharps, 1-5; sailmakers; tapestry, 18-22
Machine Needles round/set point, coarse 16-18
Threads mercerized cotton, 40; six-cord cotton, 30; waxed, 40 heavy duty; cotton-covered polyester core, heavy duty
Hems double-fold or single-fold bias binding; bound/Hong Kong finish; faced; stitched and pinked flat; overedged; seam binding
Seams plain; safety-stitched; welt; double welt
Seam Finishes book; single-ply bound; Hong Kong bound; single-ply overcast; pinked; pinked and stitched; serging/single-ply overedged; untreated/plain; single-ply zigzag
Pressing steam; safe temperature limit 325°F (164.1°C)
Care launder; dry clean
Fabric Resource Milton Sherlip Inc.

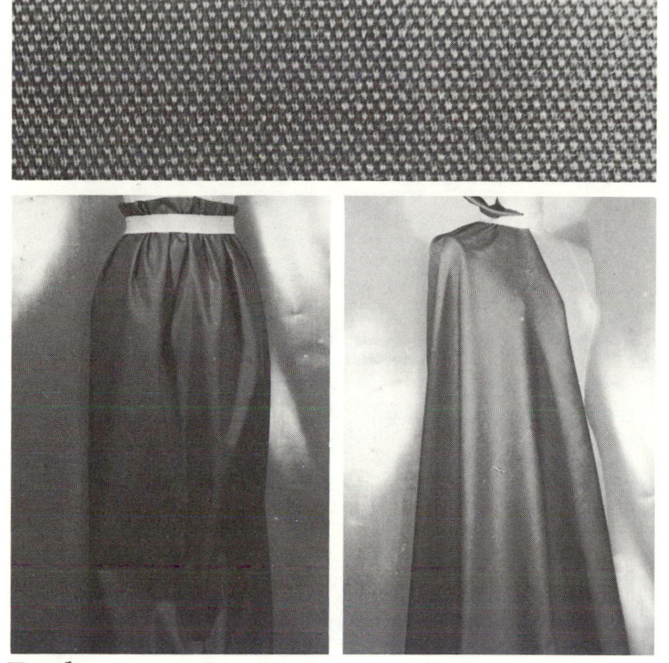

Duck

Fiber Content 50% cotton/50% polyester
Yarn Type carded or combed staple; filament fiber
Yarn Construction conventional; textured; ply
Fabric Structure closely woven plain weave
Finishing Processes calendering; mercerizing; permanent press precured
Color Application piece dyed–union dyed
Width 48 inches (121.9 cm)
Weight medium to heavy
Hand compact; crisp; nonelastic; hard
Texture coarse; papery; rough
Performance Expectations abrasion resistant; dimensional stability; durable; nonpilling; high tensile strength; crease and wrinkle resistant; wash-and-wear properties
Drapability Qualities falls into crisp cones; fullness maintains crisp effect; maintains crisp folds, creases

Recommended Construction Aids & Techniques

Hand Needles betweens, 3-5; chenilles, 18–22; cotton darners, 1-7; embroidery, 1-7; milliners, 3-7; sharps, 1-5; sailmakers; tapestry, 18–22
Machine Needles round/set point, coarse 16-18
Threads poly-core waxed, 50 all purpose; cotton-polyester blend, all purpose; cotton-covered polyester core, heavy duty; spun polyester, all purpose
Hems double-fold or single-fold bias binding; bound/Hong Kong finish; machine-stitched; overedged; seam binding
Seams flat-felled; lapped; plain; safety-stitched; welt; double welt
Seam Finishes single-ply bound; Hong Kong bound; single-ply overcast; pinked; pinked and stitched; serging/single-ply overedged; untreated/plain; single-ply zigzag
Pressing steam; safe temperature limit 325°F (164.1°C)
Care launder (wash-and-wear finish); dry clean
Fabric Resource J. P. Stevens & Co. Inc.

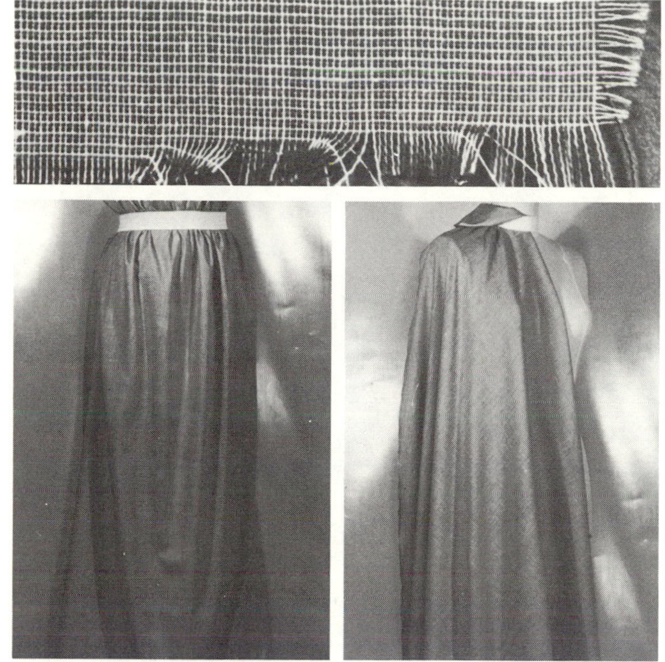

End and End Cloth (lightweight "topweight")

Fiber Content 100% cotton
Yarn Type combed staple
Yarn Construction conventional; tight twist
Fabric Structure closely woven plain weave
Finishing Processes calendering; mercerizing; shrinkage control
Color Application yarn dyed (colored and white warp and filling yarns)
Width 44-45 inches (111.8-114.3 cm)
Weight light "topweight"
Hand fine; firm; soft
Texture flat; alternating white and colored yarns in both directions create fine colored checkered appearance
Performance Expectations abrasion resistant; absorbent; antistatic; dimensional stability; subject to mildew; nonpilling; nonsnagging; high tensile strength
Drapability Qualities falls into moderate flares; accommodates fullness by pleating, gathering, elasticized shirring; fullness retains soft fall

Recommended Construction Aids & Techniques

Hand Needles betweens, 5-7; cotton darners, 8-10; embroidery, 7-10; milliners, 8-10; sharps, 9-10
Machine Needles round/set point, medium-fine 9-11
Threads mercerized cotton, 50; six-cord cotton, 60-80; cotton-covered polyester core, extra fine
Hems bound/Hong Kong finish; double fold; edge-stitched; machine-stitched; overedged
Seams flat-felled; French; lapped; plain; safety-stitched; welt; double welt
Seam Finishes single-ply bound; Hong Kong bound; edge-stitched; single-ply overcast; pinked; serging/single-ply overedged; untreated/plain; single-ply zigzag
Pressing steam; safe temperature limit 400°F (206.1°C)
Care launder
Fabric Resource Earl Glo/Erlanger Blumgart and Co. Inc.

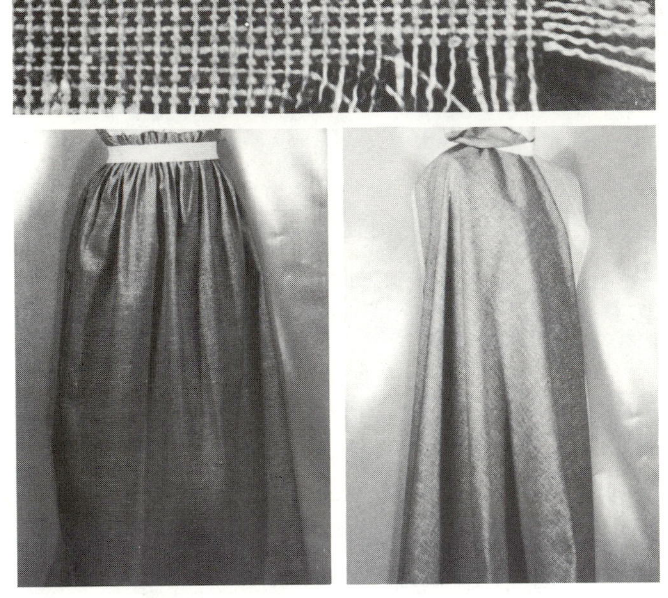

End and End Cloth (medium weight "bottomweight")

Fiber Content 65% polyester/35% cotton
Yarn Type carded staple; filament fiber
Yarn Construction conventional; textured; thick-and-thin
Fabric Structure plain weave
Finishing Processes beetling; calendering; mercerizing; stabilizing
Color Application piece dyed–cross dyed; yarn dyed (colored and white warp and filling yarns)
Width 58–60 inches (147.3–152.4 cm)
Weight medium "bottomweight"
Hand coarse; soft
Texture flat; grainy; alternating white and colored yarns in both directions create checkered appearance
Performance Expectations abrasion resistant; high covering power; durable; heat-set properties; resilient
Drapability Qualities falls into moderate flares; accommodates fullness by pleating, gathering; fullness retains soft fold, unpressed pleats

Recommended Construction Aids & Techniques

Hand Needles betweens, 3–5; cotton darners, 5–10; embroidery, 5–8; milliners, 6–8; sharps, 6–8
Machine Needles round/set point, medium 14
Threads cotton- or poly-core waxed, 50 all purpose; cotton-polyester blend, all purpose; cotton-covered polyester core, heavy duty; spun polyester, all purpose
Hems book; double-fold or single-fold bias binding; bound/Hong Kong finish; machine-stitched; overedged; seam binding
Seams flat-felled; lapped; plain; safety-stitched; welt; double welt
Seam Finishes book; single-ply or Hong Kong bound; single-ply overcast; pinked; pinked and stitched; serging/single-ply overedged; untreated/plain; single-ply zigzag
Pressing steam; safe temperature limit 325°F (164.1°C)
Care launder; dry clean
Fabric Resource **Ameritex Division**

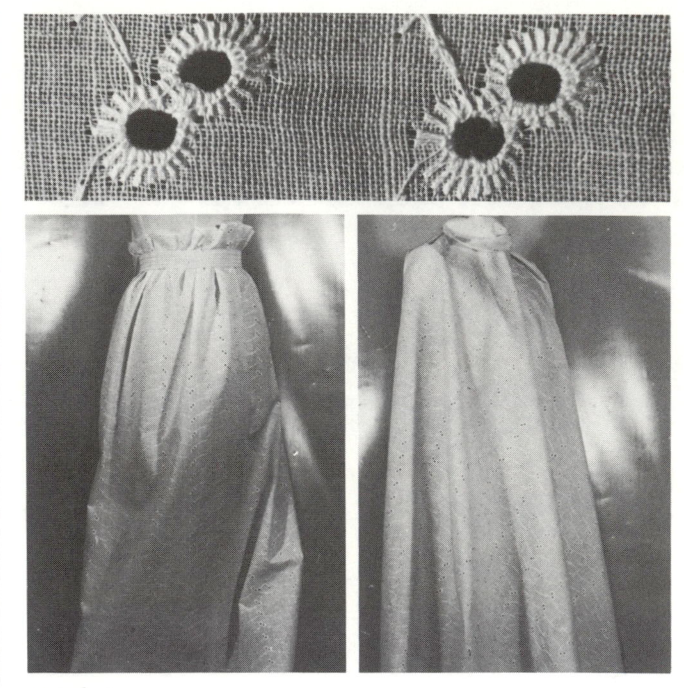

Eyelet

Fiber Content 65% polyester/35% cotton
Yarn Type combed staple; filament fiber
Yarn Construction conventional
Fabric Structure compound structure: plain weave with Schiffli embroidered design
Finishing Processes calendering; mercerizing; shrinkage control
Color Application piece dyed–union dyed; vat dyed
Width 48 inches (121.9 cm)
Weight medium
Hand fine; firm; soft to crisp depending on finish
Texture smooth flat surface with open-work design
Performance Expectations air permeable; nonpilling; resilient; wash-and-wear properties
Drapability Qualities falls into moderately soft flares; accommodates fullness by gathering; fullness retains soft fall

Recommended Construction Aids & Techniques

Hand Needles betweens, 3–5; cotton darners, 5–10; embroidery, 5–8; milliners, 6–8; sharps, 6–8
Machine Needles round/set point, medium-fine 9–11
Threads mercerized cotton, 50; six-cord cotton, 50; cotton-covered polyester core, all purpose
Hems double fold; edge-stitched and turned; machine-stitched; overedged
Seams flat-felled; French; lapped; plain; safety-stitched; welt; double welt
Seam Finishes double-stitched; double-stitched and overcast; edge-stitched; pinked; pinked and stitched; serging/single-ply overedged; untreated/plain; single-ply or double-ply zigzag
Pressing steam; safe temperature limit 325°F (164.1°C)
Care launder; permanent press cycle of dryer
Fabric Resource **Applause Fabrics Inc.**

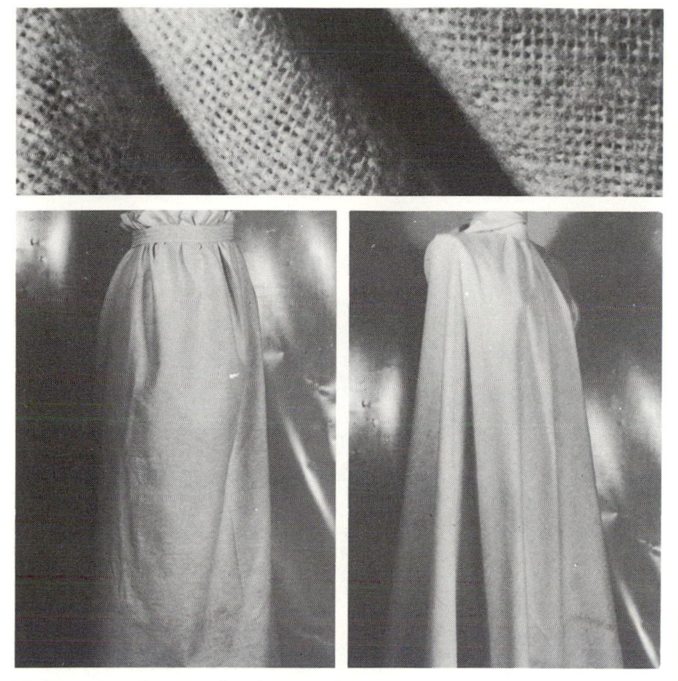

Cotton Flannel/Flannelette

Fiber Content 100% cotton
Yarn Type carded staple
Yarn Construction conventional
Fabric Structure plain weave
Finishing Processes brushing; mercerizing; napping; stabilizing
Color Application piece dyed
Width 44–45 inches (111.8–114.3 cm)
Weight light
Hand light; downy soft; supple
Texture fluffy napped face; flat back
Performance Expectations absorbent; antistatic; dimensional stability; insulating properties; nonpilling; nonsnagging; 3% maximum shrinkage; brushed face tends to abrade, wear off and flatten with wear
Drapability Qualities falls into soft flares; accommodates fullness by gathering; fullness retains soft graceful fall

Recommended Construction Aids & Techniques

Hand Needles betweens, 5–7; cotton darners, 8–10; embroidery, 7–10; milliners, 8–10; sharps, 9–10
Machine Needles round/set point, medium-fine 9–11
Threads mercerized cotton, 50; six-cord cotton, 60–80; cotton-covered polyester core, extra fine
Hems bound/Hong Kong finish; double fold; edge-stitched; machine-stitched; overedged
Seams flat-felled; lapped; plain; safety-stitched; welt; double welt
Seam Finishes single-ply bound; Hong Kong bound; edge-stitched; single-ply overcast; pinked; serging/single-ply overedged; untreated/plain; single-ply zigzag
Pressing steam; safe temperature limit 400°F (206.1°C)
Care launder
Fabric Resources **Cone Mills Corporation; Dan River Inc.**

Outing/Dommet Flannel

Fiber Content 100% cotton
Yarn Type carded staple
Yarn Construction conventional
Fabric Structure plain weave; may be twill weave
Finishing Processes brushing; mercerizing; napping; stabilizing
Color Application piece dyed
Width 36–45 inches (91.4–114.3 cm)
Weight light to medium
Hand lofty; downy soft
Texture napped face and back; opaque
Performance Expectations absorbent; antistatic; dimensional stability; insulating properties; nonpilling; nonsnagging; 2% maximum shrinkage; napped face tends to abrade, wear off and flatten with wear; soils easily
Drapability Qualities falls into moderate flares; accommodates fullness by gathering; fullness retains soft fall

Recommended Construction Aids & Techniques

Hand Needles betweens, 5–7; cotton darners, 8–10; embroidery, 7–10; milliners, 8–10; sharps, 9–10
Machine Needles round/set point, medium 11–14
Threads mercerized cotton, 50; six-cord cotton, 50; poly-core waxed, fine; cotton-covered polyester core, all purpose
Hems book; bound/Hong Kong finish; edge-stitched; machine-stitched; overedged
Seams flat-felled; lapped; plain; safety-stitched; welt; double welt
Seam Finishes book; single-ply bound; Hong Kong bound; edge-stitched; single-ply overcast; pinked; pinked and stitched; serging/single-ply overedged; untreated/plain; single-ply zigzag
Pressing steam; safe temperature limit 400°F (206.1°C)
Care launder
Fabric Resources **Cone Mills Corporation; Earl Glo/Erlanger Blumgart and Co. Inc.; Lawrence Textile Co. Inc.**

Cotton Gabardine

Fiber Content 100% polyester
Yarn Type filament fiber
Yarn Construction 2-ply; textured; high twist
Fabric Structure compactly woven warp-faced twill weave
Finishing Processes calendering; mercerizing
Color Application piece dyed–union dyed, cross dyed
Width 58-60 inches (147.3-152.4 cm)
Weight heavy
Hand compact; nonelastic; hard
Texture coarse; grainy; shiny; raised diagonal twill on face
Performance Expectations dimensional stability; durable; mildew and moth resistant; high tear strength
Drapability Qualities falls into firm wide cones; fullness maintains crisp effect; retains shape of garment

Recommended Construction Aids & Techniques

Hand Needles betweens, 3-5; chenilles 18-22; cotton darners, 1-7; embroidery, 1-7; milliners, 3-7; sharps, 1-5
Machine Needles round/set point, medium-coarse 14-16
Threads poly-core waxed, 50 all purpose; cotton-polyester blend, all purpose; cotton-covered polyester core, heavy duty; spun polyester, all purpose
Hems book; double-fold or single-fold bias binding; bound/Hong Kong finish; machine-stitched; overedged; seam binding
Seams flat-felled; lapped; plain; safety-stitched; welt; double welt
Seam Finishes book; single-ply bound; Hong Kong bound; single-ply overcast; pinked; pinked and stitched; serging/single-ply overedged; untreated/plain; single-ply zigzag
Pressing steam; safe temperature limit 325°F (164.1°C)
Care launder; dry clean
Fabric Resources **Klopman Mills; J. P. Stevens & Co. Inc.**

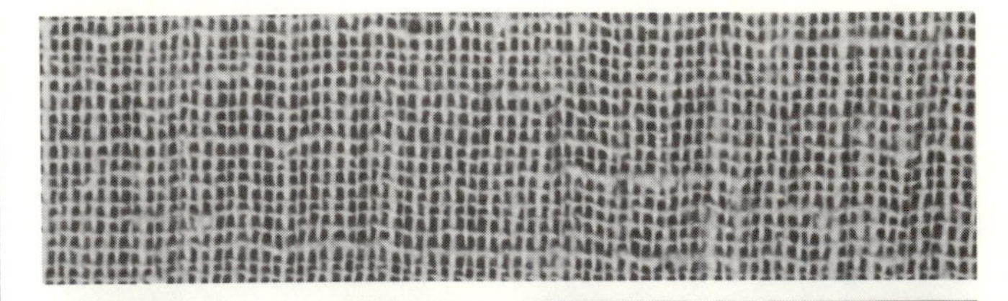

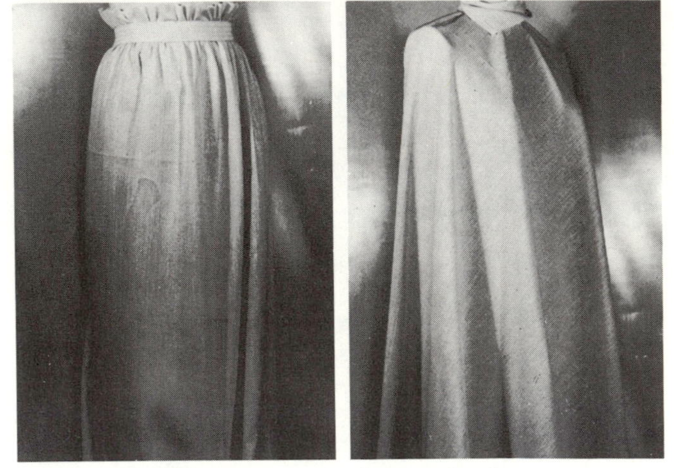

Gauze Cloth/Bunting/Cheesecloth

Fiber Content 50% cotton/50% polyester
Yarn Type carded staple; filament staple
Yarn Construction conventional; textured; irregular twist
Fabric Structure loosely woven plain weave; low thread count
Finishing Processes mercerizing; stabilizing
Color Application piece dyed–union dyed
Width 46-47 inches (116.8-119.4 cm)
Weight light to medium
Hand delicate; limp; porous; supple
Texture smooth or crinkled surface depending on finish; loose open construction
Performance Expectations air permeable; resilient; subject to shrinkage unless treated
Drapability Qualities falls into moderately soft flares; accommodates fullness by gathering; fullness retains soft graceful fall

Recommended Construction Aids & Techniques

Hand Needles betweens, 5-7; cotton darners, 8-10; embroidery, 7-10; milliners, 8-10; sharps, 9-10
Machine Needles round/set point, medium-fine 9-11
Threads mercerized cotton, 50; six-cord cotton, 60-80; cotton-covered polyester core, extra fine
Hems bound/Hong Kong finish; double fold; edge-stitched; double-edge stitched; machine-stitched; overedged
Seams flat-felled; lapped; plain; safety-stitched; taped welt; double welt
Seam Finishes self bound; double-ply bound; edge-stitched; double stitched; double-stitched and overcast; single-ply and double-ply overcast; serging/single-ply overedged
Pressing steam; safe temperature limit 325°F (164.1°C)
Care launder
Fabric Resources **Earl Glo/Erlanger Blumgart and Co. Inc.; Printsiples and Company B; Milton Sherlip Inc.**

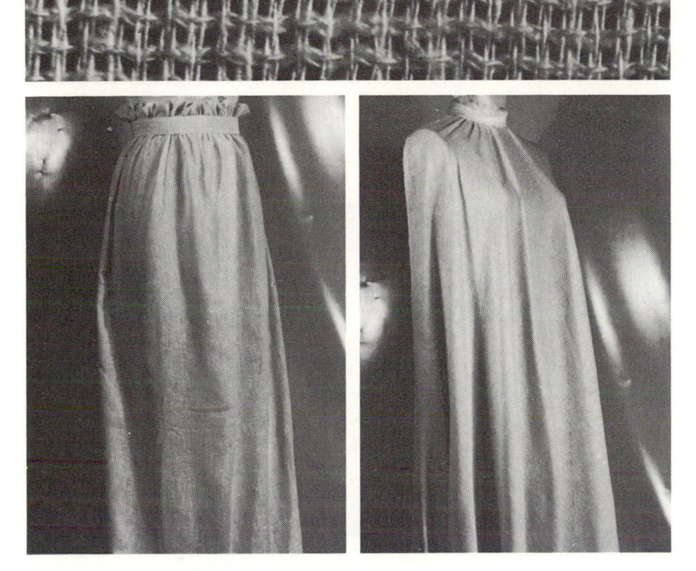

Gauze Cloth/Cotton Net

Fiber Content 50% cotton/50% polyester
Yarn Type carded or combed staple; filament staple
Yarn Construction coarse; conventional; textured
Fabric Structure plain weave variation
Finishing Processes beetling; mercerizing; shrinkage control
Color Application piece dyed
Width 44–45 inches (111.8–114.3 cm)
Weight medium
Hand harsh; pliable; soft
Texture coarse; loose open mesh effect
Performance Expectations air permeable; dimensional stability; elongation; flexible; resilient; wash-and-wear properties
Drapability Qualities falls into soft flares; accommodates fullness by gathering; fullness retains soft fall

Recommended Construction Aids & Techniques

Hand Needles betweens, 3–5; cotton darners, 5–10; embroidery, 5–8; milliners, 6–8; sharps, 6–8
Machine Needles round/set point, medium 14
Threads mercerized cotton, 50; six-cord cotton, 50; cotton- or poly-core waxed, all purpose; cotton-covered polyester core, all purpose
Hems double-fold or single-fold bias binding; bound/Hong Kong finish; faced; stitched and pinked flat; overedged; seam binding
Seams plain; safety-stitched; taped; welt; double welt
Seam Finishes single-ply bound; Hong Kong bound; single-ply overcast; pinked; pinked and stitched; serging/single-ply overedged; untreated/plain; single-ply zigzag
Pressing steam; safe temperature limit 325°F (164.1°C)
Care launder
Fabric Resources **Concord Fabrics Inc.; Printsiples and Company B**

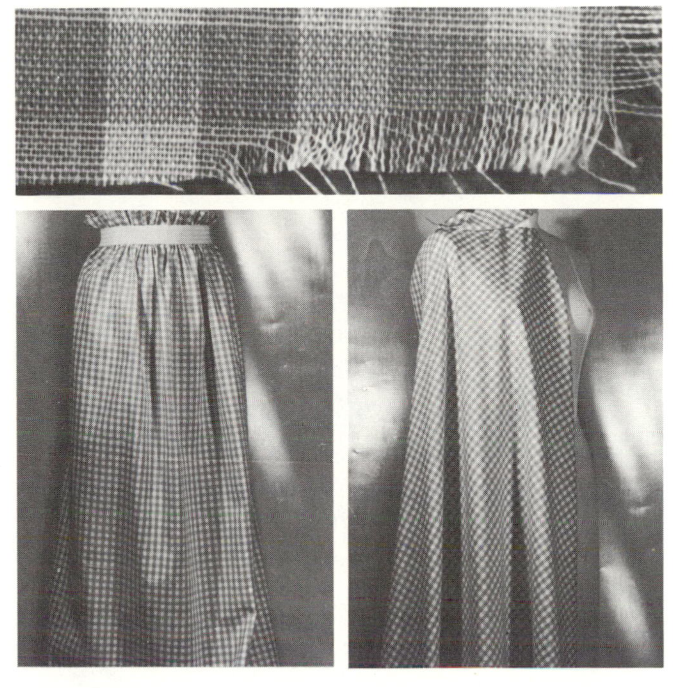

Gingham

Fiber Content 65% polyester/35% cotton
Yarn Type combed staple; filament fiber
Yarn Construction conventional; textured; high twist
Fabric Structure closely and firmly woven plain weave
Finishing Processes calendering; mercerizing; shrinkage control; sizing
Color Application piece dyed–cross dyed; yarn dyed (colored and white yarns)
Width 45 inches (114.3 cm)
Weight light to medium
Hand firm; soft to crisp depending on finish
Texture flat; smooth; check effect created by colored and white yarns
Performance Expectations abrasion resistant; dimensional stability; durable; moth resistant; 1% residual shrinkage; wash-and-wear properties
Drapability Qualities falls into moderately soft flares; accommodates fullness by pleating, gathering, elasticized shirring, smocking; fullness retains soft fall

Recommended Construction Aids & Techniques

Hand Needles betweens, 5–7; cotton darners, 8–10; embroidery, 7–10; milliners, 8–10; sharps, 9–10
Machine Needles round/set point, medium-fine 9–11
Threads cotton- or poly-core waxed, 60 fine; cotton-polyester blend, all purpose; cotton-covered polyester core, all purpose; spun polyester, all purpose
Hems bound/Hong Kong finish; double fold; edge-stitched; double-edge stitched; machine-stitched; overedged
Seams flat-felled; French; lapped; plain; safety-stitched; welt; double welt
Seam Finishes single-ply or Hong Kong bound; edge-stitched; single-ply overcast; pinked; pinked and stitched; serging/single-ply overedged; untreated/plain; single-ply zigzag
Pressing steam; safe temperature limit 325°F (164.1°C)
Care launder
Fabric Resources **Ameritex Division; Dan River Inc.**

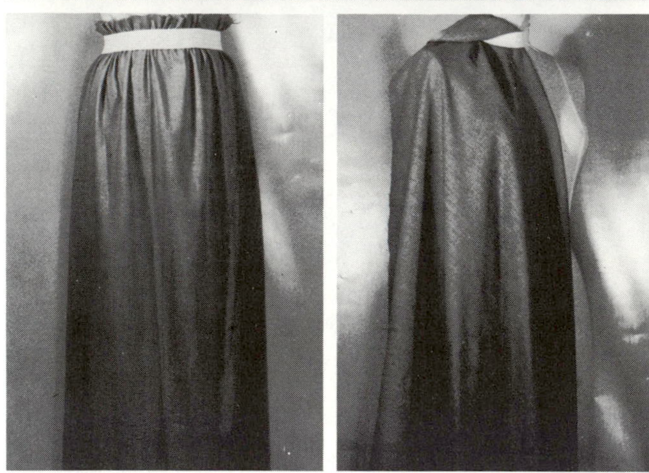

Cotton Homespun

Fiber Content 50% cotton/50% polyester
Yarn Type carded staple; filament staple
Yarn Construction conventional; textured; uneven/irregular; light twist
Fabric Structure loosely woven plain weave; low thread count
Finishing Processes beetling; mercerizing; softening; stabilizing
Color Application piece dyed-cross dyed
Width 44-45 inches (111.8-114.3 cm)
Weight medium to medium-heavy
Hand linenlike; loose; semi-soft; thick
Texture coarse; handwoven effect
Performance Expectations abrasion resistant; durable; elongation; high tensile strength; subject to shrinkage unless treated; subject to seam slippage
Drapability Qualities falls into wide cones; fullness maintains lofty effect; retains shape or silhouette of garment

Recommended Construction Aids & Techniques

Hand Needles betweens, 3-5; cotton darners, 5-10; embroidery, 5-8; milliners, 6-8; sharps, 6-8
Machine Needles round/set point, medium-coarse 14-16
Threads mercerized cotton, 40; six-cord cotton, 40; cotton- or poly-core waxed, 40 heavy duty; cotton-covered polyester core, heavy duty
Hems double-fold or single-fold bias binding; bound/Hong Kong finish; faced; stitched and pinked flat; overedged; seam binding
Seams flat-felled; lapped; plain; safety-stitched; welt; double welt
Seam Finishes single-ply or Hong Kong bound; single-ply overcast; pinked; pinked and stitched; serging/single-ply overedged; untreated/plain; single-ply zigzag
Pressing steam; safe temperature limit 325°F (164.1°C)
Care launder; dry clean
Fabric Resources Ameritex Division; Earl Glo/Erlanger Blumgart and Co. Inc.

Cotton Hopsacking

Fiber Content 100% cotton
Yarn Type carded staple
Yarn Construction conventional; ply
Fabric Structure loosely woven basket weave; low thread count
Finishing Processes beetling; calendering; softening; stabilizing
Color Application piece dyed; yarn dyed
Width 45-46 inches (114.3-116.8 cm)
Weight light to medium
Hand rough; soft; supple
Texture coarse; loose open construction
Performance Expectations absorbent; air permeable; subject to bagging and sagging; subject to snagging and seam slippage due to structure
Drapability Qualities falls into soft flares; accommodates fullness by gathering; fullness retains soft fall

Recommended Construction Aids & Techniques

Hand Needles betweens, 3-5; cotton darners, 5-10; embroidery, 5-8; milliners, 6-8; sharps, 6-8
Machine Needles round/set point, medium 11-14
Threads mercerized cotton, 50; six-cord cotton, 50; poly-core waxed, fine; cotton-covered polyester core, all purpose
Hems double-fold or single-fold bias binding; bound/Hong Kong finish; machine-stitched; overedged; seam binding
Seams flat-felled; lapped; plain; safety-stitched; taped; welt; double welt
Seam Finishes double-ply or single-ply bound; Hong Kong bound; double stitched; serging/single-ply overedged; single-ply or double-ply zigzag
Pressing steam; use pressing cloth; safe temperature limit 400°F (206.1°C)
Care launder; dry clean
Fabric Resource Ameritex Division, United Merchant and Manufacturers, Inc.

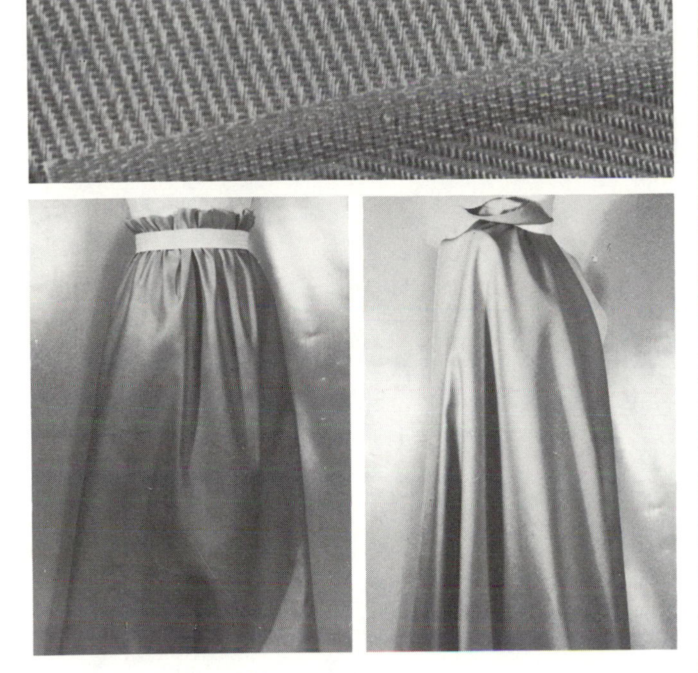

Jean Cloth/Middy Twill

Fiber Content 65% Kodel® polyester/35% cotton
Yarn Type carded or combed staple; filament fiber
Yarn Construction conventional; textured; high twist
Fabric Structure 3-harness warp-faced left-hand twill weave
Finishing Processes calendering; hard finish; mercerizing; stabilizing
Color Application piece dyed-union dyed
Width 44-45 inches (111.8-114.3 cm)
Weight light to medium
Hand compact; nonelastic; hard; harsh
Texture boardy; grainy; opaque
Performance Expectations subject to color crocking and edge abrasion; dimensional stability; durable; elongation; heat-set properties; resilient; high tensile strength
Drapability Qualities falls into firm wide flares; fullness maintains crisp effect; retains shape of garment

Recommended Construction Aids & Techniques

Hand Needles betweens, 3-5; cotton darners, 5-10; embroidery, 5-8; milliners, 6-8; sharps, 6-8
Machine Needles round/set point, medium-coarse 14-16
Threads cotton- or poly-core waxed, 50 all purpose; cotton-polyester blend, all purpose; cotton-covered polyester core, heavy duty; spun polyester, all purpose
Hems book; double-fold or single-fold bias binding; bound/Hong Kong finish; faced; stitched and pinked flat; overedged; seam binding
Seams flat-felled; lapped; plain; safety-stitched; welt; double welt
Seam Finishes book; single-ply bound; Hong Kong bound; single-ply overcast; pinked; pinked and stitched; serging/single-ply overedged; untreated/plain; single-ply zigzag
Pressing steam; safe temperature limit 325°F (164.1°C)
Care launder; dry clean
Fabric Resource Springs Mills Inc.

Khaki Cloth/Dyed Drill Cloth

Fiber Content 50% cotton/50% polyester
Yarn Type carded or combed staple; filament fiber
Yarn Construction conventional; textured; high twist
Fabric Structure twill weave
Finishing Processes permanent press precured; shrinkage control; softening
Color Application piece dyed-union dyed
Width 60 inches (152.4 cm)
Weight heavy
Hand compact; nonelastic; hard
Texture boardy; grainy; rough
Performance Expectations subject to color crocking and edge abrasion; dimensional stability; durable; elongation; heat-set properties; resilient; high tensile strength
Drapability Qualities falls into stiff wide flares; retains shape or silhouette of garment; better utilized if fitted by seaming and eliminating excess fabric

Recommended Construction Aids & Techniques

Hand Needles betweens, 3-5; chenilles, 18-22; cotton darners, 1-7; embroidery, 1-7; milliners, 3-7; sharps, 1-5; sailmakers; tapestry, 18-22
Machine Needles round/set point, medium-coarse 14-16
Threads cotton- or poly-core waxed, 50 all purpose; cotton-polyester blend, all purpose; cotton-covered polyester core, heavy duty; spun polyester, all purpose
Hems book; double-fold or single-fold bias binding; bound/Hong Kong finish; faced; stitched and pinked flat; overedged; seam binding
Seams flat-felled; plain; safety-stitched; welt; double welt
Seam Finishes book; single-ply bound; Hong Kong bound; single-ply overcast; pinked; pinked and stitched; serging/single-ply overedged; untreated/plain; single-ply zigzag
Pressing steam; safe temperature limit 325°F (164.1°C)
Care launder; dry clean
Fabric Resources Dan River Inc.; Milton Sherlip Inc.; J. P. Stevens & Co. Inc.

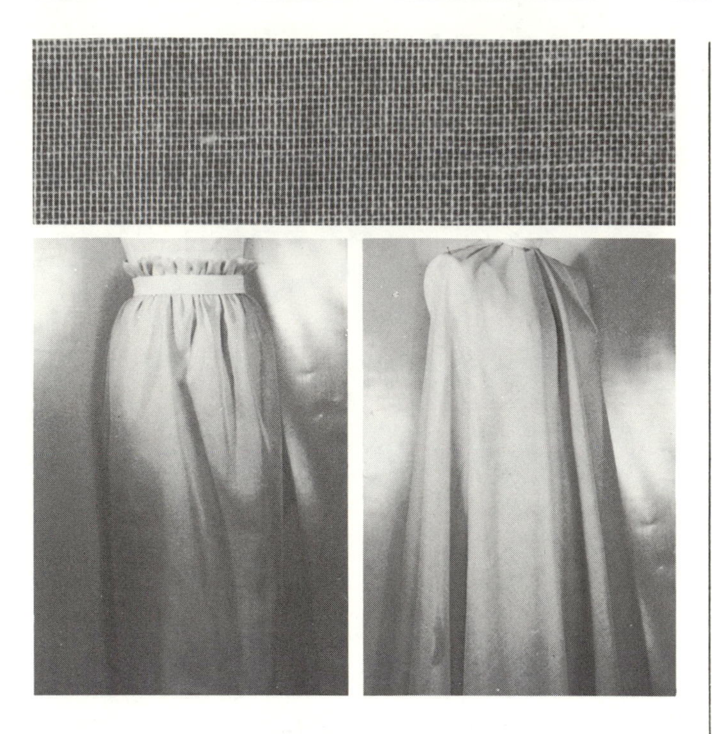

Lawn

Fiber Content 65% Kodel® polyester/35% cotton
Yarn Type combed staple; filament staple
Yarn Construction conventional; textured; high twist
Fabric Structure plain weave; medium thread count
Finishing Processes calendering; mercerizing; lightly sized
Color Application piece dyed–union dyed
Width 44–45 inches (111.8–114.3 cm)
Weight[1] light
Hand[1] delicate; soft to crisp depending on finish
Texture[1] flat; lustrous; semi-sheer; smooth
Performance Expectations abrasion resistant; air permeable; durable; high tensile strength
Drapability Qualities falls into soft flares; accommodates fullness by gathering, elasticized shirring, smocking; fullness retains soft graceful fall

Recommended Construction Aids & Techniques

Hand Needles betweens, 5–7; cotton darners, 8–10; embroidery, 7–10; milliners, 8–10; sharps, 9–10
Machine Needles round/set point, fine 9
Threads cotton- or poly-core waxed, 60 fine; cotton-polyester blend, all purpose; cotton-covered polyester core, all purpose; spun polyester, all purpose
Hems double fold; edge-stitched; double edge-stitched; horsehair; machine-stitched; hand- or machine-rolled
Seams flat-felled; French; mock French; hairline; plain; safety-stitched; tissue-stitched
Seam Finishes self bound; double-stitched and trimmed; edge-stitched; serging/single-ply overedged; untreated/plain
Pressing steam; safe temperature limit 325°F (164.1°C)
Care launder
Fabric Resources Loom Tex Corp.; Walden Textiles/Barry M. Richards Inc.

Glazed-faced Lawn

Fiber Content 100% cotton
Yarn Type combed staple
Yarn Construction conventional; fine; tight twist
Fabric Structure plain weave
Finishing Processes calendering; glazing; mercerizing; sizing
Color Application piece dyed
Width 44–45 inches (111.8–114.3 cm)
Weight light
Hand crisp; firm; light
Texture flat; glazed face; dull back; smooth
Performance Expectations abrasion resistant; absorbent; antistatic; nonpilling; nonsnagging; high tensile strength
Drapability Qualities falls into crisp flares; accommodates fullness by gathering; fullness maintains crisp effect

Recommended Construction Aids & Techniques

Hand Needles betweens, 5–7; cotton darners, 8–10; embroidery, 7–10; milliners, 8–10; sharps, 9–10
Machine Needles round/set point, fine 9
Threads mercerized cotton, 50; six-cord cotton, 80–100; cotton-covered polyester, extra fine
Hems double fold; edge-stitched; double edge-stitched; horsehair; machine-stitched; hand- or machine-rolled
Seams flat-felled; French; mock French; hairline; plain; safety-stitched; tissue-stitched
Seam Finishes self bound; double-stitched and trimmed; edge-stitched; serging/single-ply overedged; untreated/plain
Pressing steam; safe temperature limit 400°F (206.1°C)
Care launder
Fabric Resource Walden Textiles/Barry M. Richards Inc.

Madras

Fiber Content 100% cotton
Yarn Type combed staple
Yarn Construction conventional; loose twist
Fabric Structure plain weave; may be figure weaves, cords, dobbys
Finishing Processes calendering; mercerizing; stabilizing
Color Application yarn dyed
Width 39-40 inches (99.1-101.6 cm)
Weight light to medium-light
Hand fine; firm; soft
Texture flat; woven stripe, cord or check design
Performance Expectations abrasion resistant; absorbent; air permeable; elongation; high tensile strength; subject to wrinkling; damaged by chemicals and mildew; color intended to fade and create softer blends of color
Drapability Qualities falls into moderately soft flares; accommodates fullness by gathering, elasticized shirring, pleating; fullness retains soft fall

Recommended Construction Aids & Techniques

Hand Needles betweens, 5-7; cotton darners, 8-10; embroidery, 7-10; milliners, 8-10; sharps, 9-10
Machine Needles round/set point, medium-fine 9-11
Threads mercerized cotton, 50; six-cord cotton, 60-80; cotton-covered polyester, core, extra fine
Hems double fold; bound/Hong Kong finish; edge-stitched; machine-stitched; overedged
Seams flat-felled; French; lapped; plain; safety-stitched; welt; double welt
Seam Finishes single-ply bound; Hong Kong bound; edge-stitched; single-ply overcast; pinked; pinked and stitched; serging/single-ply overedged; untreated/plain; single-ply zigzag
Pressing steam; safe temperature limit 400°F (206.1°C)
Care launder
Fabric Resource **A & S Fiber Co. (sample courtesy of Bill Ditfort Inc.)**

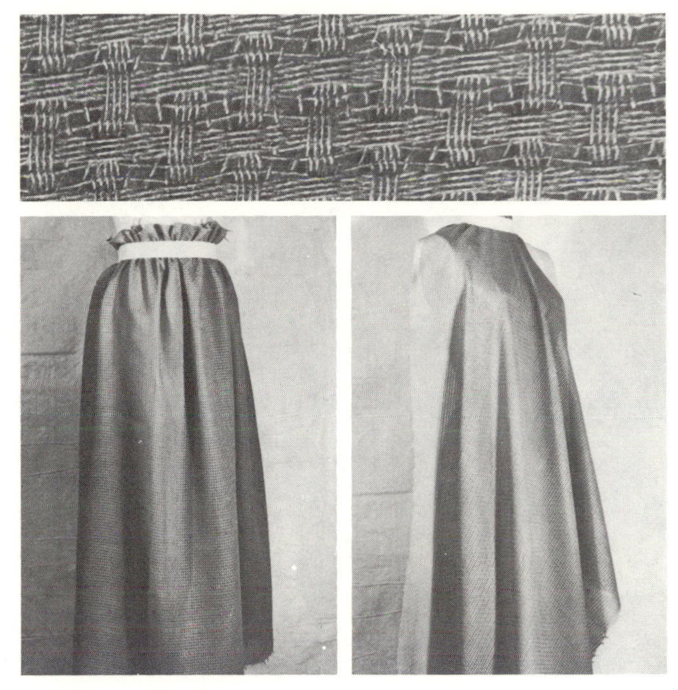

Monk's Cloth/Druid's Cloth

Fiber Content 100% cotton
Yarn Type carded staple
Yarn Construction conventional; thick
Fabric Structure 4x4 basket weave; low thread count
Finishing Processes mercerizing; singeing; stabilizing
Color Application piece dyed; yarn dyed
Width 64 inches (162.7 cm)
Weight medium to medium-heavy
Hand pliable; soft; supple
Texture coarse; high and low pattern created by loose open construction
Performance Expectations absorbent; air permeable; subject to shrinkage unless treated; subject to snagging and bagging due to weave
Drapability Qualities falls into moderately soft flares; accommodates fullness by gathering; fullness retains soft fall

Recommended Construction Aids & Techniques

Hand Needles betweens, 3-5; cotton darners, 5-10; embroidery; 5-8; milliners, 6-8; sharps, 6-8
Machine Needles round/set point, medium 14
Threads mercerized cotton, 50; six-cord cotton, 50; cotton- or poly-core waxed, all purpose; cotton-covered polyester core, all purpose
Hems single-fold or double-fold bias binding; bound/Hong Kong finish; machine-stitched
Seams plain; safety-stitched; strapped; taped
Seam Finishes single-ply bound; Hong Kong bound; double-stitched; serging/single-ply overedged
Pressing steam; use pressing cloth; safe temperature limit 400°F (206.1°C)
Care dry clean
Fabric Resource **Lawrence Textile Co. Inc.**

Muslin/Muslin Sheeting

Fiber Content 100% cotton
Yarn Type carded staple
Yarn Construction conventional; irregular coarse yarns
Fabric Structure firmly or loosely woven plain weave depending on count
Finishing Processes calendering; mercerizing; singeing; softened for soft hand; sized for crisp hand
Color Application piece dyed
Width 44-45 inches (111.8-114.3 cm)
Weight light to medium
Hand dry; firm; soft to firm depending on finish
Texture flat; dull
Performance Expectations abrasion resistant; absorbent; durable; elongation; damaged by chemicals and mildew; subject to shrinkage unless treated; high tensile strength
Drapability Qualities falls into moderately soft flares; accommodates fullness by gathering, pleating, elasticized shirring; fullness retains soft fall

Recommended Construction Aids & Techniques

Hand Needles betweens, 5-7; cotton darners, 8-10; embroidery, 7-10; milliners, 8-10; sharps, 9-10
Machine Needles round/set point, medium-fine 9-11
Threads mercerized cotton, 50; six-cord cotton, 60-80; cotton-covered polyester core, extra fine
Hems double fold; bound/Hong Kong finish; edge-stitched; machine-stitched; overedged
Seams flat-felled; lapped; plain; safety-stitched; welt; double welt
Seam Finishes single-ply bound; Hong Kong bound; edge-stitched; single-ply overcast; pinked; pinked and stitched; serging/single-ply overedged; untreated/plain; single-ply zigzag
Pressing steam; safe temperature limit 400°F (206.1°C)
Care launder
Fabric Resource Concord Fabrics Inc.

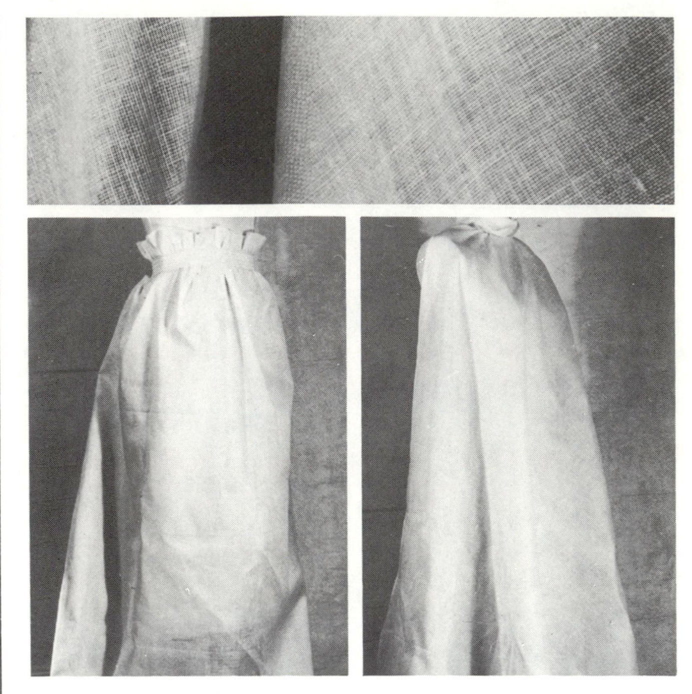

Organdy

Fiber Content 100% cotton
Yarn Type combed staple
Yarn Construction conventional; fine; tight twist
Fabric Structure plain weave
Finishing Processes calendering; glazing; mercerizing; singeing; stiffening
Color Application piece dyed
Width 45 inches (114.3 cm)
Weight light
Hand crisp; fine; firm; hard
Texture flat; sheer; smooth
Performance Expectations abrasion resistant; absorbent; air permeable; dimensional stability
Drapability Qualities falls into stiff cones; accommodates fullness by gathering, elasticized shirring, pleating; fullness maintains crisp bouffant effect

Recommended Construction Aids & Techniques

Hand Needles betweens, 6-7; cotton darners, 9-10; embroidery, 8-10; milliners, 8-10; sharps 11-12
Machine Needles round/set point, fine 9
Threads mercerized cotton, 50; six-cord cotton, 80-100; cotton-covered polyester core, extra fine
Hems double fold; edge-stitched; double edge-stitched; horsehair; machine-stitched; hand- or machine-rolled; wired
Seams flat-felled; French; mock French; hairline; plain; tissue-stitched
Seam Finishes self bound; double-stitched and trimmed; untreated/plain
Pressing steam; safe temperature limit 400°F (206.1°C)
Care launder
Fabric Resource White Rose Fabrics (sample courtesy of College Town)

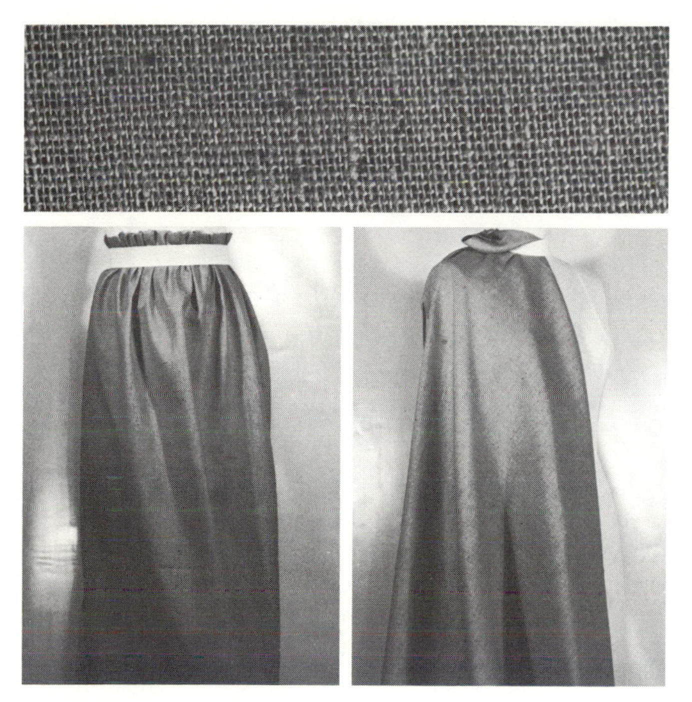

Osnaburg

Fiber Content 100% cotton
Yarn Type carded staple
Yarn Construction conventional; uneven yarns with slubs of cellulosic matter
Fabric Structure plain weave; low thread count
Finishing Processes beetling; calendering; mercerizing
Color Application piece dyed; yarn dyed
Width 44-45 inches (111.8-114.3 cm)
Weight medium to medium-heavy
Hand coarse; semi-soft; thick
Texture grainy; dark specks of cellulosic matter visible in weave
Performance Expectations abrasion resistant; absorbent; durable; elongation; subject to shrinkage unless treated; high tensile strength
Drapability Qualities falls into firm wide flares; fullness maintains lofty effect; retains shape of garment

Recommended Construction Aids & Techniques

Hand Needles betweens, 3-5; cotton darners, 5-10; embroidery, 5-8; milliners, 6-8; sharps 6-8
Machine Needles round/set point, medium 11-14
Threads mercerized cotton, 50; six-cord cotton, 50; cotton- or poly-core waxed, all purpose; cotton-covered polyester core, all purpose
Hems book; double-fold or single-fold bias binding; bound/Hong Kong finish; machine-stitched; overedged; seam binding
Seams flat-felled; lapped; plain; safety-stitched; welt; double welt
Seam Finishes single-ply bound; Hong Kong bound; single-ply overcast; pinked; pinked and stitched; serging/single-ply overedged; untreated/plain; single-ply zigzag
Pressing steam; safe temperature limit 400°F (206.1°C)
Care launder; dry clean
Fabric Resources **Concord Fabrics Inc.; M. Lowenstein & Sons, Inc.**

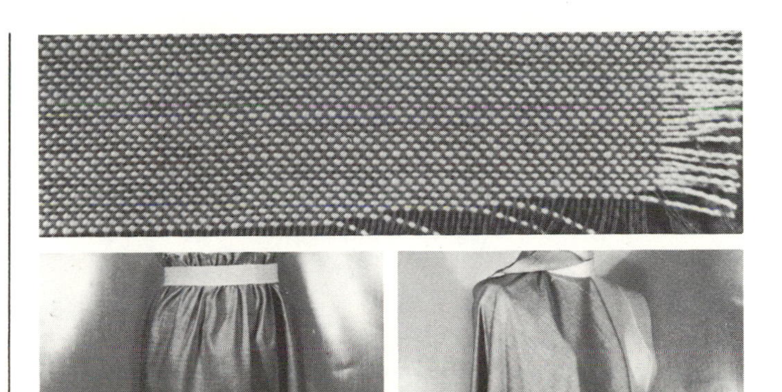

Oxford Shirting

Fiber Content 100% cotton
Yarn Type combed staple
Yarn Construction conventional; fine and thick yarns
Fabric Structure 1x2 basket weave; alternating coarse and fine yarns; count includes 88x52, 88x44
Finishing Processes calendering; mercerizing; singeing; stabilizing
Color Application yarn dyed
Width 44-45 inches (111.8-114.3 cm)
Weight light to medium-light
Hand firm; soft
Texture flat; soft luster
Performance Expectations abrasion resistant; absorbent; durable; elongation; damaged by chemicals and mildew; subject to shrinkage and wrinkling unless treated; high tensile strength
Drapability Qualities falls into soft flares; accommodates fullness by gathering, elasticized shirring, pleating; fullness retains soft fall

Recommended Construction Aids & Techniques

Hand Needles betweens, 5-7; cotton darners, 8-10; embroidery, 7-10; milliners, 8-10; sharps, 9-10
Machine Needles round/set point, medium-fine 9-11
Threads mercerized cotton, 50; six-cord cotton, 50; cotton- or poly-core waxed, fine; cotton-covered polyester core, all purpose
Hems double fold; bound/Hong Kong finish; edge-stitched; machine-stitched; overedged
Seams flat-felled; lapped; plain; safety-stitched; welt; double welt
Seam Finishes single-ply bound; Hong Kong bound; edge-stitched; single-ply overcast; pinked; pinked and stitched; serging/single-ply overedged; untreated/plain; single-ply zigzag
Pressing steam; safe temperature limit 400°F (206.1°C)
Care launder
Fabric Resource **Earl Glo/Erlanger Blumgart and Co. Inc.**

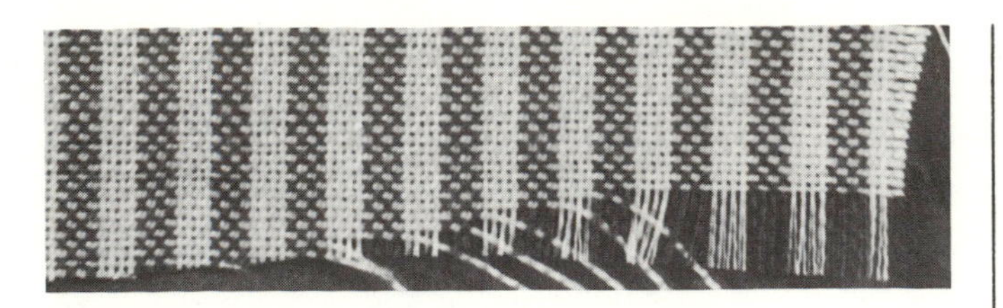

Oxford Stripe

Fiber Content 100% cotton
Yarn Type combed staple
Yarn Construction conventional; fine and thick yarns
Fabric Structure 1x2 basket weave; alternating coarse and fine yarns; count includes 88x52, 88x44
Finishing Processes calendering; mercerizing; soft luster
Color Application yarn dyed
Width 44-45 inches (111.8-114.3 cm)
Weight light to medium-light
Hand firm; soft
Texture flat; stripe pattern created by colored warp yarns
Performance Expectations abrasion resistant; absorbent; durable; elongation; subject to shrinkage and wrinkling unless treated; damaged by chemicals and mildew; high tensile strength
Drapability Qualities falls into soft flares; accommodates fullness by pleating, gathering, elasticized shirring; fullness retains soft fall

Recommended Construction Aids & Techniques

Hand Needles betweens, 5-7; cotton darners, 8-10; embroidery, 7-10; milliners, 8-10; sharps, 9-10
Machine Needles round/set point, medium-fine 9-11
Threads mercerized cotton, 50; six-cord cotton, 50; poly-core waxed, fine; cotton-covered polyester core, all purpose
Hems double fold; bound/Hong Kong finish; edge-stitched; machine-stitched; overedged
Seams flat-felled; lapped; plain; safety-stitched; welt; double welt
Seam Finishes single-ply bound; Hong Kong bound; edge-stitched; single-ply overcast; pinked; pinked and stitched; serging/single-ply overedged; untreated/plain; single-ply zigzag
Pressing steam; safe temperature limit 400°F (206.1°C)
Care launder
Fabric Resource Earl Glo/Erlanger Blumgart and Co. Inc.

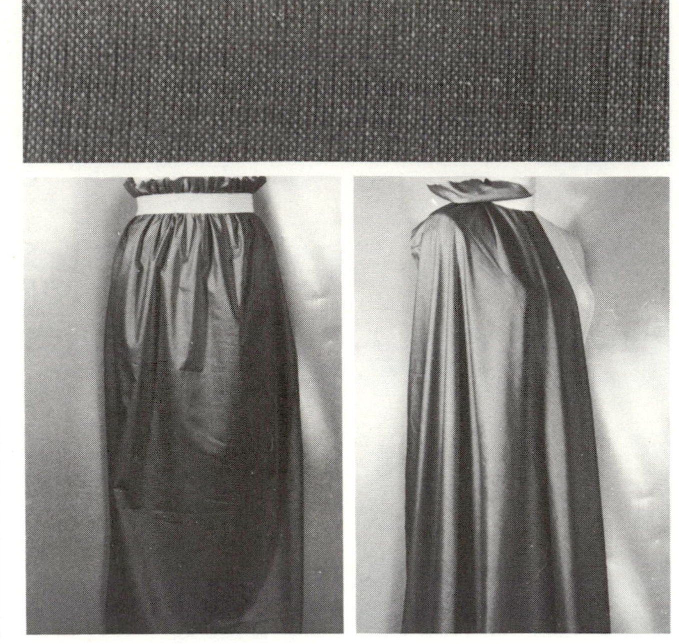

Percale

Fiber Content 100% cotton
Yarn Type carded staple
Yarn Construction conventional; fine; high twist
Fabric Structure closely woven plain weave; 90x90 or finer thread count
Finishing Processes calendering; mercerizing; sizing
Color Application piece dyed
Width 44-45 inches (111.8-114.3 cm)
Weight[1] light to medium-light
Hand[1] dry; firm; soft to firm depending on finish
Texture[1] flat; smooth
Performance Expectations abrasion resistant; absorbent; durable; elongation; damaged by chemicals and mildew; high tensile strength; subject to wrinkling
Drapability Qualities falls into soft flares; accommodates fullness by pleating, gathering, elasticized shirring; fullness retains soft fall

Recommended Construction Aids & Techniques

Hand Needles betweens, 5-7; cotton darners, 8-10; embroidery, 7-10; milliners, 8-10; sharps, 9-10
Machine Needles round/set point, medium-fine 9-11
Threads mercerized cotton, 50; six-cord cotton, 50; cotton- or poly-core waxed, fine; cotton-covered polyester core, all purpose
Hems double fold; bound/Hong Kong finish; edge-stitched; double edge-stitched; machine-stitched
Seams flat-felled; French; lapped; plain; safety-stitched; welt; double welt
Seam Finishes single-ply bound; Hong Kong bound; edge-stitched; single-ply overcast; pinked; pinked and stitched; serging/single-ply overedged; untreated/plain; single-ply zigzag
Pressing steam; safe temperature limit 400°F (206.1°C)
Care launder
Fabric Resource Printsiples and Company B

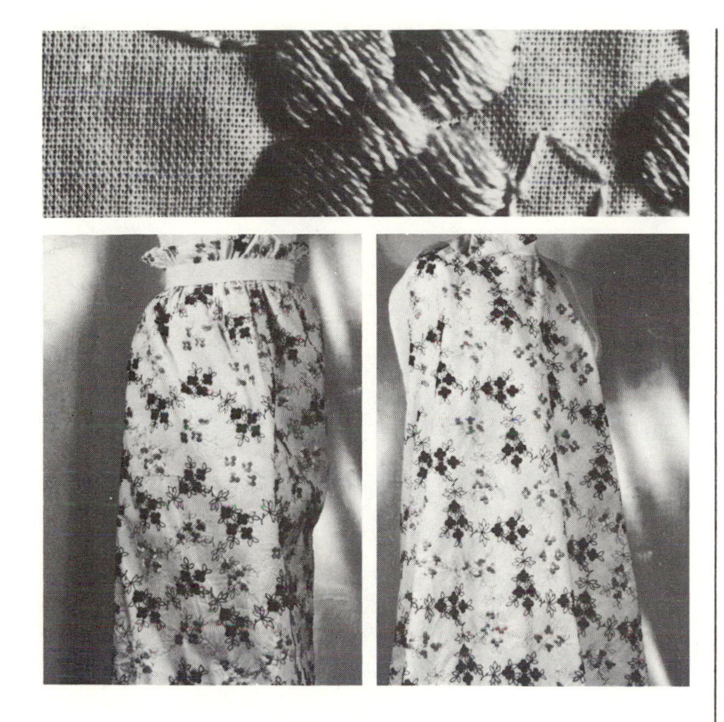

Percale with Yarn Embroidery

Fiber Content 65% cotton/35% polyester
Yarn Type filament fiber; staple
Yarn Construction conventional; textured
Fabric Structure Schiffli embroidery on plain weave base
Finishing Processes calendering; compressive shrinkage; mercerizing
Color Application base: piece dyed–union dyed; embroidery yarn: skein or packaged dyed
Width 45 inches (114.3 cm)
Weight medium-light
Hand cotton; rough
Texture base: coarse; embroidery yarn forms high and low raised pattern on surface
Performance Expectations air permeable; durable; elongation; high tensile strength; subject to wrinkling and creasing; raised embroidered design subject to snagging
Drapability Qualities falls into wide cones; fullness maintains crisp effect; retains shape of garment

Recommended Construction Aids & Techniques

Hand Needles beading, 13–16; betweens, 4–6; embroidery, 6–8; milliners, 7–9; sharps, 6–8
Machine Needles round/set point, medium-fine 9–11
Threads silk, industrial size A; poly-core waxed, fine; cotton-covered polyester core, all purpose; spun polyester, all purpose; nylon/Dacron, monocord A
Hems bound/Hong Kong finish; single-fold bias binding; faced; machine-stitched; overedged
Seams plain; safety-stitched; tissue-stitched
Seam Finishes single-ply bound; Hong Kong bound; edge-stitched; single-ply overcast; pinked and stitched; serging/single-ply overedged; untreated/plain
Pressing use pressing cloth; safe temperature limit 325°F (164.1°C)
Care launder; dry clean
Fabric Resource **Embroidery Council of America**

Plissé

Fiber Content 100% cotton
Yarn Type carded or combed staple
Yarn Construction conventional
Fabric Structure plain weave
Finishing Processes mercerizing; chemical process of crinkling and crepeing
Color Application piece dyed
Width 40 inches (101.6 cm)
Weight medium-light
Hand crisp; springy; thin
Texture allover crinkled/blistered effect
Performance Expectations absorbent; antistatic; nonpilling; texture tends to flatten after repeated launderings; high tensile strength
Drapability Qualities falls into crisp flares; accommodates fullness by gathering, elasticized shirring; fullness maintains crisp effect

Recommended Construction Aids & Techniques

Hand Needles betweens, 5–7; cotton darners, 8–10; embroidery, 7–10; milliners, 8–10; sharps, 9–10
Machine Needles round/set point, medium-fine 9–11
Threads mercerized cotton, 50; six-cord cotton, 50; cotton- or poly-core waxed, fine; cotton-covered polyester core, all purpose
Hems double fold; bound/Hong Kong finish; edge-stitched; machine-stitched; overedged
Seams flat-felled; French; lapped; plain; safety-stitched; welt
Seam Finishes single-ply bound; Hong Kong bound; edge-stitched; single-ply overcast; pinked; pinked and stitched; serging/single-ply overedged; untreated/plain; single-ply zigzag
Pressing steam; safe temperature limit 400°F (206.1°C)
Care launder
Fabric Resource **Ameritex Division, United Merchants and Manufacturers, Inc.**

Polished Cotton

Fiber Content 50% cotton/50% polyester
Yarn Type combed staple; filament staple
Yarn Construction conventional; textured; tight twist
Fabric Structure closely woven plain weave; may also be twill weave
Finishing Processes calendering; glazing; mercerizing; stabilizing
Color Application piece dyed–union dyed
Width 42 inches (106.7 cm)
Weight medium-light
Hand firm; soft; thin
Texture lustrous face; dull back; smooth
Performance Expectations abrasion resistant; dimensional stability; durable; elongation; high tensile strength; polished face tends to dull and flatten after repeated launderings
Drapability Qualities falls into moderately soft flares; accommodates fullness by pleating, gathering, elasticized shirring; fullness retains moderately soft fall

Recommended Construction Aids & Techniques

Hand Needles betweens, 5–7; cotton darners, 8–10; embroidery, 7–10; milliners, 8–10; sharps, 9–10
Machine Needles round/set point, medium-fine 9–11
Threads mercerized cotton, 50; six-cord cotton, 50; cotton- or poly-core waxed, fine; cotton-covered polyester core, all purpose
Hems double fold; bound/Hong Kong finish; edge-stitched; double edge-stitched; machine-stitched; overedged
Seams flat-felled; lapped; plain; safety-stitched; welt; double welt
Seam Finishes single-ply bound; Hong Kong bound; edge-stitched; single-ply overcast; pinked; pinked and stitched; serging/single-ply overedged; untreated/plain; single-ply zigzag
Pressing steam; safe temperature limit 325°F (164.1°C)
Care launder
Fabric Resource **Concord Fabrics Inc.**

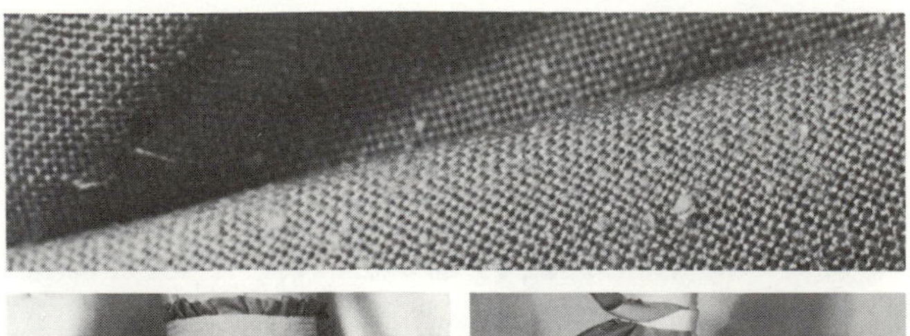

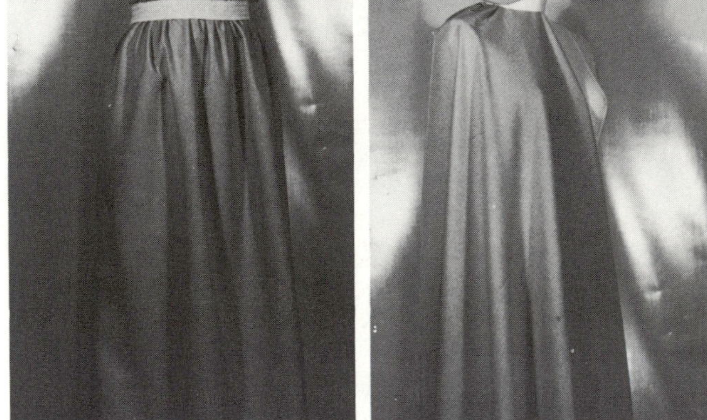

Cotton Pongee/Silk Noil Cotton

Fiber Content cotton/polyester blend
Yarn Type carded staple; filament fiber
Yarn Construction conventional; textured; seed and noil novelty yarns
Fabric Structure plain weave; heavier novelty yarns in filling
Finishing Processes mercerizing; schreiner calendering; stabilizing
Color Application piece dyed–union dyed
Width 44–45 inches (111.8–114.3 cm)
Weight medium
Hand coarse; light; porous; soft
Texture irregular slubbed yarn with raised noils in filling
Performance Expectations abrasion resistant; dimensional stability; durable; elongation; high tensile strength; noils tend to wear off or dislodge after repeated launderings
Drapability Qualities falls into soft flares; accommodates fullness by gathering, elasticized shirring; fullness retains soft fall

Recommended Construction Aids & Techniques

Hand Needles betweens, 5–7; cotton darners, 8–10; embroidery, 7–10; milliners, 8–10; sharps, 9–10
Machine Needles round/set point, medium 11–14
Threads mercerized cotton, 50; six-cord cotton, 60–80; cotton-covered polyester core, extra fine
Hems bound/Hong Kong finish; edge-stitched; machine-stitched; overedged
Seams flat-felled; lapped; plain; safety-stitched; welt; double welt
Seam Finishes single-ply bound; Hong Kong bound; edge-stitched; single-ply overcast; pinked; pinked and stitched; serging/single-ply overedged; untreated/plain; single-ply zigzag
Pressing steam; safe temperature limit 325°F (164.1°C)
Care launder
Fabric Resources **Concord Fabrics Inc.; Dan River Inc.; Milton Sherlip Inc.**

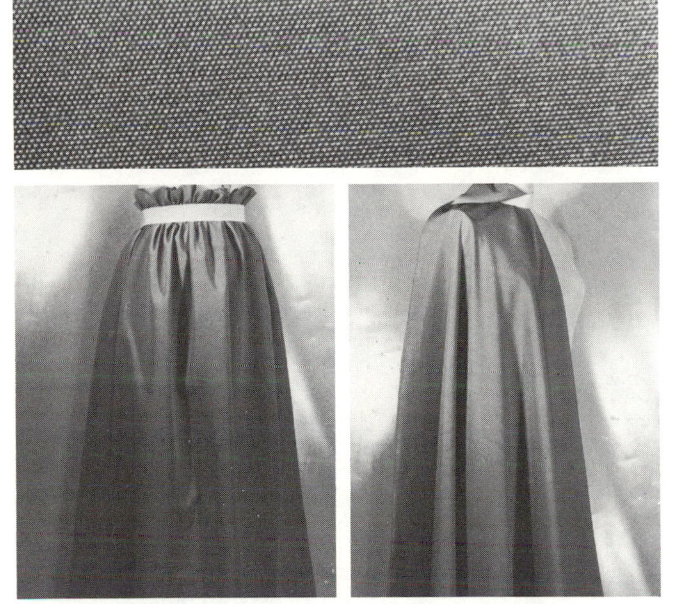

Cotton Poplin (soft finish)

Fiber Content 100% polyester
Yarn Type filament fiber
Yarn Construction textured; high twist
Fabric Structure plain weave; additional filling yarns create rib in crosswise direction
Finishing Processes calendering; mercerizing; shrinkage control; singeing; softening
Color Application piece dyed; yarn dyed
Width 59–60 inches (149.9–152.4 cm)
Weight medium
Hand nonelastic; firm; scroopy; soft
Texture papery; fine crosswise rib
Performance Expectations subject to color crocking and edge abrasion; dimensional stability; durable; elongation; heat-set properties; chemical and mildew resistant; high tensile strength
Drapability Qualities falls into firm wide flares; fullness maintains crisp effect; maintains crisp folds, creases, pleats

Recommended Construction Aids & Techniques

Hand Needles betweens, 5–7; cotton darners, 8–10; embroidery, 7–10; milliners, 8–10; sharps, 9–10
Machine Needles round/set point, medium 11–14
Threads cotton- or poly-core waxed, 60 fine; cotton-polyester blend; all purpose; cotton-covered polyester core, all purpose; spun polyester, all purpose
Hems book; double-fold or single-fold bias binding; bound/Hong Kong finish; machine-stitched; overedged; seam binding
Seams flat-felled; lapped; plain; safety-stitched; welt; double welt
Seam Finishes book; single-ply bound; Hong Kong bound; single-ply overcast; pinked; pinked and stitched; serging/single-ply overedged; untreated/plain; single-ply zigzag
Pressing steam; safe temperature limit 325°F (164.1°C)
Care launder; dry clean
Fabric Resource Reeves Brothers Inc.

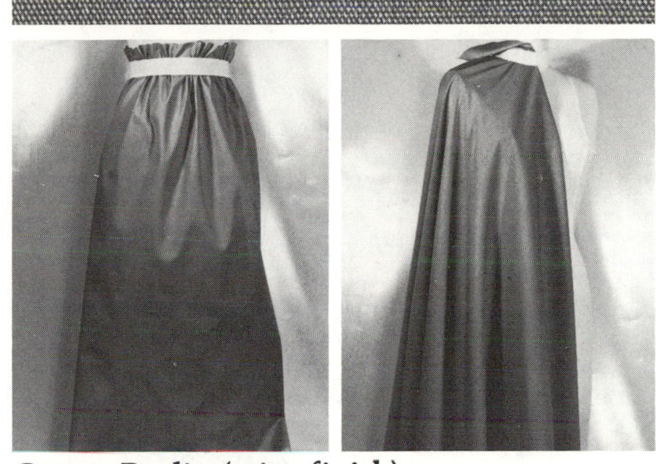

Cotton Poplin (crisp finish)

Fiber Content 65% polyester/35% cotton
Yarn Type combed staple; filament fiber
Yarn Construction conventional; textured; high twist; fine warp yarns/heavy filling yarns
Fabric Structure tightly woven plain weave; heavier filling yarns create rib in crosswise direction
Finishing Processes flame retardant; mildew proof; sizing; Zepel® water/stain repellent
Color Application piece dyed–union dyed
Width 44–45 inches (111.8–114.3 cm)
Weight medium to medium-heavy
Hand nonelastic; firm; scroopy; stiff
Texture boardy; fine crosswise rib
Performance Expectations subject to color crocking and edge abrasion; dimensional stability; durable; elongation; heat-set properties; resilient; high tensile strength; water and stain resistant
Drapability Qualities falls into stiff wide cones; maintains crisp folds, creases, pleats; retains shape or silhouette of garment

Recommended Construction Aids & Techniques

Hand Needles betweens, 3–5; cotton darners, 5–10; embroidery, 5–8; milliners, 6–8; sharps, 6–8
Machine Needles round/set point, medium 11–14
Threads poly-core waxed, 50 all purpose; cotton-polyester blend, all purpose; cotton-covered polyester core, heavy duty; spun polyester, all purpose
Hems book; double-fold or single-fold bias binding; bound/Hong Kong finish; faced; stitched and pinked flat; overedged; seam binding
Seams flat-felled; lapped; plain; safety-stitched; welt; double welt
Seam Finishes book; single-ply bound; Hong Kong bound; single-ply overcast; pinked; pinked and stitched; serging/single-ply overedged; untreated/plain; single-ply zigzag
Pressing steam; safe temperature limit 325°F (164.1°C)
Care dry clean
Fabric Resource Reeves Brothers Inc.

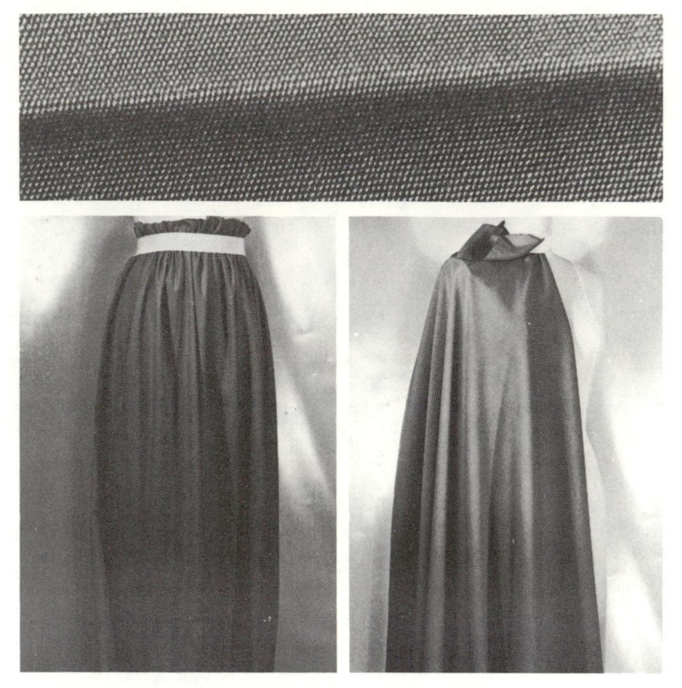

Cotton Poplin (heavyweight)

Fiber Content 50% Avril® polyester/50% polyester
Yarn Type filament fiber
Yarn Type textured; high twist
Fabric Structure plain weave; additional filling yarns create rib in crosswise direction
Finishing Processes beetling; calendering; mercerizing; singeing; stabilizing
Color Application piece dyed–union dyed
Width 44–45 inches (111.8–114.3 cm)
Weight heavy
Hand compact; scroopy; semi-soft; solid; thick
Texture papery; pronounced crosswise rib
Performance Expectations subject to color crocking and edge abrasion; dimensional stability; durable; elongation; heat-set properties; chemical and mildew resistant; resilient; high tensile strength
Drapability Qualities falls into crisp wide flares; fullness maintains crisp effect; maintains crisp folds, creases, pleats

Recommended Construction Aids & Techniques

Hand Needles betweens, 3–5; cotton darners, 5–10; embroidery, 5–8; milliners, 6–8; sharps, 6–8
Machine Needles round/set point, medium 11–14
Threads cotton- or poly-core waxed, 50 all purpose; cotton-polyester blend, all purpose; cotton-covered polyester core, heavy duty; spun polyester, all purpose
Hems book; double-fold or single-fold bias binding; bound/Hong Kong finish; faced; stitched and pinked flat; overedged; seam binding
Seams flat-felled; lapped; plain; safety-stitched; welt; double welt
Seam Finishes book; single-ply bound; Hong Kong bound; single-ply overcast; pinked; pinked and stitched; serging/single-ply overedged; untreated/plain; single-ply zigzag
Pressing steam; safe temperature limit 325°F (164.1°C)
Care dry clean
Fabric Resource **Concord Fabrics Inc.**

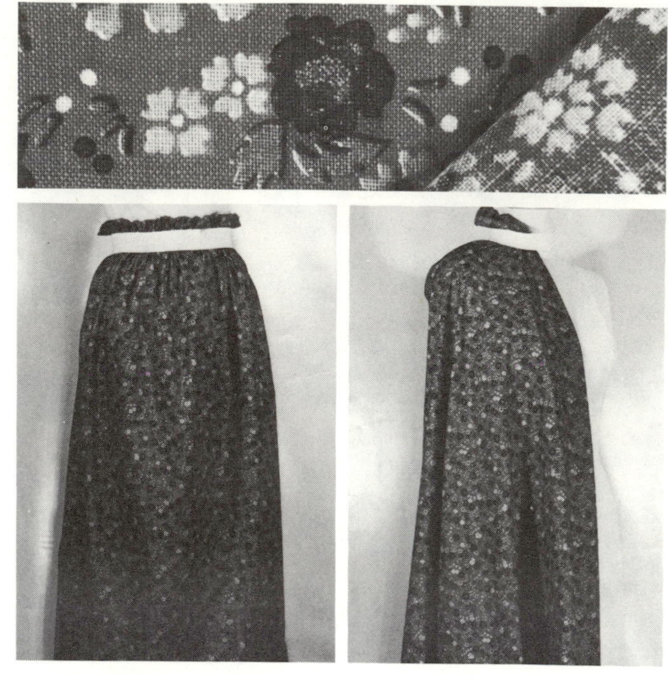

Printcloth[2]

Fiber Content cotton/polyester blend
Yarn Type carded staple; filament fiber
Yarn Construction conventional; thin; high twist
Fabric Structure plain weave; low to medium thread count
Finishing Processes calendering; mercerizing; sizing
Color Application printing: direct, discharge, resist, roller methods
Width 44–45 inches (111.8–114.3 cm)
Weight medium-light to light
Hand thin; soft to firm depending on finish
Texture flat; printed design on surface
Performance Expectations abrasion resistant; durable; elongation; nonsnagging; subject to shrinkage unless treated
Drapability Qualities falls into moderately soft flares; accommodates fullness by pleating, gathering, elasticized shirring; fullness retains soft fall

Recommended Construction Aids & Techniques

Hand Needles betweens, 5–7; cotton darners, 8–10; embroidery, 7–10; milliners, 8–10; sharps, 9–10
Machine Needles round/set point, medium-fine 9–11
Threads mercerized cotton, 50; six-cord cotton, 50; cotton- or poly-core waxed, fine; cotton-covered polyester core, all purpose
Hems double fold; bound/Hong Kong finish; edge-stitched; machine-stitched; overedged
Seams flat-felled; French; lapped; plain; safety-stitched; welt; double welt
Seam Finishes single-ply bound; Hong Kong bound; edge-stitched; pinked; pinked and stitched; single-ply overcast; serging/single-ply overedged; untreated/plain; single-ply zigzag
Pressing steam; safe temperature limit 325°F (164.1°C)
Care launder
Fabric Resources **M. Lowenstein & Sons, Inc.; Marcus Brothers Textile Inc.; Peter Pan Fabrics (sample courtesy of Angel Martinez)**

Calico Printcloth[3]

Fiber Content cotton/polyester blend
Yarn Type carded staple; filament fiber
Yarn Construction conventional; thin; high twist
Fabric Structure plain weave; low to medium thread count
Finishing Processes calendering; mercerizing; sizing
Color Application printing: direct, discharge, resist, roller methods
Width 44–45 inches (111.8–114.3 cm)
Weight medium-light to light
Hand thin; soft to firm depending on finish
Texture flat; printed design on surface
Performance Expectations abrasion resistant; durable; elongation; nonsnagging; subject to shrinkage unless treated
Drapability Qualities falls into moderately soft flares; accommodates fullness by pleating, gathering, elasticized shirring; fullness retains soft fall

Recommended Construction Aids & Techniques

Hand Needles betweens, 5–7; cotton darners, 8–10; embroidery, 7–10; milliners, 8–10; sharps, 9–10
Machine Needles round/set point, medium-fine 9–11
Threads mercerized cotton, 50; six-cord cotton, 50; cotton- or poly-core waxed, fine; cotton-covered polyester core, all purpose
Hems double fold; bound/Hong Kong finish; edge-stitched; machine-stitched; overedged
Seams flat-felled; French; lapped; plain; safety-stitched; welt; double welt
Seam Finishes single-ply bound; Hong Kong bound; edge-stitched; pinked; pinked and stitched; single-ply overcast; serging/single-ply overedged; untreated/plain; single-ply zigzag
Pressing steam; safe temperature limit 325°F (164.1°C)
Care launder
Fabric Resources **M. Lowenstein & Sons, Inc.; Marcus Brothers Textile Inc.; Peter Pan Fabrics (sample courtesy of Angel Martinez)**

Cotton Ratiné

Fiber Content 85% rayon/15% acetate
Yarn Type filament fiber
Yarn Construction bulked; novelty seed and slub yarns; textured
Fabric Structure loosely woven plain weave
Finishing Processes beetling; mercerizing; stabilizing
Color Application piece dyed–cross dyed; yarn dyed
Width 44–45 inches (111.8–114.3 cm)
Weight medium-heavy
Hand coarse; loose; medium-hard; thick
Texture slub yarn and seed noil create rough surface
Performance Expectations elongation; nonpilling; subject to abrasion and snagging; subject to shrinkage unless treated
Drapability Qualities falls into wide cones; fullness maintains lofty effect; retains shape or silhouette of garment

Recommended Construction Aids & Techniques

Hand Needles betweens, 3–5; cotton darners, 5–10; embroidery, 5–8; milliners, 6–8; sharps, 6–8
Machine Needles round/set point, medium-coarse 14–16
Threads cotton- or poly-core waxed, 50 all purpose; cotton-polyester blend, all purpose; cotton-covered polyester core, heavy duty; spun polyester, all purpose
Hems double fold or single-fold bias binding; bound/Hong Kong finish; faced; stitched and pinked flat; overedged; seam binding
Seam plain; safety-stitched; taped; welt; double welt
Seam Finishes single-ply bound; Hong Kong bound; single-ply overcast; pinked; pinked and stitched; serging/single-ply overedged; single-ply zigzag
Pressing steam; safe temperature limit 350°F (178.1°C)
Care dry clean
Fabric Resource **DeMarco Fabrics (sample courtesy of College Town)**

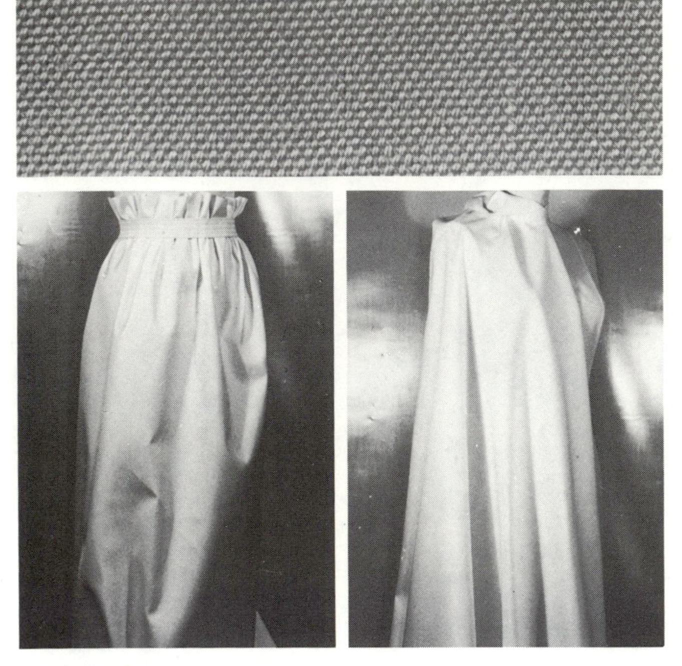

Sailcloth

Fiber Content 85% polyester/15% cotton
Yarn Type carded or combed staple; filament fiber
Yarn Construction conventional; textured
Fabric Structure closely woven plain weave
Finishing Processes calendering; crease and wrinkle resistant; mercerizing
Color Application piece dyed–union dyed
Width 45 inches (114.3 cm)
Weight medium to medium-heavy
Hand compact; crisp; rigid; thick
Texture grainy; rough
Performance Expectations abrasion resistant; dimensional stability; durable; heat-set properties; resilient; high tensile strength; crease and wrinkle resistant
Drapability Qualities falls into firm wide cones; fullness maintains crisp effect; retains shape of garment

Recommended Construction Aids & Techniques

Hand Needles betweens, 3–5; cotton darners, 5–10; embroidery, 5–8; milliners, 6–8; sharps, 6–8
Machine Needles round/set point, medium-coarse 14–16
Threads cotton- or poly-core waxed, 50 all purpose; cotton-polyester blend, all purpose; cotton-covered polyester core, heavy duty; spun polyester, all purpose
Hems book; double-fold or single-fold bias binding; bound/Hong Kong finish; machine-stitched; overedged; seam binding
Seams flat-felled; lapped; plain; safety-stitched; welt; double welt
Seam Finishes book; single-ply bound; Hong Kong bound; single-ply overcast; pinked; pinked and stitched; serging/single-ply overedged; untreated/plain; single-ply zigzag
Pressing steam; safe temperature limit 325°F (164.1°C)
Care launder
Fabric Resources **Ameritex Division, United Merchants and Manufacturers, Inc.; Hargro Fabrics Inc.; Loom Tex Corp.**

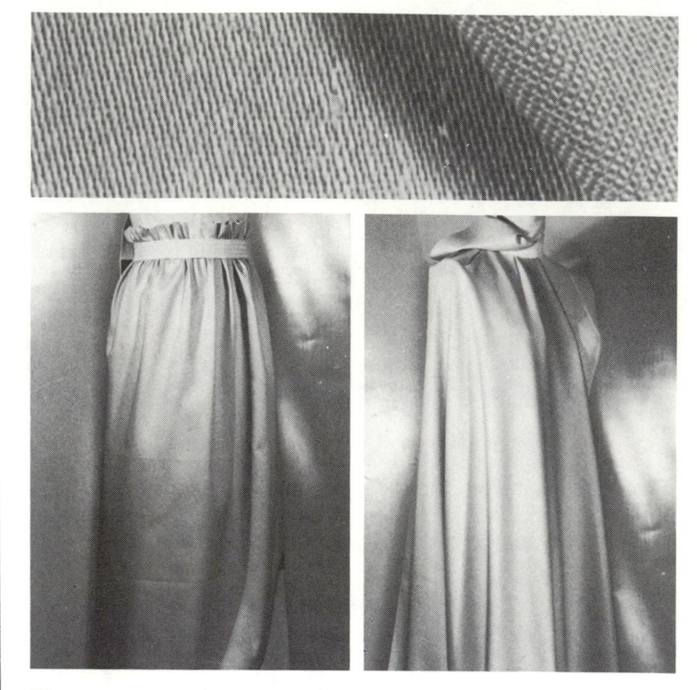

Sateen

Fiber Content 50% cotton/50% polyester
Yarn Type combed staple; filament fiber
Yarn Construction conventional; textured; high twist
Fabric Structure filling face satin weave
Finishing Processes schreiner calendering; crease and wrinkle resistant; mercerizing; permanent press precured
Color Application piece dyed–union dyed
Width 60 inches (152.4 cm)
Weight medium to medium-heavy
Hand compact; firm; satiny soft; solid
Texture smooth; soft lustrous face; dull back
Performance Expectations dimensional stability; durable; elongation; high tensile strength; subject to edge abrasion and snagging; wash-and-wear properties; crease and wrinkle resistant
Drapability Qualities falls into wide cones; fullness maintains lofty effect; retains shape of garment

Recommended Construction Aids & Techniques

Hand Needles betweens, 3–5; cotton darners, 5–10; embroidery, 5–8; milliners, 6–8; sharps, 6–8
Machine Needles round/set point, medium 11–14
Threads cotton- or poly-core waxed, 50 all purpose; cotton-polyester blend, all purpose; cotton-covered polyester core, heavy duty; spun polyester, all purpose
Hems book; double-fold or single-fold bias binding; bound/Hong Kong finish; machine-stitched; overedged; seam binding
Seams flat-felled; lapped; plain; safety-stitched; welt; double welt
Seam Finishes book; single-ply bound; Hong Kong bound; single-ply overcast; pinked; pinked and stitched; serging/single-ply overedged; untreated/plain; single-ply zigzag
Pressing steam; safe temperature limit 325°F (164.1°C)
Care launder; dry clean
Fabric Resource **J. P. Stevens & Co. Inc.**

Cotton Seersucker (even stripe)

Fiber Content 100% cotton; cotton/polyester blend
Yarn Type combed staple; filament fiber
Yarn Construction conventional; high twist
Fabric Structure plain weave; alternating crinkled stripes permanently woven by weaving warp slack and tight
Finishing Processes caustic acid for crinkled effect; Russpress® permanent press precured; shrinkage control
Color Application yarn dyed
Width 44-45 inches (111.8-114.3 cm)
Weight medium-light; medium; medium-heavy
Hand crisp; nonelastic; firm
Texture stripe effect created by alternating crinkled and smooth surface
Performance Expectations abrasion resistant; absorbent; antistatic; dimensional stability; durable; nonpilling; nonsnagging; resilient; high tensile strength; wash-and-wear properties
Drapability Qualities falls into crisp flares; accommodates fullness by pleating, gathering, elasticized shirring; fullness maintains crisp effect

Recommended Construction Aids & Techniques

Hand Needles betweens, 3-5; cotton darners, 5-10; embroidery, 5-8; milliners, 6-8; sharps, 6-8
Machine Needles round/set point, medium-fine 9-11
Threads mercerized cotton, 50; six-cord cotton, 50; cotton- or poly-core waxed, fine; cotton-covered polyester core, all purpose
Hems book; double fold; bound/Hong Kong finish; edge-stitched; machine-stitched; overedged
Seams flat-felled; lapped; plain; safety-stitched; welt; double welt
Seam Finishes book; single-ply or Hong Kong bound; edge-stitched; single-ply and overcast; pinked; pinked and stitched; serging/single-ply overedged; untreated/plain; single-ply zigzag
Pressing steam; safe temperature limit 325°F (164.1°C)
Care launder
Fabric Resources **Drexler Associates Inc.; Earl Glo/Erlanger Blumgart and Co. Inc.; Russell Corporation**

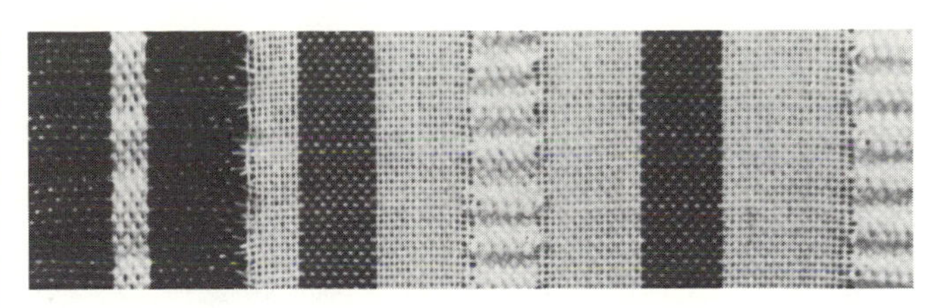

Cotton Seersucker (novelty stripe)

Fabric Resource Russell Corporation

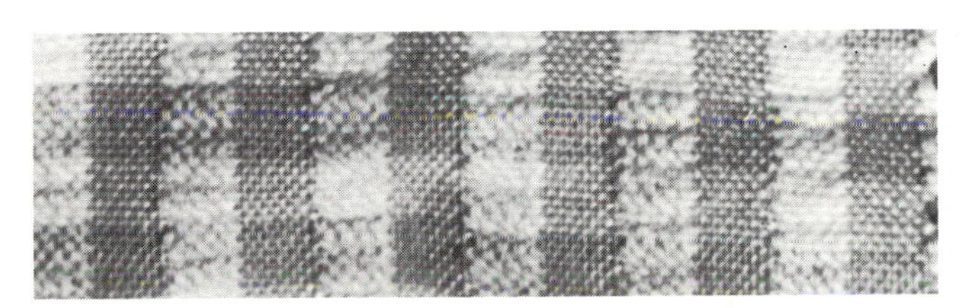

Cotton Seersucker (checkered effect)

Fabric Resource Russell Corporation

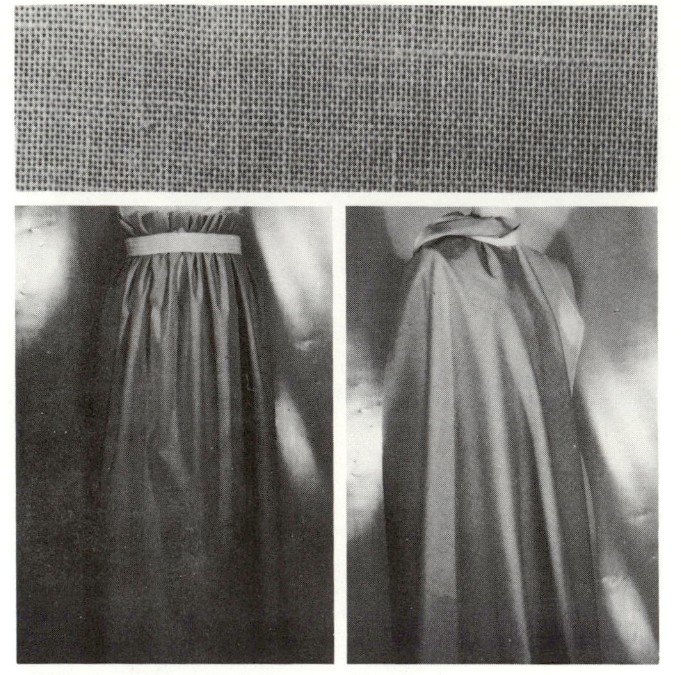

Sheeting/Muslin Sheeting

Fiber Content 50% cotton/50% polyester
Yarn Type carded staple; filament fiber
Yarn Construction conventional; textured; slightly irregular, heavier filling yarn
Fabric Structure plain weave; low thread count
Finishing Processes calendering; mercerizing; softly sized
Color Application piece dyed–union dyed
Width 68–69 inches (172.7–175.3 cm)
Weight[1] medium-light to medium
Hand[1] firm; soft to crisp depending on finish
Texture[1] flat; coarser than muslin; noticeable irregular yarns in structure
Performance Expectations abrasion resistant; durable; elongation; nonsnagging; subject to shrinkage unless treated
Drapability Qualities falls into moderately soft flares; accommodates fullness by gathering, elasticized shirring, pleating; fullness retains moderately soft fall

Recommended Construction Aids & Techniques

Hand Needles betweens, 5–7; cotton darners, 8–10; embroidery, 7–10; milliners, 8–10; sharps, 9–10
Machine Needles round/set point, medium-fine 9–11
Threads mercerized cotton, 50; six-cord cotton, 50; cotton- or poly-core waxed, all purpose; cotton-covered polyester core, all purpose
Hems single-fold bias binding; bound/Hong Kong finish; double fold; edge-stitched; machine-stitched; overedged
Seams flat-felled; lapped; plain; safety-stitched; welt; double welt
Seam Finishes single-ply bound; Hong Kong bound; edge-stitched; single-ply overcast; pinked; pinked and stitched; serging/single-ply overedged; untreated/plain; single-ply zigzag
Pressing steam; safe temperature limit 325°F (164.1°C)
Care launder
Fabric Resource M. Lowenstein & Sons, Inc.

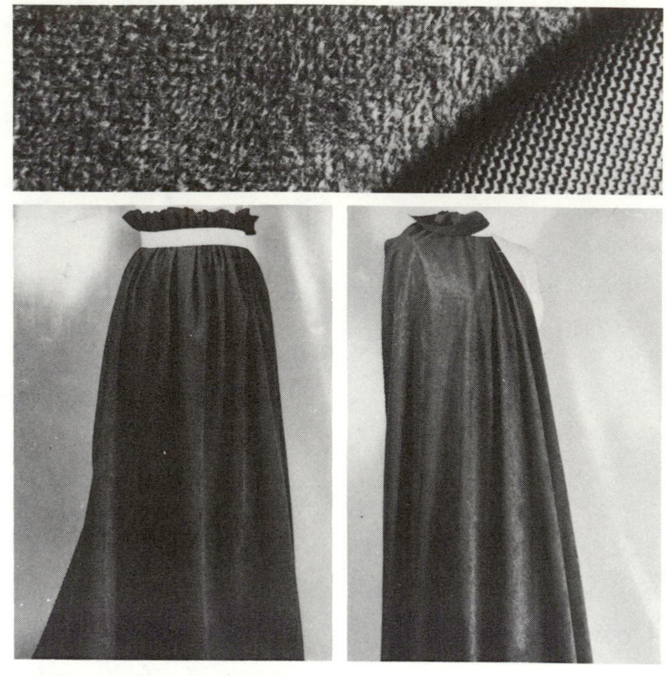

Cotton Suede Cloth

Fiber Content 100% polyester
Yarn Type filament fiber
Yarn Construction textured
Fabric Structure pile face; knit back
Finishing Processes heat-set stabilizing; sanding/suedeing
Color Application piece dyed; solution dyed; yarn dyed
Width 62 inches (157.5 cm)
Weight medium-light to medium
Hand downy soft; supple
Texture napped face; plain back; opaque; velvety
Performance Expectations subject to color crocking and edge abrasion; dimensional stability; elongation; heat-set properties; moth repellent; resilient; high tensile strength; subject to static buildup
Drapability Qualities falls into soft flares; accommodates fullness by gathering, elasticized shirring, pleating; fullness retains soft graceful fall

Recommended Construction Aids & Techniques

Hand Needles betweens, 5–7; cotton darners, 8–10; embroidery, 7–10; milliners, 8–10; sharps, 9–10
Machine Needles round/set point, medium 11–14
Threads cotton- or poly-core waxed, 60 fine; cotton-polyester blend, all purpose; cotton-covered polyester core, all purpose; spun polyester, all purpose
Hems book; double-fold or single-fold bias binding; bound/Hong Kong finish; machine-stitched; overedged; seam binding
Seams flat-felled; lapped; plain; safety-stitched; welt; double welt
Seam Finishes book; single-ply bound; Hong Kong bound; single-ply overcast; pinked; pinked and stitched; serging/single-ply overedged; untreated/plain; single-ply zigzag
Pressing steam; safe temperature limit 325°F (164.1°C)
Care launder
Fabric Resources Gloversville Mills Inc.; Private Collections Fabrics, Ltd.; Springs Mills Inc.

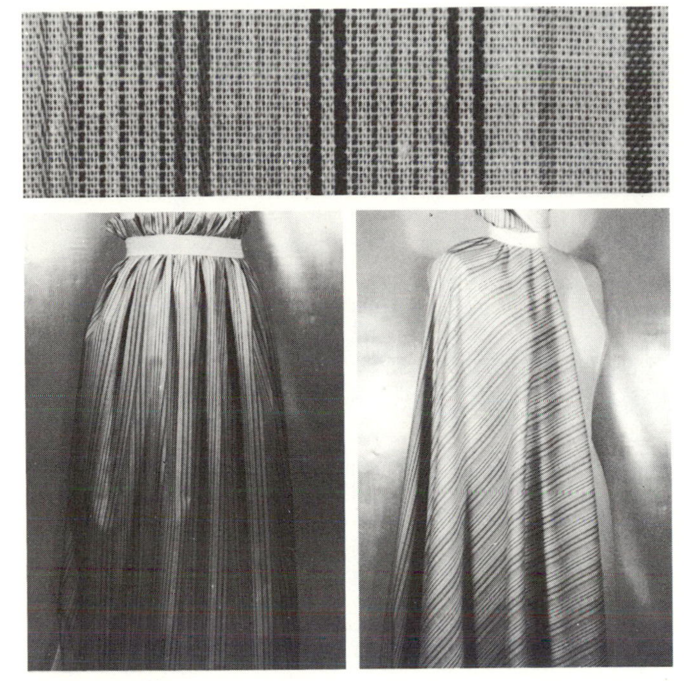

Tattersall

Fiber Content 65% Kodel® polyester/35% cotton
Yarn Type combed staple; filament fiber
Yarn Construction conventional; textured
Fabric Structure plain weave
Finishing Processes calendered for slight luster or polish on face; mercerizing; softening
Color Application piece dyed-cross dyed; yarn dyed
Width 60 inches (152.4 cm)
Weight light
Hand fine; firm; soft
Texture flat; smooth; checkered or plaid patterned design
Performance Expectations abrasion resistant; dimensional stability; durable; elongation; nonsnagging; high tensile strength
Drapability Qualities falls into moderately soft flares; accommodates fullness by gathering, elasticized shirring, pleating; fullness retains soft fall

Recommended Construction Aids & Techniques

Hand Needles betweens, 5–7; cotton darners, 8–10; embroidery, 7–10; milliners, 8–10; sharps, 9–10
Machine Needles round/set point, medium-fine 9–11
Threads cotton- or poly-core waxed, 60 fine; cotton-polyester blend, all purpose; cotton-covered polyester core, all purpose; spun polyester, all purpose
Hems bound/Hong Kong finish; edge-stitched; double edge-stitched; machine-stitched; overedged
Seams flat-felled; French; lapped; plain; safety-stitched; welt; double welt
Seam Finishes single-ply bound; Hong Kong bound; edge-stitched; single-ply overcast; pinked; pinked and stitched; serging/single-ply overedged; untreated/plain; single-ply zigzag
Pressing steam; safe temperature limit 325°F (164.1°C)
Care launder
Fabric Resource **Russell Corporation**

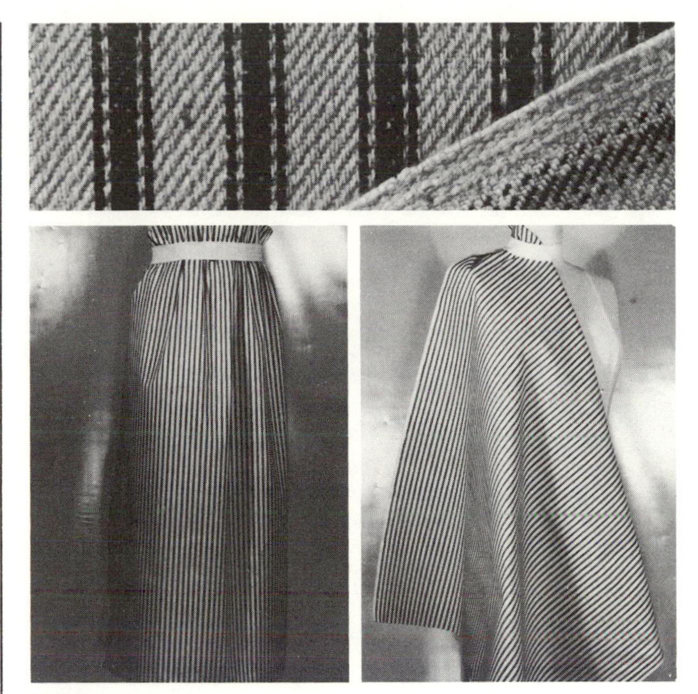

Ticking

Fiber Content 100% cotton
Yarn Type carded or combed staple
Yarn Construction conventional; high twist
Fabric Structure closely woven twill weave; may also be plain weave
Finishing Processes beetling; calendering; mercerizing; stabilizing
Color Application yarn dyed (colored warp/white filling)
Width 45 inches (114.3 cm)
Weight medium-heavy to heavy
Hand compact; crisp; heavy; rigid
Texture grainy; rough; evenly spaced stripe pattern in lengthwise direction
Performance Expectations abrasion resistant; absorbent; durable; elongation; high tensile and bursting strength; subject to shrinkage unless treated
Drapability Qualities falls into stiff wide cones; maintains crisp folds, creases; retains shape or silhouette of garment

Recommended Construction Aids & Techniques

Hand Needles betweens, 3–5; chenilles, 18–22; cotton darners, 1–7; embroidery, 1–7; milliners, 3–7; sharps, 1–5; sailmakers; tapestry, 18–22
Machine Needles round/set point, coarse 16–18
Threads mercerized cotton, 40; six-cord cotton, 30; cotton- or poly-core waxed, 40 heavy duty; cotton-covered polyester core, heavy duty
Hems double-fold or single-fold bias binding; bound/Hong Kong finish; faced; stitched and pinked flat; overedged; seam binding
Seams plain; safety-stitched; welt; double welt
Seam Finishes single-ply or Hong Kong bound; single-ply overcast; pinked; pinked and stitched; serging/single-ply overedged; untreated/plain; single-ply zigzag
Pressing steam; safe temperature limit 400°F (206.1°C)
Care launder; dry clean
Fabric Resource **Milton Sherlip Inc.**

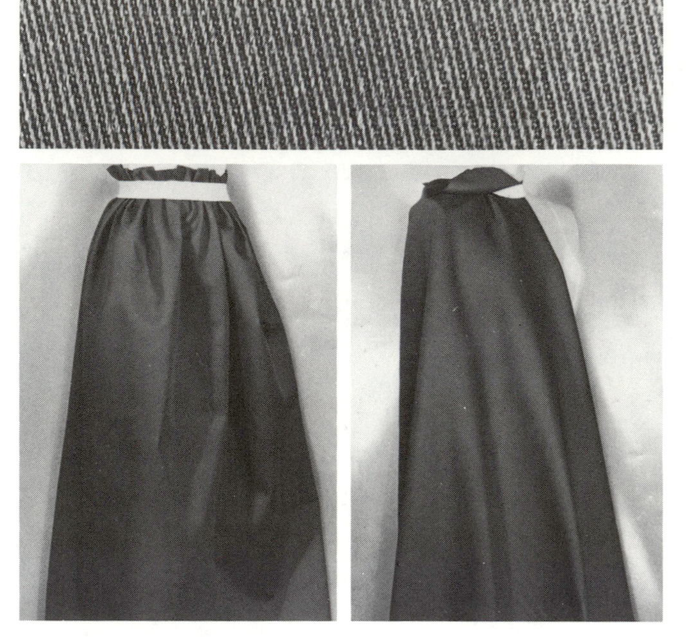

Cotton Tricotine

Fiber Content 65% polyester/35% cotton
Yarn Type combed staple; filament fiber
Yarn Construction conventional; textured; high twist
Fabric Structure twill weave; double twill line
Finishing Processes calendering; clear finish; Zepel® water/stain repellent
Color Application piece dyed–union dyed
Width 59–60 inches (149.9–152.4 cm)
Weight medium-heavy
Hand compact; crisp; harsh
Texture grainy; harsh; double twill line visible on face
Performance Expectations subject to color crocking and edge abrasion; dimensional stability; durable; elongation; heat-set properties; resilient; high tensile strength; water and stain resistant
Drapability Qualities falls into crisp wide cones; maintains crisp folds, creases; retains shape or silhouette of garment

Recommended Construction Aids & Techniques

Hand Needles betweens, 3–5; cotton darners, 5–10; embroidery, 5–8; milliners, 6–8; sharps, 6–8
Machine Needles round/set point, coarse 16–18
Threads cotton- or poly-core waxed, 50 all purpose; cotton-polyester blend, all purpose; cotton-covered polyester core, heavy duty; spun polyester, all purpose
Hems book; double-fold or single-fold bias binding; bound/Hong Kong finish; machine-stitched; overedged; seam binding
Seams flat-felled; lapped; plain; safety-stitched; welt; double welt
Seam Finishes book; single-ply bound; Hong Kong bound; single-ply overcast; pinked; pinked and stitched; serging/single-ply overedged; untreated/plain; single-ply zigzag
Pressing steam; safe temperature limit 325°F (164.1°C)
Care dry clean
Fabric Resource **Reeves Brothers Inc.**

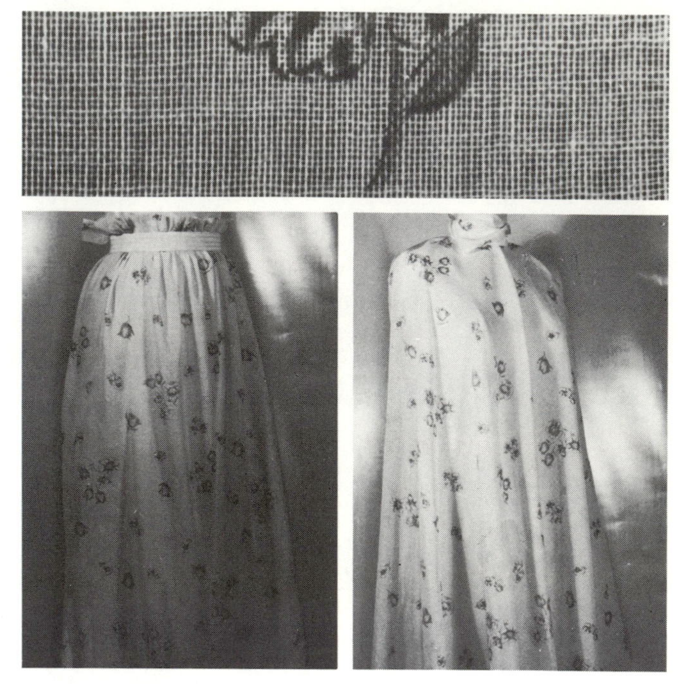

Cotton Voile

Fiber Content 65% polyester/35% cotton
Yarn Type combed staple; filament fiber
Yarn Construction conventional; textured; high twist
Fabric Structure loosely woven plain weave
Finishing Processes calendering; mercerizing; sizing; softening
Color Application piece dyed–union dyed
Width 44–45 inches (111.8–114.3 cm)
Weight[1] light
Hand[1] delicate; soft; supple
Texture[1] flat; semi-sheer; smooth
Performance Expectations abrasion resistant; air permeable; dimensional stability; elongation; nonsnagging; high tensile strength
Drapability Qualities falls into soft languid flares; accommodates fullness by gathering, elasticized shirring; fullness retains soft graceful fall

Recommended Construction Aids & Techniques

Hand Needles betweens, 6–7; cotton darners, 9–10; embroidery, 8–10; milliners, 8–10; sharps, 11–12
Machine Needles round/set point, fine 9
Threads cotton- or poly-core waxed, 60 fine; cotton-polyester blend, all purpose; cotton-covered polyester core, all purpose; spun polyester, all purpose
Hems double fold; edge-stitched; double edge-stitched; horsehair; machine-stitched; hand- or machine-rolled; wired
Seams flat-felled; French; mock French; hairline; plain; safety-stitched; tissue-stitched
Seam Finishes self bound; double-stitched and trimmed; edge-stitched; serging/single-ply overedged; untreated/plain
Pressing steam; safe temperature limit 325°F (164.1°C)
Care launder
Fabric Resource **Printsiples and Company B**

Crinkled Cotton Voile

Fiber Content 65% polyester/35% cotton
Yarn Type combed staple; filament fiber
Yarn Construction conventional; textured; high twist
Fabric Structure loosely woven plain weave
Finishing Processes calendering; heat-set crinkle; mercerizing; slack mercerization (stretching)
Color Application piece dyed-union dyed
Width 40 inches (101.6 cm)
Weight light
Hand delicate; soft; supple
Texture crinkled; semi-sheer
Performance Expectations abrasion resistant; air permeable; dimensional stability; elongation; nonsnagging; high tensile strength
Drapability Qualities falls into moderately soft flares; accommodates fullness by gathering, elasticized shirring; fullness retains soft fall

Recommended Construction Aids & Techniques

Hand Needles betweens, 6–7; cotton darners, 9–10; embroidery, 8–10; milliners, 8–10; sharps, 11–12
Machine Needles round/set point, fine 9
Threads cotton- or poly-core waxed, 60 fine; cotton-polyester blend, all purpose; cotton-covered polyester core, all purpose; spun polyester, all purpose
Hems double fold; edge-stitched; horsehair; machine-stitched; hand- or machine-rolled
Seams flat-felled; French; mock French; hairline; plain; safety-stitched; tissue-stitched
Seam Finishes self bound; double-stitched and trimmed; edge-stitched; serging/single-ply overedged; untreated/plain
Pressing steam; safe temperature limit 325°F (164.1°C)
Care launder
Fabric Resource **Milton Sherlip Inc.**

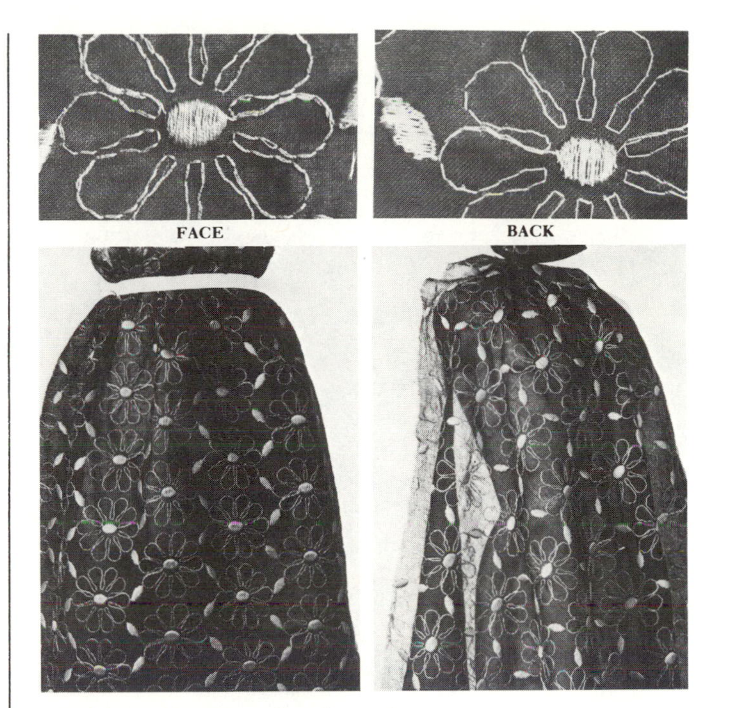

Voile with Yarn Embroidery

Fiber Content cotton/polyester blend
Yarn Type combed staple; filament fiber
Yarn Construction conventional
Fabric Structure Schiffli embroidery design on plain weave base
Finishing Processes calendering; mercerizing; stabilizing
Color Application piece dyed (base); yarn dyed (embroidery yarn)
Width 44-45 inches (111.8–114.3 cm)
Weight light
Hand crisp; firm; thin
Texture base: smooth; embroidery yarn forms high and low raised pattern on surface
Performance Expectations air permeable; nonpilling; embroidery tends to abrade and snag; wash-and-wear properties
Drapability Qualities falls into firm flares; accommodates fullness by gathering, elasticized shirring; fullness maintains crisp effect

Recommended Construction Aids & Techniques

Hand Needles betweens, 5–7; cotton darners, 8–10; embroidery, 7–10; milliners, 8–10; sharps, 9–10
Machine Needles round/set point, medium-fine 9–11
Threads mercerized cotton, 50; six-cord cotton, 50; cotton- or poly-core waxed, all purpose; cotton-covered polyester core, all purpose
Hems double fold; edge-stitched; machine-stitched; overedged
Seams flat-felled; French; lapped; plain; safety-stitched; welt; double welt
Seam Finishes double-stitched; double-stitched and overcast; edge-stitched; pinked; pinked and stitched; serging/single-ply overedged; untreated/plain; single-ply or double-ply zigzag
Pressing steam; use pressing cloth to protect weave; safe temperature limit 325°F (164.1°C)
Care launder; drip dry; permanent press cycle of dryer
Fabric Resource **Lawrence Textile Co. Inc.**

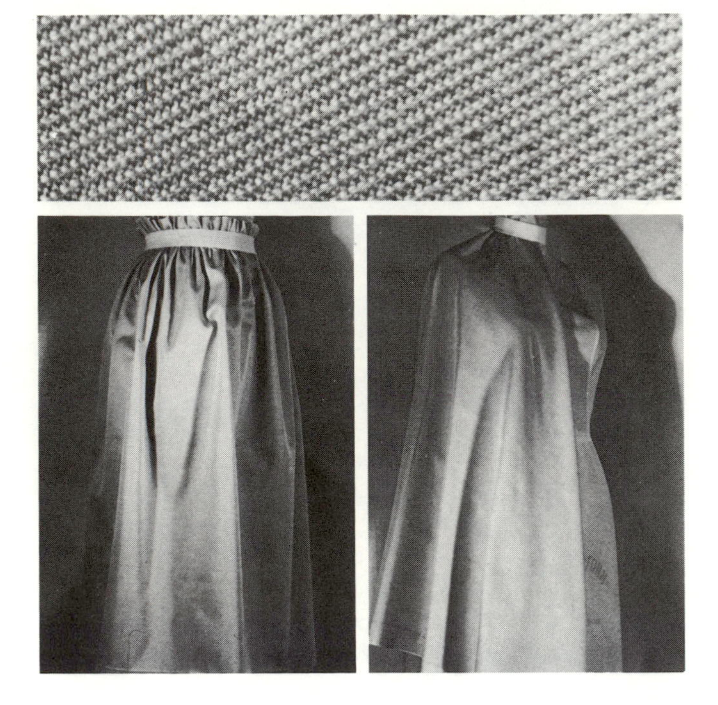

Cotton Whipcord

Fiber Content 100% cotton
Yarn Type combed staple
Yarn Construction conventional; high twist
Fabric Structure left-hand warp twill weave
Finishing Processes calendering; mercerizing; napping; preshrunk
Color Application yarn dyed
Width 56–58 inches (142.2–147.3 cm)
Weight medium to medium heavy
Hand firm; hard
Texture rough; prominent steep diagonal weave on face; slight nap on back
Performance Expectations abrasion resistant; absorbent; durable; high tensile strength
Drapability Qualities falls into firm flares; fullness maintains crisp effect; retains shape of garment

Recommended Construction Aids & Techniques

Hand Needles betweens, 3–5; cotton darners, 5–10; embroidery, 5–8; milliners, 6–8; sharps, 6–8
Machine Needles round/set point, medium-fine 9–11
Threads mercerized cotton, 50; six-cord cotton, 50; cotton- or poly-core waxed, all purpose; cotton-covered polyester core, all purpose
Hems book; single-fold bias binding; bound/Hong Kong finish; machine-stitched; overedged; seam binding
Seams flat-felled; lapped; plain; safety-stitched; welt; double welt
Seam Finishes single-ply bound; Hong Kong bound; edge-stitched; single-ply and overcast; pinked; pinked and stitched; serging/single-ply overedged; untreated/plain; single-ply zigzag
Pressing steam; safe temperature limit 400°F (206.1°C)
Care launder; dry clean
Fabric Resource Auburn Fabrics Inc.

Novelty Cottons

Novelty cotton is a term used to describe a large variety of new or unusual fabrics made with cotton or cotton-blended fiber yarns. Novelty cottons may be composed of one or more shades, tones or colors, may utilize any novelty yarn or combination of yarns, may be made in any combination or variety of weave structure, and may vary in:

1. unusual surface texture and pattern;
2. internal or external finishes;
3. soft to stiff hand.

Some manufacturers may refer to their novelty cottons as *Cotton Fancies* and to their line of plaids, tweeds, surface patterns, florals or finishes and/or other types of decorative cottons as *Fancies*.

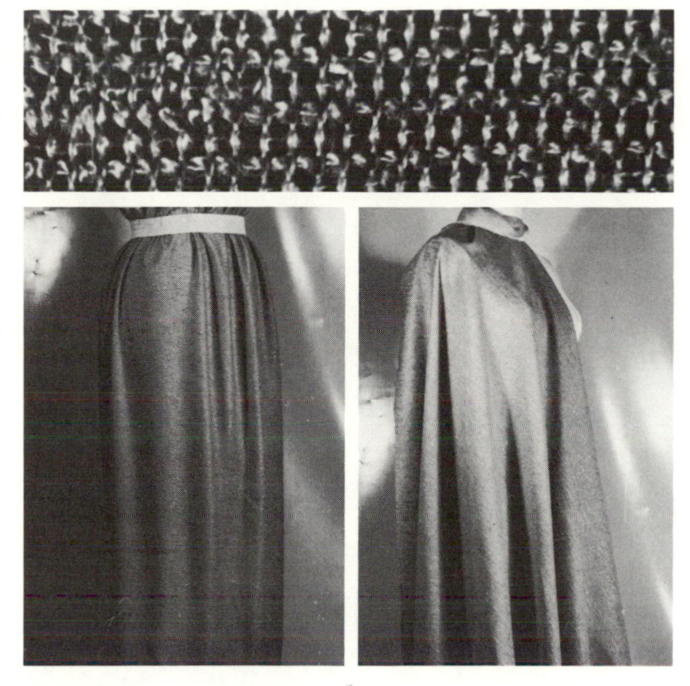

Bouclé Yarn Cotton

Fiber Content cotton/nylon/polyester blend
Yarn Type carded staple; filament fiber
Yarn Construction novelty bouclé yarn (filling); textured yarn (warp)
Fabric Structure leno weave
Finishing Processes calendering; mercerizing; stabilizing
Color Application piece dyed–cross dyed; yarn dyed
Width 44–45 inches (111.8–114.3 cm)
Weight medium
Hand loose; pliable; silky; soft
Texture lustrous; bouclé yarn creates looped surface
Performance Expectations air permeable; elasticity; flexible; nonpilling; resilient
Drapability Qualities falls into soft flares; accommodates fullness by gathering; fullness retains soft fall

Recommended Construction Aids & Techniques

Hand Needles betweens, 3–6; cotton darners, 7–10; embroidery, 6–10; milliners, 6–10; sharps, 3–6
Machine Needles round/set point, fine 9
Threads mercerized cotton, 50; six-cord cotton, 50; cotton- or poly-core waxed, all purpose; cotton-covered polyester core, all purpose
Hems double-fold or single-fold bias binding; bound/Hong Kong finish; stitched and pinked flat; overedged; seam binding
Seams plain; safety-stitched; taped; welt; double welt
Seam Finishes single-ply bound; Hong Kong bound; single-ply overcast; pinked; pinked and stitched; serging/single-ply overedged; single-ply zigzag
Pressing steam; safe temperature limit 325°F (164.1°C)
Care launder; dry clean
Fabric Resource **Concord Fabrics Inc.**

Dobby Weave Cotton

Fiber Content 65% Kodel® polyester/35% cotton
Yarn Type combed staple; filament fiber
Yarn Construction conventional; fine; textured; high twist
Fabric Structure dobby weave
Finishing Processes mercerizing; Russpress® permanent press precured
Color Application piece dyed–union dyed; yarn dyed
Width 48 inches (121.9 cm)
Weight light
Hand fine; firm; soft; springy
Texture flat; smooth; uneven light-reflecting surface created by weave
Performance Expectations abrasion resistant; dimensional stability; durable; flexible; nonpilling; resilient; 1% residual shrinkage; high tensile strength; wash-and-wear properties
Drapability Qualities falls into moderately soft flares; accommodates fullness by pleating, gathering, elasticized shirring; fullness retains soft fall

Recommended Construction Aids & Techniques

Hand Needles betweens, 5–7; cotton darners, 8–10; embroidery, 7–10; milliners, 8–10; sharps, 9–10
Machine Needles round/set point, medium-fine 9-11
Threads cotton- or poly-core waxed, 60 fine; cotton-polyester blend, all purpose; cotton-covered polyester core, all purpose; spun polyester, all purpose
Hems double fold; edge-stitched; machine-stitched; overedged
Seams flat-felled; French; lapped; plain; safety-stitched; welt; double welt
Seam Finishes double-stitched; double-stitched and overcast; edge-stitched; pinked; pinked and stitched; serging/single-ply overedged; untreated/plain; single-ply zigzag
Pressing steam; safe temperature limit 325°F (164.1°C)
Care launder; drip dry; permanent press cycle of dryer
Fabric Resource **Russell Corporation**

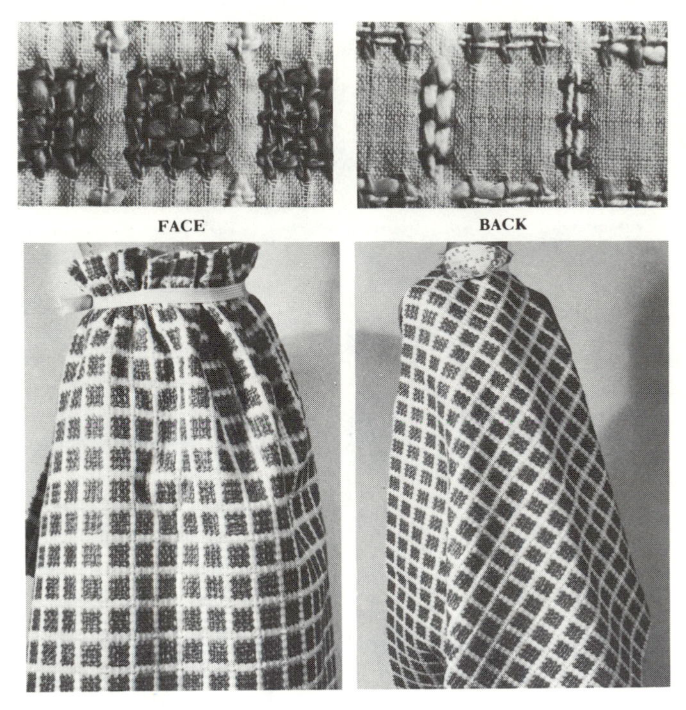

Double Weave Cotton/Double Cloth

Fiber Content 100% cotton
Yarn Type carded or combed staple
Yarn Construction conventional; fine; high twist for base and thick slack twist for double weave effect
Fabric Structure double cloth weave
Finishing Processes mercerizing; stabilizing
Color Application yarn dyed
Width 45 inches (114.3 cm)
Weight heavy
Hand rough; semi-soft; thick
Texture rough raised surface created by weave
Performance Expectations dimensional stability; durable; nonpilling; resilient; subject to snagging due to weave
Drapability Qualities falls into firm wide cones; fullness maintains lofty effect; retains shape or silhouette of garment

Recommended Construction Aids & Techniques

Hand Needles betweens, 3-5; cotton darners, 1-7; embroidery, 1-7; milliners, 3-7; sharps, 1-5
Machine Needles round/set point, medium 14
Threads mercerized cotton, 40; six-cord cotton, 30; cotton- or poly-core waxed, 40 heavy duty; cotton-covered polyester core, heavy duty
Hems double-fold or single-fold bias binding; bound/ Hong Kong finish; faced; stitched and pinked flat; overedged; seam binding
Seams plain; safety-stitched; welt; double welt
Seam Finishes single-ply bound; Hong Kong bound; single-ply overcast; pinked; pinked and stitched; serging/single-ply overedged; untreated/plain; single-ply zigzag
Pressing steam; use pressing cloth to protect weave; safe temperature limit 400°F (206.1°C)
Care launder; dry clean
Fabric Resource **Lawrence Textile Co. Inc.**

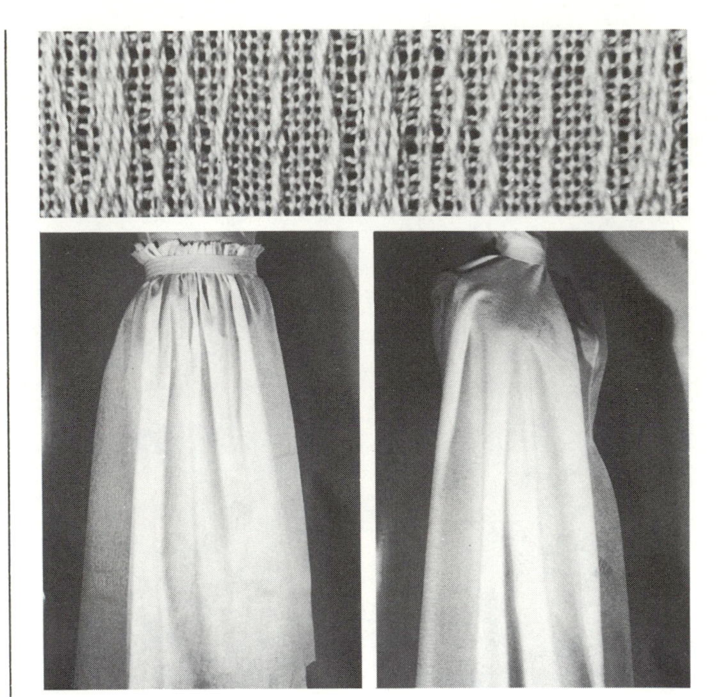

Jacquard Weave Cotton

Fiber Content 100% cotton
Yarn Type combed staple
Yarn Construction conventional; thick-and-thin; high twist
Fabric Structure Jacquard weave
Finishing Processes calendering; mercerizing; stabilizing
Color Application piece dyed; yarn dyed
Width 45 inches (114.3 cm)
Weight medium-light
Hand firm; hard; rough
Texture high and low patterned surface created by weave
Performance Expectations absorbent; antistatic; durable; nonpilling; high tensile strength
Drapability Qualities falls into moderately soft flares; accommodates fullness by gathering, pleating; fullness retains soft fall

Recommended Construction Aids & Techniques

Hand Needles betweens, 5-7; cotton darners, 8-10; embroidery, 7-10; milliners, 8-10; sharps, 9-10
Machine Needles round/set point, medium-fine 11
Threads mercerized cotton, 50; six-cord cotton, 50; cotton- or poly-core waxed, fine; cotton-covered polyester core, all purpose
Hems double fold; edge-stitched; machine-stitched; overedged
Seams flat-felled; French; lapped; plain; safety-stitched; welt; double welt
Seam Finishes double-stitched; double-stitched and overcast; edge-stitched; pinked; pinked and stitched; serging/single-ply overedged; untreated/plain; single-ply and double-ply zigzag
Pressing steam; safe temperature limit 400°F (206.1°C)
Care launder
Fabric Resource **Auburn Fabrics Inc.**

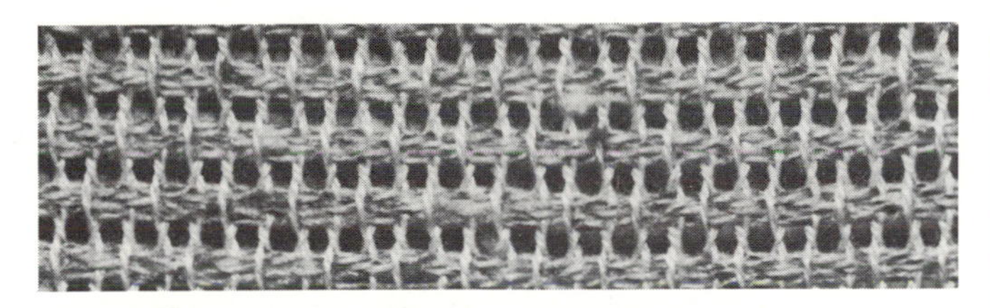

Leno Weave Cotton (lightweight)

Fiber Content 65% polyester/35% cotton
Yarn Type combed staple; filament fiber
Yarn Construction conventional; textured; high twist
Fabric Structure leno weave
Finishing Processes calendering; mercerizing; stabilizing
Color Application piece-dyed–union dyed
Width 44–45 inches (111.8–114.3 cm)
Weight light
Hand firm; semi-soft
Texture loose open weave; textured areas
Performance Expectations abrasion resistant; dimensional stability; durable; flexible; nonpilling; resilient; high tensile strength; wash-and-wear properties
Drapability Qualities falls into moderately soft flares; accommodates fullness by gathering, elasticized shirring, pleating; fullness maintains moderately crisp effect

Recommended Construction Aids & Techniques

Hand Needles betweens, 3–6; cotton darners, 7–10; embroidery, 6–10; milliners, 6–10; sharps, 6–8
Machine Needles round/set point, medium-fine 9–11
Threads mercerized cotton, 50; six-cord cotton, 60–80; cotton- or poly-core waxed, 60 fine; cotton-covered polyester core, extra fine
Hems double fold; edge-stitched; machine-stitched; overedged
Seams flat-felled; French; lapped; plain; safety-stitched; taped; welt; double welt
Seam Finishes double-stitched; double-stitched and trimmed; edge-stitched; pinked; pinked and stitched; serging/single-ply overedged; single-ply or double-ply zigzag
Pressing steam; safe temperature limit 325°F (164.1°C)
Care launder; drip dry; permanent press cycle of dryer
Fabric Resource **Lawrence Textile Co. Inc.**

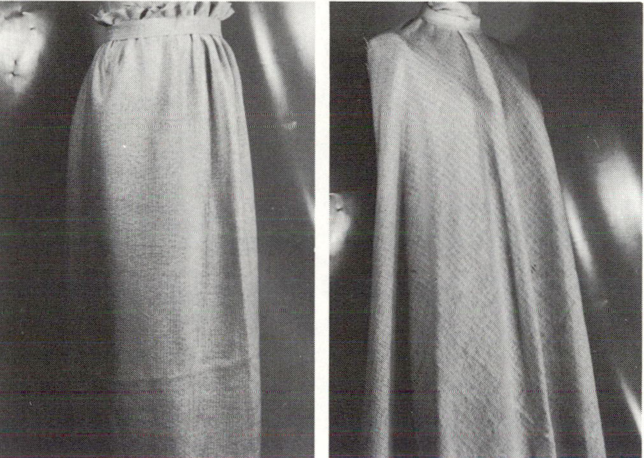

Leno Weave Cotton (heavyweight)

Fiber Content 100% cotton
Yarn Type carded staple
Yarn Construction conventional; low twist
Fabric Structure leno weave
Finishing Processes beetling; calendering; mercerizing
Color Application piece dyed
Width 45-46 inches (114.3–116.8 cm)
Weight heavy
Hand coarse; porous; soft; thick
Texture loose open weave; rough
Performance Expectations absorbent; air permeable; resilient; subject to shrinkage; subject to stretching and bagging out of shape
Drapability Qualities falls into wide cones; fullness maintains bouffant effect; retains shape or silhouette of garment

Recommended Construction Aids & Techniques

Hand Needles betweens, 3–5; cotton darners, 1–7; embroidery, 1–7; milliners, 3–8; sharps, 1–6
Machine Needles round/set point, medium-coarse 14–16
Threads mercerized cotton, heavy duty; six-cord cotton, 40; cotton-covered polyester core, heavy duty
Hems double-fold or single-fold bias binding; bound/Hong Kong finish; faced; stitched and pinked flat; overedged; seam binding
Seams plain; safety-stitched; welt; double welt
Seam Finishes single-ply bound; Hong Kong bound; single-ply overcast; pinked; pinked and stitched; serging/single-ply overedged; single-ply zigzag
Pressing steam; safe temperature limit 400°F (206.1°C)
Care launder; dry clean
Fabric Resource **Truemark Discount Fabrics Inc.**

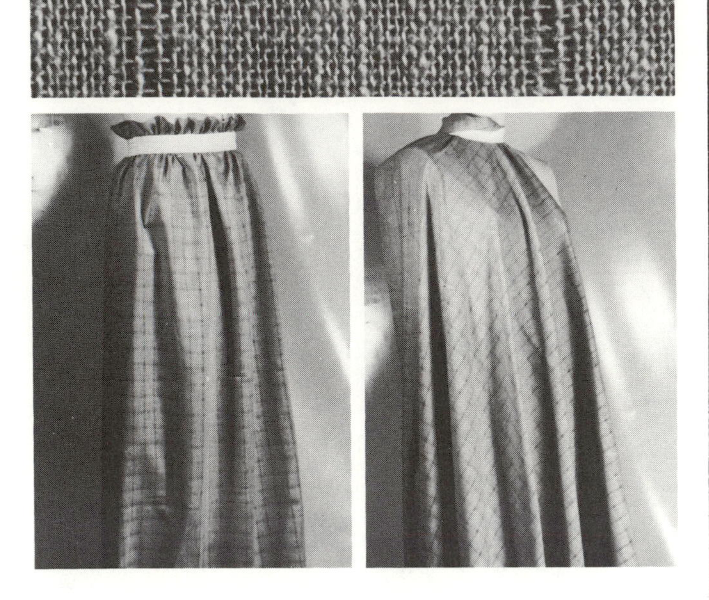

Skip Denting—Open Effect Cotton

Fiber Content 100% cotton
Yarn Type carded or combed staple
Yarn Construction conventional
Fabric Structure plain weave variation; novelty skip denting process
Finishing Processes mercerizing; shrinkage control; wrinkle resistant
Color Application piece dyed
Width 44-45 inches (111.8-114.3 cm)
Weight light
Hand loose; semi-soft; thin
Texture flat; grainy, open effect created by weave
Performance Expectations absorbent; air permeable; antistatic; flexible; nonpilling; subject to snagging due to weave
Drapability Qualities falls into moderately soft flares; accommodates fullness by gathering, elasticized shirring; fullness retains soft fall

Recommended Construction Aids & Techniques

Hand Needles betweens, 5-7; cotton darners, 8-10; embroidery, 7-10; milliners, 8-10; sharps, 9-10
Machine Needles round/set point, medium-fine 9-11
Threads mercerized cotton, 50; six-cord cotton, 60-80; cotton-covered polyester core, extra fine
Hems double fold; edge-stitched; machine-stitched; overedged
Seams flat-felled; French; lapped; plain; safety-stitched; welt; double welt
Seam Finishes double-stitched; double-stitched and overcast; edge-stitched; pinked; pinked and stitched; serging/single-ply overedged; untreated/plain; single-ply or double-ply zigzag
Pressing steam; safe temperature limit 400°F (206.1°C)
Care launder
Fabric Resources Concord Fabrics Inc.; Earl Glo/Erlanger Blumgart and Co. Inc.

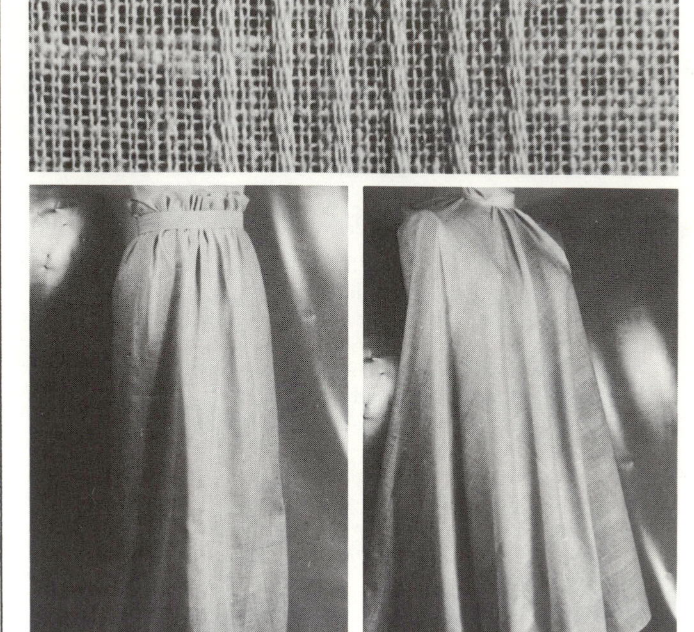

Skip Denting—Corded Effect Cotton

Fiber Content 100% cotton
Yarn Type carded or combed staple
Yarn Construction conventional; fine tightly twisted yarn for cord
Fabric Structure plain weave variation; novelty skip denting with cording
Finishing Processes calendering; mercerizing; stabilizing
Color Application piece dyed
Width 44-45 inches (111.8-114.3 cm)
Weight light
Hand porous; semi-soft; thin
Texture coarse; raised cord and weave create uneven surface
Performance Expectations absorbent; air permeable; antistatic; nonpilling
Drapability Qualities falls into moderately soft flares; accommodates fullness by gathering, elasticized shirring; fullness retains soft fall

Recommended Construction Aids & Techniques

Hand Needles betweens, 5-7; cotton darners, 8-10; embroidery, 7-10; milliners, 8-10; sharps, 9-10
Machine Needles round/set point, medium-fine 9-11
Threads mercerized cotton, 50; six-cord cotton, 60-80; cotton-covered polyester core, extra fine
Hems double fold; edge-stitched; machine-stitched; overedged
Seams flat-felled; French; lapped; plain; safety-stitched; welt; double welt
Seam Finishes double-stitched; double-stitched and overcast; edge-stitched; pinked; pinked and stitched; serging/single-ply overedged; untreated/plain; single-ply or double-ply zigzag
Pressing steam; safe temperature limit 400°F (206.1°C)
Care launder
Fabric Resource Iselin Jefferson Co. Inc. (sample courtesy of Jonan Enterprise)

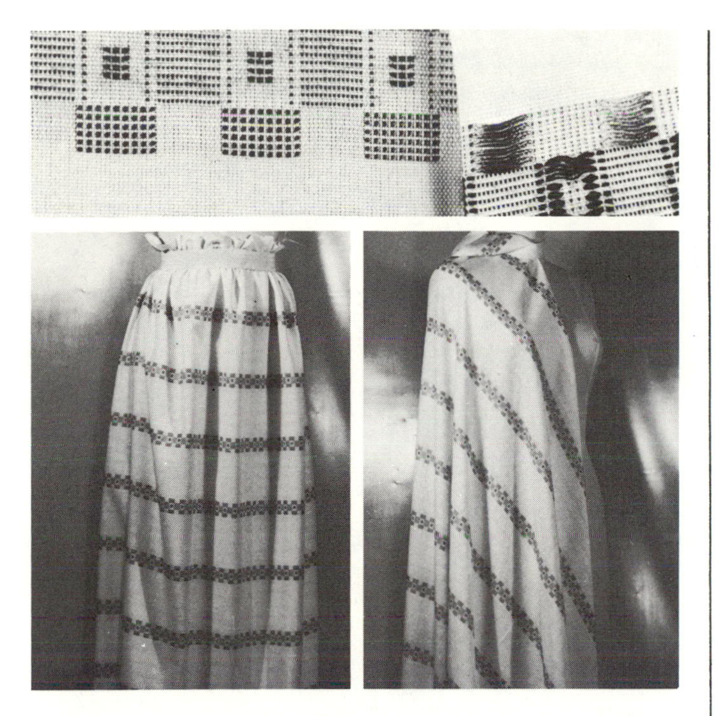

Spot/Unclipped Spot/Unclipped Dot Cotton (lightweight)

Fiber Content cotton/polyester blend
Yarn Type carded or combed staple; filament fiber
Yarn Construction conventional; textured; fine; high twist
Fabric Structure spot or swivel figure weave
Finishing Processes calendering; mercerizing; stabilizing
Color Application yarn dyed
Width 45 inches (114.3 cm)
Weight light to medium
Hand firm; soft; springy; thin
Texture smooth and pattern design; thread floats on back
Performance Expectations abrasion resistant; durable; flexible; nonpilling; nonsnagging; resilient; thread floats tend to snag; wash-and-wear properties
Drapability Qualities falls into soft flares; accommodates fullness by pleating, gathering, elasticized shirring; fullness retains soft fall

Recommended Construction Aids & Techniques

Hand Needles betweens, 3–6; cotton darners, 7–10; embroidery, 6–10; milliners, 6–10; sharps, 6–10
Machine Needles round/set point, medium-fine 9–11
Threads mercerized cotton, 50; six-cord cotton, 50; cotton- or poly-core waxed, fine; cotton-covered polyester core, all purpose
Hems double fold; edge-stitched; machine-stitched; overedged
Seams flat-felled; French; lapped; plain; safety-stitched; welt; double welt
Seam Finishes double-stitched; double-stitched and overcast; edge-stitched; pinked; pinked and stitched; serging/single-ply overedged; untreated/plain; single-ply or double-ply zigzag
Pressing steam; safe temperature limit 325°F (164.1°C)
Care launder
Fabric Resource **Lawrence Textile Co. Inc.**

Spot/Unclipped Spot/Unclipped Dot Cotton (heavyweight)

Fiber Content cotton/polyester blend
Yarn Type carded or combed staple; filament fiber
Yarn Construction conventional; ply; textured
Fabric Structure spot or swivel figure weave
Finishing Processes mercerizing; stabilizing
Color Application piece dyed; yarn dyed
Width 45 inches (114.3 cm)
Weight heavy
Hand semi-crisp; nonelastic; thick
Texture thread floats on face create uneven surface
Performance Expectations dimensional stability; durable; nonpilling; resilient; thread floats tend to snag
Drapability Qualities falls into firm wide cones; fullness maintains lofty effect; retains shape or silhouette of garment

Recommended Construction Aids & Techniques

Hand Needles betweens, 3–5; cotton darners, 1–7; embroidery, 1–7; milliners, 3–8; sharps, 1–6
Machine Needles round/set point, medium 11–14
Threads mercerized cotton, heavy; six-cord cotton, 40; cotton- or poly-core waxed, heavy duty; cotton-covered polyester core, heavy duty
Hems double-fold or single-fold bias binding; bound/Hong Kong finish; faced; stitched and pinked flat; overedged; seam binding
Seams plain; safety-stitched; welt; double welt
Seam Finishes single-ply bound; Hong Kong bound; single-ply overcast; pinked; pinked and stitched; serging/single-ply overedged; untreated/plain; single-ply zigzag
Pressing steam; use pressing cloth to protect weave; safe temperature limit 325°F (164.1°C)
Care launder; dry clean
Fabric Resource **Lawrence Textile Co. Inc.**

Spot/Clip Spot/Clip Dot "Eyelash" Cotton

Fiber Content cotton/polyester blend
Yarn Type carded or combed staple; filament fiber
Yarn Construction conventional; ply; textured
Fabric Structure spot or swivel figure weave
Finishing Processes mercerizing; stabilizing
Color Application piece dyed; yarn dyed
Width 45 inches (114.3 cm)
Weight heavy
Hand semi-crisp; nonelastic; thick
Texture thread floats on face create uneven surface
Performance Expectations dimensional stability; durable; nonpilling; resilient; thread floats tend to snag
Drapability Qualities falls into firm wide cones; fullness maintains lofty effect; retains shape or silhouette of garment

Recommended Construction Aids & Techniques

Hand Needles betweens, 3-5; cotton darners, 1-7; embroidery, 1-7; milliners, 3-8; sharps, 1-6
Machine Needles round/set point, medium 11-14
Threads mercerized cotton, heavy; six-cord cotton, 40; cotton- or poly-core waxed, heavy duty; cotton-covered polyester core, heavy duty
Hems double-fold or single-fold bias binding; bound/Hong Kong finish; faced; stitched and pinked flat; overedged; seam binding
Seams plain; safety-stitched; welt; double welt
Seam Finishes single-ply bound; Hong Kong bound; single-ply overcast; pinked; pinked and stitched; serging/single-ply overedged; untreated/plain; single-ply zigzag
Pressing steam; use pressing cloth to protect weave; safe temperature limit 325°F (164.1°C)
Care launder; dry clean
Fabric Resource American Silk Mills Corp.

Spot/Clip Spot/Clip Dot Cotton

Fiber Content 100% cotton
Yarn Type carded or combed staple
Yarn Construction conventional
Fabric Structure spot or swivel figure weave; double denting process
Finishing Processes clipping; mercerizing; shearing
Color Application piece dyed
Width 45 inches (114.3 cm)
Weight light
Hand fine; soft to crisp depending on finish
Texture clipped yarns create uneven surface; open effect created by weave
Performance Expectations absorbent; air permeable; nonpilling; open weave tends to snag; clip spot tends to abrade and dislodge with wear and repeated launderings
Drapability Qualities falls into moderately soft flares; accommodates fullness by gathering, elasticized shirring; fullness maintains moderately soft effect

Recommended Construction Aids & Techniques

Hand Needles betweens, 5-7; cotton darners, 8-10; embroidery, 7-10; milliners, 8-10; sharps, 9-10
Machine Needles round/set point, medium-fine 9-11
Threads mercerized cotton, 50; six-cord cotton, 50; cotton-covered polyester core, fine
Hems double fold; edge-stitched; machine-stitched; overedged
Seams flat-felled; French; lapped; plain; safety-stitched
Seam Finishes double-stitched; double-stitched and overcast; edge-stitched; pinked; pinked and stitched; serging/single-ply overedged; untreated/plain; single-ply or double-ply zigzag
Pressing steam; safe temperature limit 400°F (206.1°C)
Care launder
Fabric Resource Lawrence Textile Co. Inc.

Reverse Twill Weave Cotton

Fiber Content 100% cotton
Yarn Type combed staple
Yarn Construction conventional; high twist
Fabric Structure twill weave; reverse twill pattern
Finishing Processes calendering; mercerizing; stabilizing
Color Application piece dyed; yarn dyed
Width 56–58 inches (142.2–147.3 cm)
Weight medium
Hand firm; semi-soft
Texture uneven ridge created by diagonal weave
Performance Expectations abrasion resistant; absorbent; antistatic; durable; nonsnagging; high tensile and bursting strength
Drapability Qualities falls into soft flares; fullness retains soft fall; retains shape of garment

Recommended Construction Aids & Techniques

Hand Needles betweens, 3–5; cotton darners, 5–10; embroidery, 5–8; milliners, 6–8; sharps, 6–8
Machine Needles round/set point, medium 14
Threads mercerized cotton, 50; six-cord cotton, 50; cotton-covered polyester core, all purpose
Hems book; double fold; edge-stitched; machine-stitched; overedged
Seams flat-felled; French; lapped; plain; safety-stitched; welt; double welt
Seam Finishes book; single-ply bound; Hong Kong bound; edge-stitched; single-ply overcast; pinked; pinked and stitched; serging/single-ply overedged; untreated/plain; single-ply zigzag
Pressing steam; safe temperature limit 400°F (206.1°C)
Care launder
Fabric Resource **Auburn Fabrics Inc.**

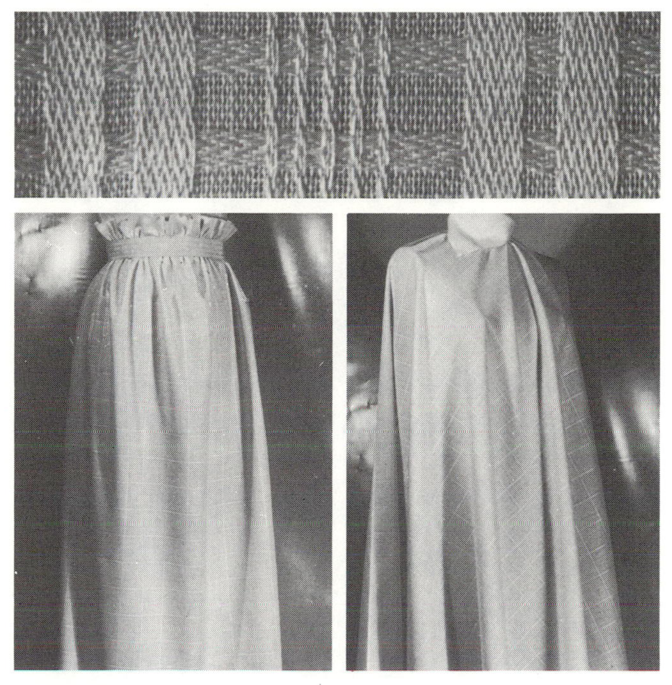

White-on-White Cotton

Fiber Content 65% polyester/35% cotton
Yarn Type combed staple; filament fiber; filament staple
Yarn Construction conventional; textured; high twist
Fabric Structure novelty weave; dobby weave
Finishing Processes mercerizing; permanent press; shrinkage control
Color Application piece dyed–union dyed; yarn dyed
Width 44–45 inches (111.8–114.3 cm)
Weight light
Hand firm; soft; thin
Texture light-reflecting high and low pattern created by woven stripe and check
Performance Expectations abrasion resistant; dimensional stability; durable; nonsnagging; high tensile strength; wash-and-wear properties
Drapability Qualities falls into moderately soft flares; accommodates fullness by pleating, gathering; fullness retains moderately soft fall

Recommended Construction Aids & Techniques

Hand Needles betweens, 5–7; cotton darners, 8–10; embroidery, 7–10; milliners, 8–10; sharps, 9–10
Machine Needles round/set point, medium-fine 9–11
Threads mercerized cotton, 50; six-cord cotton, 50; cotton-polyester blend, all purpose; cotton-covered polyester core, all purpose; spun polyester, all purpose
Hems double fold; edge-stitched; machine-stitched; overedged
Seams flat-felled; French; lapped; plain; safety-stitched; welt; double welt
Seam Finishes double-stitched; double-stitched and overcast; edge-stitched; pinked; pinked and stitched; serging/single-ply overedged; untreated/plain; single-ply or double-ply zigzag
Pressing steam; safe temperature limit 325°F (164.1°C)
Care launder
Fabric Resources **Earl Glo/Erlanger Blumgart and Co. Inc.; Russell Corporation**

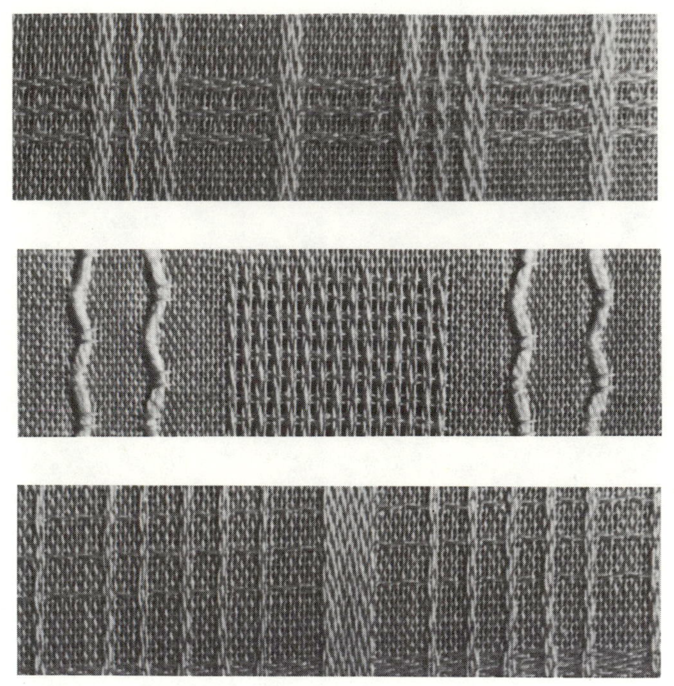

White-on-White Cotton Variation

Fiber Content 65% polyester/35% cotton
Yarn Type combed staple; filament fiber; filament staple
Yarn Construction conventional; textured; high twist
Fabric Structure novelty weave; dobby weave
Finishing Processes mercerizing; permanent press; shrinkage control
Color Application piece dyed-union dyed; yarn dyed
Width 44-45 inches (111.8-114.3 cm)
Weight light
Hand firm; soft; thin
Texture light-reflecting high and low pattern created by woven stripe and check
Performance Expectations abrasion resistant; dimensional stability; durable; nonsnagging; high tensile strength; wash-and-wear properties
Drapability Qualities falls into moderately soft flares; accommodates fullness by pleating, gathering; fullness retains moderately soft fall

Recommended Construction Aids & Techniques

Hand Needles betweens, 5-7; cotton darners, 8-10; embroidery, 7-10; milliners, 8-10; sharps, 9-10
Machine Needles round/set point, medium-fine 9-11
Threads mercerized cotton, 50; six-cord cotton, 50; cotton-polyester blend, all purpose; cotton-covered polyester core, all purpose; spun polyester, all purpose
Hems double fold; edge-stitched; machine-stitched; overedged
Seams flat-felled; French; lapped; plain; safety-stitched; welt; double welt
Seam Finishes double-stitched; double-stitched and overcast; edge-stitched; pinked; pinked and stitched; serging/single-ply overedged; untreated/plain; single-ply or double-ply zigzag
Pressing steam; safe temperature limit 325°F (164.1°C)
Care launder
Fabric Resources Klopman Mills; Earl Glo/Erlanger Blumgart and Co. Inc.; Russell Corporation

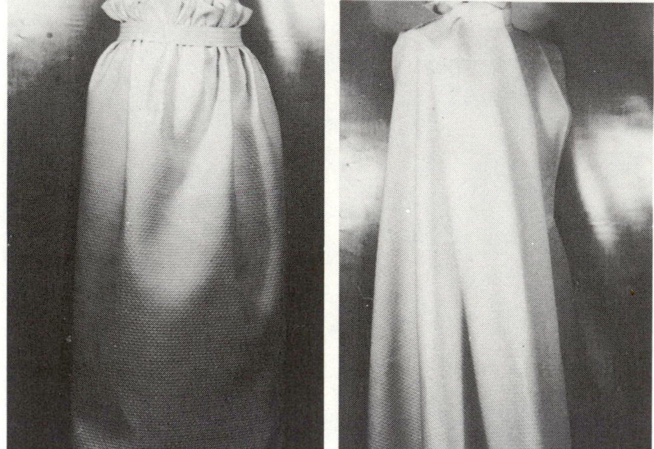

Bull's Eye Piqué

Fiber Content 65% Kodel® polyester/35% cotton
Yarn Type combed staple; filament fiber
Yarn Construction conventional; 2-ply (face)/heavy single-ply (back); textured
Fabric Structure piqué weave; dobby attachment
Finishing Processes calendering, mercerizing; Russpress® permanent press precured
Color Application piece dyed-union dyed
Width 44-45 inches (111.8-114.3 cm)
Weight medium-heavy
Hand coarse; crisp; nonelastic; firm
Texture weave creates uneven surface
Performance Expectations durable; elongation; resilient; 2-3% residual shrinkage; high tensile strength; damaged by bleaching agents; wash-and-wear properties; wrinkle resistant
Drapability Qualities falls into crisp flares; fullness maintains crisp effect; retains shape or silhouette of garment

Recommended Construction Aids & Techniques

Hand Needles betweens, 3-5; cotton darners, 5-10; embroidery, 5-8; milliners, 6-8; sharps, 6-8
Machine Needles round/set point, medium 14
Threads mercerized cotton, 50; six-cord cotton, 50; cotton-or poly-core waxed, all purpose; cotton-covered polyester core, all purpose
Hems double fold or single-fold bias binding; bound/Hong Kong finish; machine stitched; overedged; seam binding
Seams flat-felled; lapped; plain; safety-stitched; welt; double welt
Seam Finishes single-ply bound; Hong Kong bound; single-ply overcast; pinked; pinked and stitched; serging/single-ply overedged; untreated/plain; single-ply zigzag
Pressing steam; safe temperature limit 325°F (164.1°C)
Care launder; *do not* use bleaching agents
Fabric Resource Russell Corporation

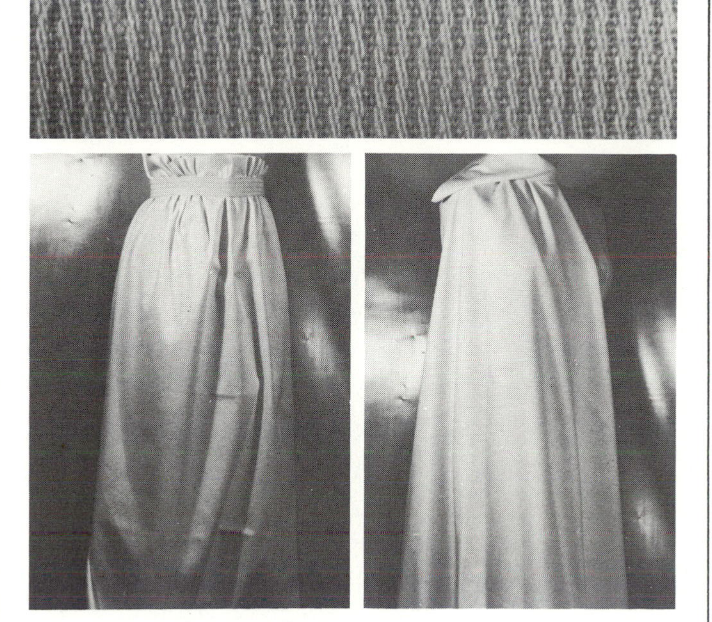

Corded Piqué/Piqué Cord

Fiber Content 50% cotton/50% polyester
Yarn Type combed staple; filament fiber; filament staple
Yarn Construction conventional; 2-ply (face); high twist (back); textured
Fabric Structure piqué weave; dobby attachment
Finishing Processes bleaching; mercerizing; shrinkage control
Color Application piece dyed–union dyed
Width 44–45 inches (111.8–114.3 cm)
Weight medium
Hand crisp; firm; thick
Texture fine lengthwise rib creates uneven surface
Performance Expectations durable; elongation; resilient; 2–3% residual shrinkage; high tensile strength; damaged by bleaching agents; wash-and-wear properties; wrinkle resistant
Drapability Qualities falls into crisp flares; fullness maintains crisp effect; retains shape or silhouette of garment

Recommended Construction Aids & Techniques

Hand Needles betweens, 3–5; cotton darners, 5–10; embroidery, 5–8; milliners, 6–8; sharps, 6–8
Machine Needles round/set point, medium 14
Threads mercerized cotton, 50; six-cord cotton, 50; cotton- or poly-core waxed, all purpose; cotton-covered polyester core, all purpose
Hems double- or single-fold bias binding; bound/Hong Kong finish; machine-stitched; overedged; seam binding
Seams flat-felled; lapped; plain; safety-stitched; welt; double welt
Seam Finishes single-ply or Hong Kong bound; single-ply overcast; pinked; pinked and stitched; serging/single-ply overedged; untreated/plain; single-ply zigzag
Pressing steam; safe temperature limit 325°F (164.1°C)
Care launder; *do not* use bleaching agents
Fabric Resources **Concord Fabrics Inc.; Earl Glo/Erlanger Blumgart and Co. Inc.; Loom Tex Corp.; Russell Corporation**

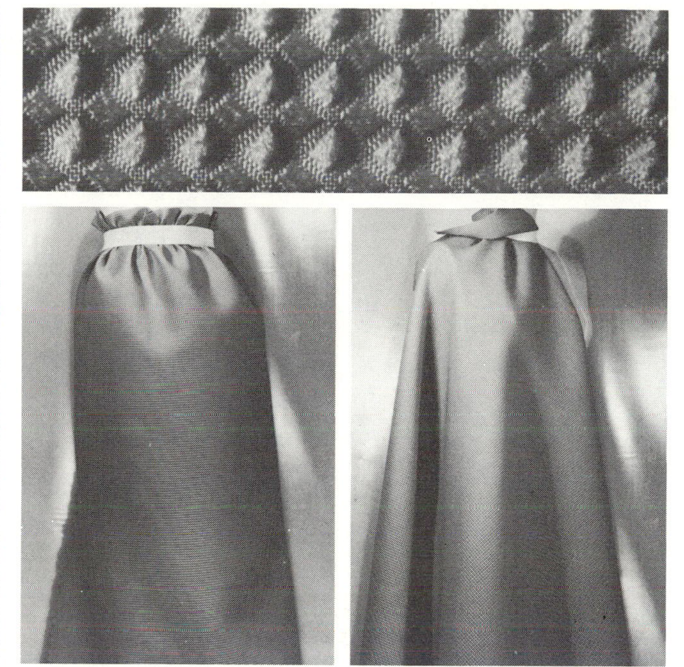

Embossed Piqué

Fiber Content 100% polyester
Yarn Type filament fiber
Yarn Construction fine; textured; slack twist
Fabric Structure piqué weave; dobby or Jacquard attachment
Finishing Processes heat-set embossing; thermoplastic molding
Color Application piece dyed; solution dyed; yarn dyed
Width 44–45 inches (111.8–114.3 cm)
Weight medium-light
Hand soft; spongy
Texture finish creates deeply raised and recessed face and back
Performance Expectations subject to color crocking and edge abrasion; dimensional stability; durable; elongation; heat-set properties; resilient; high tensile strength
Drapability Qualities falls into wide cones; fullness maintains bouffant effect; retains shape or silhouette of garment

Recommended Construction Aids & Techniques

Hand Needles betweens, 3–5; cotton darners, 5–10; embroidery, 5–8; milliners, 6–8; sharps, 6–8
Machine Needles round/set point, medium 14
Threads cotton- or poly-core waxed, 60 fine; cotton-polyester blend, all purpose; cotton-covered polyester core, all purpose; spun polyester, all purpose
Hems double-fold or single-fold bias binding; bound/Hong Kong finish; machine-stitched; overedged; seam binding
Seams plain; safety-stitched; welt; double welt
Seam Finishes single-ply bound; Hong Kong bound; single-ply overcast; pinked; pinked and stitched; serging/single-ply overedged; untreated/plain; single-ply zigzag
Pressing steam; safe temperature limit 325°F (164.1°C)
Care launder
Fabric Resource **Lawrence Textile Co. Inc.**

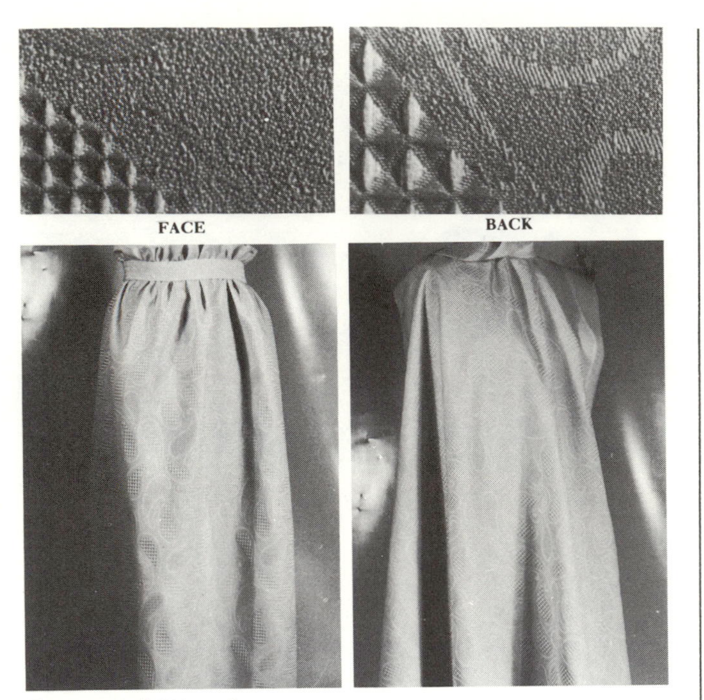

Jacquard Piqué

Fiber Content polyester/rayon blend
Yarn Type filament fiber
Yarn Construction fine; textured; slack twist
Fabric Structure piqué weave; Jacquard weave
Finishing Processes stabilizing
Color Application piece dyed-cross dyed; yarn dyed
Width 44-45 inches (111.8-114.3 cm)
Weight medium
Hand soft; springy
Texture weave creates uneven surface
Performance Expectations dimensional stability; elongation; resilient; high tensile strength; subject to snagging and edge abrasion
Drapability Qualities falls into wide cones; fullness maintains lofty effect; retains shape or silhouette of garment

Recommended Construction Aids & Techniques

Hand Needles betweens, 3-5; cotton darners, 5-10; embroidery, 5-8; milliners, 6-8; sharps, 6-8
Machine Needles round/set point, medium 11-14
Threads cotton- or poly-core waxed, 60 fine; cotton-polyester blend, all purpose; cotton-covered polyester core, all purpose; spun polyester, all purpose
Hems double-fold or single-fold bias binding; bound/Hong Kong finish; machine-stitched; overedged; seam binding
Seams flat-felled; lapped; plain; safety-stitched; welt; double welt
Seam Finishes single-ply bound; Hong Kong bound; single-ply overcast; pinked; pinked and stitched; serging/single-ply overedged; untreated/plain; single-ply zigzag
Pressing steam; safe temperature limit 325°F (164.1°C)
Care launder
Fabric Resource **Truemark Discount Fabrics Inc.**

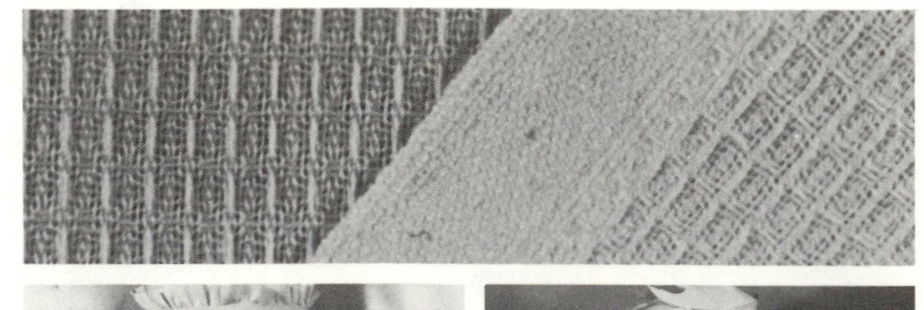

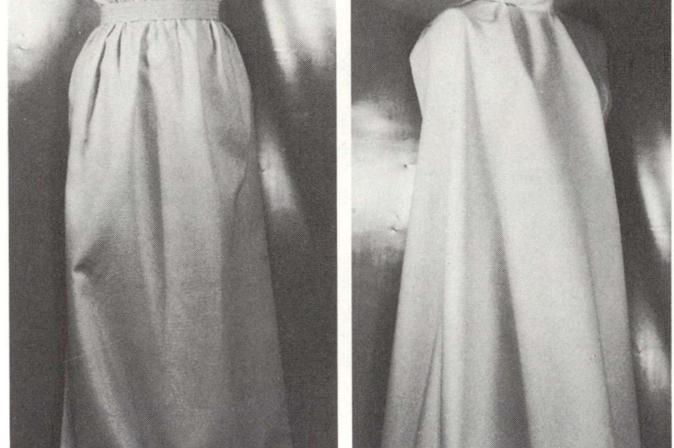

Waffle Piqué

Fiber Content 65% Kodel® polyester/35% cotton
Yarn Type combed staple; filament fiber
Yarn Construction conventional; textured; 2-ply (face)/heavy single-ply (back)
Fabric Structure piqué weave; dobby attachment
Finishing Processes mercerizing; permanent press/wash and wear; stabilizing
Color Application piece dyed-union dyed
Width 44-45 inches (111.8-114.3 cm)
Weight medium
Hand crisp; nonelastic; firm; rough
Texture weave creates waffle-patterned design
Performance Expectations durable; elongation; resilient; 2-3% residual shrinkage; high tensile strength; damaged by bleaching agents; wash-and-wear properties; wrinkle resistant
Drapability Qualities falls into crisp flares; fullness maintains crisp effect; retains shape or silhouette of garment

Recommended Construction Aids & Techniques

Hand Needles betweens, 3-5; cotton darners, 5-10; embroidery, 5-8; milliners, 6-8; sharps, 6-8
Machine Needles round/set point, medium 14
Threads mercerized cotton, 50; six-cord cotton, 50; cotton- or poly-core waxed, all purpose; cotton-covered polyester core, all purpose
Hems double-fold or single-fold bias binding; bound/Hong Kong finish; machine-stitched; overedged; seam binding
Seams flat-felled; lapped; plain; safety-stitched; welt; double welt
Seam Finishes single-ply bound; Hong Kong bound; single-ply overcast; pinked; pinked and stitched; serging/single-ply overedged; untreated/plain; single-ply zigzag
Pressing steam; safe temperature limit 325°F (164.1°C)
Care launder; *do not* use bleaching agents
Fabric Resource **Springmaid Fabrics/Springs Mills Inc.**

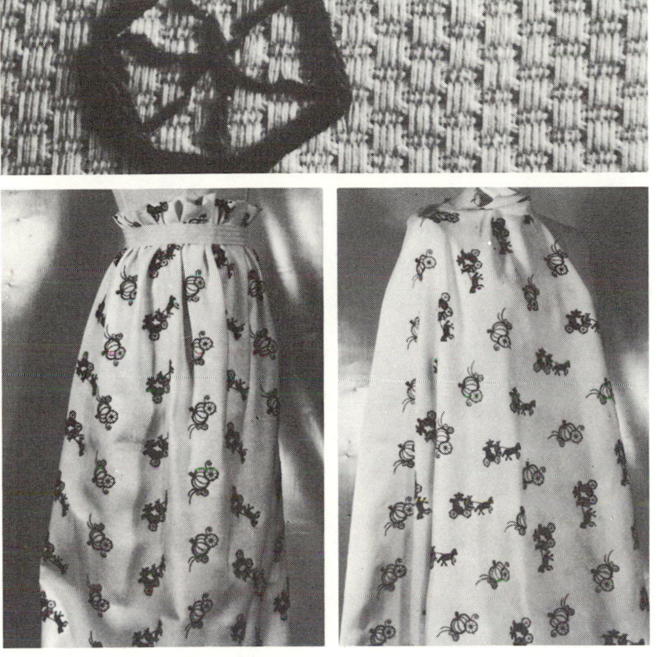

Piqué with Yarn Embroidery

Fiber Content 65% polyester/35% cotton
Yarn Type combed staple; filament fiber
Yarn Construction textured; light twist
Fabric Structure Schiffli embroidery on piqué weave base
Finishing Processes compressive shrinkage; heat-set; singeing
Color Application base: piece dyed–union dyed; embroidery yarn: skein or packaged dyed
Width 45 inches (114.3 cm)
Weight medium-heavy
Hand coarse; crisp; thick
Texture base: rough; embroidery yarn forms high and low raised pattern surface
Performance Expectations dimensional stability; durable; elongation; heat-set properties; resilient; high tensile strength
Drapability Qualities falls into wide cones; fullness maintains crisp bouffant effect; retains shape or silhouette of garment

Recommended Construction Aids & Techniques

Hand Needles betweens, 3–5; embroidery, 1–9; milliners, 3–7; sharps, 1–5
Machine Needles round/set point, medium 14
Threads poly-core waxed, 50; cotton-polyester blend, all purpose; cotton-covered polyester core, heavy duty; spun polyester, heavy duty; nylon/Dacron, twist A
Hems double-fold or single-fold bias binding; bound/ Hong Kong finish; faced; overedged; seam binding
Seams plain; safety-stitched
Seam Finishes single-ply bound; Hong Kong bound; single-ply overcast; serging/single-ply overedged; untreated/plain
Pressing use pressing cloth; safe temperature limit 325°F (164.1°C)
Care launder; dry clean
Fabric Resource Embroidery Council of America

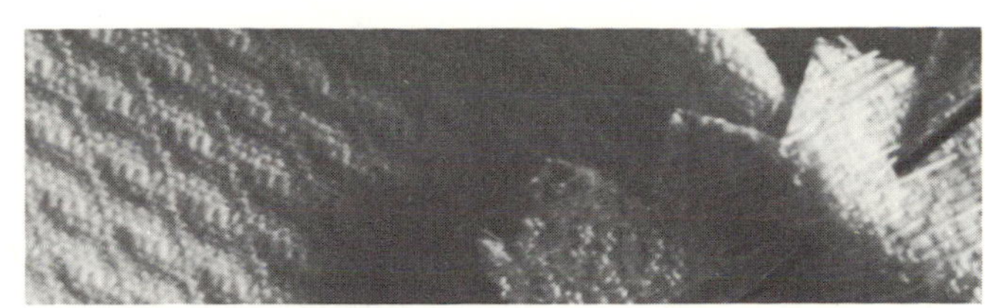

Bird's Eye Piqué

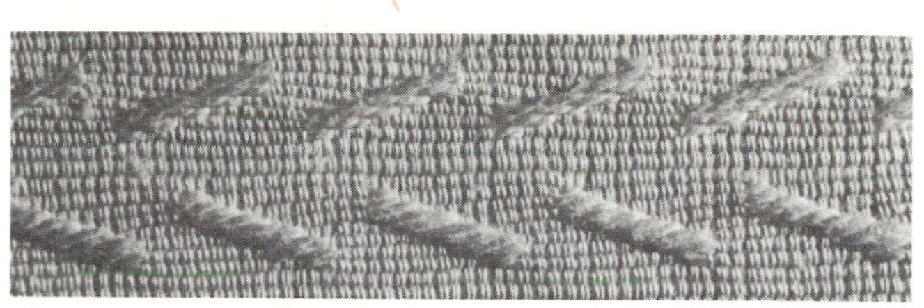

Crow's Foot Piqué

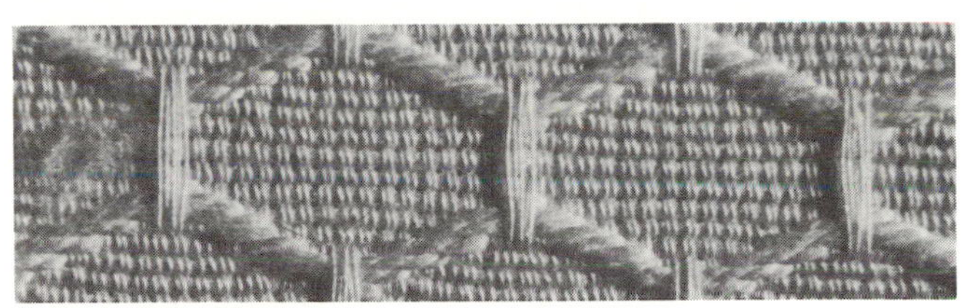

Diamond Piqué

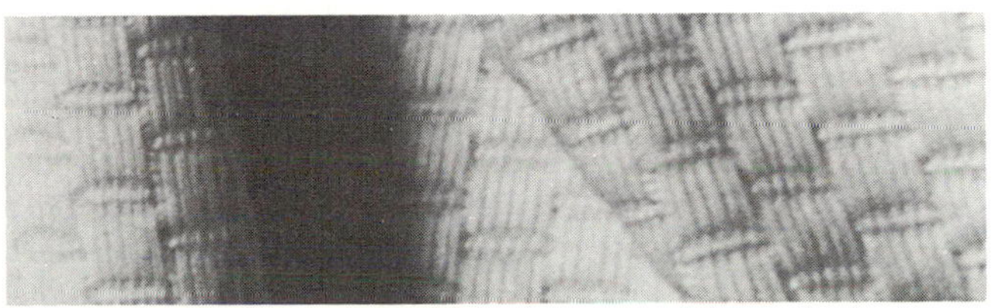

Ladder Piqué

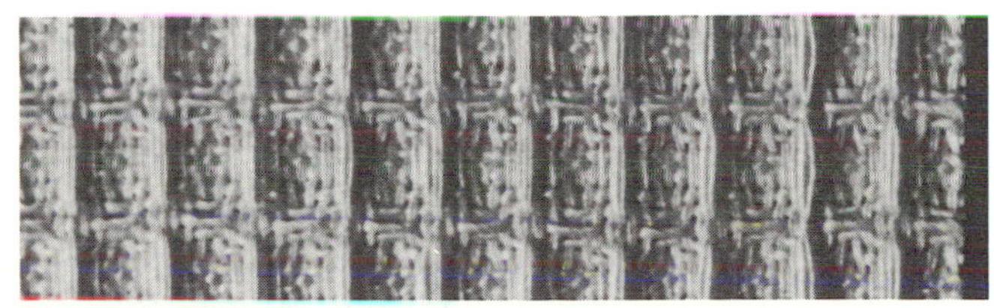

Square Piqué

Fabric Resource Russell Corporation

2 ~ Linen/Linen-type Fabrics

Art Linen
Bisso Linen / Altar Linen
Butcher Linen
  Butcher Linen (soft finish)
  Butcher Linen (crisp finish)
Cambric / Batiste Linen / Linen Lawn
  Cambric / Batiste Linen / Linen Lawn (soft finish)
  Cambric / Batiste Linen / Linen Lawn (crisp finish)
Crash Linen
  Crash Linen (soft finish)
  Crash Linen (crisp finish)
Dobby
  Linen Dobby
Dress Linen
  Dress Linen (soft finish)
  Dress Linen (crisp finish)
  Dress Linen (novelty/open weave)
Handkerchief Linen
  Handkerchief Linen (soft finish)
  Handkerchief Linen (crisp finish)
  Handkerchief Linen with Cord
Homespun
  Linen Homespun
Suiting
  Linen Suiting

Linen and linen-type fabrics, as listed in this unit, refer to fabrics generically named and classified as linen fabrics which were originally made of one-hundred percent natural flax fibers. This unit includes linen and linen-type fabrics which may be made of:

- Natural flax fibers
- Man-made fibers simulating flax
- Man-made fiber blends
- Natural and man-made fiber blends
- Mixed or combined yarns of different fiber origins

Regardless of the fiber content, linen and linen-type fabrics maintain or simulate the same outward characteristics of the originally generically named fabric. The following characteristics remain the same:

- Appearance
- Surface texture and interest
- Hand or feel
- Yarn type and construction
- Fabric structure
- Drapability qualities
- Weight (usually)

Although the original outward characteristics of the fabric are maintained, the fiber content used *changes* the performance, thread selection, pressing and care factors of the fabric. Each fiber has its own particular characteristics and properties.

There are variations within the basic generically named fabric. Variations may be obtained by changing, altering, modifying or combining:

- Yarn type
- Yarn construction
- Yarn count and size
- Yarns in warp and filling
- Usually weave structure
- Finishing processes

By combining or varying the components the textile designer can create new fabrics. When one component is changed, the fabric is changed. The newly designed or produced fabrics receive a new trade name. The new fabrics or newly named fabrics presented each season are usually placed into categories which may refer to their hand or feel, texture, surface appearance, structure or weight. Different textile companies have their own trade names or trademarks for the fabrics they produce.

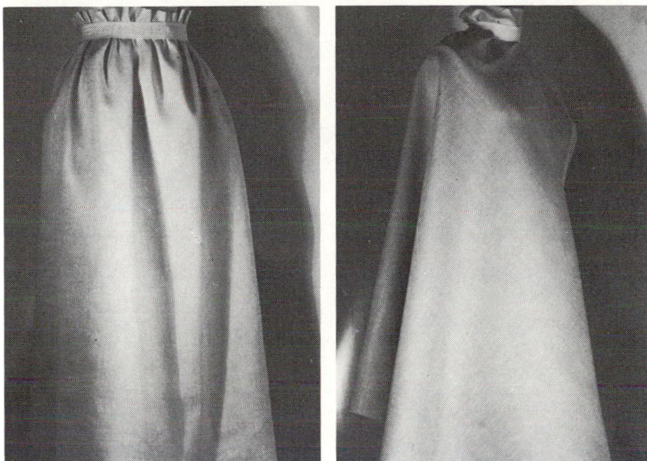

Art Linen

Fiber Content 100% rayon
Yarn Type filament fiber
Yarn Construction conventional; cylindrical; hard twist
Fabric Structure closely woven plain weave; balanced construction
Finishing Processes beetling; crease resistance; softening
Color Application piece dyed; solution dyed; yarn dyed
Width 56-58 inches (142.2-147.3 cm)
Weight medium-heavy
Hand compact; nonelastic; firm; hard
Texture flat; lustrous; smooth
Performance Expectations subject to edge abrasion; dimensional stability; durable; nonpilling; nonsnagging; soil resistant; high tensile strength
Drapability Qualities falls into wide cones; fullness maintains lofty effect; retains shape or silhouette of garment

Recommended Construction Aids & Techniques

Hand Needles betweens, 3-5; cotton darners, 1-7; embroidery, 1-7; milliners, 3-8; sharps, 1-6
Machine Needles round/set point, medium 11
Threads mercerized cotton, 50; silk, industrial size A-B; six-cord cotton, 40; cotton-covered polyester core, all purpose
Hems double-fold or single-fold bias binding; bound/ Hong Kong finish; stitched and pinked flat; overedged; seam binding
Seams flat-felled; plain; safety-stitched; welt
Seam Finishes single-ply or double-ply bound; Hong Kong bound; single-ply overcast; pinked and stitched; serging/single-ply overedged; single-ply zigzag
Pressing steam; safe temperature limit 350°F (178.1°C)
Care launder; dry clean
Fabric Resource **Auburn Fabrics Inc.**

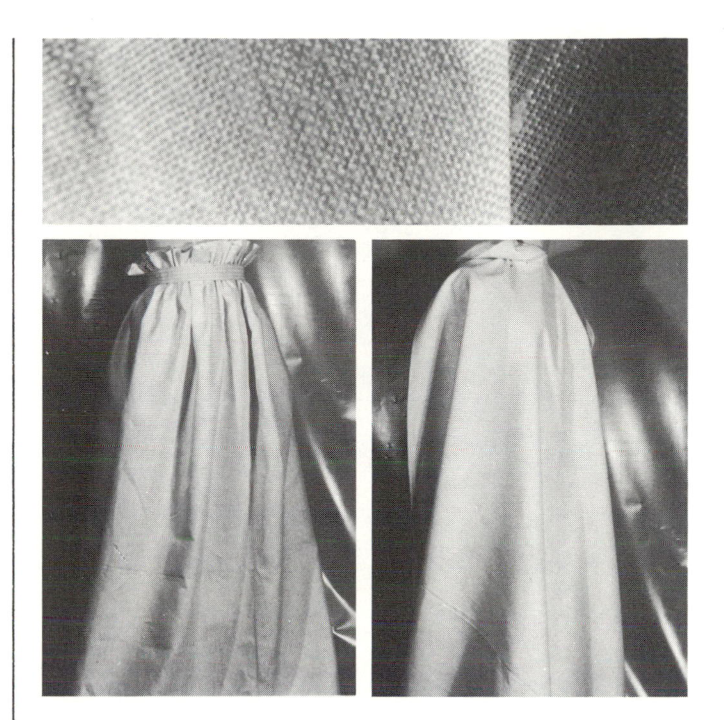

Bisso Linen/Altar Linen

Fiber Content 100% flax
Yarn Type line fiber
Yarn Construction conventional; fine; round; tight twist
Fabric Structure closely woven plain weave; high thread count
Finishing Processes bleaching; crease resistance; mercerizing; scouring; schreiner calendering
Color Application piece dyed; yarn dyed
Width 72 inches (182.9 cm)
Weight medium-light
Hand crisp; fine; firm
Texture polished face; smooth
Performance Expectations subject to flex abrasion; absorbent; antistatic; durable; nonpilling; nonsnagging; high tensile strength
Drapability Qualities falls into firm crisp flares; fullness maintains crisp effect; retains crisp folds, creases

Recommended Construction Aids & Techniques

Hand Needles betweens, 3-6; cotton darners, 7-10; embroidery, 6-10; milliners, 6-10; sharps, 3-6
Machine Needles round/set point, medium-fine 9-11
Threads mercerized cotton, 50; silk, industrial size A-B; six-cord cotton, 50; cotton-covered polyester core, extra fine
Hems edge-stitched; machine-stitched; double-fold machine-stitched; double edge-stitched
Seams French; lapped; plain; safety-stitched
Seam Finishes self-bound; single-ply bound; edge-stitched; pinked; pinked and stitched
Pressing steam; safe temperature limit 450°F (234.1°C)
Care launder
Fabric Resource **Hamilton Adams Imports, Ltd./ Moygashel®**

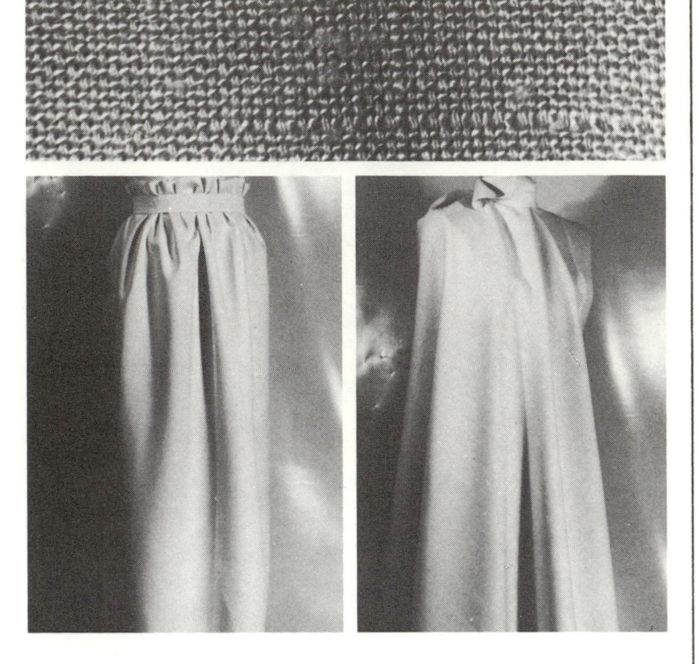

Butcher Linen (soft finish)

Fiber Content 100% rayon
Yarn Type filament fiber; filament staple
Yarn Construction conventional; thick-and-thin
Fabric Structure plain weave
Finishing Processes beetling; mercerizing; softening
Color Application piece dyed; yarn dyed
Width 56–58 inches (142.2–147.3 cm)
Weight medium
Hand compact; grainy; semi-soft
Texture coarse; irregular yarn
Performance Expectations abrasion resistant; dimensional stability; nonpilling; nonsnagging; high tensile strength
Drapability Qualities falls into moderately firm flares; fullness maintains lofty effect; retains shape or silhouette of garment

Recommended Construction Aids & Techniques

Hand Needles betweens, 3–5; cotton darners, 1–7; embroidery, 1–7; milliners, 3–8; sharps, 1–6
Machine Needles round/set point, medium 11–14
Threads mercerized cotton, 50; silk, industrial size A-B; six-cord cotton, 40; cotton-covered polyester core, all purpose
Hems double-fold or single-fold bias binding; bound/Hong Kong finish; stitched and pinked flat; overedged; seam binding
Seams flat-felled; plain; safety-stitched; welt
Seam Finishes single-ply or double-ply bound; Hong Kong bound; single-ply overcast; pinked and stitched; serging/single-ply overedged
Pressing steam; safe temperature limit 350°F (178.1°C)
Care launder; dry clean
Fabric Resource Auburn Fabrics Inc.

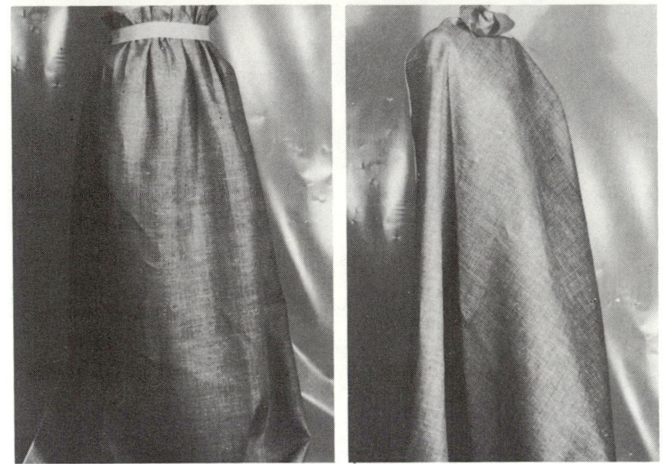

Butcher Linen (crisp finish)

Fiber Content 100% flax
Yarn Type line fiber
Yarn Construction conventional; thick-and-thin
Fabric Structure plain weave
Finishing Processes beetling; crease resistance; mercerizing; scouring
Color Application jig dyed; molten metal dyed
Width 54 inches (137.2 cm)
Weight medium to medium-heavy
Hand firm; hard; harsh; scratchy
Texture coarse; grainy
Performance Expectations subject to edge abrasion; absorbent; antistatic; durable; nonpilling; high tensile strength
Drapability Qualities falls into crisp wide cones; fullness maintains bouffant effect; retains shape or silhouette of garment

Recommended Construction Aids & Techniques

Hand Needles betweens, 3–5; cotton darners, 17; embroidery, 1–7; milliners, 3–8; sharps, 1–6
Machine Needles round/set point, medium 11–14
Threads mercerized cotton, heavy duty; silk, industrial size A-B; six-cord cotton, 40; cotton-covered polyester core, heavy duty
Hems double-fold or single-fold bias binding; bound/Hong Kong finish; stitched and pinked flat; overedged; seam binding
Seams flat-felled; safety-stitched welt; plain
Seam Finishes single-ply or double-ply bound; Hong Kong bound; single-ply overcast; pinked and stitched; serging/single-ply overedged; single-ply zigzag
Pressing steam; safe temperature limit 450°F (234.1°C)
Care launder; dry clean
Fabric Resource Hamilton Adams Imports, Ltd./Moygashel®

Cambric/Batiste Linen/Linen Lawn (soft finish)

Fiber Content 50% polyester/50% rayon
Yarn Type filament fiber; filament staple
Yarn Construction conventional; fine
Fabric Structure closely woven twill weave; high thread count
Finishing Processes beetling; friction calendering; mercerizing; softening
Color Application piece dyed-union dyed
Width 44-45 inches (111.8-114.3 cm)
Weight[1] light
Hand[1] firm; pliable; soft
Texture[1] flat; slightly glossed face; smooth
Performance Expectations abrasion resistant; nonsnagging; high tensile strength
Drapability Qualities falls into soft flares; accommodates fullness by gathering, elasticized shirring, smocking; fullness retains soft fall

Recommended Construction Aids & Techniques

Hand Needles betweens, 3-16; cotton darners, 7-10; embroidery, 6-10; milliners, 6-10; sharps, 3-6
Machine Needles round/set point, medium-fine 9-11
Threads mercerized cotton, 50; silk, industrial size A-B; six-cord cotton, 50; cotton-covered polyester core, extra fine
Hems edge-stitched; double edge-stitched; machine-stitched; double fold machine-stitched
Seams flat-felled; French; mock French; lapped; plain
Seam Finishes self bound; edge-stitched; pinked; pinked and stitched
Pressing steam; safe temperature limit 325°F (164.1°C)
Care launder
Fabric Resource **Printsiples and Company B**

Cambric/Batiste Linen/Linen Lawn (crisp finish)

Fiber Content 50% Avril® polyester/50% Avril® rayon
Yarn Type filament fiber; filament staple
Yarn Construction conventional; fine; high twist
Fabric Structure plain weave
Finishing Processes friction calendering; mercerizing; resin sizing
Color Application piece dyed-union dyed; solution dyed
Width 44-45 inches (111.8-114.3 cm)
Weight[1] light
Hand[1] crisp; fine; firm
Texture[1] flat; slightly glossed face; smooth
Performance Expectations abrasion resistant; nonsnagging; high tensile strength
Drapability Qualities falls into crisp flares; accommodates fullness by pleating, gathering; fullness maintains crisp effect

Recommended Construction Aids & Techniques

Hand Needles betweens, 3-16; cotton darners, 7-10; embroidery, 6-10; milliners, 6-10; sharps, 3-6
Machine Needles round/set point, medium-fine, 9-11
Threads mercerized cotton, 50; silk, industrial size A-B; six-cord cotton, 50; cotton-covered polyester core, extra fine
Hems edge-stitched; double edge-stitched; machine-stitched; double fold machine-stitched
Seams flat-felled; French; mock French; lapped; plain
Seam Finishes self bound; edge-stitched; pinked; pinked and stitched
Pressing steam; safe temperature limit 325°F (164.1°C)
Care launder; dry clean
Fabric Resource **Cameo Fabrics Inc.**

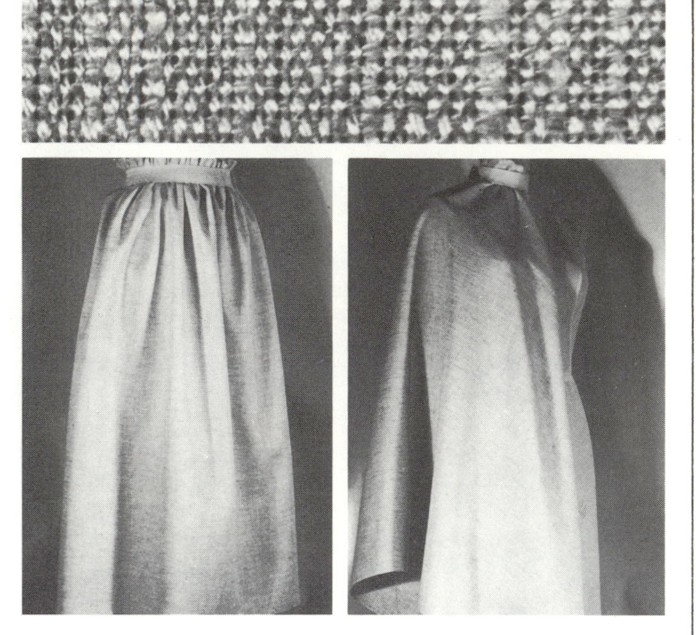

Crash Linen (soft finish)

Fiber Content 58% rayon/34% polyester/8% flax
Yarn Type filament fiber; line fiber
Yarn Construction conventional; irregular yarns in a variety of thicknesses; uneven; slack twist
Fabric Structure plain weave
Finishing Processes beetling; crease resistance; mercerizing
Color Application piece dyed–union dyed; yarn dyed
Width 58–60 inches (147.3–152.4 cm)
Weight medium to heavy
Hand compact; rough; soft
Texture coarse; yarn creates uneven surface
Performance Expectations abrasion resistance; dimensional stability; nonpilling; nonsnagging; high tensile strength
Drapability Qualities falls into moderately soft flares; fullness retains moderately soft fall; retains shape of garment

Recommended Construction Aids & Techniques

Hand Needles betweens, 3–5; cotton darners, 1–7; embroidery, 1–7; milliners, 3–8; sharps, 1–6
Machine Needles round/set point, medium 11–14
Threads mercerized cotton, 50; silk, industrial size A-B; six-cord cotton, 40; cotton-covered polyester core, all purpose
Hems double-fold or single-fold bias binding; bound/Hong Kong finish; stitched and pinked flat; overedged; seam binding
Seams flat-felled; plain; safety-stitched; welt
Seam Finishes single-ply or double-ply bound; Hong Kong bound; single-ply overcast; pinked and stitched; serging/single-ply overedged; single-ply zigzag
Pressing steam; safe temperature limit 325°F (164.1°C)
Care launder; dry clean
Fabric Resource **Auburn Fabrics Inc.**

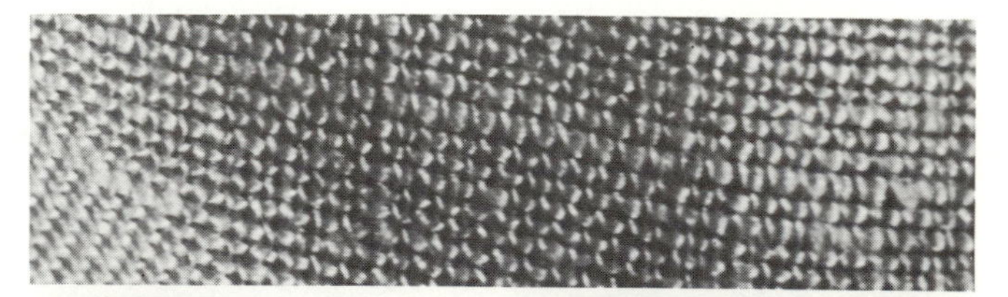

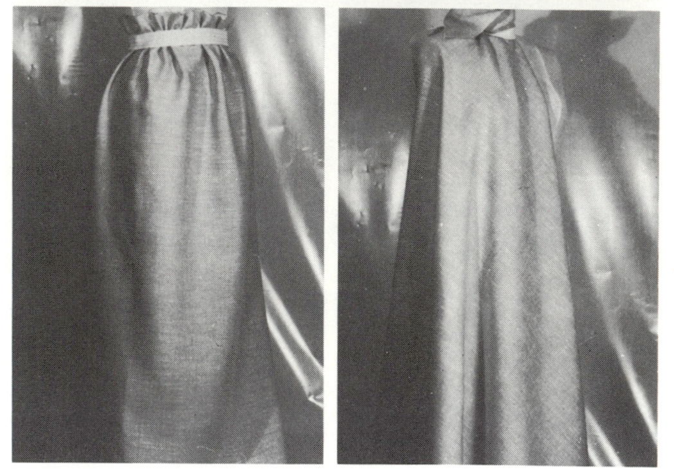

Crash Linen (crisp finish)

Fiber Content 100% flax
Yarn Type line fiber
Yarn Construction conventional; irregular yarns in a variety of thicknesses; slack twist
Fabric Structure plain weave
Finishing Processes beetling; crease resistance; mercerizing; scouring
Color Application jig dyed; molten metal dyed
Width 45 inches (114.3 cm)
Weight medium to heavy
Hand compact; crisp; nonelastic; scratchy
Texture coarse; harsh; yarn creates uneven surface
Performance Expectations subject to edge abrasion; absorbent; antistatic; durable; inflexible; nonpilling; nonsnagging; high tensile strength
Drapability Qualities falls into stiff wide cones; fullness maintains bouffant effect; retains shape or silhouette of garment

Recommended Construction Aids & Techniques

Hand Needles betweens, 3–5; cotton darners, 1–7; embroidery, 11–14; milliners, 3–8; sharps, 1–6
Machine Needles round/set point, medium 11–14
Threads mercerized cotton, heavy duty; silk, industrial size A-B; six-cord cotton, 40; cotton-covered polyester core, all purpose
Hems double-fold or single-fold bias binding; bound/Hong Kong finish; stitched and pinked flat; overedged; seam binding
Seams flat-felled; plain; safety-stitched; welt
Seam Finishes single-ply or double-ply bound; Hong Kong bound; single-ply overcast; pinked and stitched; serging/single-ply overedged; single-ply zigzag
Pressing steam; safe temperature limit 450°F (234.1°C)
Care launder; dry clean
Fabric Resource **Hamilton Adams Imports, Ltd./Moygashel®**

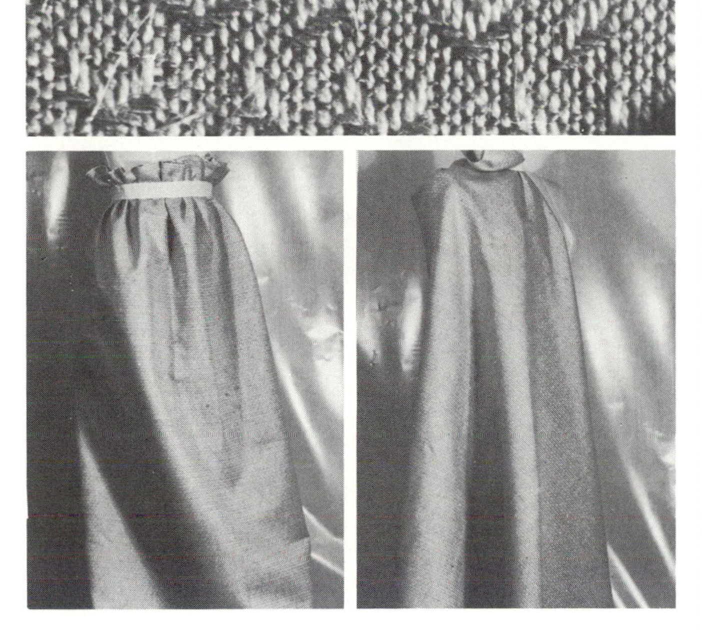

Linen Dobby

Fiber Content 100% flax
Yarn Type line fiber
Yarn Construction conventional; slack twist
Fabric Structure novelty figure weave on dobby loom
Finishing Processes beetling; crease resistance; mercerzing; scouring
Color Application yarn dyed
Width 54 inches (137.2 cm)
Weight medium to heavy
Hand compact; soft to crisp depending on finish; thick
Texture coarse; raised pattern
Performance Expectations subject to edge abrasion; absorbent; antistatic; nonpilling
Drapability Qualities falls into firm wide cones; fullness maintains bouffant effect; retains shape or silhouette of garment

Recommended Construction Aids & Techniques

Hand Needles betweens, 3-5; cotton darners, 1-7; embroidery, 1-7; milliners, 3-8; sharps, 1-6
Machine Needles round/set point, medium 11-14
Threads mercerized cotton, heavy duty; silk, industrial size A-B; six-cord cotton, 40; cotton-covered polyester core, all purpose
Hems double-fold or single-fold bias binding; bound/Hong Kong finish; stitched and pinked flat; overedged; seam binding
Seams flat-felled; plain; safety-stitched; welt
Seam Finishes single-ply or double-ply bound; Hong Kong bound; single-ply overcast; pinked and stitched; serging/single-ply overedged; single-ply zigzag
Pressing steam; safe temperature limit 450°F (234.1°C)
Care dry clean
Fabric Resource **Hamilton Adams Imports, Ltd./Moygashel®**

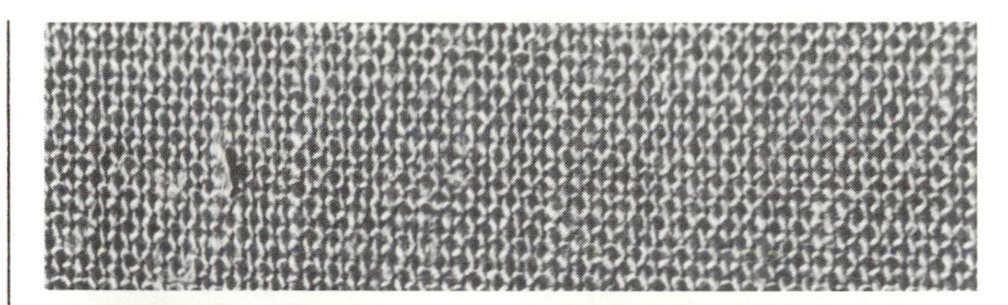

Dress Linen (soft finish)

Fiber Content 55% Dacron® polyester/35% rayon/10% flax
Yarn Type filament fiber; filament staple; line fiber
Yarn Construction conventional
Fabric Structure plain weave
Finishing Processes beetling; crease resistance; mercerizing; softening
Color Application piece dyed-union dyed
Width 44-45 inches (111.8-114.3 cm)
Weight medium
Hand compact; semi-soft
Texture coarse; grainy
Performance Expectations abrasion resistant; dimensional stability; high tensile strength
Drapability Qualities falls into moderately firm flares; accommodates fullness by pleating, gathering; fullness maintains crisp effect

Recommended Construction Aids & Techniques

Hand Needles betweens, 3-5; cotton darners, 1-7; embroidery, 1-7; milliners, 3-8; sharps, 1-6
Machine Needles round/set point, medium 11-14
Threads mercerized cotton, 50; silk, industrial size A-B; six-cord cotton, 40; cotton-covered polyester core, all purpose
Hems double-fold or single-fold bias binding; bound/Hong Kong finish; stitched and pinked flat; overedged; seam binding
Seams flat-felled; plain; safety-stitched; welt
Seam Finishes single-ply or double-ply bound; Hong Kong bound; single-ply overcast; pinked and stitched; serging/single-ply overedged; single-ply zigzag
Pressing steam; safe temperature limit 350°F (178.1°C)
Care launder; dry clean
Fabric Resource **Bloomsburg Mills**

Dress Linen (crisp finish)

Fiber Content 100% flax
Yarn Type line fiber
Yarn Construction conventional; cylindrical; thick
Fabric Structure plain weave
Finishing Processes bleaching; crease resistance; mercerizing; scouring
Color Application jig dyed; molten metal dyed
Width 45 inches (114.3 cm)
Weight medium-light to medium-heavy
Hand semi-crisp; firm
Texture coarse; grainy
Performance Expectations subject to edge abrasion; absorbent; antistatic; dimensional stability; durable; nonpilling; nonsnagging; high tensile strength
Drapability Qualities falls into moderately firm flares; fullness maintains lofty effect; retains shape or silhouette of garment

Recommended Construction Aids & Techniques

Hand Needles betweens, 3-8; cotton darners, 1-7; embroidery, 1-7; milliners, 3-8; sharps, 1-6
Machine Needles round/set point, medium 11-14
Threads mercerized cotton, 50; silk, industrial size A-B; six-cord cotton, 40; cotton-covered polyester core, all purpose
Hems double-fold or single-fold bias binding; bound/Hong Kong finish; stitched and pinked flat; overedged; seam binding
Seams flat-felled; plain; safety-stitched; welt
Seam Finishes single-ply or double-ply bound; Hong Kong bound; single-ply overcast; pinked and stitched; serging/single-ply overedged; single-ply zigzag
Pressing steam; safe temperature limit 450°F (234.1°C)
Care launder; dry clean
Fabric Resource **Hamilton Adams Imports, Ltd./Moygashel®**

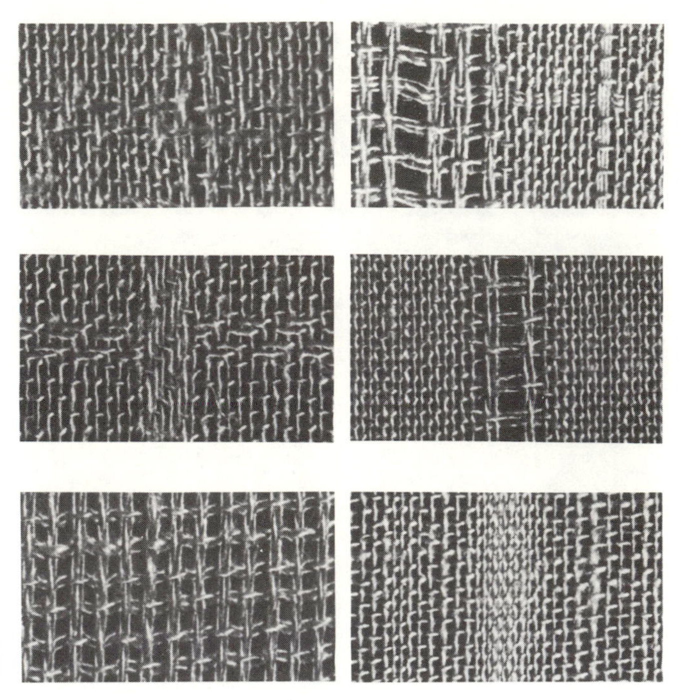

Dress Linen (novelty/open weave)

Fiber Content 55% Dacron® polyester/35% rayon/10% flax
Yarn Type filament fiber; filament staple; line fiber
Yarn Construction conventional
Fabric Structure novelty weave; leno weave; gauze weave
Finishing Processes beetling; crease resistance; mercerizing; softening
Color Application piece dyed-union dyed
Width 44-45 inches (111.8-114.3 cm)
Weight medium
Hand semi-crisp
Texture weave creates open effect and uneven surface
Performance Expectations abrasion resistant; subject to snagging due to weave
Drapability Qualities falls into moderately firm flares; accommodates fullness by pleating, gathering; fullness maintains crisp effect

Recommended Construction Aids & Techniques

Hand Needles betweens, 3-5; cotton darners, 1-7; embroidery, 1-7; milliners, 3-8; sharps, 1-6
Machine Needles round/set point, medium 11-14
Threads mercerized cotton, 50; silk, industrial size A-B; six-cord cotton, 40; cotton-covered polyester core, all purpose
Hems double-fold or single-fold bias binding; bound/Hong Kong finish; stitched and pinked flat; overedged; seam binding
Seams flat-felled; plain; safety-stitched; welt
Seam Finishes single-ply or double-ply bound; Hong Kong bound; single-ply overcast; pinked and stitched; serging/single-ply overedged; single-ply zigzag
Pressing steam; safe temperature limit 350°F (178.1°C)
Care launder; dry clean
Fabric Resource **Bloomsburg Mills**

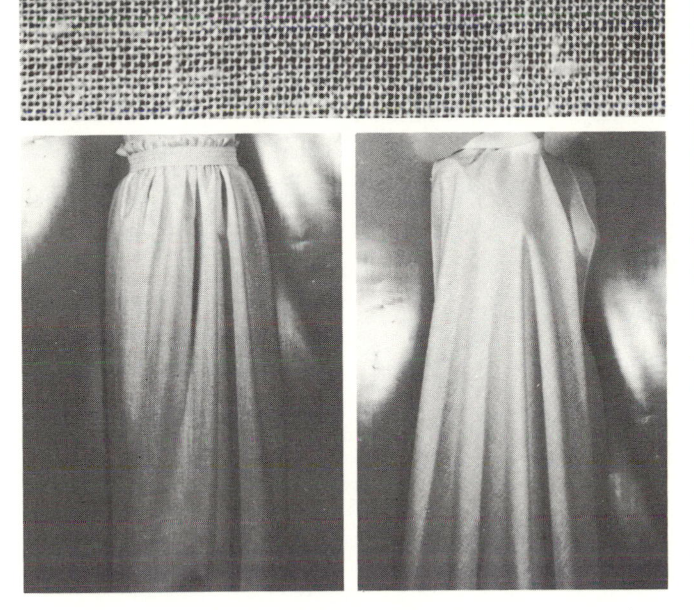

Handkerchief Linen (soft finish)

Fiber Content 50% Trevira® polyester/50% rayon
Yarn Type filament fiber
Yarn Construction conventional; fine yarns with slight irregularities
Fabric Structure plain weave
Finishing Processes calendering; mercerizing; shrinkage control; softening
Color Application piece dyed-union dyed
Width 44–45 inches (111.8–114.3 cm)
Weight[1] light
Hand[1] fine; firm; soft
Texture[1] coarse; irregularly spaced slight nub
Performance Expectations abrasion resistant; dimensional stability; durable; nonsnagging; high tensile strength
Drapability Qualities falls into soft flares; accommodates fullness by gathering, elasticized shirring, smocking; fullness retains soft fall

Recommended Construction Aids & Techniques

Hand Needles betweens, 3–6; cotton darners, 7–10; embroidery, 6–10; milliners, 6–10; sharps, 3–6
Machine Needles round/set point, medium-fine 9–11
Threads mercerized cotton, 50; silk, industrial size A-B; six-cord cotton, 50; cotton-covered polyester core, extra fine
Hems edge-stitched; double edge-stitched; machine-stitched; double fold machine-stitched
Seams French; lapped; plain; safety-stitched
Seam Finishes self bound; double-stitched and trimmed; edge-stitched; pinked; pinked and stitched
Pressing steam; safe temperature limit 325°F (164.1°C)
Care launder; dry clean
Fabric Resources **Cameo Fabrics Inc.; Hargro Fabrics Inc.**

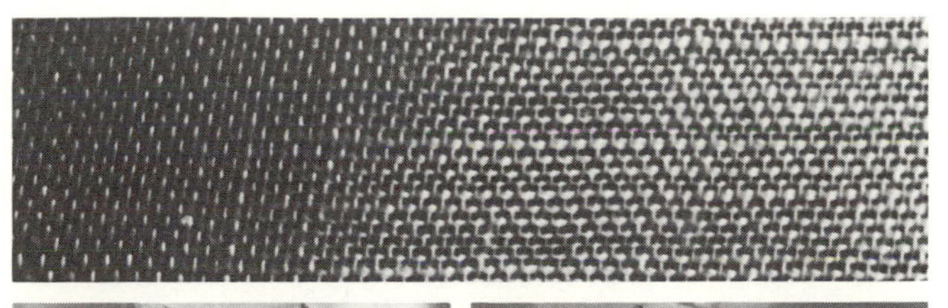

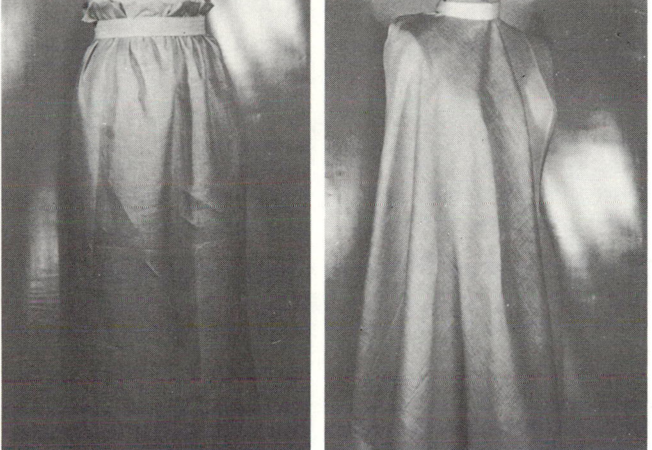

Handkerchief Linen (crisp finish)

Fiber Content 100% flax
Yarn Type line fiber
Yarn Construction conventional; fine yarns with slight irregularities
Fabric Structure plain weave
Finishing Processes bleaching; crease resistance; mercerizing; scouring; sizing
Color Application jig dyed; molten metal dyed
Width 36 inches (91.4 cm)
Weight[1] light
Hand[1] crisp; fine; firm; harsh
Texture[1] coarse; irregularly spaced slight nub
Performance Expectations subject to edge abrasion; absorbent; antistatic; durable; nonpilling, nonsnagging; high tensile strength
Drapability Qualities falls into stiff cones; accommodates fullness by pleating, gathering, elasticized shirring; fullness maintains crisp effect

Recommended Construction Aids & Techniques

Hand Needles betweens, 3–6; cotton darners, 7–10; embroidery, 6–10; milliners, 6–10; sharps, 3–6
Machine Needles round/set point, medium-fine 9–11
Threads mercerized cotton, 50; silk, industrial size A-B; six-cord cotton, 50; cotton-covered polyester core, extra fine
Hems edge-stitched; double edge-stitched; machine-stitched; double machine-stitched; hand- or machine-rolled
Seams French; lapped; plain; safety-stitched
Seam Finishes self bound; double-stitched and trimmed; edge-stitched; pinked; pinked and stitched
Pressing steam; safe temperature limit 450°F (234.1°C)
Care launder; dry clean
Fabric Resource **Hamilton Adams Import, Ltd./ Moygashel®**

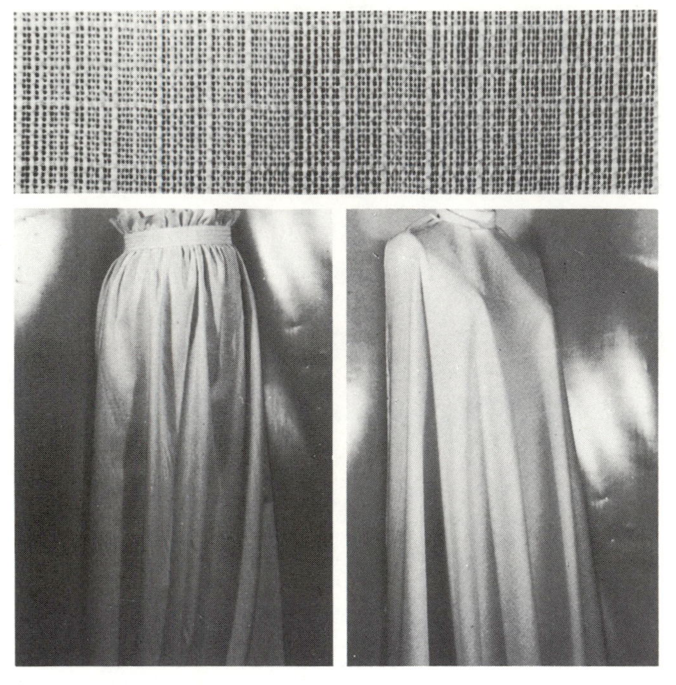

Handkerchief Linen with Cord

Fiber Content 65% Kodel® polyester/35% cotton
Yarn Type combed staple; filament fiber; filament staple
Yarn Construction conventional; fine; thick yarns for cord
Fabric Structure plain weave
Finishing Processes mercerizing; Russpress® permanent press precured
Color Application piece dyed-union dyed
Width 44-45 inches (111.8-114.3 cm)
Weight[1] light
Hand[1] semi-crisp; fine; firm
Texture[1] lengthwise and/or crosswise corded surface
Performance Expectations abrasion resistant; durable; permanent press; 1% residual shrinkage; high tensile strength; wrinkle resistant
Drapability Qualities falls into soft flares; accommodates fullness by gathering, elasticized shirring, smocking; fullness retains soft fall

Recommended Construction Aids & Techniques

Hand Needles betweens, 3-6; cotton darners, 7-10; embroidery, 6-10; milliners, 6-10; sharps, 3-6
Machine Needles round/set point, medium-fine 9-11
Threads mercerized cotton, 50; silk, industrial size A-B; six-cord cotton, 50; cotton-covered polyester core, extra fine
Hems edge-stitched; double edge-stitched; machine-stitched; double fold machine-stitched
Seams French; lapped; plain; safety-stitched
Seam Finishes self bound; double-stitched and trimmed; edge-stitched; pinked; pinked and stitched
Pressing steam; safe temperature limit 325°F (164.1°C)
Care launder
Fabric Resource **Russell Corporation**

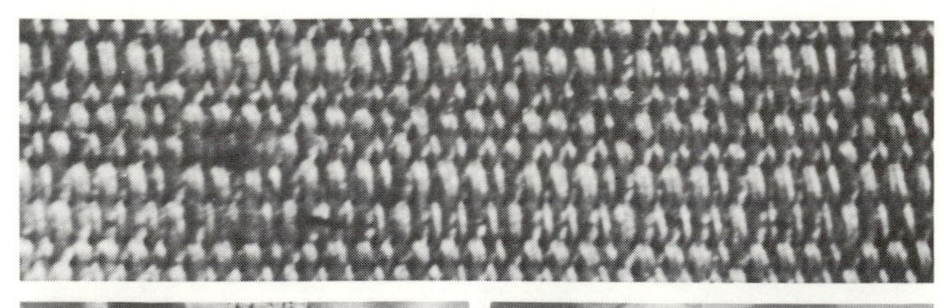

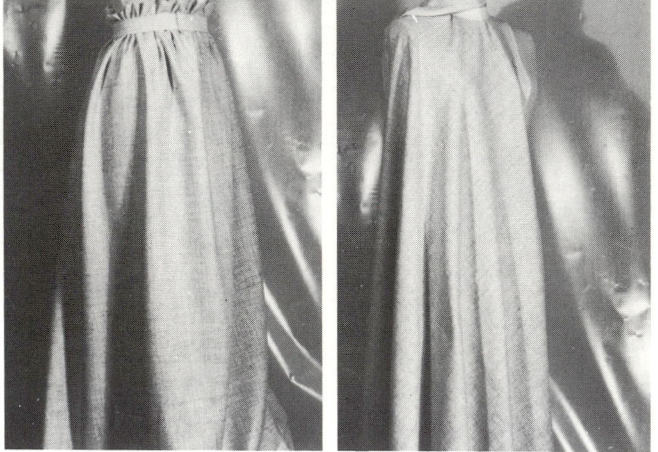

Linen Homespun[2]

Fiber Content flax/nylon blend
Yarn Type filament; line fiber
Yarn Construction conventional; ply; slack twist
Fabric Structure loosely woven plain weave variation
Finishing Processes beetling; bleaching; crease resistance; mercerizing; scouring
Color Application bleached; piece dyed-union dyed; yarn dyed
Width 60 inches (152.4 cm)
Weight medium to medium-heavy
Hand hard; harsh; porous
Texture coarse; loose; rough
Performance Expectations absorbent; high tensile strength
Drapability Qualities falls into wide cones; accommodates fullness by gathering; fullness maintains lofty effect

Recommended Construction Aids & Techniques

Hand Needles betweens, 3-5; cotton darners, 1-7; embroidery, 1-7; milliners, 3-8; sharps, 1-6
Machine Needles round/set point, medium 11-14
Threads mercerized cotton, heavy duty; silk, industrial size A-B; six-cord cotton, 40; cotton-covered polyester core, all purpose
Hems double-fold or single-fold bias binding; bound/Hong Kong finish; stitched and pinked flat; overedged; seam binding
Seams flat-felled; plain; safety-stitched; welt
Seam Finishes single-ply or double-ply bound; Hong Kong bound; single-ply overcast; pinked and stitched; serging/single-ply overedged
Pressing steam; safe temperature limit 350°F (178.1°C)
Care dry clean
Fabric Resource **Hamilton Adams Imports, Ltd./Moygashel®**

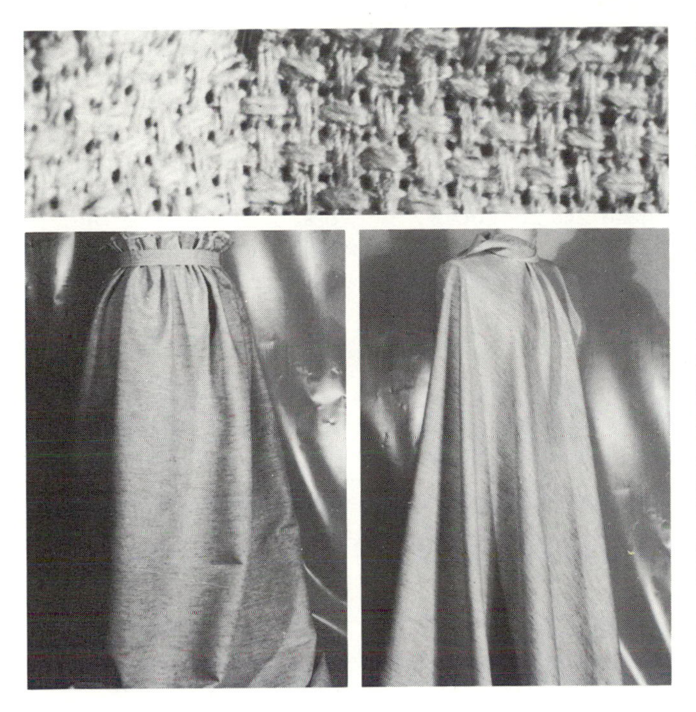

Linen Suiting

Fiber Content flax/nylon blend
Yarn Type filament fiber; line fiber
Yarn Construction conventional; slack twist
Fabric Structure plain weave; fine warp/heavy filling
Finishing Processes bleaching; crease resistance; mercerizing; scouring
Color Application jig dyed; molten metal dyed; yarn dyed
Width 64 inches (162.7 cm)
Weight medium to heavy
Hand compact; firm; soft to crisp depending on finish; scratchy
Texture coarse; harsh; yarn creates uneven surface
Performance Expectations absorbent; durable; nonpilling; nonsnagging; soil resistant; high tensile strength
Drapability Qualities falls into crisp wide flares; fullness maintains crisp bouffant effect; retains shape or silhouette of garment

Recommended Construction Aids & Techniques

Hand Needles betweens, 3-5; cotton darners, 1-7; embroidery, 1-7; milliners, 3-8; sharps, 1-6
Machine Needles round/set point, medium 11-14
Threads mercerized cotton, heavy duty; silk, industrial size A-B; six-cord cotton, 40; cotton-covered polyester core, all purpose
Hems double-fold or single-fold bias binding; bound/ Hong Kong finish; stitched and pinked flat; overedged; seam binding
Seams flat-felled; plain; safety-stitched; welt
Seam Finishes single-ply or double-ply bound; Hong Kong bound; single-ply overcast; pinked and stitched; serging/single-ply overedged; single-ply zigzag
Pressing steam; safe temperature limit 350°F (178.1°C)
Care dry clean
Fabric Resource **Hamilton Adams Imports, Ltd./ Moygashel®**

3 ~ Burlap/Burlap-type Fabrics

Natural Burlap
Simulated Burlap

Burlap and burlap-type fabrics, as listed in this unit, refer to fabrics generically named and classified as burlap fabrics which were originally made of one-hundred percent natural jute fibers. This unit includes burlap and burlap-type fabrics which may be made of:

- Natural jute fibers
- Man-made fibers simulating jute
- Man-made fiber blends
- Natural and man-made fiber blends
- Mixed or combined yarns of different fiber origins

Regardless of the fiber content, burlap and burlap-type fabrics maintain or simulate the same outward characteristics of the original generically named fabric. The following characteristics remain the same:

- Appearance
- Surface texture and interest
- Hand or feel
- Yarn type and construction
- Fabric structure
- Drapability qualities
- Weight (usually)

Although the original outward characteristics of the fabrics are maintained, the fiber content used *changes* the performance, thread selection, pressing and care factors of the fabric. Each fiber has its own particular characteristics and properties.

There are variations within the generically named fabric. Variations may be obtained by changing, altering, modifying or combining:

- Yarn type
- Yarn construction
- Yarn count and size
- Yarns in warp and filling
- Usual weave structure
- Finishing processes

By combining or varying the components the textile designer can create new fabrics. When one component is changed, the fabric is changed. The newly designed or produced fabric receives a new trade name. The new fabrics or newly named fabrics presented each season are usually placed into categories which refer to their hand or feel, texture, surface appearance, structure or weight. Different textile companies have their own trade names or trademarks for the fabrics they produce.

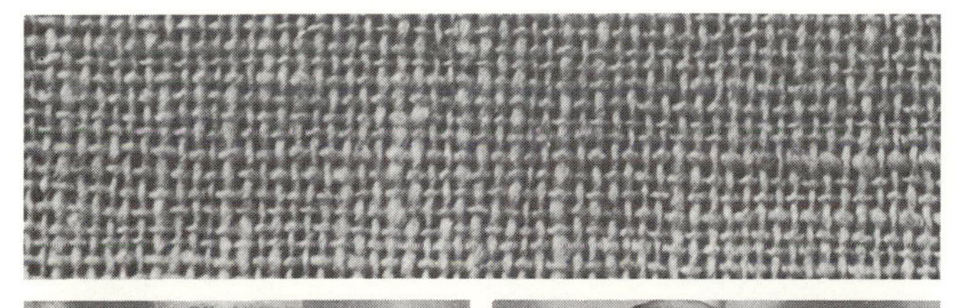

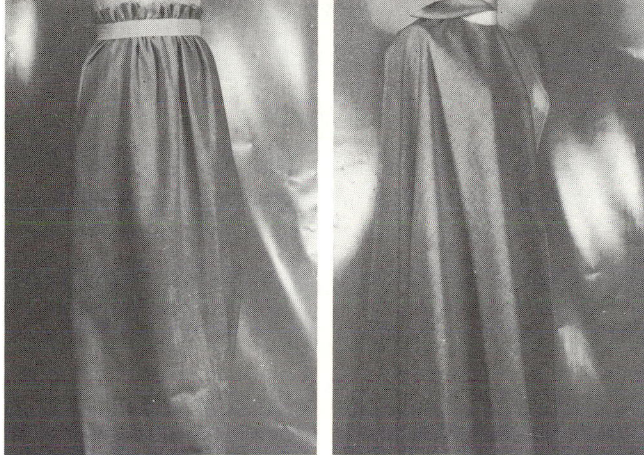

Natural Burlap

Fiber Content 100% jute
Yarn Type line fiber
Yarn Construction conventional; thick; heavy; slack twist
Fabric Structure plain weave
Finishing Processes bleaching; scouring; stabilizing
Color Application natural fiber color; may be yarn dyed or piece dyed
Width 36-45 inches (91.4-114.3 cm)
Weight medium-heavy
Hand harsh; scratchy; stiff
Texture loose; rough
Performance Expectations subject to flex abrasion; sunlight resistant; subject to limpness and bagging after cleaning, snagging, seam slippage
Drapability Qualities falls into stiff wide cones; fullness maintains bouffant effect; better utilized if fitted loosely by seaming and eliminating excess fabric

Recommended Construction Aids & Techniques

Hand Needles betweens, 3-5; chenilles, 12-18; cotton darners, 1-7; embroidery, 1-7; milliners, 3-7; tapestry, 18-22
Machine Needles round/set point, medium-coarse 14-16
Threads mercerized cotton, 40; six-cord cotton, 40; waxed, 40 heavy duty; cotton-covered polyester core, heavy duty
Hems double-fold or single-fold bias binding; bound/ Hong Kong finish; double machine-stitched
Seams double-stitched; plain
Seam Finishes single-ply bound; Hong Kong bound; strapped; taped
Pressing steam; safe temperature limit 375°F (192.1°C)
Care dry clean
Fabric Resource **Lawrence Textile Co. Inc.**

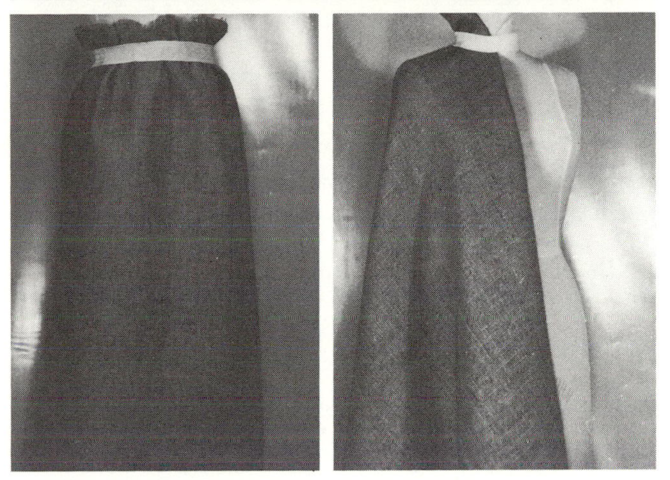

Simulated Burlap

Fiber Content 50% cotton/50% polyester
Yarn Type filament fiber; filament staple; staple
Yarn Construction conventional; thick-and-thin; slack twist
Fabric Structure plain weave
Finishing Processes beetling; mercerizing; stabilizing
Color Application piece dyed-union dyed
Width 44-45 inches (111.8-114.3 cm)
Weight medium-heavy
Hand dry; grainy; hard
Texture loose; rough
Performance Expectations subject to edge abrasion; durable; subject to shrinkage unless treated, bagging, snagging
Drapability Qualities falls into wide cones; fullness maintains lofty effect; retains shape or silhouette of garment

Recommended Construction Aids & Techniques

Hand Needles betweens, 3-5; cotton darners, 5-10; embroidery, 5-8; milliners, 6-8; sharps, 6-8
Machine Needles round/set point, medium-coarse 14-16
Threads mercerized cotton, 40; six-cord cotton, 40; waxed, 40 heavy duty; cotton-covered polyester core, heavy duty
Hems double-fold or single-fold bias binding; bound/ Hong Kong finish; overedged; seam binding
Seams plain; safety-stitched; welt; double welt
Seam Finishes single-ply bound; Hong Kong bound; pinked and stitched; serging/single-ply overedged; single-ply zigzag
Pressing steam; safe temperature limit 325°F (164.1°C)
Care launder
Fabric Resource **Printsiples and Company B**

4 – Silk/Silk-type Fabrics

- Barathea
  - Barathea
  - Barathea Suiting
- Bengaline
  - Silk Bengaline
- Broadcloth
  - Silk Broadcloth
  - Printed Silk Broadcloth
- Brocade
  - Silk Brocade
- Canvas
  - Silk Canvas
- Charvet
- Chiffon
  - Silk Chiffon
  - Sequin-embroidered Chiffon (pavé design)
  - Sequin-embroidered Chiffon (stripe design)
- China Silk
- Damask
  - Silk Damask (thick textured weave)
  - Silk Damask (thin flat weave)
- Faille
  - Silk Faille
  - Moiré Faille (close up only)
  - Tissue Faille
- Gazar
- Georgette
  - Silk Georgette
  - Double Georgette
  - Triple Georgette
  - Openweave Georgette / Georgette Leno
  - Sequin-embroidered Georgette
- Habutai
- Honan
  - Silk Honan / Silk Pongee
  - Simulated Silk Honan
  - Silk Pongee Dobby
- Homespun
  - Silk Homespun / Silk Burlap
- Hopsacking
  - Silk Hopsacking
- Marquisette
- Matelassé
  - Silk Matelassé
- Mousseline de Soie / Silk Gauze
- Mull
- Ninon
- Organza
  - Organza
  - Organza with Yarn Embroidery
  - Organza with Yarn Embroidery & Flat Appliqué
  - Organza with Yarn Embroidery & Raised Appliqué
- Ottoman
  - Silk Ottoman
  - Silk Ottoman Coating
- Peau de Soie
- Poplin
  - Silk Poplin
- Radium Silk
- Rep / Repp
  - Silk Repp
- Seersucker
  - Silk Seersucker
- Serge
  - Silk Serge
- Shantung
  - Shantung (soft finish)
  - Shantung (crisp finish)
  - Shantung Faille
  - Shantung Georgette
- Sharkskin
  - Silk Sharkskin
- Suiting
  - Silk Suiting
- Surah
  - Surah
  - Surah Suiting
- Tussah / Wild Silk

Silk and silk-type fabrics, as listed in this unit, refer to fabrics generically named and classified as silk fabrics which were originally made of one-hundred percent natural silk fibers. This unit includes silk and silk-type fabrics which may be made of:

- Natural silk fibers
- Man-made fibers simulating silk
- Man-made fiber blends
- Natural and man-made fiber blends
- Mixed or combined yarns of different fiber origins

Regardless of the fiber content, silk and silk-type fabrics maintain or simulate the same outward characteristics of the original generically named fabric. The following characteristics remain the same:

- Appearance
- Surface texture and interest
- Hand or feel
- Yarn type and construction
- Fabric structure
- Drapability qualities
- Weight (usually)

Although the original outward characteristics of the fabrics are maintained, the fiber content used *changes* the performance, thread selection, pressing and care factors of the fabric. Each fiber has its own particular characteristics and properties.

There are variations within the basic generically named fabrics. Variations may be obtained by changing, altering, modifying or combining:

- Yarn type
- Yarn construction
- Yarn count and size
- Yarns in warp and filling
- Usual weave structure
- Finishing processes

By combining or varying the components the textile designer can create new fabrics. When one component is changed, the fabric is changed. The newly designed or produced fabrics receive a new trade name. The new fabrics or newly named fabrics presented each season are usually placed into categories which may refer to their hand or feel, texture, surface appearance, structure or weight. Different textile companies have their own trade names or trademarks for the fabrics they produce.

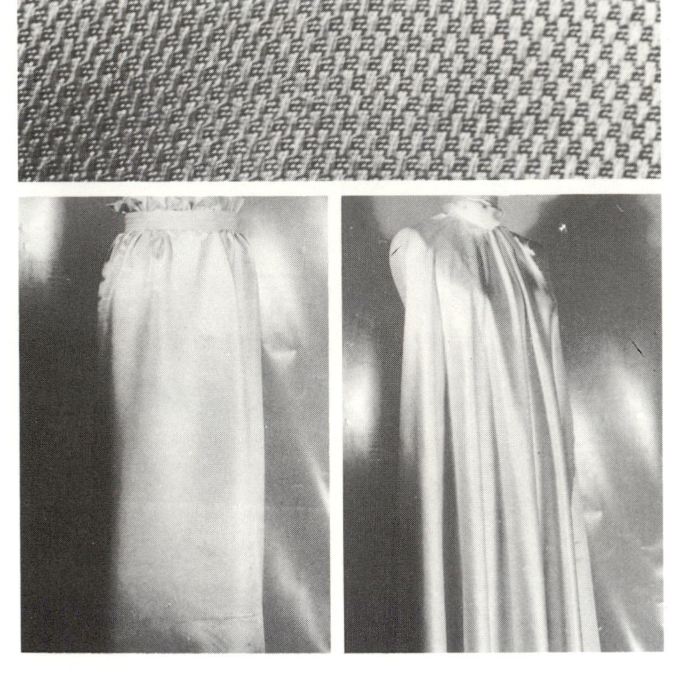

Barathea

Fiber Content 100% silk
Yarn Type natural filament fiber
Yarn Construction conventional; fine; slack twist
Fabric Structure broken twill weave variation
Finishing Processes calendering; softening; stabilizing
Color Application piece dyed; yarn dyed
Width 44-45 inches (111.8-114.3 cm)
Weight light
Hand limp; soft; supple
Texture pebbly; weave creates broken rib effect
Performance Expectations subject to abrasion; absorbent; poor coverage; dimensional stability; poor elongation; flexible; nonpilling
Drapability Qualities falls into soft flares; accommodates fullness by gathering, elasticized shirring, smocking; fullness retains soft fall

Recommended Construction Aids & Techniques

Hand Needles beading, 10-13; betweens, 5-7; milliners, 8-10; sharps, 8-10
Machine Needles round/set point, medium-fine 9-11
Threads mercerized cotton, 50; silk, industrial size A; cotton-covered polyester core, extra fine
Hems double-fold bias binding; bound/Hong Kong finish; edge-stitched; horsehair; machine-stitched; hand- or machine-rolled; seam binding
Seams flat-felled; French; plain; tissue-stitched
Seam Finishes self bound; double-ply bound; double-stitched; double-stitched and overcast; edge-stitched; mock French; double-ply overcast; double-ply zigzag
Pressing steam; safe temperature limit 300°F (150.1°C)
Care dry clean
Fabric Resource American Silk Mills Corp.

Barathea Suiting

Fiber Content 100% Silk
Yarn Type natural filament fiber
Yarn Construction conventional; slack twist
Fabric Structure broken twill weave variation
Finishing Processes calendering; stabilizing; stiffening
Color Application piece dyed; yarn dyed
Width 58 inches (147.3 cm)
Weight medium-light
Hand firm; soft
Texture grainy; uneven surface due to weave
Performance Expectations subject to abrasion; absorbent; poor coverage; dimensional stability; poor elongation; flexible; nonpilling; resilient
Drapability Qualities falls into moderately soft flares; accommodates fullness by pleating, gathering; fullness maintains a soft fold, unpressed pleat

Recommended Construction Aids & Techniques

Hand Needles betweens, 3-4; milliners, 6-8; sharps, 4-8
Machine Needles round/set point, medium-fine 11
Threads mercerized cotton, 50; silk, industrial size A-B; cotton-covered polyester core, all purpose
Hems book; double-fold or single-fold bias binding; bound/Hong Kong finish; edge-stitched; machine-stitched; seam binding
Seams flat-felled; French; false French; plain; safety-stitched; top stitched; welt; double welt
Seam Finishes self bound; single-ply or double-ply bound; double-stitched; double-stitched and overcast; serging/single-ply overedged; double-ply zigzag
Pressing steam; safe temperature limit 300°F (150.1°C)
Care dry clean
Fabric Resource American Silk Mills Corp.

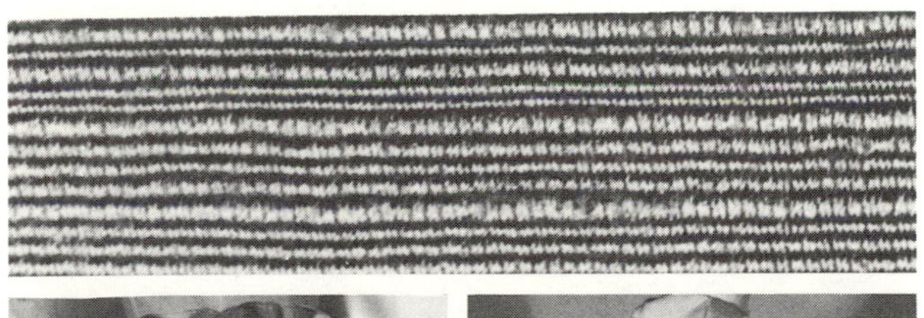

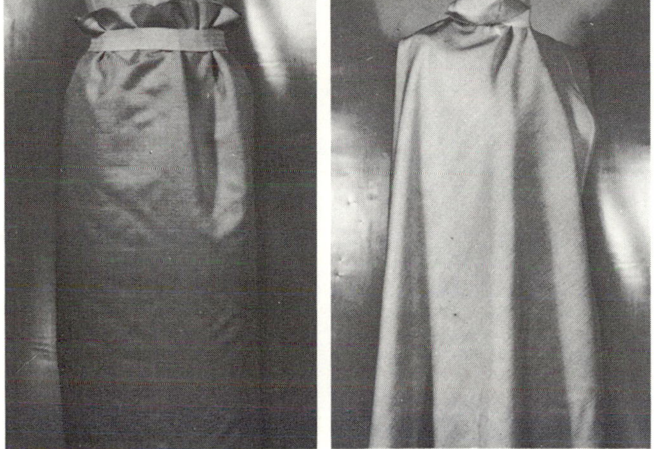

Silk Bengaline

Fiber Content 100% silk
Yarn Type natural filament fiber
Yarn Construction conventional; thick-and-thin
Fabric Structure firmly woven plain weave; heavy horizontal rib
Finishing Processes calendering; stiffening
Color Application piece dyed; yarn dyed
Width 36 inches (91.4 cm)
Weight medium-heavy
Hand nonelastic; firm; soft
Texture coarse; raised crosswise rib; slubbed
Performance Expectations absorbent; dimensional stability; poor elongation; flexible; resilient; subject to abrasion and seam slippage
Drapability Qualities falls into crisp wide flares; fullness maintains crisp effect; retains shape of garment

Recommended Construction Aids & Techniques

Hand Needles cotton darners, 9-10; milliners, 3-6; sharps, 1-4
Machine Needles round/set point, medium 11-14
Threads mercerized cotton, 40; silk, industrial size A-B-C; cotton-covered polyester core, heavy duty
Hems book; double-fold or single-fold bias binding; bound/Hong Kong finish; edge-stitched; faced; stitched and pinked; overedged; seam binding
Seams plain; safety-stitched; top-stitched; welt; double welt
Seam Finishes single-ply bound; Hong Kong bound; single-ply overcast; pinked; pinked and stitched; serging/single-ply overedged; single-ply zigzag
Pressing steam; safe temperature limit 300°F (150.1°C)
Care dry clean
Fabric Resource **Jack Larsen Incorporated**

Silk Broadcloth

Fiber Content 100% silk
Yarn Type natural filament fiber
Yarn Construction conventional; high twist
Fabric Structure closely woven plain weave
Finishing Processes calendering; singeing; sizing
Color Application piece dyed; yarn dyed
Width 36 inches (91.4 cm)
Weight light
Hand fine; soft; supple
Texture flat; smooth
Performance Expectations absorbent; air permeable; poor coverage; dimensional stability; poor elongation; flexible; nonpilling; nonsagging
Drapability Qualities falls into soft flares; accommodates fullness by gathering, elasticized shirring, smocking; fullness retains soft fall

Recommended Construction Aids & Techniques

Hand Needles beading, 10-13; betweens, 5-7; milliners, 8-10; sharps, 8-10
Machine Needles round/set point, medium-fine 9-11
Threads mercerized cotton, 50; silk, industrial size A; cotton-covered polyester core, extra fine
Hems bound/Hong Kong finish; double fold; edge-stitched; double edge-stitched; horsehair; machine-stitched; hand- or machine-rolled; seam binding
Seams flat-felled; French; false French; mock French; plain; tissue-stitched
Seam Finishes self bound; double-ply bound; double-stitched; double-stitched and overcast; edge-stitched; mock French; double-ply overcast; double-ply zigzag
Pressing steam; safe temperature limit 300°F (150.1°C)
Care dry clean
Fabric Resources **American Silk Mills Corp.; Jack Larsen Incorporated**

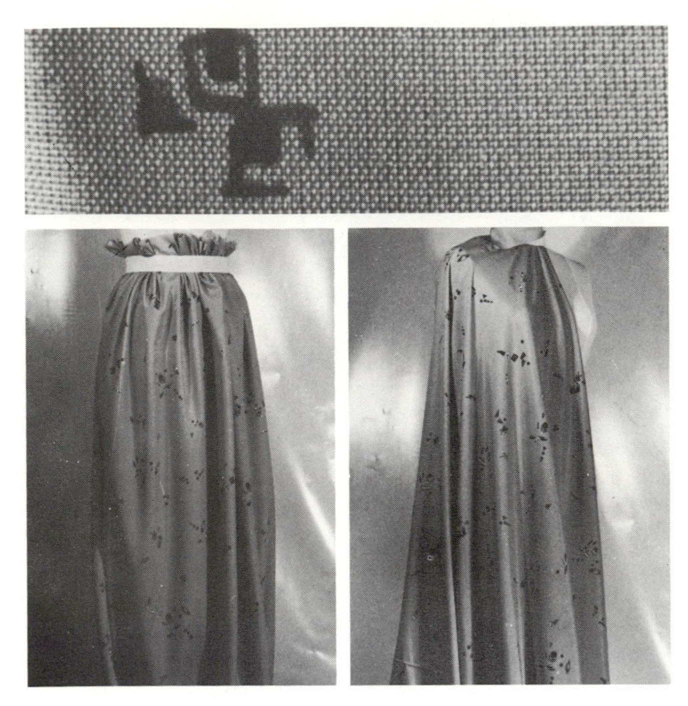

Printed Silk Broadcloth

Fiber Content 100% polyester
Yarn Type filament fiber; filament staple
Yarn Construction conventional; slack twist
Fabric Structure closely woven plain weave
Finishing Processes calendering; singeing; stabilizing
Color Application piece dyed; silk-screen printed
Width 64 inches (162.7 cm)
Weight light
Hand fine; soft
Texture flat; smooth
Performance Expectations subject to color crocking and edge abrasion; dimensional stability; durable; elongation; heat-set properties; resilient; high tensile strength
Drapability Qualities falls into soft flares; accommodates fullness by gathering, elasticized shirring, smocking; fullness retains soft fall

Recommended Construction Aids & Techniques

Hand Needles beading, 10–13; betweens, 5–7; milliners, 8–10; sharps, 8–10
Machine Needles round/set point, medium-fine 9–11
Threads poly-core waxed, 60; cotton-polyester blend, fine; cotton-covered polyester core, extra fine: spun polyester, fine
Hems bound/Hong Kong finish; double fold; edge-stitched; horsehair; machine-stitched; hand- or machine-rolled; seam binding
Seams flat-felled; French; false French; mock French; plain; tissue-stitched
Seam Finishes self-bound; double-ply bound; double-stitched; double-stitched and overcast; edge-stitched; mock French; double-ply overcast; single-ply zigzag
Pressing steam; safe temperature limit 325°F (164.1°C)
Care launder; dry clean
Fabric Resource **Loomskill/Gallery Screen Print**

Silk Brocade

Fiber Content 91% silk/9% polyester
Yarn Type natural and man-made filament fibers
Yarn Construction conventional; heavy and fine yarns; low twist
Fabric Structure figure weave
Finishing Processes calendering; stiffening; stretching
Color Application yarn dyed
Width 40 inches (101.6 cm)
Weight heavy
Hand nonelastic; soft; thick
Texture raised patterned design on face; floats on back
Performance Expectations subject to abrasion; absorbent; dimensional stability; poor elongation; flexible; non-pilling; resilient; subject to snagging due to weave
Drapability Qualities falls into stiff wide cones; retains shape or silhouette of garment; better utilized if fitted by seaming and eliminating excess fabric

Recommended Construction Aids & Techniques

Hand Needles cotton darners, 9–10; milliners, 3–6; sharps, 1–4
Machine Needles round/set point, medium 11–14
Threads mercerized cotton, 40; silk, industrial size A-B-C; cotton-covered polyester core, heavy duty
Hems book; double-fold or single-fold bias binding; bound/Hong Kong finish; edge-stitched; faced; overedged; stitched and pinked; seam binding
Seams plain; safety-stitched; top-stitched; welt; double welt
Seam Finishes single-ply bound; Hong Kong bound; single-ply overcast; pinked; pinked and stitched; serging/single-ply overedged; single-ply zigzag
Pressing steam; safe temperature limit 300°F (150.1°C)
Care dry clean
Fabric Resource **Jack Larsen Incorporated**

Silk Canvas

Fiber Content 70% silk/30% polyester
Yarn Type natural and man-made filament fibers
Yarn Construction conventional; low twist
Fabric Structure twill weave
Finishing Processes bleaching; calendering; stretching
Color Application piece dyed–union dyed/cross dyed; yarn dyed
Width 40 inches (101.6 cm)
Weight heavy
Hand compact; firm; hard; thick
Texture grainy; harsh; rough
Performance Expectations dimensional stability; durable; flexible; resilient; subject to snagging due to weave
Drapability Qualities falls into stiff wide cones; retains shape or silhouette of garment; better utilized if fitted by seaming and eliminating excess fabric

Recommended Construction Aids & Techniques

Hand Needles cotton darners, 9–10; milliners, 3–6; sharps, 1–4
Machine Needles round/set point, medium-coarse 14–16
Threads mercerized cotton, 40; silk, industrial size A-B-C; cotton-covered polyester core, heavy duty
Hems book; double-fold or single-fold bias binding; bound/Hong Kong finish; edge-stitched; faced; overedged; stitched and pinked; seam binding
Seams plain; safety-stitched; top-stitched; welt; double welt
Seam Finishes single-ply bound; Hong Kong bound; single-ply overcast; pinked; pinked and stitched; serging/single-ply overedged; single-ply zigzag
Pressing steam; safe temperature limit 300°F (150.1°C)
Care dry clean
Fabric Resource **Jack Larsen Incorporated**

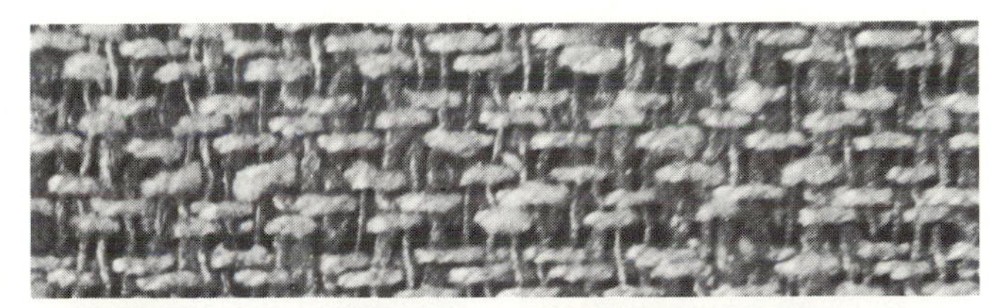

Charvet

Fiber Content 100% silk
Yarn Type natural filament fiber; noils
Yarn Construction novelty; textured
Fabric Structure irregular twill weave
Finishing Processes singeing; stretching
Color Application yarn dyed; warp and filling contrasting colors
Width 58–60 inches (147.3–152.4 cm)
Weight medium
Hand loose; pliable
Texture coarse; rough; protruding seed noil
Performance Expectations absorbent; elasticity; elongation; flexible; moth resistant; subject to snagging due to weave and seam and yarn slippage; wrinkle resistant
Drapability Qualities falls into moderately soft flares; accommodates fullness by gathering; fullness retains soft fall

Recommended Construction Aids & Techniques

Hand Needles cotton darners, 9–10; milliners, 3–6; sharps, 1–4
Machine Needles round/set point, medium 11–14
Threads mercerized cotton, 50; silk, industrial size A-B; cotton-covered polyester core, all purpose
Hems book; double-fold or single-fold bias binding; bound/Hong Kong finish; edge-stitched; machine-stitched; seam binding
Seams flat-felled; French; false-French; plain; safety-stitched; taped; top-stitched; welt; double welt
Seam Finishes double-ply bound; Hong Kong bound; edge-stitched; single-ply overcast; serging/single-ply overedged; single-ply zigzag
Pressing steam; safe temperature limit 300°F (150.1°C)
Care dry clean
Fabric Resource **Auburn Fabrics Inc.**

Silk Chiffon

Fiber Content 100% silk
Yarn Type natural filament fiber
Yarn Construction conventional; high twist
Fabric Structure plain weave; same size yarns for warp and filling
Finishing Processes calendering; degumming; singeing; softening
Color Application piece dyed; yarn dyed
Width 45 inches (114.3 cm)
Weight fine to lightweight
Hand soft; supple; thin
Texture crepey; flat; sheer
Performance Expectations absorbent; air permeable; dimensional stability; flexible; nonpilling
Drapability Qualities falls into soft languid flares and ripples; flows and is fluid; fullness retains soft graceful fall

Recommended Construction Aids & Techniques

Hand Needles beading, 10–13; betweens, 5–7; milliners, 10; sharps, 8–10
Machine Needles round/set point, fine 9
Threads mercerized cotton, 50; silk, industrial size A; cotton-covered polyester core, extra fine
Hems double fold; horsehair; machine-stitched; hand- or machine-rolled; wired; edge-stitched only; double edge-stitched
Seams French; mock French; hairline; plain; tissue-stitched
Seam Finishes self bound; mock French; double-stitched; double-stitched and trimmed
Pressing steam; safe temperature limit 300°F (150.1°C)
Care dry clean
Fabric Resource **Kabat Textile Corp.**

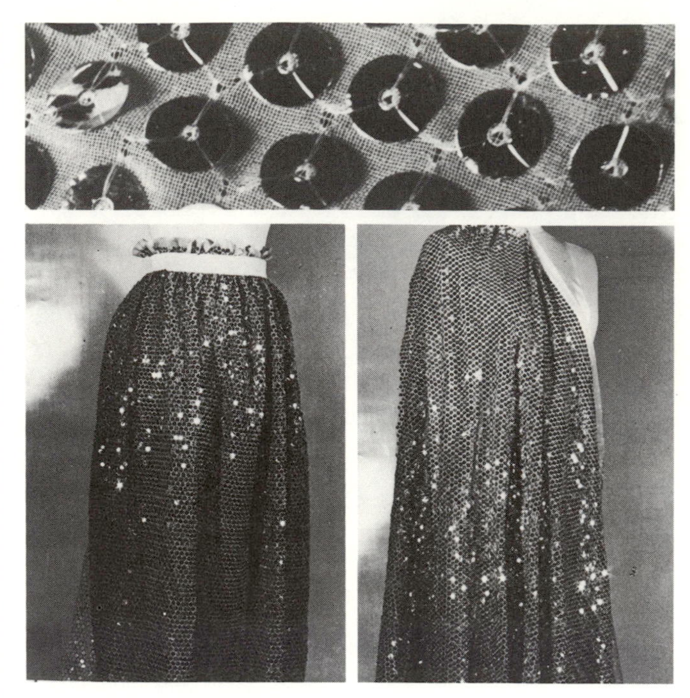

Sequin-embroidered Chiffon (pavé design)

Fiber Content 100% nylon; plastic paillettes
Yarn Type filament fiber
Yarn Construction textured
Fabric Structure Schiffli embroidery on plain weave base
Finishing Processes calendering; flame retardant; singeing
Color Application piece dyed; solution dyed; yarn dyed
Width 45 inches (114.3 cm)
Weight light
Hand base: delicate, supple; paillettes: allover/pavé design, rough, scratchy
Texture sheer base; pailletes form high and low raised pattern on surface
Performance Expectations air permeable; dimensional stability; durable; elasticity; elongation; resilient; high tensile strength; paillettes subject to abrasion, breaking, snagging
Drapability Qualities falls into soft supple flares; accommodates fullness by gathering; fullness retains soft fall

Recommended Construction Aids & Techniques

Hand Needles beading, 10–13; betweens, 7; embroidery, 9–10; sharps, 9–10
Machine Needles round/set point, fine 9
Threads silk, industrial size A; spun polyester, fine; nylon/Dacron, monocord A
Hems bound/Hong Kong finish; net binding; faced; horsehair
Seams hairline; plain; tissue-stitched; overlap and hand-stitch sequins on seamline after stitching
Seam Finishes net bound; Hong Kong bound; double-stitched; double-stitched and trimmed; untreated/plain; remove sequins in seam allowance
Pressing use pressing cloth; safe temperature limit 300°F (150.1°C)
Care dry clean only
Fabric Resource **Sequin International Corp.**

Sequin-embroidered Chiffon (stripe design)

Fiber Content 100% nylon; plastic paillettes
Yarn Type filament fiber
Yarn Construction creped; textured
Fabric Structure Schiffli embroidery on plain weave base
Finishing Processes calendering; flame retardant; heat-set stabilizing
Color Application piece dyed; solution dyed; yarn dyed
Width 45 inches (114.3 cm)
Weight light
Hand base: delicate, supple; paillettes: rough, scratchy
Texture sheer base; pailettes form high and low raised pattern on surface
Performance Expectations air permeable; elasticity; elongation; dimensional stability; durable; resilient; high tensile strength; paillettes subject to abrasion, breaking, snagging
Drapability Qualities falls into soft supple flares; accommodates fullness by gathering; fullness retains soft fall

Recommended Construction Aids & Techniques

Hand Needles beading, 10–13; betweens, 7; embroidery, 9–10; sharps, 9–10
Machine Needles round/set point, fine 9
Threads silk, industrial size A; spun polyester, fine; nylon/Dacron, monocord A
Hems bound/Hong Kong finish; net binding; faced; horsehair
Seams hairline; plain; tissue-stitched; overlap and hand-stitch sequins on seamline after stitching
Seam Finishes net bound; Hong Kong bound; double-stitched; double-stitched and trimmed; untreated/plain; remove sequins in seam allowance
Pressing use pressing cloth; safe temperature limit 300°F (150.1°C)
Care dry clean only
Fabric Resource **Sequin International Corp.**

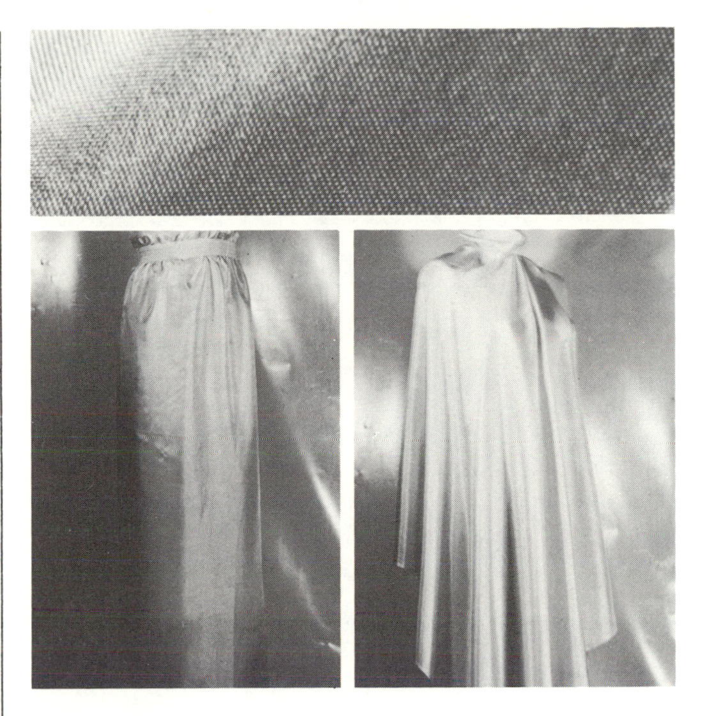

China Silk

Fiber Content 100% silk
Yarn Type natural filament fiber
Yarn Construction conventional; fine
Fabric Structure plain weave
Finishing Processes calendering; singeing; sizing; stretching
Color Application piece dyed
Width 36 inches (91.4 cm)
Weight fine to lightweight
Hand limp; sleazy; soft
Texture slippery; smooth
Performance Expectations absorbent; poor elongation; flexible; subject to seam and yarn slippage
Drapability Qualities falls into soft languid flares; accommodates fullness by gathering; fullness retains soft graceful fall

Recommended Construction Aids & Techniques

Hand Needles beading, 10–13; betweens, 5–7; milliners, 10; sharps, 8–10
Machine Needles round/set point, fine 9
Threads silk, industrial size A; cotton-covered polyester core, extra fine; mercerized cotton, 50
Hems bound/Hong Kong finish; double fold; edge-stitched; double edge-stitched; horsehair; machine-stitched; hand- or machine-rolled; seam binding
Seams flat-felled; French; mock French; plain; tissue-stitched
Seam Finishes self bound; double-ply or single-ply bound; double-stitched; double-stitched and trimmed; mock French
Pressing steam; safe temperature limit 300°F (150.1°C)
Care dry clean
Fabric Resource **American Silk Mills Corp.**

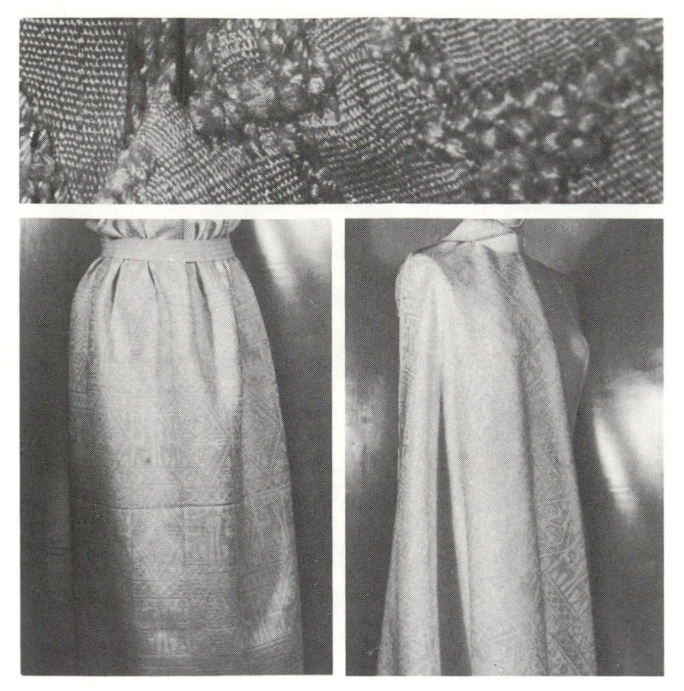

Silk Damask (thick textured weave)

Fiber Content 100% silk
Yarn Type natural filament fiber
Yarn Construction conventional; thick and thin; low twist
Fabric Structure figure weave; double cloth; ground and pattern formed by warp and filling satin weave
Finishing Processes calendering; degumming; stretching
Color Application yarn dyed
Width 42-43 inches (106.7-109.2 cm)
Weight medium to medium-heavy
Hand firm; soft; thick
Texture reversible; high and low areas; weave and yarn create shiny and dull contrasting surface
Performance Expectations absorbent; dimensional stability; poor elongation; flexible; nonpilling; resilient; subject to snagging due to weave
Drapability Qualities falls into stiff wide flares; fullness maintains crisp, lofty effect; retains shape or silhouette of garment

Recommended Construction Aids & Techniques

Hand Needles cotton darners, 9-10; milliners, 3-6; sharps, 1-4
Machine Needles round/set point, medium 11-14
Threads mercerized cotton, 40; silk, industrial size A-B-C; cotton-covered polyester core, heavy duty
Hems book; double-fold or single-fold bias binding; bound/Hong Kong finish; edge-stitched; faced; overedged; stitched and pinked; seam binding
Seams plain; safety-stitched; top-stitched; welt; double welt
Seam Finishes single-ply bound; Hong Kong bound; single-ply overcast; pinked; pinked and stitched; serging/single-ply overedged; single-ply zigzag
Pressing steam; safe temperature limit 300°F (150.1°C)
Care dry clean
Fabric Resources **Jack Larsen Incorporated**

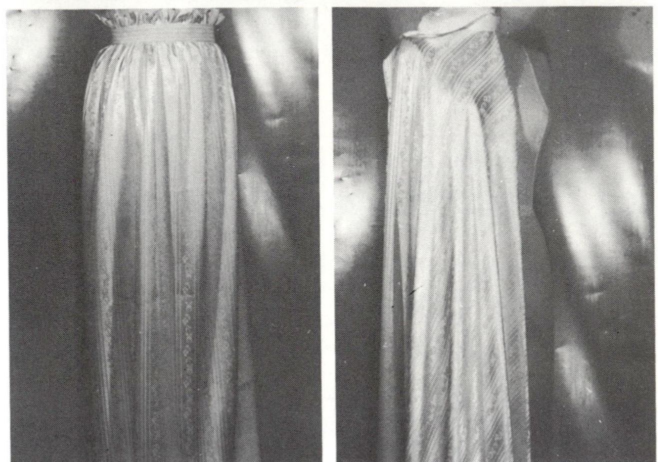

Silk Damask (thin flat weave)

Fiber Content 70% Arnel® acetate/30% polyester
Yarn Type filament fiber
Yarn Construction conventional; low twist; textured
Fabric Structure satin weave stripe; Jacquard weave design
Finishing Processes calendering; softening; stabilizing
Color Application piece dyed-cross dyed; yarn dyed
Width 44-45 inches (111.8-114.3 cm)
Weight light
Hand firm; soft; supple
Texture smooth with slightly raised patterned areas; weave and yarns create light-reflecting qualities
Performance Expectations subject to abrasion; dimensional stability; elongation; heat-set properties
Drapability Qualities falls into soft flares; accommodates fullness by pleating, gathering; fullness retains soft fall

Recommended Construction Aids & Techniques

Hand Needles beading, 10-13; betweens, 5-7; milliners, 8-10; sharps, 8-10
Machine Needles round/set point, medium 11
Threads poly-core-waxed, 60; cotton-polyester blend, fine; cotton-covered polyester core, extra fine; spun polyester, fine
Hems book; double-fold or single-fold bias binding; bound/Hong Kong finish; edge-stitched; machine-stitched; seam binding
Seams flat-felled; French; false French; safety-stitched; top-stitched; welt; double welt
Seam Finishes single-ply or double-ply bound; Hong Kong bound; edge-stitched; single-ply overcast; pinked and stitched; serging/single-ply overedged; single-ply zigzag
Pressing steam; safe temperature limit 250°F (122.1°C)
Care hand washable; dry clean
Fabric Resource **Hargro Fabrics Inc.**

Silk Faille

Fiber Content 37% silk/63% nylon
Yarn Type natural and man-made filament fibers
Yarn Construction conventional; thin warp/heavy filling; slack twist
Fabric Structure plain weave; heavy filling yarns form rib in crosswise direction
Finishing Processes ciré calendering; singeing; stretching
Color Application piece dyed–cross dyed
Width 50 inches (127 cm)
Weight medium-light
Hand compact; firm; soft
Texture fine; raised crosswise rib; slippery
Performance Expectations abrasion resistant; dimensional stability; durable; flexible
Drapability Qualities falls into firm flares; accommodates fullness by gathering; fullness maintains crisp effect

Recommended Construction Aids & Techniques

Hand Needles betweens, 3–4; milliners, 6–8; sharps, 4–8
Machine Needles round/set point, medium-fine 9–11
Threads poly-core waxed, 60; cotton-polyester blend, fine; cotton-covered polyester core, extra fine; spun polyester, fine
Hems book; double-fold or single-fold bias binding; bound/Hong Kong finish; edge-stitched; machine-stitched; seam binding
Seams flat-felled; French; false-French; plain; safety-stitched; top-stitched; welt; double welt
Seam Finishes single-ply or double-ply bound; Hong Kong bound; edge-stitched; single-ply overcast; pinked and stitched; serging/single-ply overedged; single-ply zigzag
Pressing steam; safe temperature limit 300°F (150.1°C)
Care dry clean
Fabric Resource **American Silk Mills Corp.**

Moiré Faille

Fiber Content 37% silk/63% nylon
Yarn Type natural and man-made filament fiber
Yarn Construction conventional; thin warp/heavy filling; slack twist
Fabric Structure plain weave; heavy filling yarns form rib in crosswise direction
Finishing Processes moiré calendering; singeing; stretching
Color Application piece dyed–cross dyed
Width 50 inches (127 cm)
Weight medium-light
Hand compact; firm; soft
Texture wavy reflecting pattern due to finishing process
Performance Expectations abrasion resistant; dimensional stability; durable; flexible
Drapability Qualities falls into firm flares; accommodates fullness by gathering; fullness maintains crisp effect

Recommended Construction Aids & Techniques

Hand Needles betweens, 3–4; milliners, 6–8; sharps, 4–8
Machine Needles round/set point, medium-fine 9–11
Threads poly-core waxed, 60; cotton-polyester blend, fine; cotton-covered polyester core, extra fine; spun polyester, fine
Hems book; double-fold or single-fold bias binding; bound/Hong Kong finish; edge-stitched; machine-stitched; seam binding
Seams flat-felled; French; false-French; plain; safety-stitched; top-stitched; welt; double welt
Seam Finishes single-ply or double-ply bound; Hong Kong bound; edge-stitched; single-ply overcast; pinked and stitched; serging/single-ply overedged; single-ply zigzag
Pressing steam; safe temperature limit 300°F (150.1°C)
Care dry clean
Fabric Resource **American Silk Mills Corp./Bucol**

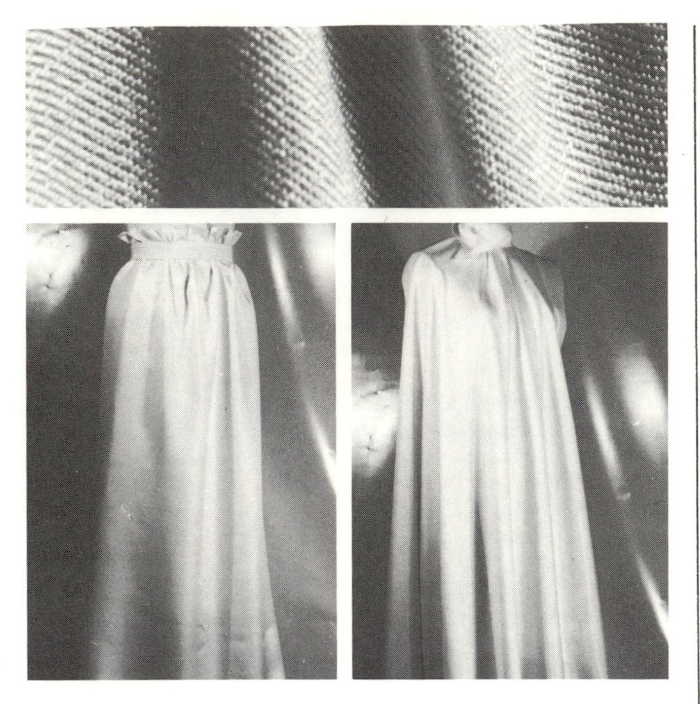

Tissue Faille

Fiber Content 100% polyester
Yarn Type filament fiber
Yarn Construction conventional; fine; slack twist
Fabric Structure plain weave; additional filling yarns form rib in crosswise direction
Finishing Processes calendering; softening; stabilizing
Color Application piece dyed; solution dyed; yarn dyed
Width 44-45 inches (111.8-114.3 cm)
Weight light
Hand firm; soft; supple
Texture grainy; noticeable crosswise rib
Performance Expectations subject to color crocking and edge abrasion; dimensional stability; durable; elongation; heat-set properties; resilient; high tensile strength
Drapability Qualities falls into moderately soft flares; accommodates fullness by pleating, gathering, elasticized shirring; fullness retains soft fall

Recommended Construction Aids & Techniques

Hand Needles beading, 10-13; betweens, 5-7; milliners, 8-10; sharps, 8-10
Machine Needles round/set point, medium-fine 9-11
Threads poly-core waxed, 60; cotton-polyester blend, fine; cotton-covered polyester core, extra fine; spun polyester, fine
Hems bound/Hong Kong finish; double fold; edge-stitched; horsehair; machine-stitched; hand- or machine-rolled; seam binding
Seams flat-felled; French; mock French; plain; safety-stitched; top-stitched; welt; double welt
Seam Finishes self bound; double-ply bound; double-stitched; double-stitched and overcast; edge-stitched; mock French; double-ply overcast; double-ply zigzag
Pressing steam; safe temperature limit 325°F (164.1°C)
Care launder; dry clean
Fabric Resource Kabat Textile Corp.

Gazar

Fiber Content 100% silk
Yarn Type natural filament fiber
Yarn Construction conventional; high twist
Fabric Structure plain weave; double yarns interlaced as one
Finishing Processes calendering; permanent stiffening; stretching
Color Application piece dyed; yarn dyed
Width 47 inches (119.4 cm)
Weight fine to lightweight
Hand crisp; delicate; firm
Texture flat; semi-sheer; smooth
Performance Expectations absorbent; air permeable; dimensional stability; poor elongation; flexible; nonpilling
Drapability Qualities falls into crisp cones; accommodates fullness by gathering; fullness maintains crisp bouffant effect

Recommended Construction Aids & Techniques

Hand Needles beading, 10-13; betweens, 5-7; milliners, 10; sharps, 8-10
Machine Needles round/set point, fine 9
Threads mercerized cotton, 50; silk, industrial size A; cotton-covered polyester core, extra fine
Hems double fold; edge-stitched; edge-stitched only; double edge-stitched; horsehair; machine-stitched; hand- or machine-rolled; wired
Seams French; mock French; hairline; plain; tissue-stitched
Seam Finishes self bound; double-stitched; double-stitched and trimmed; mock French; serging/single-ply overedged
Pressing steam; safe temperature limit 300°F (150.1°C)
Care dry clean
Fabric Resource Sormani Co. Inc./Taroni

Silk Georgette

Fiber Content 100% polyester
Yarn Type filament fiber
Yarn Construction creped; textured
Fabric Structure plain weave
Finishing Processes calendering; creping; singeing; stabilizing
Color Application piece dyed; solution dyed; yarn dyed
Width 44–45 inches (111.8–114.3 cm)
Weight fine to lightweight
Hand soft; supple; thin
Texture crepey; grainy; sheer
Performance Expectations air permeable; dimensional stability; durable; elongation; heat-set properties; resilient; high tensile strength
Drapability Qualities falls into soft languid flares and ripples; flows and is fluid; fullness retains soft graceful fall

Recommended Construction Aids & Techniques

Hand Needles beading, 10–13; betweens, 5–7; milliners, 10; sharps, 8–10
Machine Needles round/set point, fine 9
Threads poly-core waxed, 60; cotton-polyester blend, fine; cotton-covered polyester core, extra fine; spun polyester, fine
Hems double fold; edge-stitched only; double edge-stitched; horsehair; machine-stitched; hand- or machine-rolled; wired
Seams French; mock French; hairline; plain; tissue-stitched
Seam Finishes self bound; mock French; double-stitched; double-stitched and trimmed
Pressing steam; safe temperature limit 325°F (164.1°C)
Care launder; dry clean
Fabric Resource **Kabat Textile Corp.**

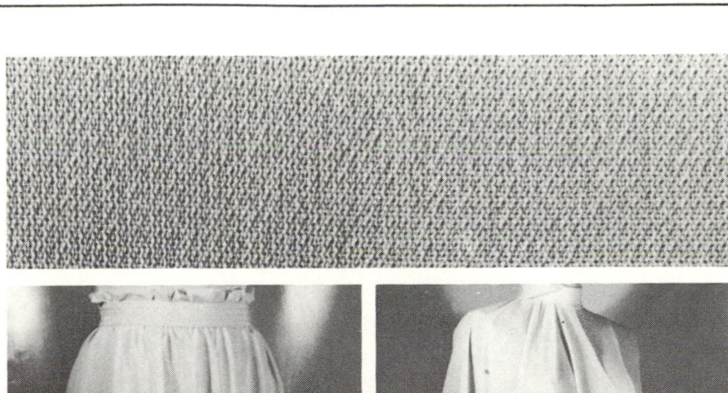

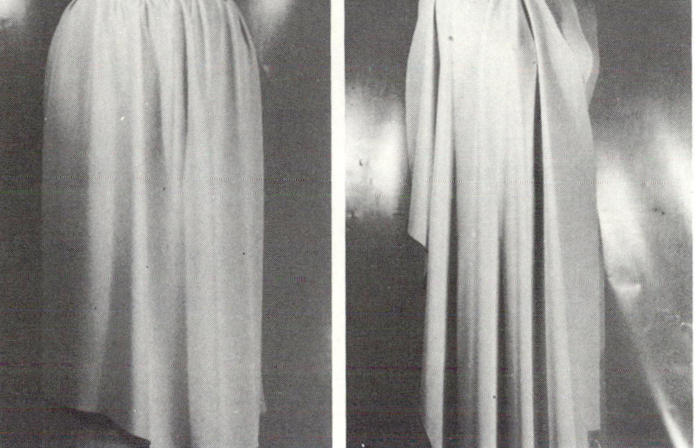

Double Georgette

Fiber Content 100% polyester
Yarn Type filament fiber
Yarn Construction textured
Fabric Structure plain weave
Finishing Processes calendering; creping; singeing; stabilizing
Color Application piece dyed; solution dyed; yarn dyed
Width 44–45 inches (111.8–114.3 cm)
Weight fine to lightweight
Hand delicate; soft; supple
Texture flat; semi-sheer
Performance Expectations air permeable; dimensional stability; durable; elongation; heat-set properties; resilient; high tensile strength
Drapability Qualities falls into soft languid flares and ripples; flows and is fluid; fullness retains soft graceful fall

Recommended Construction Aids & Techniques

Hand Needles beading, 10–13; betweens, 5–7; milliners, 10; sharps, 8–10
Machine Needles round/set point, fine 9
Threads poly-core waxed, 60; cotton-polyester blend, fine; cotton-covered polyester core, extra fine; spun polyester, fine
Hems double fold; edge-stitched only; double edge-stitched; horsehair; machine-stitched; hand- or machine-rolled; wired
Seams French; mock French; hairline; plain; tissue-stitched
Seam Finishes self bound; double-stitched; double-stitched and trimmed; mock French
Pressing steam; safe temperature limit 325°F (164.1°C)
Care launder; dry clean
Fabric Resource **Private Collections Fabrics Ltd.**

Triple Georgette

Fiber Content 100% silk
Yarn Type natural filament fiber
Yarn Construction conventional; high twist
Fabric Structure plain weave
Finishing Processes calendering; creping; singeing; stretching
Color Application piece dyed; yarn dyed
Width 45 inches (114.3 cm)
Weight light
Hand fine; soft; supple; thin
Texture flat; smooth; light admitting
Performance Expectations absorbent; dimensional stability; poor elongation; flexible; nonpilling
Drapability Qualities falls into soft languid flares and ripples; flows and is fluid; fullness retains soft graceful fall

Recommended Construction Aids & Techniques

Hand Needles beading, 10–13; betweens, 5–7; milliners, 8–10; sharps, 8–10
Machine Needles round/set point, fine 9
Threads mercerized cotton, 50; silk, industrial size A; cotton-covered polyester core, extra fine
Hems bound/Hong Kong finish; double fold; edge-stitched only; double edge-stitched; horsehair; machine-stitched; hand- or machine-rolled
Seams French; mock French; hairline; plain; tissue-stitched
Seam Finishes self bound; double-ply bound; double-stitched; double-stitched and overcast; edge-stitched; mock French; double-ply overcast; double-ply zigzag
Pressing steam; safe temperature limit 300°F (150.1°C)
Care dry clean
Fabric Resource **American Silk Mills Corp.**

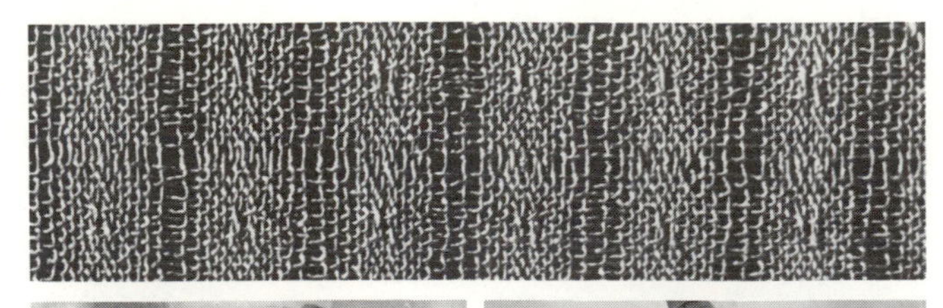

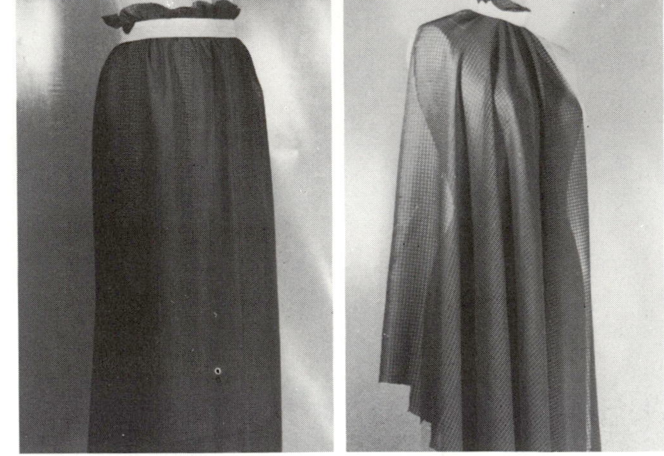

Openweave Georgette/Georgette Leno

Fiber Content 100% polyester
Yarn Type filament fiber
Yarn Construction high twist; textured
Fabric Structure plain weave variation; skip denting
Finishing Processes singeing; stiffening; stabilizing
Color Application piece dyed; yarn dyed
Width 44–45 inches (111.8–114.3 cm)
Weight light
Hand sandy; soft; supple
Texture crepey; grainy; open areas created by weave; semi-sheer
Performance Expectations air permeable; dimensional stability; durable; elongation; heat-set properties; resilient; subject to snagging due to weave
Drapability Qualities falls into soft languid flares and ripples; flows and is fluid; fullness retains soft graceful fall

Recommended Construction Aids & Techniques

Hand Needles beading, 10–13; betweens, 5–7; milliners, 10; sharps, 8–10
Machine Needles round/set point, fine 9
Threads poly-core waxed, 60; cotton-polyester blend, fine; cotton-covered polyester core, extra fine; spun polyester, fine
Hems bound/Hong Kong finish; edge-stitched; edge-stitched only; double edge-stitched; double fold; horsehair; machine-stitched; hand- or machine-rolled; seam binding
Seams French; mock French; hairline; plain; tissue-stitched
Seam Finishes self bound; double-stitched; double-stitched and trimmed; mock French
Pressing steam; safe temperature limit 325°F (164.1°C)
Care launder; dry clean
Fabric Resource **Private Collections Fabrics Ltd.**

Sequin-embroidered Georgette

Fiber Content 100% polyester; plastic paillettes
Yarn Type filament fiber
Yarn Construction creped; textured
Fabric Structure Schiffli embroidery on plain weave base
Finishing Processes compressive shrinkage; flame retardant; singeing
Color Application piece dyed; solution dyed; yarn dyed
Width 45 inches (114.3 cm)
Weight light
Hand base: delicate, supple; paillettes: rough, scratchy
Texture sheer base; paillettes form high and low raised pattern on surface
Performance Expectations air permeable; dimensional stability; durable; elongation; resilient; high tensile strength; paillettes subject to abrasion, breaking, snagging
Drapability Qualities falls into soft supple flares; accommodates fullness by gathering; fullness retains soft fall

Recommended Construction Aids & Techniques

Hand Needles beading, 10–13; betweens, 7; embroidery, 9–10; sharps, 9–10
Machine Needles round/set point, fine 9
Threads silk, industrial size A; spun polyester, fine; nylon/Dacron, monocord A
Hems bound/Hong Kong finish; net binding; faced; horsehair
Seams hairline; plain; tissue-stitched; overlap and hand-stitch sequins on seamline after stitching
Seam Finishes net bound; Hong Kong bound; double-stitched and trimmed; untreated/plain; remove sequins in seam allowance
Pressing use pressing cloth; safe temperature limit 300°F (150.1°C)
Care dry clean only
Fabric Resource **Sequin International Corp.**

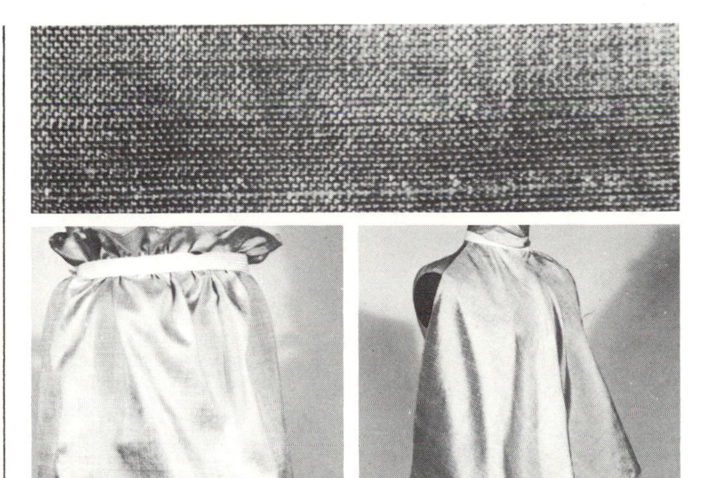

Habutai

Fiber Content 100% silk
Yarn Type natural filament fiber
Yarn Construction conventional; natural irregularity of doupioni yarn
Fabric Structure plain weave
Finishing Processes calendering; degumming; stretching
Color Application piece dyed
Width 40 inches (101.6 cm)
Weight light
Hand fine; firm; soft
Texture rough, uneven nubby yarns irregularly spaced in crosswise direction
Performance Expectations absorbent; dimensional stability; poor elongation; flexible; nonpilling; nonsnagging
Drapability Qualities falls into moderately soft flares; accommodates fullness by pleating, gathering, elasticized shirring; fullness retains soft fall

Recommended Construction Aids & Techniques

Hand Needles beading, 10–13; betweens, 5–7; milliners, 8–10; sharps, 8–10
Machine Needles round/set point, medium-fine 9–11
Threads mercerized cotton, 50; silk, industrial size A; cotton-covered polyester core, extra fine
Hems bound/Hong Kong finish; double fold; edge-stitched; horsehair; machine-stitched; hand- or machine-rolled; seam binding
Seams flat-felled; French; mock French; plain; tissue-stitched
Seam Finishes self bound; double-ply bound; double-stitched; double-stitched and overcast; edge-stitched; mock French; double-ply overcast; double-ply zigzag
Pressing steam; safe temperature limit 300°F (150.1°C)
Care dry clean
Fabric Resources **American Silk Mills Corp.; Jack Larsen Incorporated**

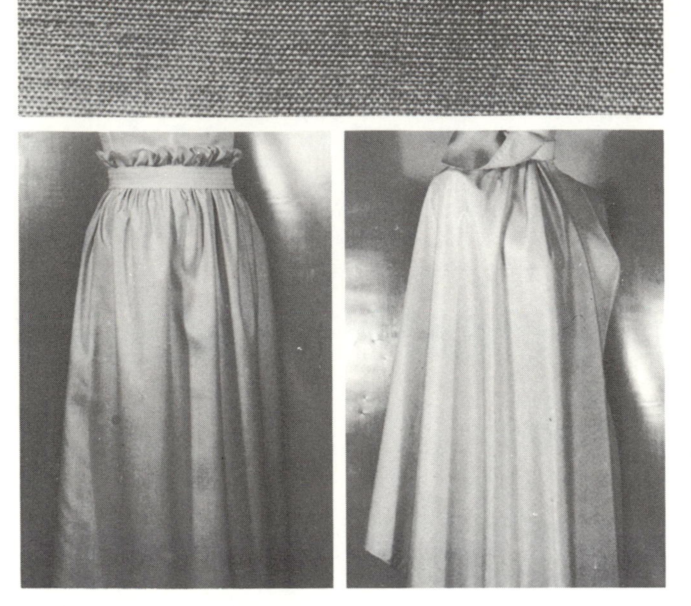

Silk Honan/Silk Pongee

Fiber Content 100% wild silk
Yarn Type natural filament fiber
Yarn Construction conventional; low twist; natural slub and irregularity of wild silk yarn
Fabric Structure plain weave
Finishing Processes bleaching; calendering; degumming; sizing
Color Application piece dyed; yarn dyed
Width 45 inches (114.3 cm)
Weight light
Hand fine; firm; soft
Texture slight slub in lengthwise and crosswise directions
Performance Expectations absorbent; dimensional stability; poor elongation; flexible; nonpilling; nonsnagging
Drapability Qualities falls into moderately soft flares; accommodates fullness by pleating, gathering, elasticized shirring; fullness retains soft fall

Recommended Construction Aids & Techniques

Hand Needles beading, 10–13; betweens, 5–7; milliners, 8–10; sharps, 8–10
Machine Needles round/set point, medium-fine 9–11
Threads mercerized cotton, 50; silk, industrial size A; cotton-covered polyester core, extra fine
Hems bound/Hong Kong finish; double fold; edge-stitched; horsehair; machine-stitched; hand- or machine-rolled; seam binding
Seams flat-felled; French; false French; mock French; plain; tissue-stitched
Seam Finishes self bound; double-ply bound; double-stitched; double-stitched and overcast; edge-stitched; mock French; double-ply overcast; double-ply zigzag
Pressing steam; safe temperature limit 300°F (150.1°C)
Care dry clean
Fabric Resource Jack Larsen Incorporated

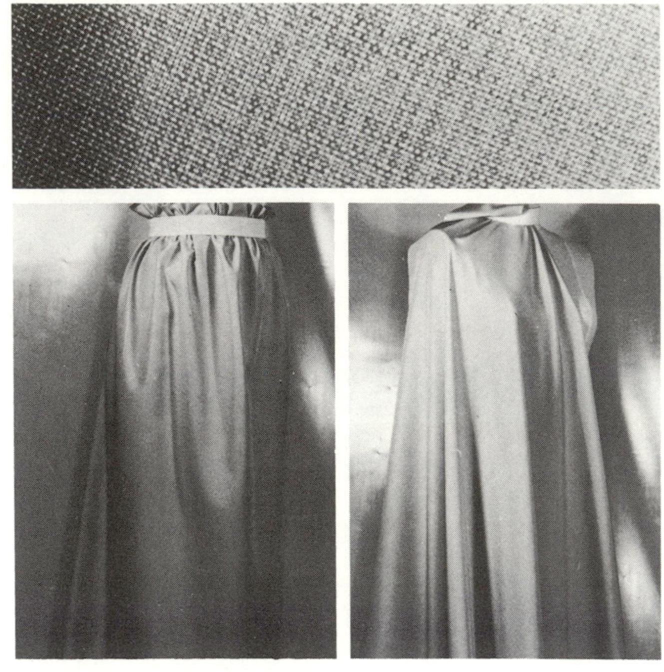

Simulated Silk Honan

Fiber Content 100% polyester
Yarn Type filament fiber
Yarn Construction textured; thick-and-thin
Fabric Structure plain weave
Finishing Processes calendering; delustering; stabilizing
Color Application piece dyed; solution dyed; yarn dyed
Width 54 inches (137.2 cm)
Weight light
Hand fine; firm; soft; springy
Texture thicker yarns regularly spaced in lengthwise and crosswise directions
Performance Expectations subject to abrasion; dimensional stability; durable; elongation; heat-set properties; resilient; high tensile strength
Drapability Qualities falls into moderately soft flares; accommodates fullness by pleating, gathering, elasticized shirring; fullness retains soft fall

Recommended Construction Aids & Techniques

Hand Needles beading, 10–13; betweens, 5–7; milliners, 8–10; sharps, 8–10
Machine Needles round/set point; medium-fine 9–11
Threads poly-core waxed, 60; cotton-polyester blend, fine; cotton-covered polyester core, extra fine; spun polyester, fine
Hems bound/Hong Kong finish; double fold; edge-stitched; horsehair; machine-stitched; hand- or machine-rolled; seam binding
Seams flat-felled; French; false French; mock French; plain; tissue-stitched
Seam Finishes self bound; double-ply bound; double-stitched; double-stitched and overcast; edge-stitched; mock French; double-ply overcast; double-ply zigzag
Pressing steam; safe temperature limit 325°F (164.1°C)
Care launder; dry clean
Fabric Resource Klopman Mills

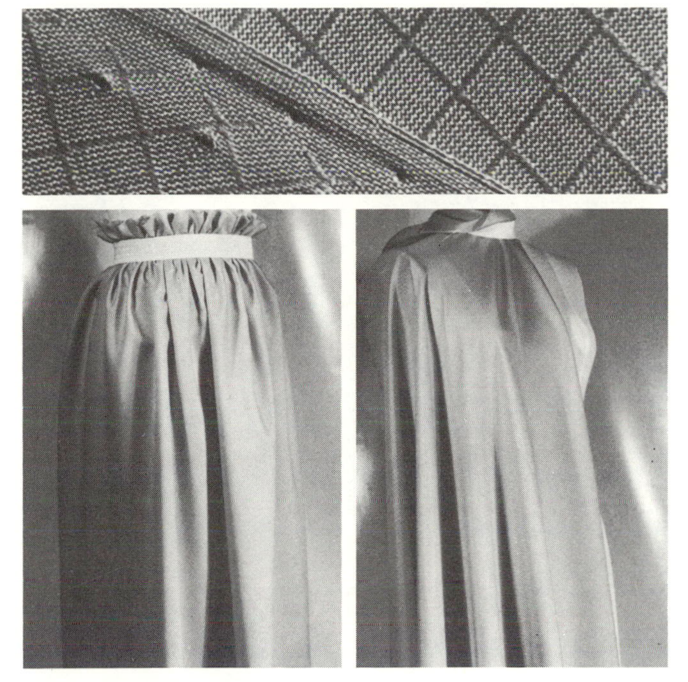

Silk Pongee Dobby

Fiber Content 100% polyester
Yarn Type filament fiber
Yarn Construction textured
Fabric Structure dobby weave
Finishing Processes calendering; delustering; stabilizing
Color Application piece dyed; solution dyed; yarn dyed
Width 45 inches (114.3 cm)
Weight light
Hand firm; soft; springy
Texture slippery; smooth; light-reflecting yarns visible in design
Performance Expectations subject to abrasion; dimensional stability; durable; elongation; heat-set properties; resilient; high tensile strength
Drapability Qualities falls into moderately soft flares; accommodates fullness by pleating, gathering, elasticized shirring; fullness retains soft fall

Recommended Construction Aids & Techniques

Hand Needles beading, 10–13; betweens, 5–7; milliners, 8–10; sharps, 8–10
Machine Needles round/set point, medium-fine 9–11
Threads poly-core waxed, 60; cotton-polyester blend, fine; cotton-covered polyester core, extra fine; spun polyester, fine
Hems bound/Hong Kong finish; double fold; edge-stitched; horsehair; machine-stitched; hand- or machine-rolled; seam binding
Seams flat-felled; French; false French; mock French; plain; tissue-stitched
Seam Finishes self bound; double-ply bound; double-stitched; double-stitched and overcast; edge-stitched; mock French; double-ply overcast; double-ply zigzag
Pressing steam; safe temperature limit 325°F (164.1°C)
Care launder; dry clean
Fabric Resource **Bloomsburg Mills**

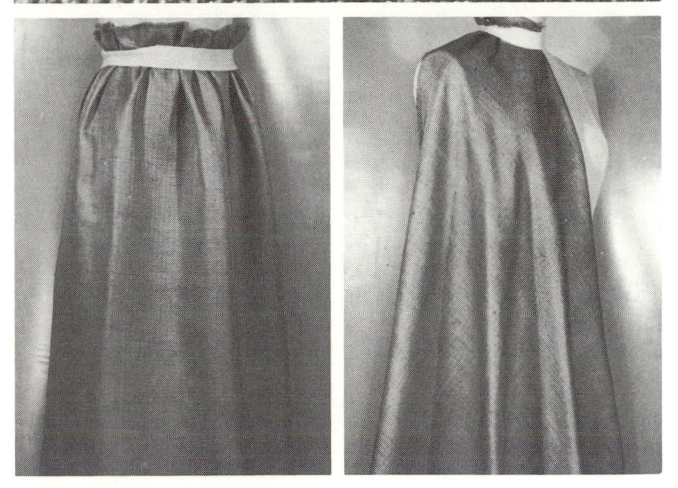

Silk Homespun/Silk Burlap

Fiber Content 100% silk
Yarn Type natural filament fiber
Yarn Construction conventional; thick; slack twist
Fabric Structure plain weave
Finishing Processes bleaching; degumming; sizing; stretching
Color Application piece dyed; yarn dyed
Width 43 inches (109.2 cm)
Weight heavy
Hand soft; spongy; thick
Texture coarse; rough
Performance Expectations dimensional stability; flexible; poor elongation; moth resistant; nonpilling; resilient; subject to snagging due to weave
Drapability Qualities falls into wide cones; retains shape or silhouette of garment; better utilized if fitted by seaming and eliminating excess fabric

Recommended Construction Aids & Techniques

Hand Needles cotton darners, 9–10; milliners, 3–6; sharps, 1–4
Machine Needles round/set point, medium-coarse 14–16
Threads mercerized cotton, 40; silk, industrial size A-B-C; cotton-covered polyester core, regular
Hems book; double-fold or single-fold bias binding; bound/Hong Kong finish; edge-stitched; faced; overedged; stitched and pinked; seam binding
Seams plain; safety-stitched; top-stitched; welt; double welt
Seam Finishes single-ply bound; Hong Kong bound; single-ply overcast; pinked; pinked and stitched; serging/single-ply overedged; single-ply zigzag
Pressing steam; safe temperature limit 300°F (150.1°C)
Care dry clean
Fabric Resource **Jack Larsen Incorporated**

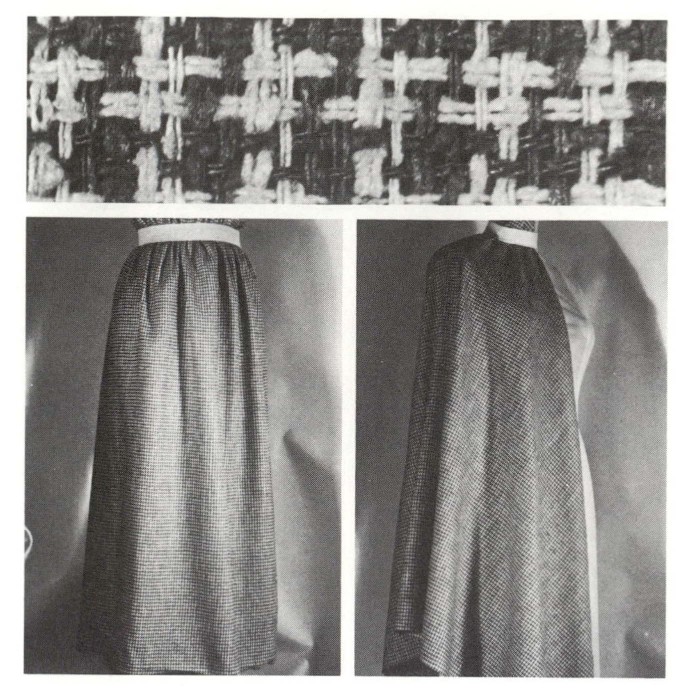

Silk Hopsacking

Fiber Content 100% silk
Yarn Type natural filament fiber
Yarn Construction conventional; low twist
Fabric Structure 2x2 basket weave
Finishing Processes calendering; sizing; stretching
Color Application piece dyed; yarn dyed
Width 47–48 inches (119.4–121.9 cm)
Weight medium
Hand loose; pliable; semi-soft
Texture coarse; grainy
Performance Expectations absorbent; dimensional stability; poor elongation; flexible; moth resistant; nonpilling; resilient; subject to snagging due to weave
Drapability Qualities falls into moderately soft flares; accommodates fullness by gathering; fullness retains soft fall

Recommended Construction Aids & Techniques

Hand Needles cotton darners, 9–10; milliners, 3–6; sharps, 1–4
Machine Needles round/set point, medium 11–14
Threads mercerized cotton, 50; silk, industrial size A-B; cotton-covered polyester core, regular
Hems book; double-fold or single-fold bias binding; bound/Hong Kong finish; edge-stitched; faced; overedged; stitched and pinked; seam binding
Seams plain; safety-stitched; top-stitched; welt; double welt
Seam Finishes single-ply bound; Hong Kong bound; single-ply overcast; pinked; pinked and stitched; serging/single-ply overedged; single-ply zigzag
Pressing steam; safe temperature limit 300°F (150.1°C)
Care dry clean
Fabric Resources **American Silk Mills Corp.; Auburn Fabrics Inc.**

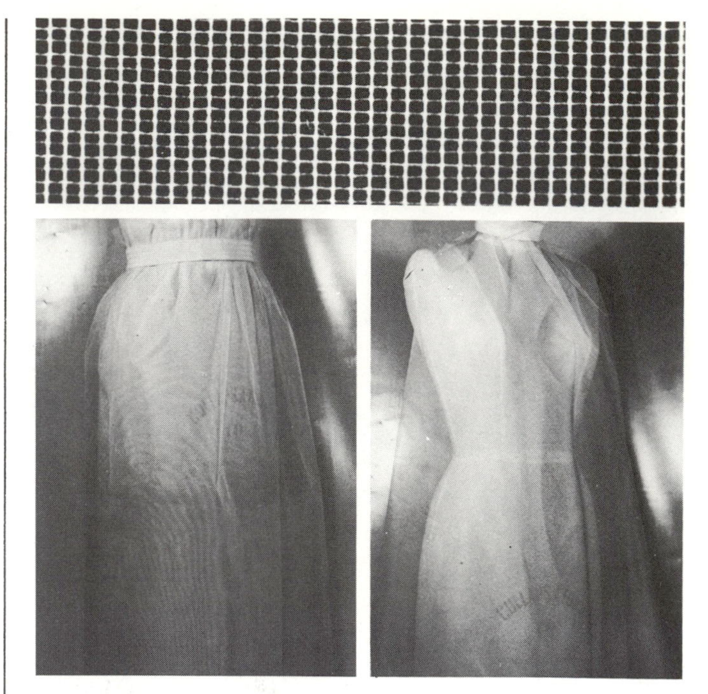

Marquisette

Fiber Content 100% nylon
Yarn Type filament fiber
Yarn Construction monofilament
Fabric Structure leno weave
Finishing Processes calendering; delustering; stabilizing; stiffening
Color Application piece dyed; solution dyed
Width 52 inches (132.1 cm)
Weight fine to lightweight
Hand crisp; delicate
Texture open mesh; papery
Performance Expectations abrasion resistant; air permeable; dimensional stability; durable; high elongation; heat-set properties; high tensile strength
Drapability Qualities falls into stiff cones; accommodates fullness by gathering; fullness maintains crisp bouffant effect

Recommended Construction Aids & Techniques

Hand Needles beading, 10–13; betweens, 5–7; milliners, 10; sharps, 8–10
Machine Needles round/set point, fine 9
Threads poly-core waxed, 60; cotton-polyester blend, fine; cotton-covered polyester core, extra fine; spun polyester, fine
Hems double fold; edge-stitched only; double edge-stitched; horsehair; machine-stitched; hand- or machine-rolled; wired
Seams French; mock French; hairline; plain; tissue-stitched
Seam Finishes self bound; double-stitched; double-stitched and trimmed; mock French; untreated/plain
Pressing steam; safe temperature limit 350°F (178.1°C)
Care launder
Fabric Resource **Kortex Associates, Inc.**

Matelassé Silk

Fiber Content 100% silk
Yarn Type natural filament fiber
Yarn Construction conventional; high twist
Fabric Structure double weave; additional filling yarn on back
Finishing Processes degumming; stabilizing; stretching
Color Application piece dyed; yarn dyed
Width 36 inches (91.4 cm)
Weight medium
Hand soft; spongy; springy; thick
Texture blistered; puffed design creates uneven surface
Performance Expectations subject to abrasion; absorbent; dimensional stability; flexible; resilient
Drapability Qualities falls into wide cones; fullness maintains lofty bouffant effect; retains shape or silhouette of garment

Recommended Construction Aids & Techniques

Hand Needles betweens, 3–4; milliners, 6–8; sharps, 4–8
Machine Needles round/set point, medium-fine 9–11
Threads mercerized cotton, 50; silk, industrial size A-B; cotton-covered polyester core, all purpose
Hems book; double-fold or single-fold bias binding; bound/Hong Kong finish; edge-stitched; machine-stitched; seam binding
Seams flat-felled; plain; safety-stitched; top-stitched; welt; double welt
Seam Finishes single-ply or double-ply bound; Hong Kong bound; edge-stitched; single-ply overcast; pinked and stitched; serging/single-ply overedged; single-ply zigzag
Pressing steam; safe temperature limit 300°F (150.1°C)
Care dry clean
Fabric Resources **American Silk Mills Corp.; Sormani Co. Inc./Taroni**

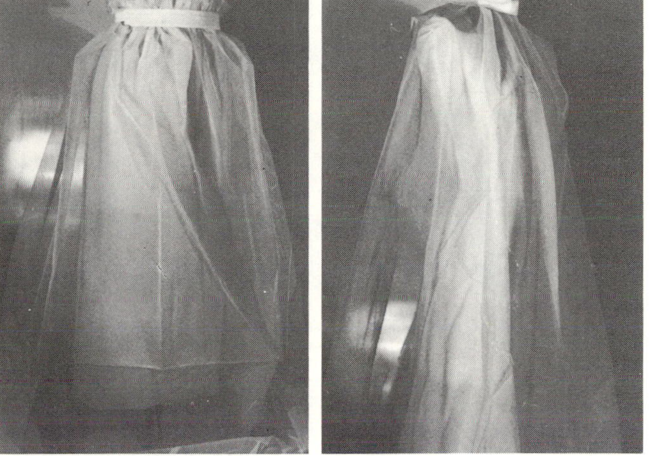

Mousseline de Soie/Silk Gauze

Fiber Content 100% silk
Yarn Type natural filament fiber
Yarn Construction conventional; fine; high twist
Fabric Structure gauze weave
Finishing Processes calendering; singeing; permanent stiffening
Color Application piece dyed
Width 69-70 inches (175.3-177.8 cm)
Weight fine to lightweight
Hand crisp; delicate; firm
Texture flat; sheer
Performance Expectations absorbent; air permeable; dimensional stability; poor elongation; flexible; nonpilling; subject to snagging due to weave
Drapability Qualities falls into firm cones; accommodates fullness by gathering, elasticized shirring; fullness maintains crisp effect

Recommended Construction Aids & Techniques

Hand Needles beading, 10–13; betweens, 5–7; milliners, 10; sharps, 8–10
Machine Needles round/set point, fine 9
Threads mercerized cotton, 50; silk, industrial size A; cotton-covered polyester core, extra fine
Hems double fold; edge-stitched only; double edge-stitched; horsehair; machine-stitched; hand- or machine-rolled; wired
Seams French; mock French; hairline; plain; tissue-stitched
Seam Finishes self bound; double-stitched; double-stitched and trimmed; mock French
Pressing steam; safe temperature limit 300°F (150.1°C)
Care dry clean
Fabric Resource **Sormani Co. Inc./Taroni**

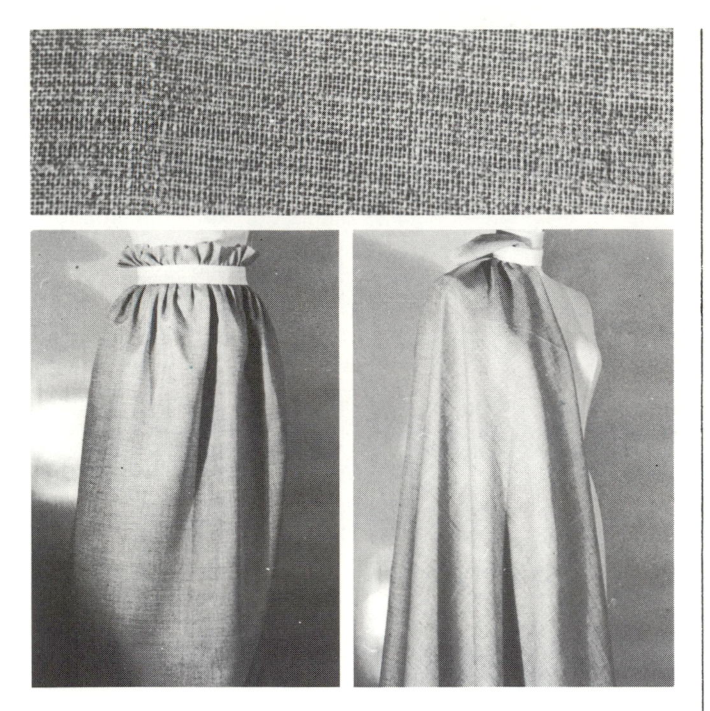

Mull

Fiber Content 100% silk
Yarn Type natural filament fiber
Yarn Construction conventional; hard twist
Fabric Structure loosely woven plain weave
Finishing Processes bleaching; calendering; degumming
Color Application natural silk color; yarn dyed
Width 56 inches (142.2 cm)
Weight medium-light
Hand loose; pliable; soft
Texture grainy; scratchy
Performance Expectations absorbent; dimensional stability; poor elongation; flexible; moth resistant; nonpilling
Drapability Qualities falls into moderately soft flares; accommodates fullness by pleating, gathering, elasticized shirring; fullness retains soft fall

Recommended Construction Aids & Techniques

Hand Needles beading, 10–13; betweens, 5–7; milliners, 8–10; sharps, 8–10
Machine Needles round/set point, medium-fine 9–11
Threads mercerized cotton, 50; silk, industrial size A; cotton-covered polyester core, extra fine
Hems bound/Hong Kong finish; double fold; edge-stitched; horsehair; machine-stitched; hand- or machine-rolled; seam binding
Seams flat-felled; French; false French; mock French; plain; tissue-stitched
Seam Finishes self bound; double-ply bound; double-stitched; double-stitched and overcast; edge-stitched; mock French; double-ply overcast; double-ply zigzag
Pressing steam; safe temperature limit 300°F (150.1°C)
Care dry clean
Fabric Resource Sormani Co. Inc./Taroni

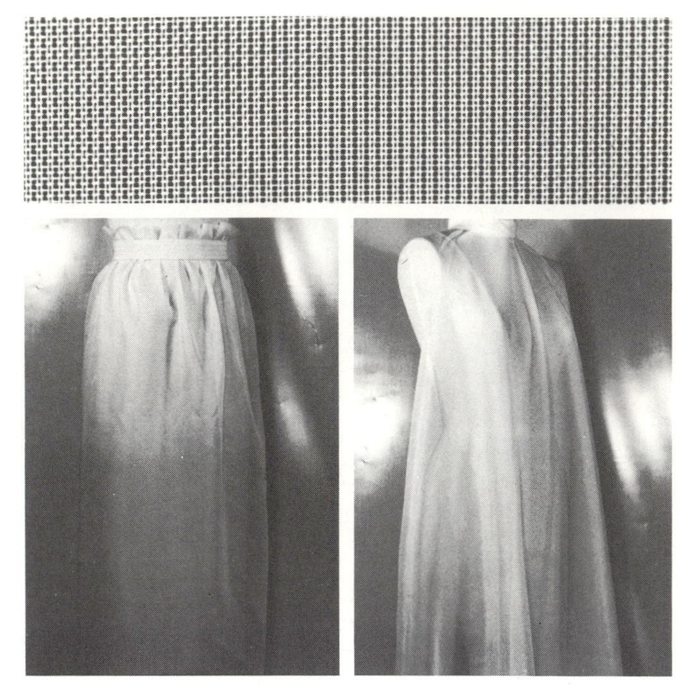

Ninon

Fiber Content 100% polyester
Yarn Type filament fiber
Yarn Construction conventional; fine; high twist
Fabric Structure plain weave; open effect
Finishing Processes calendering; stabilizing; permanent stiffening
Color Application piece dyed; solution dyed; yarn dyed
Width 45 inches (114.3 cm)
Weight fine to lightweight
Hand firm; thin; soft to crisp depending on finish
Texture flat; open weave; sheer; smooth
Performance Expectations abrasion resistant; air permeable; dimensional stability; durable; high elongation; heat-set properties; high tensile strength
Drapability Qualities falls into soft cones; accommodates fullness by gathering; fullness retains soft fall

Recommended Construction Aids & Techniques

Hand Needles beading, 10–13; betweens, 5–7; milliners, 10; sharps, 8–10
Machine Needles round/set point, fine 9
Threads poly-core waxed, 60; cotton-polyester blend, fine; cotton-covered polyester core, extra fine; spun polyester, fine
Hems double fold; edge-stitched only; double edge-stitched; horsehair; machine-stitched; hand- or machine-rolled; wired
Seams French; mock French; hairline; plain; tissue-stitched
Seam Finishes self bound; double-stitched; double-stitched and trimmed; mock French
Pressing steam; safe temperature limit 325°F (164.1°C)
Care launder
Fabric Resources Kortex Associates Inc.; Lawrence Textile Co. Inc.

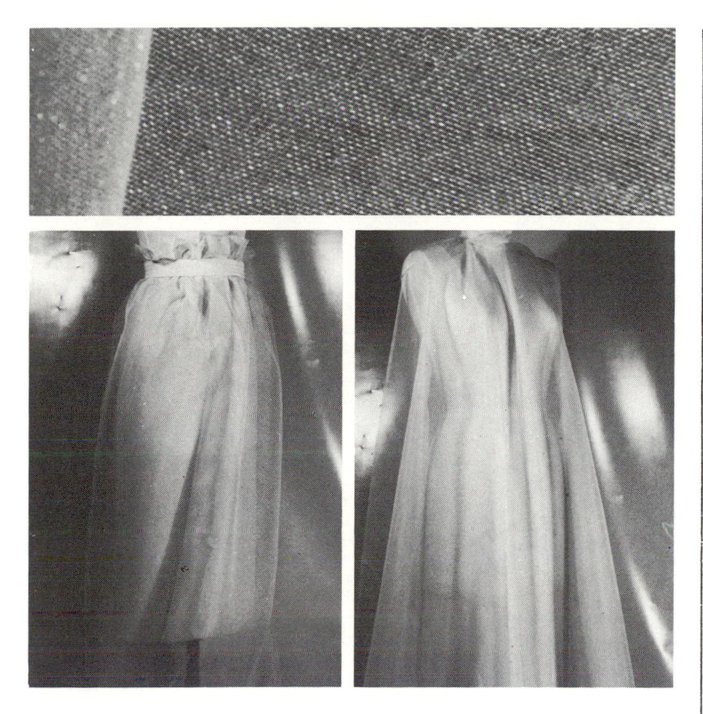

Organza

Fiber Content 100% polyester
Yarn Type filament fiber
Yarn Construction fine; textured
Fabric Structure loosely woven plain weave
Finishing Processes calendering; stabilizing; permanent stiffening
Color Application piece dyed; solution dyed; yarn dyed
Width 44–45 inches (111.8–114.3 cm)
Weight fine to lightweight
Hand crisp; delicate; firm
Texture flat; sheer; smooth
Performance Expectations air permeable; dimensional stability; durable; elongation; heat-set properties; resilient; high tensile strength
Drapability Qualities falls into crisp cones; accommodates fullness by gathering; fullness maintains crisp effect

Recommended Construction Aids & Techniques

Hand Needles beading, 10–13; betweens, 5–7; milliners, 10; sharps, 8–10
Machine Needles round/set point, fine 9
Threads poly-core waxed, 60; cotton-polyester blend, fine; cotton-covered polyester core, extra fine; spun polyester, fine
Hems double fold; edge-stitched only; double edge-stitched; horsehair; machine-stitched; hand- or machine-rolled; wired
Seams French; mock French; hairline; plain; tissue-stitched
Seam Finishes self bound; double-stitched; double stitched and trimmed; mock French
Pressing steam; safe temperature limit 325°F (164.1°C)
Care launder; dry clean
Fabric Resource Kabat Textile Corp.

Organza with Yarn Embroidery

Fiber Content 100% nylon
Yarn Type filament fiber
Yarn Construction textured
Fabric Structure Schiffli embroidery on plain weave base
Finishing Processes calendering; flame retardant; stiffening
Color Application base: piece dyed; embroidery yarn: skein or packaged dyed
Width 45 inches (114.3 cm)
Weight light
Hand semi-crisp; fine; wiry
Texture base: sheer; embroidery yarn forms high and low raised pattern on surface
Performance Expectations air permeable; dimensional stability; durable; elasticity; elongation; heat-set properties; resilient; high tensile strength; raised embroidered design subject to snagging
Drapability Qualities falls into moderately crisp flares; accommodates fullness by gathering, elasticized shirring; fullness maintains crisp effect

Recommended Construction Aids & Techniques

Hand Needles beading, 10–13; betweens, 7; embroidery, 9–10; sharps, 9–10
Machine Needles round/set point, fine 9
Threads silk, industrial size A; spun polyester, fine; nylon/Dacron, monocord A
Hems bound/Hong Kong finish; net binding; faced; horsehair; scalloped edge may be used as the hem edge
Seams hairline; plain; tissue-stitched
Seam Finishes net bound; Hong Kong bound; double-stitched; double-stitched and trimmed; untreated/plain
Pressing use pressing cloth; safe temperature limit 350°F (178.1°C)
Care launder; dry clean
Fabric Resource Embroidery Council of America

Organza with Yarn Embroidery & Flat Appliqué

Fiber Content 100% silk
Yarn Type natural filament fiber
Yarn Construction conventional; fine; tight twist
Fabric Structure Schiffli embroidery and raised appliqué on plain weave base
Finishing Processes calendering; decating; singeing; stiffening
Color Application base: piece dyed; embroidery yarn: skein or packaged dyed; appliqué: printed
Width 45 inches (114.3 cm)
Weight fine to lightweight
Hand crisp; delicate; thin
Texture base: sheer; flat, smooth appliqué and embroidery form high and low raised pattern on surface
Performance Expectations absorbent; air permeable; dimensional stability; poor elongation; flexible; raised yarn and appliqué subject to abrasion and snagging
Drapability Qualities falls into crisp cones; accommodates fullness by gathering; fullness maintains bouffant effect

Recommended Construction Aids & Techniques

Hand Needles beading, 10–13; betweens, 7; embroidery, 9–10; sharps, 9–10
Machine Needles round/set point, fine 9
Threads silk, industrial size A; spun polyester, fine; nylon/Dacron, monocord A
Hems bound/Hong Kong finish; edge-stitched; faced; horsehair; machine-stitched; hand- or machine-rolled
Seams hairline; plain; tissue-stitched; join appliqué at seamline
Seam Finishes single-ply or Hong Kong bound; double-stitched; double-stitched and trimmed; untreated/plain; remove excess appliqué in seam allowance
Pressing use pressing cloth; safe temperature limit 300°F (150.1°C)
Care dry clean
Fabric Resource Schiffli Lace & Embroidery Manufacturers Association (sample courtesy of Lita Konde)

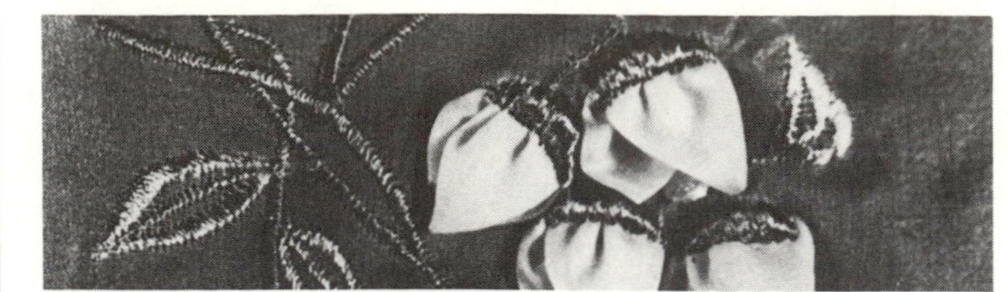

DESIGN ON FABRIC DOES NOT LEND ITSELF TO BIAS DRAPE.

Organza with Yarn Embroidery & Raised Appliqué

Fiber Content 100% silk
Yarn Type natural filament fiber
Yarn Construction conventional; fine; tight twist
Fabric Structure Schiffli embroidery and appliqué on plain weave base
Finishing Processes calendering; decating; singeing; stiffening
Color Application base: piece dyed; embroidery yarn: skein or packaged dyed; appliqué: pre-dyed fabric
Width 45 inches (114.3 cm)
Weight light
Hand crisp; delicate; fine
Texture base: sheer; raised appliqué and embroidery form three-dimensional design on surface
Performance Expectations absorbent; air permeable; dimensional stability; poor elongation; flexible; raised appliqué and yarn subject to abrasion and snagging
Drapability Qualities falls into crisp cones; accommodates fullness by gathering; fullness maintains bouffant effect

Recommended Construction Aids & Techniques

Hand Needles beading, 10–13; betweens, 7; embroidery, 9–10; sharps, 9–10
Machine Needles round/set point, fine 9
Threads silk, industrial size A; spun polyester, fine; nylon/Dacron, monocord A
Hems Hong Kong bound; edge-stitched; faced; horsehair; machine-stitched; hand- or machine-rolled
Seams hairline; plain; tissue stitched; join appliqué at seamline
Seam Finishes single-ply or Hong Kong bound; double-stitched; double-stitched and trimmed; untreated/plain; remove excess appliqué in seam allowance
Pressing use pressing cloth; safe temperature limit 300°F (150.1°C)
Care dry clean
Fabric Resource Schiffli Lace & Embroidery Manufacturers Association (sample courtesy of Lita Konde)

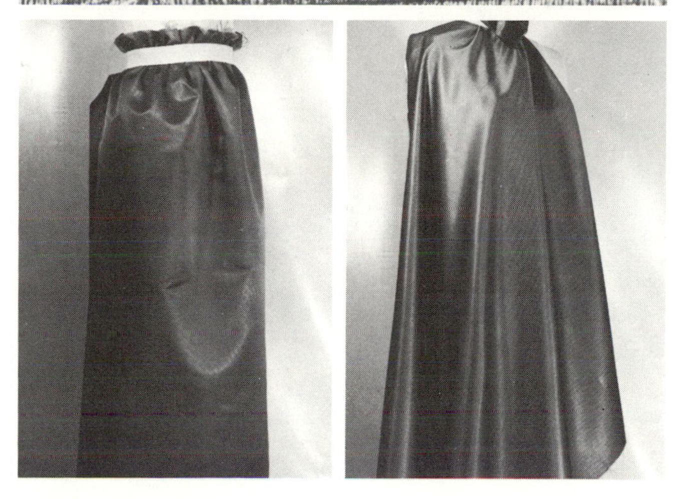

Silk Ottoman

Fiber Content 100% silk
Yarn Type natural filament fiber
Yarn Construction conventional; low twist
Fabric Structure plain weave; fine warp and multiple filling yarns form rib in crosswise direction
Finishing Processes ciré calendering; lustering; stretching
Color Application piece dyed; yarn dyed
Width 54 inches (137.2 cm)
Weight medium
Hand firm; soft
Texture pronounced wide, flat rib in crosswise direction; slippery
Performance Expectations subject to abrasion; absorbent; dimensional stability; poor elongation; flexible; nonpilling
Drapability Qualities falls into moderately soft cones; accommodates fullness by gathering; fullness retains moderately soft fall

Recommended Construction Aids & Techniques

Hand Needles betweens, 3–4; milliners, 6–8; sharps, 4–8
Machine Needles round/set point, medium 11
Threads mercerized cotton, 50; silk, industrial size A-B; cotton-covered polyester core, all purpose
Hems book; double-fold or single-fold bias binding; bound/Hong Kong finish; edge-stitched; machine-stitched; seam binding
Seams flat-felled; false French; plain; safety-stitched; top-stitched; welt; double welt
Seam Finishes single-ply or double-ply bound; Hong Kong bound; edge-stitched; single-ply overcast; pinked and stitched; serging/single-ply overedged
Pressing steam; safe temperature limit 300°F (150.1°C)
Care dry clean
Fabric Resource **American Silk Mills Corp.**

Silk Ottoman Coating

Fiber Content acetate/mylar
Yarn Type filament fiber; metallic fiber
Yarn Construction metallic film; textured
Fabric Structure plain weave; ply warp and multiple filling yarns form rib in crosswise direction
Finishing Processes calendering; sizing; stabilizing
Color Application piece dyed-cross dyed; yarn dyed
Width 45 inches (114.3 cm)
Weight heavy
Hand compact; hard; thick
Texture pronounced wide, flat rib in crosswise direction
Performance Expectations subject to abrasion; dimensional stability; elongation; flexible; heat-set properties; mildew resistant; moth repellent
Drapability Qualities falls into stiff wide cones; fullness maintains crisp lofty effect; retains shape or silhouette of garment

Recommended Construction Aids & Techniques

Hand Needles cotton darners, 9–10; milliners, 3–6; sharps, 1–4
Machine Needles round/set point, medium 11–14
Threads poly-core waxed, 50; cotton-polyester blend, all purpose; cotton-covered polyester core, regular; spun polyester, all purpose
Hems book; double-fold or single-fold bias binding; bound/Hong Kong finish; edge-stitched; faced; overedged; stitched and pinked; seam binding
Seams plain; safety-stitched; top-stitched; welt; double welt
Seam Finishes single-ply bound; Hong Kong bound; single-ply overcast; pinked; pinked and stitched; serging/single-ply overedged; single-ply zigzag
Pressing steam; safe temperature limit 250°F (122.1°C)
Care dry clean
Fabric Resource **Lawrence Textile Co. Inc.**

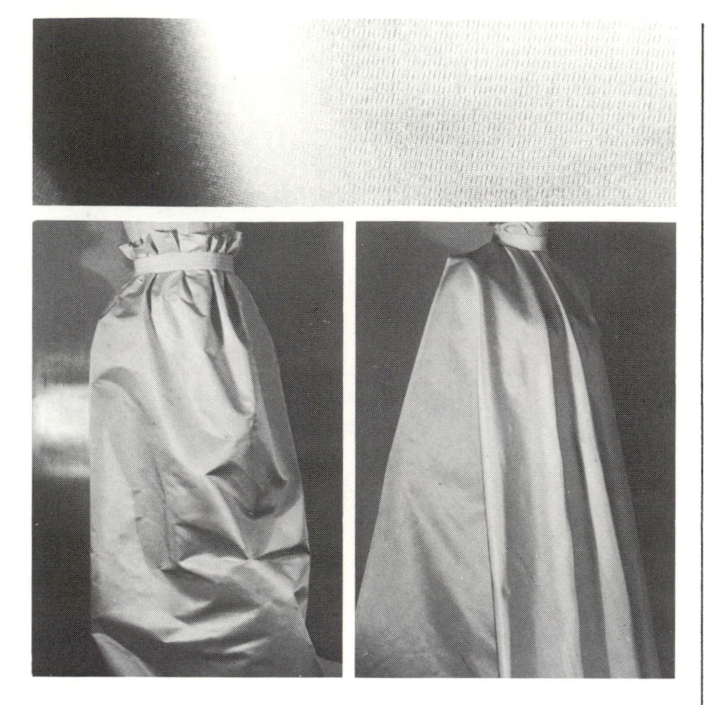

Peau de Soie

Fiber Content 100% silk
Yarn Type natural filament fiber
Yarn Construction conventional; slack twist
Fabric Structure 8-shaft satin weave variation
Finishing Processes ciré calendering; lustering; stretching
Color Application piece dyed; yarn dyed
Width 48 inches (121.9 cm)
Weight medium-heavy
Hand compact; semi-crisp
Texture shiny face; dull back; satiny smooth; opaque
Performance Expectations absorbent; dimensional stability; flexible; nonpilling
Drapability Qualities falls into wide cones; fullness maintains crisp lofty effect; retains shape or silhouette of garment

Recommended Construction Aids & Techniques

Hand Needles betweens, 3–4; milliners, 6–8; sharps, 4–8
Machine Needles round/set point, medium 11
Threads mercerized cotton, 40; silk, industrial size A-B-C; cotton-covered polyester core, heavy duty
Hems book; double-fold or single-fold bias binding; bound/Hong Kong finish; edge-stitched; machine-stitched; seam binding
Seams flat-felled; false French; plain; safety-stitched; welt; double welt
Seam Finishes double-ply bound; Hong Kong bound; edge stitched; single-ply overcast; pinked and stitched; serging/single-ply overedged; single-ply zigzag
Pressing steam; safe temperature limit 300°F (150.1°C)
Care dry clean
Fabric Resource Sormani Co. Inc./Taroni

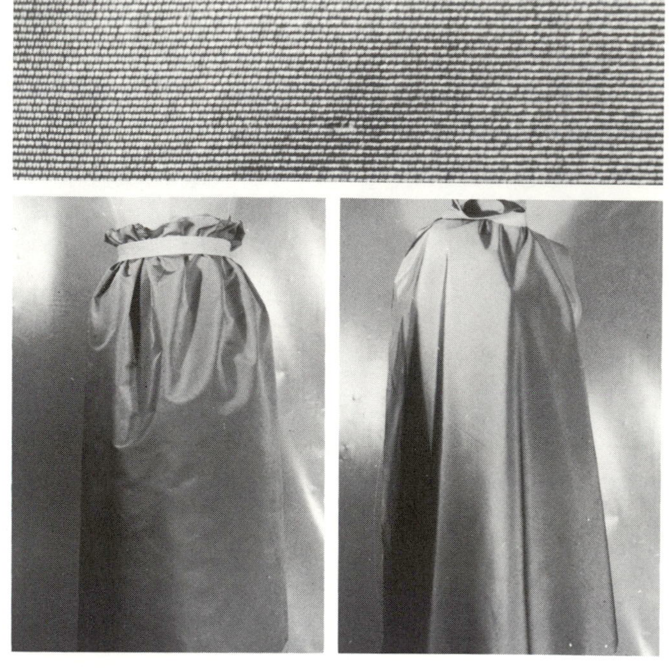

Silk Poplin

Fiber Content 100% silk
Yarn Type natural filament fiber
Yarn Construction conventional; fine; slack twist
Fabric Structure plain weave; multiple filling yarns form rib in crosswise direction
Finishing Processes calendering; singeing; water repellent
Color Application piece dyed; yarn dyed
Width 50–52 inches (127–132.1 cm)
Weight medium-light
Hand firm; rigid; scroopy
Texture harsh; papery; fine even crosswise rib
Performance Expectations absorbent; dimensional stability; flexible; water resistant due to finish
Drapability Qualities falls into crisp folds; accommodates fullness by gathering; fullness maintains crisp effect

Recommended Construction Aids & Techniques

Hand Needles betweens, 3–4; milliners, 6–8; sharps, 4–8
Machine Needles round/set point, medium 11–14
Threads mercerized cotton, 50; silk, industrial size A-B; cotton-covered polyester core, all purpose
Hems book; double-fold or single-fold bias binding; bound/Hong Kong finish; edge-stitched; machine-stitched; seam binding
Seams flat-felled; French; false French; plain; safety-stitched; top-stitched; welt; double welt
Seam Finishes single-ply or double-ply bound; Hong Kong bound; edge-stitched; single-ply overcast; pinked and stitched; serging/single-ply overedged; single-ply zigzag
Pressing steam; safe temperature limit 300°F (150.1°C)
Care dry clean
Fabric Resource Lafitte Inc.

Radium Silk

Fiber Content 100% silk
Yarn Type natural filament fiber
Yarn Construction textured; high twist
Fabric Structure compactly woven plain weave
Finishing Processes calendering; degumming; singeing; stretching
Color Application piece dyed
Width 42 inches (106.7 cm)
Weight light
Hand fine; firm; soft; thin
Texture flat; even surface
Performance Expectations absorbent; air permeable; dimensional stability; flexible
Drapability Qualities falls into soft flares and ripples; accommodates fullness by pleating, gathering, elasticized shirring; fullness retains soft fall

Recommended Construction Aids & Techniques

Hand Needles beading, 10–13; betweens, 5–7; milliners, 8–10; sharps, 8–10
Machine Needles round/set point, medium-fine 9–11
Threads mercerized cotton, 50; silk, industrial size A; cotton-covered polyester core, extra fine
Hems bound/Hong Kong finish; double fold; edge-stitched; horsehair; machine-stitched; hand- or machine-rolled; seam binding
Seams flat-felled; French; false French; mock French; plain; tissue-stitched
Seam Finishes self bound; double-ply bound; double-stitched; double-stitched and overcast; edge-stitched; mock French; double-ply overcast; double-ply zigzag
Pressing steam; safe temperature limit 300°F (150.1°C)
Care dry clean
Fabric Resource American Silk Mills Corp.

Silk Repp

Fiber Content 100% silk
Yarn Type natural filament fiber
Yarn Construction conventional; slubbed doupioni yarns
Fabric Structure firmly woven plain weave
Finishing Processes calendering; stiffening; stretching
Color Application piece dyed; yarn dyed
Width 40–52 inches (101.6–132.1 cm)
Weight medium-heavy
Hand compact; rigid; thick
Texture harsh; prominent rounded crosswise rib
Performance Expectations absorbent; dimensional stability; flexible; nonpilling; resilient
Drapability Qualities falls into firm wide cones; fullness maintains bouffant effect; retains shape or silhouette of garment

Recommended Construction Aids & Techniques

Hand Needles cotton darners, 9–10; milliners, 3–6; sharps, 1–4
Machine Needles round/set point, medium 14
Threads mercerized cotton, 40; silk, industrial size A-B-C; cotton-covered polyester core, heavy duty
Hems book; double-fold or single-fold bias binding; bound/Hong Kong finish; edge-stitched; faced; overedged; stitched and pinked; seam binding
Seams plain; safety stitched; top stitched; welt; double welt
Seam Finishes single-ply bound; Hong Kong bound; single-ply overcast; pinked; pinked and stitched; serging/single-ply overedged; single-ply zigzag
Pressing steam; safe temperature limit 300°F (150.1°C)
Care dry clean
Fabric Resources American Silk Mills Corp.; Jack Larsen Incorporated

Silk Seersucker

Fiber Content 100% silk
Yarn Type natural filament fiber
Yarn Construction conventional; fine
Fabric Structure plain weave; alternately tight and slack woven
Finishing Processes caustic acid bath to produce crinkled areas; calendering; singeing; sizing
Color Application piece dyed; yarn dyed
Width 45 inches (114.3 cm)
Weight light
Hand semi-crisp; pliable; thin
Texture stripe effect created by alternating crinkled and smooth surface
Performance Expectations absorbent; air permeable; dimensional stability; flexible; nonpilling
Drapability Qualities falls into moderately soft flares; accommodates fullness by gathering, elasticized shirring, smocking; fullness retains soft fall

Recommended Construction Aids & Techniques

Hand Needles beading, 10–13; betweens, 5–7; milliners, 8–10; sharps, 8–10
Machine Needles round/set point, fine 9
Threads mercerized cotton, 50; silk, industrial size A; cotton-covered polyester core, extra fine
Hems book; double-fold or single-fold bias binding; bound/Hong Kong finish; double fold; edge-stitched; machine-stitched; hand- or machine-rolled; seam binding
Seams flat-felled; French; safety-stitched; tissue-stitched; top-stitched; welt; double welt
Seam Finishes single-ply or double-ply bound; Hong Kong bound; edge-stitched; single-ply or double-ply overcast; pinked and stitched; serging/single-ply overedged; single-ply or double-ply zigzag
Pressing steam; safe temperature limit 300°F (150.1°C)
Care dry clean
Fabric Resource **American Silk Mills Corp.**

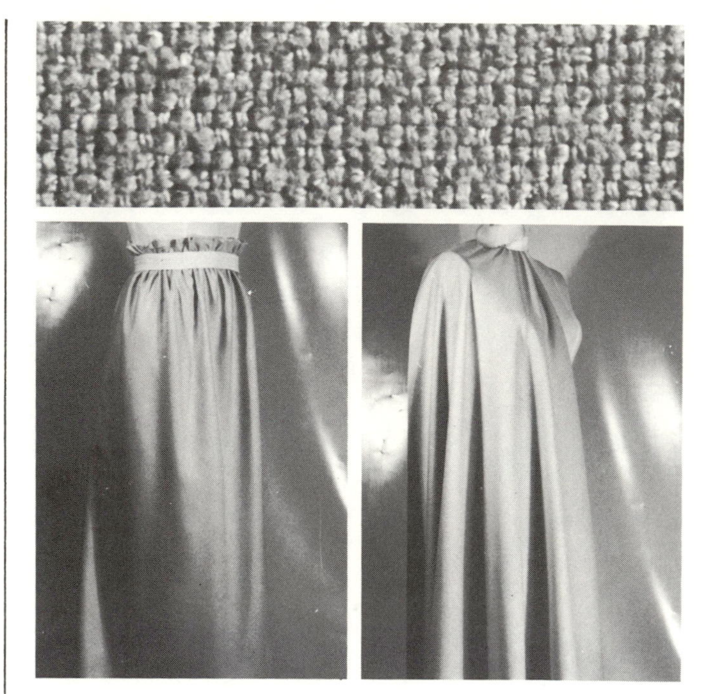

Silk Serge

Fiber Content 100% silk
Yarn Type natural filament fiber
Yarn Construction conventional; multiple yarn; slack twist
Fabric Structure plain weave manipulated to resemble basket weave
Finishing Processes calendering; softening; stabilizing
Color Application piece dyed; yarn dyed
Width 42 inches (106.7 cm)
Weight medium-heavy
Hand clammy; semi-soft
Texture loose; pebbly; woolly
Performance Expectations absorbent; dimensional stability; flexible; nonpilling; subject to bagging and sagging due to structure
Drapability Qualities falls into soft wide flares; accommodates fullness by gathering; fullness retains moderately soft fall

Recommended Construction Aids & Techniques

Hand Needles cotton darners, 9–10; milliners, 3–6; sharps, 1–4
Machine Needles round/set point, medium-coarse 14–16
Threads mercerized cotton, 40; silk, industrial size A-B-C; cotton-covered polyester core, heavy duty
Hems book; double-fold or single-fold bias binding; bound/Hong Kong finish; edge-stitched; faced; overedged; stitched and pinked; seam binding
Seams plain; safety-stitched; top-stitched; welt; double welt
Seam Finishes single-ply bound; Hong Kong bound; single-ply overcast; pinked; pinked and stitched; serging/single-ply overedged; single-ply zigzag
Pressing steam; safe temperature limit 300°F (150.1°C)
Care dry clean
Fabric Resource **American Silk Mills Corp.**

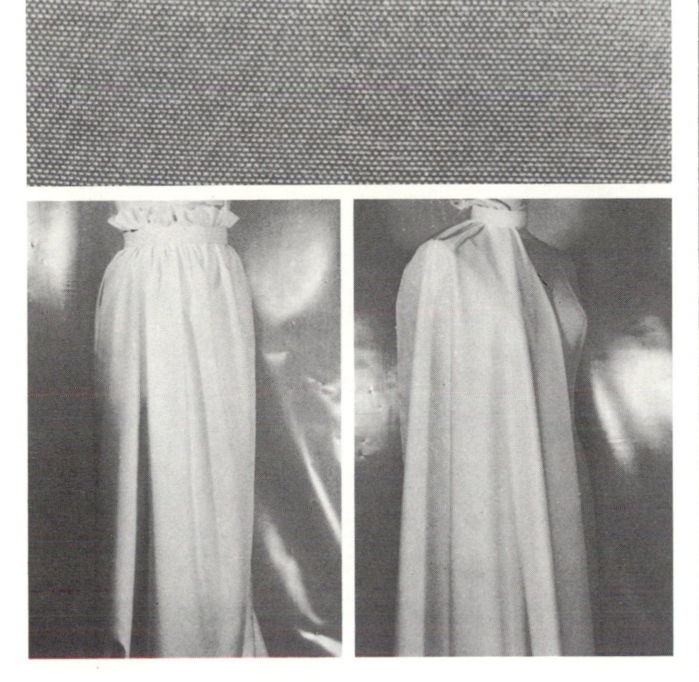

Shantung (soft finish)

Fiber Content 100% Qiana® nylon
Yarn Type filament fiber
Yarn Construction conventional; elongated irregular slub filling yarn; low twist
Fabric Structure plain weave
Finishing Processes calendering; delustering; stabilizing
Color Application piece dyed; solution dyed; yarn dyed
Width 46–47 inches (116.8–119.4 cm)
Weight medium-light
Hand firm; semi-soft; supple; thin
Texture irregular slub yarn in crosswise direction
Performance Expectations abrasion resistant; dimensional stability; durable; high elongation; heat-set properties; resilient; high tensile strength
Drapability Qualities falls into soft flares; accommodates fullness by pleating, gathering; fullness retains soft fall

Recommended Construction Aids & Techniques

Hand Needles beading, 10–13; betweens, 5–7; milliners, 8–10; sharps, 8–10
Machine Needles round/set point, medium 11
Threads poly-core waxed, 60; cotton-polyester blend, fine; cotton-covered polyester core, extra fine; spun polyester, fine
Hems book; double-fold or single-fold bias binding; bound/Hong Kong finish; edge-stitched; machine-stitched; seam binding
Seams flat-felled; French; false French; plain; safety-stitched; top-stitched; welt; double welt
Seam Finishes single-ply or double-ply bound; Hong Kong bound; edge-stitched; single-ply overcast; pinked and stitched; serging/single-ply overedged; single-ply zigzag
Pressing steam; safe temperature limit 350°F (178.1°C)
Care launder; dry clean
Fabric Resource **Klopman Mills**

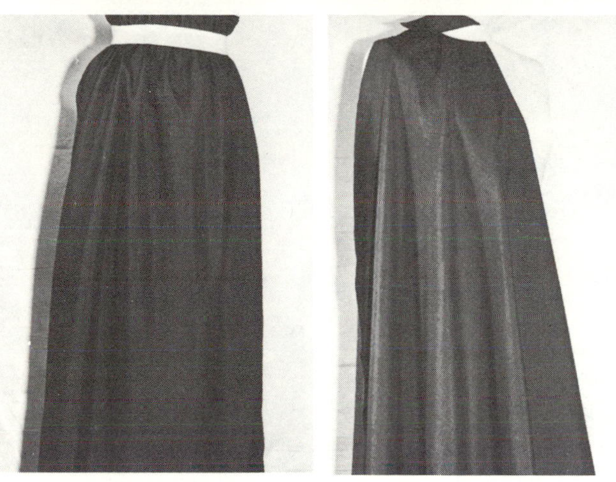

Shantung (crisp finish)

Fiber Content 100% silk
Yarn Type natural filament fiber
Yarn Construction conventional; slubbed doupioni filling yarns
Fabric Structure plain weave
Finishing Processes calendering; stiffening; stretching
Color Application piece dyed; yarn dyed
Width 45 inches (114.3 cm)
Weight medium-light
Hand semi-crisp; firm; thin
Texture thick irregular slub yarn in crosswise direction
Performance Expectations absorbent; dimensional stability; flexible; nonpilling
Drapability Qualities falls into moderately crisp flares; accommodates fullness by pleating, gathering; fullness retains moderately crisp fall

Recommended Construction Aids & Techniques

Hand Needles betweens, 3–4; milliners, 6–8; sharps, 4–8
Machine Needles round/set point, medium 11
Threads mercerized cotton, 50; silk, industrial size A-B; cotton-covered polyester core, all purpose
Hems book; double-fold or single-fold bias binding; bound/Hong Kong finish; edge-stitched; machine-stitched; seam binding
Seams flat-felled; French; false French; plain; safety-stitched; top-stitched; welt; double welt
Seam Finishes single-ply or double-ply bound; Hong Kong bound; edge stitched; single-ply overcast; pinked and stitched; serging/single-ply overedged; single-ply zigzag
Pressing steam; safe temperature limit 300°F (150.1°C)
Care dry clean
Fabric Resource **American Silk Mills Corp.**

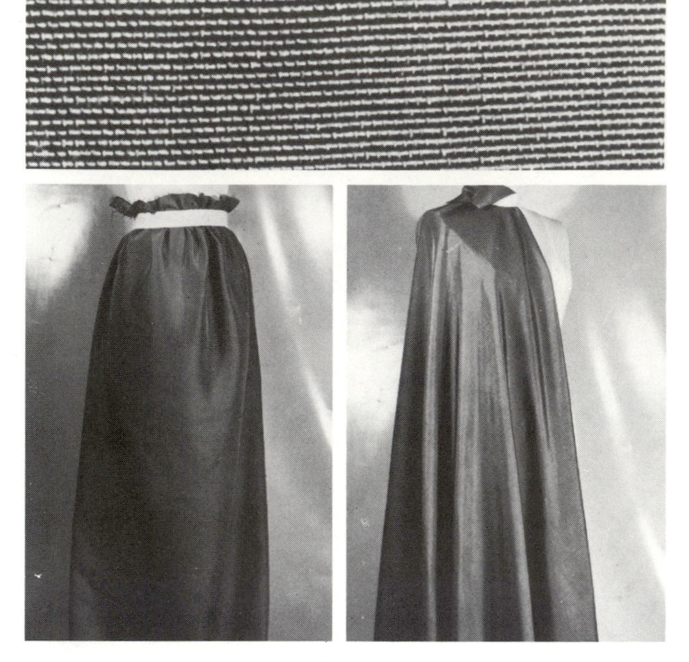

Shantung Faille

Fiber Content 100% silk
Yarn Type natural filament fiber
Yarn Construction conventional; slack twist
Fabric Structure plain weave; multiple filling yarns form rib in crosswise direction
Finishing Processes calendering; singeing; stiffening
Color Application piece dyed; yarn dyed
Width 44-45 inches (111.8-114.3 cm)
Weight medium
Hand crisp; nonelastic; firm; scroopy
Texture fine crosswise rib
Performance Expectations absorbent; dimensional stability; poor elongation; flexible; nonpilling
Drapability Qualities falls into moderately crisp flares; accommodates fullness by gathering; fullness retains crisp effect

Recommended Construction Aids & Techniques

Hand Needles beading, 10-13; betweens, 5-7; milliners, 8-10; sharps, 8-10
Machine Needles round/set point, medium 11
Threads mercerized cotton, 50; silk, industrial size A-B; cotton-covered polyester core, all purpose
Hems book; double-fold or single-fold bias binding; bound/Hong Kong finish; edge-stitched; machine-stitched; seam binding
Seams flat-felled; French; false French; plain; safety-stitched; top-stitched; welt; double welt
Seam Finishes single-ply or double-ply bound; Hong Kong bound; edge-stitched; single-ply overcast; pinked and stitched; serging/single-ply overedged; single-ply zigzag
Pressing steam; safe temperature limit 300°F (150.1°C)
Care dry clean
Fabric Resource **American Silk Mills Corp.**

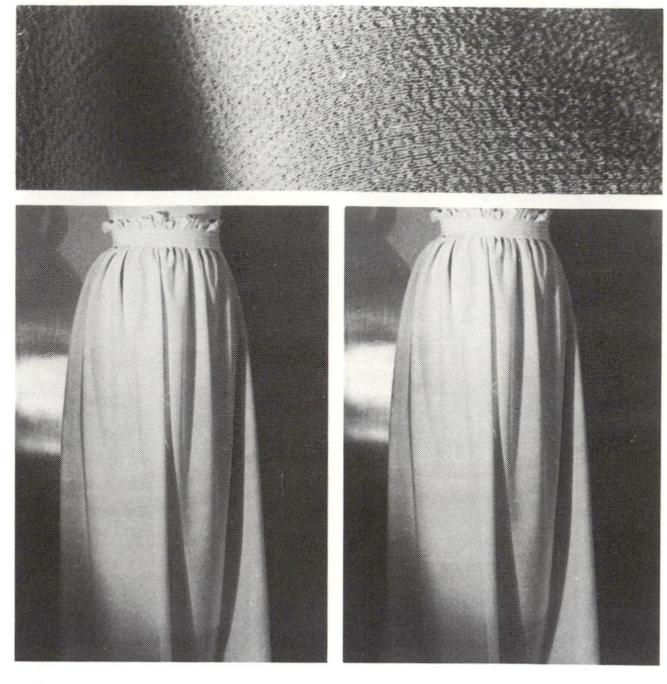

Shantung Georgette

Fiber Content 100% silk
Yarn Type natural filament fiber
Yarn Construction creped; tight twist
Fabric Structure plain weave; alternating left- and right-hand twisted yarns
Finishing Processes singeing; stretching; stiffening
Color Application piece dyed; yarn dyed
Width 47 inches (119.4 cm)
Weight fine to lightweight
Hand dry; harsh; sandy; thin
Texture crepey; irregularly spaced thick slub yarn in crosswise direction; transparent
Performance Expectations absorbent; air permeable; dimensional stability; flexible; nonpilling; subject to snagging
Drapability Qualities falls into soft flares; accommodates fullness by pleating, gathering, elasticized shirring; fullness retains soft graceful fall

Recommended Construction Aids & Techniques

Hand Needles beading, 10-13; betweens, 5-7; milliners, 10; sharps, 8-10
Machine Needles round/set point, fine 9
Threads mercerized cotton, 50; silk, industrial size A; cotton-covered polyester core, extra fine
Hems bound/Hong Kong finish; double fold; edge-stitched; double edge-stitched; horsehair; machine-stitched; hand- or machine-rolled; seam binding
Seams flat-felled; French; mock French; tissue-stitched
Seam Finishes self bound; double-ply bound; double-stitched; double-stitched and overcast; edge-stitched; mock French; double-ply overcast; double-ply zigzag
Pressing steam; safe temperature limit 300°F (150.1°C)
Care dry clean
Fabric Resource **Sormani Co. Inc./Taroni**

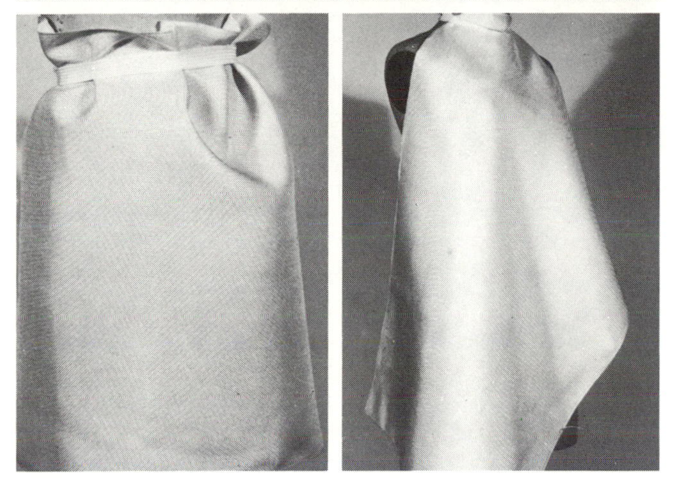

Silk Sharkskin

Fiber Content 100% silk
Yarn Type natural filament fiber
Yarn Construction conventional; slack twist
Fabric Structure rib weave variation; multiple yarns alternately woven with single yarn
Finishing Processes calendering; singeing; stiffening
Color Application piece dyed; yarn dyed
Width 36 inches (91.4 cm)
Weight medium-heavy
Hand coarse; nonelastic; rigid
Texture boardy; harsh; rough
Performance Expectations absorbent; dimensional stability; poor elongation; flexible; nonpilling; subject to abrasion and snagging
Drapability Qualities falls into crisp wide cones; fullness maintains crisp effect; retains shape or silhouette of garment

Recommended Construction Aids & Techniques

Hand Needles cotton darners, 9–10; milliners, 3–6; sharps, 1–4
Machine Needles round/set point, medium 14
Threads mercerized cotton, 50; silk, industrial size A-B; cotton-covered polyester core, all purpose
Hems book; double-fold or single-fold bias binding; bound/Hong Kong finish; edge-stitched; faced; overedged; stitched and pinked; seam binding
Seams plain; safety-stitched; top-stitched; welt; double welt
Seam Finishes single-ply bound; Hong Kong bound; single-ply overcast; pinked; pinked and stitched; serging/single-ply overedged; single-ply zigzag
Pressing steam; safe temperature limit 300°F (150.1°C)
Care dry clean
Fabric Resource **Jack Larsen Incorporated**

Silk Suiting

Fiber Content 40% silk/60% wool
Yarn Type natural filament fiber; worsted wool staple
Yarn Construction conventional
Fabric Structure plain weave; fine warp/heavier filling
Finishing Processes calendering; singeing; stabilizing
Color Application piece dyed–cross dyed; yarn dyed
Width 54-55 inches (137.2–139.7 cm)
Weight medium-heavy
Hand compact; rigid; sandy
Texture harsh; rough
Performance Expectations absorbent; flexible; resilient; wrinkle resistant
Drapability Qualities falls into firm wide cones; fullness maintains crisp bouffant effect; retains shape or silhouette of garment

Recommended Construction Aids & Techniques

Hand Needles cotton darners, 9-10; milliners, 3-6; sharps, 1-4
Machine Needles round/set point, medium 14
Threads mercerized cotton, 40; silk, industrial size A-B-C; cotton-covered polyester core, heavy duty
Hems book; double-fold or single-fold bias binding; bound/Hong Kong finish; edge-stitched; faced; overedged; stitched and pinked; seam binding
Seams plain; safety-stitched; top-stitched; welt; double welt
Seam Finishes single-ply bound; Hong Kong bound; single-ply overcast; pinked; pinked and stitched; serging/single-ply overedged; single-ply zigzag
Pressing steam; safe temperature limit 300°F (150.1°C)
Care dry clean
Fabric Resource **Lafitte Inc.**

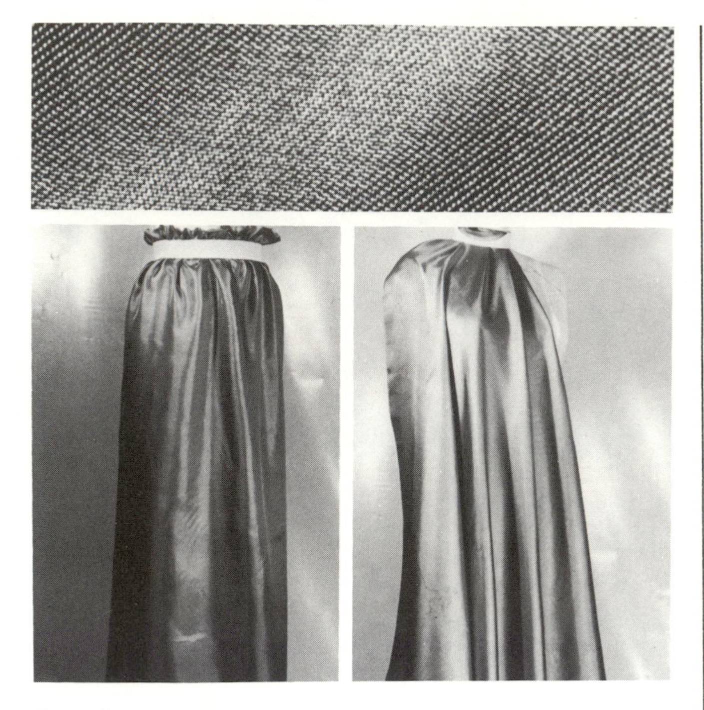

Surah

Fiber Content 100% silk
Yarn Type natural filament fiber
Yarn Construction conventional; fine; slack twist
Fabric Structure twill weave
Finishing Processes ciré calendering; degumming; singeing; stretching
Color Application piece dyed; yarn dyed
Width 44–45 inches (111.8–114.3 cm)
Weight light
Hand fine; soft; supple
Texture flat; slippery; smooth
Performance Expectations absorbent; dimensional stability; poor elongation; flexible; nonpilling
Drapability Qualities falls into soft flares; accommodates fullness by pleating, gathering, elasticized shirring; fullness retains soft fall

Recommended Construction Aids & Techniques

Hand Needles beading, 10–13; betweens, 5–7; milliners, 8–10; sharps, 8–10
Machine Needles round/set point, medium 11
Threads mercerized cotton, 50; silk, industrial size A; cotton-covered polyester core, extra fine
Hems bound/Hong Kong finish; double fold; edge-stitched; double edge-stitched; horsehair; machine-stitched; hand- or machine-rolled; seam binding
Seams flat-felled; French; false French; mock French; plain; tissue-stitched
Seam Finishes self bound; double-ply bound; double-stitched; double-stitched and overcast; edge-stitched; mock French; double-ply overcast; double-ply zigzag
Pressing steam; safe temperature limit 300°F (150.1°C)
Care dry clean
Fabric Resource American Silk Mills Corp.

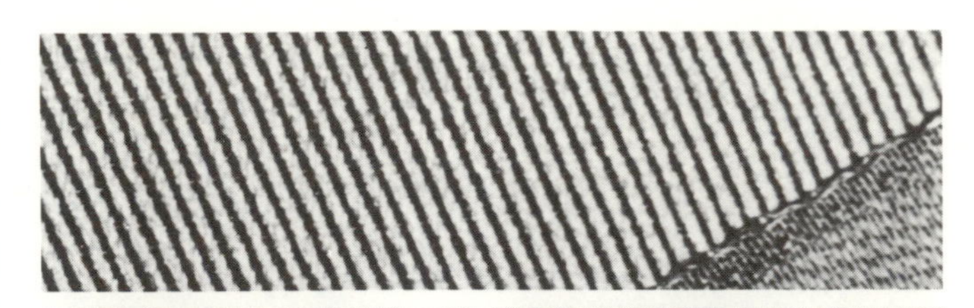

Surah Suiting

Fiber Content 100% silk
Yarn Type natural filament fiber
Yarn Construction conventional; slack twist
Fabric Structure twill weave
Finishing Processes calendering; singeing; stretching
Color Application piece dyed; yarn dyed
Width 48–49 inches (121.9–124.5 cm)
Weight medium
Hand nonelastic; firm; soft
Texture flat; slippery; smooth
Performance Expectations subject to abrasion; absorbent; dimensional stability; poor elongation; flexible; nonpilling
Drapability Qualities falls into soft wide cones; fullness maintains lofty effect; retains shape of garment

Recommended Construction Aids & Techniques

Hand Needles betweens, 3–4; milliners, 6–8; sharps, 4–8
Machine Needles round/set point, medium 14
Threads mercerized cotton, 50; silk, industrial size A-B; cotton-covered polyester core, all purpose
Hems book; double-fold or single-fold bias binding; bound/Hong Kong finish; edge-stitched; machine-stitched; seam binding
Seams flat-felled; French; false French; plain; safety-stitched; top-stitched; welt; double welt
Seam Finishes single-ply or double-ply bound; Hong Kong bound; edge-stitched; single-ply overcast; pinked and stitched; serging/single-ply overedged; single-ply zigzag
Pressing steam; safe temperature limit 300°F (150.1°C)
Care dry clean
Fabric Resource American Silk Mills Corp.

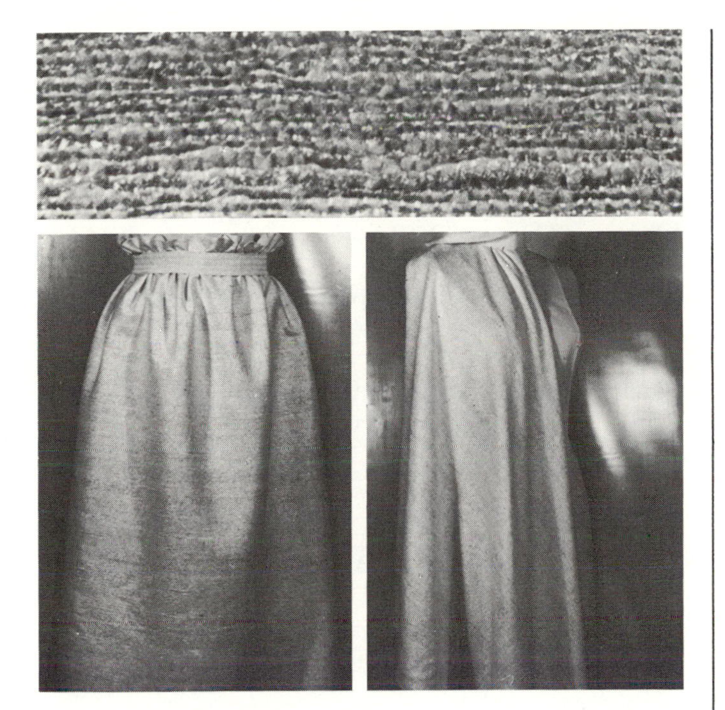

Tussah/Wild Silk

Fiber Content 100% wild silk
Yarn Type natural filament fiber
Yarn Construction conventional; low twist; heavy
Fabric Structure plain weave
Finishing Processes bleaching; degumming; stiffening
Color Application bleached to natural color
Width 50 inches (127 cm)
Weight medium-heavy
Hand coarse; springy; thick
Texture rough; irregular yarns create uneven surface; opaque
Performance Expectations subject to abrasion; absorbent; dimensional stability; flexible; subject to snagging and seam slippage due to weave
Drapability Qualities falls into wide cones; fullness maintains lofty effect; retains shape of garment

Recommended Construction Aids & Techniques

Hand Needles cotton darners, 9–10; milliners, 3–6; sharps, 1–4
Machine Needles round/set point, medium 14
Threads mercerized cotton, 40; silk, industrial size A-B-C; cotton-covered polyester core, heavy duty
Hems book; double-fold or single-fold bias binding; bound/Hong Kong finish; edge-stitched; faced; overedged; stitched and pinked; seam binding
Seams plain; safety-stitched; top-stitched; welt; double welt
Seam Finishes single-ply bound; Hong Kong bound; single-ply overcast; pinked; pinked and stitched; serging/single-ply overedged; single-ply zigzag
Pressing steam; safe temperature limit 300°F (150.1°C)
Care dry clean
Fabric Resource **Jack Larsen Incorporated**

5 – Crepe/Crepe-type Fabrics

Albatross Cloth
- Albatross Crepe

Bark Cloth
- Bark Crepe

Canton Crepe

Chiffon
- Crepe Chiffon

Crepe-backed Satin / Satin-backed Crepe

Crepe de Chine
- Crepe de Chine
- Crepe de Chine Dobby

Crepon / Crinkle Crepe / Plissé Crepe

Faille
- Crepe Faille

Georgette
- Crepe Georgette

Flat Crepe

Lingerie / "French" Crepe
- Lingerie / "French" Crepe (lightweight)
- Lingerie / "French" Crepe (heavy texture)

Matelassé
- Crepe Matelassé

Meteor Crepe / Satin-faced Chiffon

Pebbly / Mossy Crepe / Sand Crepe®

Crepe is a term applied to fabrics characterized by a pebbly or crinkly surface texture. Natural, man-made or combination fiber blends may be used to produce crepe fabrics. Crepe or crepe-type fabrics vary from:

- Smooth to rough hand
- Fine to pronounced surface texture
- Sheer to opaque density
- Light to heavy weight

Crepe or crepe-type fabrics may be constructed in plain, twill, satin or jacquard weave structures. Crepe and crepe-type fabrics are achieved by:

- Twisting yarns
- Plying or texturizing yarns
- Using crepe yarn in warp, filling or both directions
- Weaving
- Chemical or mechanical finishing action
- Embossing

True crepe fabric is made with tightly twisted crepe yarns in the warp, filling or in both directions.

Crepe-effect fabric is produced by chemical or mechanical finishes to provide the characteristic pebbly or crinkly surface.

Albatross Crepe

Fiber Content 100% polyester
Yarn Type filament fiber
Yarn Construction creped; textured
Fabric Structure plain weave
Finishing Processes creping; softening; stabilizing
Color Application piece dyed; solution dyed; yarn dyed
Width 44–45 inches (111.8–114.3 cm)
Weight fine to lightweight
Hand delicate; harsh; supple
Texture openweave; pebbly; scratchy; semi-sheer
Performance Expectations dimensional stability; durable; elongation; heat-set properties; resilient; high tensile strength
Drapability Qualities falls into soft flares; accommodates fullness by pleating, gathering, elasticized shirring; fullness retains soft graceful fall

Recommended Construction Aids & Techniques

Hand Needles beading, 10–13; betweens, 5–7; milliners, 10; sharps, 8–10
Machine Needles round/set point, fine 9
Threads poly-core waxed, 60; cotton-polyester blend, fine; cotton-covered polyester core, extra fine; spun polyester, fine
Hems double fold; edge-stitched only; double edge-stitched; horsehair; machine-stitched; hand- or machine-rolled; wired
Seams French; mock French; hairline; plain; tissue-stitched
Seam Finishes self bound; double-stitched; double-stitched and trimmed; mock French
Pressing steam; safe temperature limit 325°F (164.1°C)
Care launder; dry clean
Fabric Resource **Bloomsburg Mills**

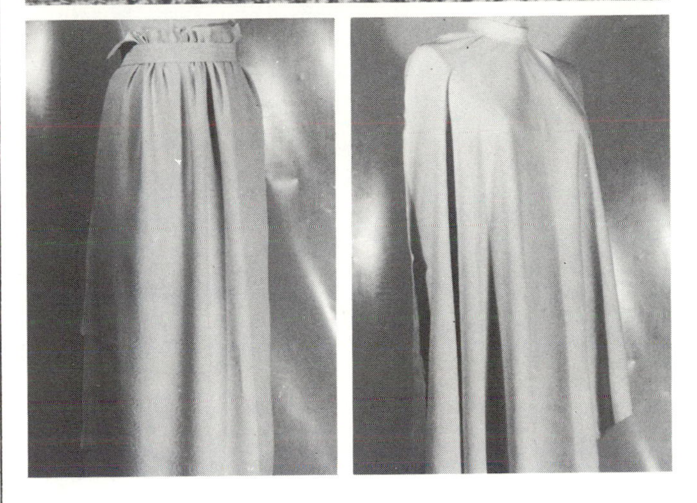

Bark Crepe

Fiber Content 100% acetate
Yarn Type filament fiber
Yarn Construction creped; textured
Fabric Structure plain weave
Finishing Processes heat-set texture; stabilizing
Color Application piece dyed; solution dyed; yarn dyed
Width 32 inches (81.3 cm)
Weight medium-heavy
Hand harsh; thick
Texture high and low wavy pattern; finish creates rough "tree bark" appearance and uneven surface
Performance Expectations subject to abrasion; dimensional stability; elongation; flexible; heat-set properties; nonpilling; subject to snagging due to weave
Drapability Qualities falls into crisp cones; fullness maintains lofty effect; retains shape of garment

Recommended Construction Aids & Techniques

Hand Needles betweens, 3–4; milliners, 6–8; sharps, 4–8
Machine Needles round/set point, medium 11–14
Threads poly-core waxed, 50; cotton-polyester blend, all purpose; cotton-covered polyester core, all purpose; spun polyester, all purpose
Hems book; double-fold or single-fold bias binding; bound/Hong Kong finish; edge-stitched; faced; overedged; pinked and stitched; seam binding
Seams plain; safety-stitched; top-stitched; welt; double welt
Seam Finishes single-ply bound; Hong Kong bound; single-ply overcast; pinked; pinked and stitched; serging/single-ply overedged; single-ply zigzag
Pressing steam; safe temperature limit 250°F (122.1°C)
Care dry clean
Fabric Resource **Lawrence Textile Co. Inc.**

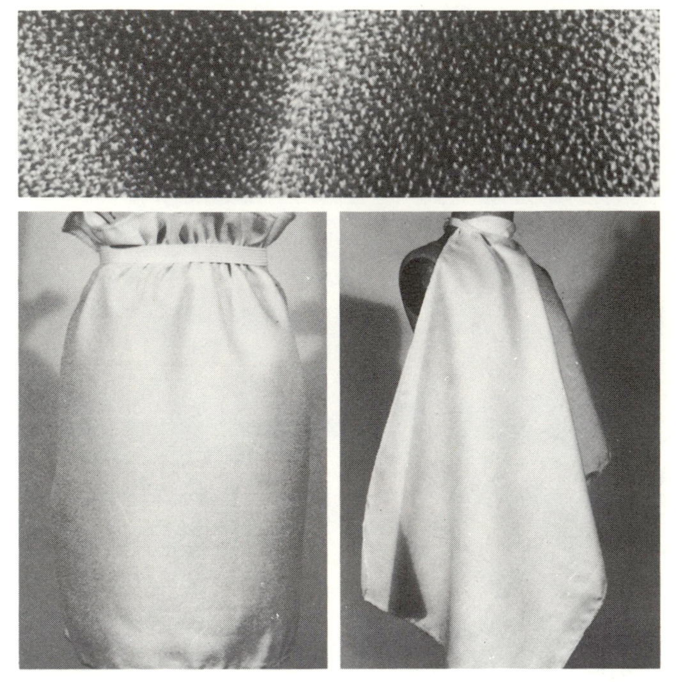

Canton Crepe

Fiber Content 100% silk
Yarn Type natural filament fiber
Yarn Construction creped; fine warp yarns/heavier filling yarns
Fabric Structure plain weave; alternating Z- and S-twisted yarns for warp and filling
Finishing Processes bleaching; calendering; stabilizing
Color Application piece dyed; yarn dyed
Width 36 inches (91.4 cm)
Weight medium-heavy
Hand pliable; soft; stretches in crosswise direction
Texture crinkly; pebbly; fine crosswise rib
Performance Expectations absorbent; dimensional stability; flexible; nonpilling
Drapability Qualities falls into wide cones; accommodates fullness by gathering; fullness retains soft fall

Recommended Construction Aids & Techniques

Hand Needles betweens, 3–4; milliners, 6–8; sharps, 4–8
Machine Needles round/set point, medium 11
Threads mercerized cotton, 50; silk, industrial size A-B; cotton-covered polyester core, all purpose
Hems book; double-fold or single-fold bias binding; bound/Hong Kong finish; edge-stitched; faced; overedged; stitched and pinked; seam binding
Seams flat-felled; French; false French; plain; safety-stitched; top-stitched; welt; double welt
Seam Finishes single-ply or double-ply bound; Hong Kong bound; edge-stitched; single-ply overcast; pinked and stitched; serging/single-ply overedged; single-ply zigzag
Pressing steam; safe temperature limit 300°F (150.1°C)
Care dry clean
Fabric Resource Jack Larsen Incorporated

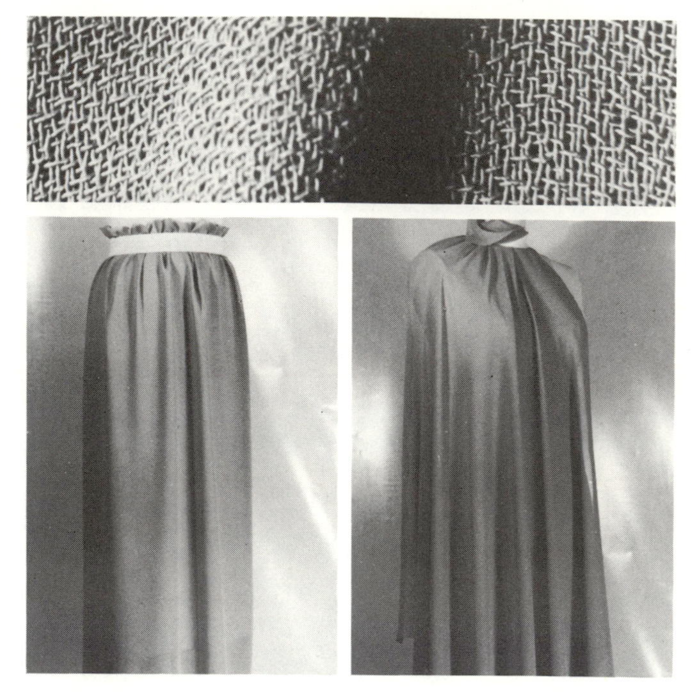

Crepe Chiffon

Fiber Content 100% polyester
Yarn Type filament fiber
Yarn Construction creped; fine; textured
Fabric Structure plain weave
Finishing Processes creping; softening; stabilizing
Color Application piece dyed; solution dyed; yarn dyed
Width 44–45 inches (111.8–114.3 cm)
Weight light
Hand semi-soft; supple; thin
Texture slightly sandy; semi-sheer; softer than georgette
Performance Expectations dimensional stability; durable; elongation; heat-set properties; resilient; high tensile strength
Drapability Qualities falls into soft flares and ripples; accommodates fullness by gathering, elasticized shirring; fullness retains soft graceful fall

Recommended Construction Aids & Techniques

Hand Needles beading, 10–13; betweens, 5–7; milliners, 8–10; sharps, 8–10
Machine Needles round/set point, fine 9
Threads poly-core waxed, 60; cotton-polyester blend, fine; cotton-covered polyester core, extra fine; spun polyester, fine
Hems double fold; edge-stitched only; double edge-stitched; horsehair; machine-stitched; hand- or machine-rolled; wired
Seams French; mock French; hairline; plain; tissue-stitched
Seam Finishes self bound; double-ply bound; double-stitched; double-stitched and overcast; edge-stitched; mock French; double-ply overcast; double-ply zigzag
Pressing steam; safe temperature limit 325°F (164.1°C)
Care launder; dry clean
Fabric Resource Private Collections Fabrics Ltd.

Crepe-backed Satin/Satin-backed Crepe

Fiber Content 87% acetate/13% nylon
Yarn Type filament fiber
Yarn Construction conventional; hard twist for crepe face; low twist for satin face
Fabric Structure satin weave; crepe-twisted filling yarns
Finishing Processes calendering; stabilizing; stretching
Color Application piece dyed–cross dyed; yarn dyed
Width 44–45 inches (111.8–114.3 cm)
Weight medium-heavy
Hand pliable; soft
Texture reversible; smooth face; pebbly back
Performance Expectations subject to abrasion; dimensional stability; elongation; flexible; heat-set properties; subject to snagging due to weave
Drapability Qualities falls into soft flares; accommodates fullness by pleating, gathering, elasticized shirring; fullness retains soft graceful fall

Recommended Construction Aids & Techniques

Hand Needles beading, 10–13; betweens, 5–7; milliners, 8–10; sharps, 8–10
Machine Needles round/set point, medium-fine 9–11
Threads poly-core waxed, 60; cotton-polyester blend, fine; spun polyester, fine
Hems bound/Hong Kong finish; double fold; edge-stitched; double edge-stitched; horsehair; machine-stitched; hand- or machine-rolled; seam binding
Seams flat-felled; French; false French; plain; safety-stitched; top-stitched; welt; double welt
Seam Finishes single-ply or double-ply bound; Hong Kong bound; edge-stitched; single-ply overcast; pinked and stitched; serging/single-ply overedged; single-ply zigzag
Pressing steam; safe temperature limit 250°F (122.1°C)
Care dry clean
Fabric Resources **Hargro Fabrics Inc.; Private Collections Fabrics Ltd.; Springs Mills Inc.**

Crepe de Chine

Fiber Content 100% acetate; may be of Qiana® nylon or Fortrel® polyester
Yarn Type filament fiber
Yarn Construction fine; textured
Fabric Structure plain weave; alternating Z- and S-twisted yarns for warp and filling
Finishing Processes calendering; stretching
Color Application piece dyed–cross dyed; yarn dyed
Width 46–47 inches (116.8–119.4 cm)
Weight light to medium-light
Hand fine; firm; soft
Texture crepey; slippery
Performance Expectations dimensional stability; elongation; flexible; heat-set properties; mildew and moth resistant; subject to yarn slippage
Drapability Qualities falls into soft flares and ripples; accommodates fullness by pleating, shirring; elasticized shirring; fullness retains soft graceful fall

Recommended Construction Aids & Techniques

Hand Needles beading, 10–13; betweens, 5–7; milliners, 8–10; sharps, 8–10
Machine Needles round/set point, medium-fine 9–11
Threads poly-core waxed, 60; cotton-polyester blend, fine; cotton-covered polyester core, extra fine; spun polyester, fine
Hems bound/Hong Kong finish; double fold; edge-stitched; double edge-stitched; horsehair; machine-stitched; hand- or machine-rolled; seam binding
Seams flat-felled; French; mock French; tissue-stitched
Seam Finishes self bound; double-ply bound; double-stitched; double-stitched and overcast; edge-stitched; mock French; double-ply overcast; double-ply zigzag
Pressing steam; safe temperature limit 250°F (122.1°C)
Care launder; dry clean
Fabric Resources **Hargro Fabrics Inc.; Klopman Mills; Loom Tex Corp.; Private Collections Fabrics Ltd.**

Crepe de Chine Dobby

Fiber Content 65% triacetate/35% nylon
Yarn Type filament fiber
Yarn Construction textured
Fabric Structure dobby weave
Finishing Processes heat-set stabilizing; stretching
Color Application piece dyed; solution dyed; yarn dyed
Width 44–45 inches (111.8–114.3 cm)
Weight light
Hand fine; firm; soft
Texture slightly crepey
Performance Expectations dimensional stability; durable; elongation; heat-set properties; high tensile strength
Drapability Qualities falls into soft flares and ripples; accommodates fullness by pleating, shirring, elasticized shirring; fullness retains soft graceful fall

Recommended Construction Aids & Techniques

Hand Needles beading, 10–13; betweens, 5–7; milliners, 8–10; sharps, 8–10
Machine Needles round/set point, medium-fine 9–11
Threads poly-core waxed, 60; cotton-polyester blend, fine; cotton-covered polyester core, extra fine; spun polyester, fine
Hems bound/Hong Kong finish; double fold; edge-stitched; horsehair; machine-stitched; hand- or machine-rolled; seam binding
Seams flat-felled; French; mock French; tissue-stitched
Seam Finishes self bound; double-ply bound; double-stitched; double-stitched and overcast; edge-stitched; mock French; double-ply overcast; double-ply zigzag
Pressing steam; safe temperature limit 350°F (178.1°C)
Care launder; dry clean
Fabric Resource Earl Glo/Erlanger Blumgart and Co. Inc.

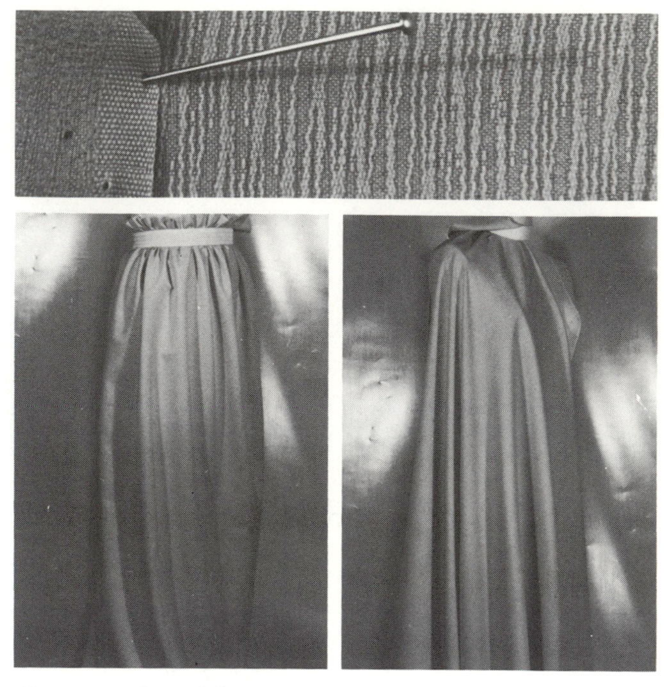

Crepon/Crinkle Crepe/Plissé Crepe

Fiber Content 100% polyester
Yarn Type filament fiber
Yarn Construction conventional; creped; textured; high and low twist
Fabric Structure plain weave; alternating conventional and crepe yarns
Finishing Processes caustic bath and heat-set crinkle; stabilizing
Color Application piece dyed; solution dyed
Width 45 inches (114.3 cm)
Weight light
Hand pliable; soft; thin
Texture crinkled or blistered stripe effect in lengthwise direction
Performance Expectations dimensional stability; durable; elongation; heat-set properties; resilient; high tensile strength
Drapability Qualities falls into moderate flares; accommodates fullness by gathering, elasticized shirring; fullness retains soft fall

Recommended Construction Aids & Techniques

Hand Needles beading, 10–13; betweens, 5–7; milliners, 8–10; sharps, 8–10
Machine Needles round/set point, medium-fine 9–11
Threads poly-core waxed, 60; cotton-polyester blend, fine; cotton-covered polyester core, extra fine; spun polyester, fine
Hems book; double-fold or single-fold bias binding; bound/Hong Kong finish; edge-stitched; machine-stitched; seam binding
Seams flat-felled; French; false French; safety-stitched; top-stitched; welt; double welt
Seam Finishes single-ply or double-ply bound; Hong Kong bound; edge-stitched; single-ply overcast; pinked and stitched; serging/single-ply overedged; single-ply zigzag
Pressing steam; safe temperature limit 325°F (164.1°C)
Care launder; dry clean
Fabric Resources Kabat Textile Corp.; Private Collections Fabrics Ltd.; Springs Mills Inc.

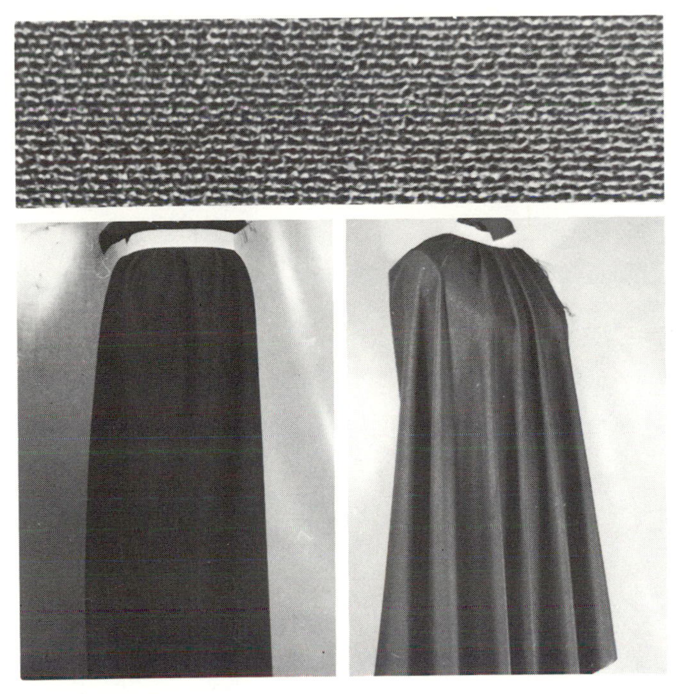

Crepe Faille

Fiber Content 100% silk
Yarn Type natural filament fiber
Yarn Construction conventional; creped
Fabric Structure plain weave; additional filling yarns form rib in crosswise direction
Finishing Processes stabilizing; stiffening; stretching
Color Application piece dyed; yarn dyed
Width 43-44 inches (109.2-111.8 cm)
Weight medium
Hand compact; crisp; nonelastic; harsh
Texture dull appearance; fine but pronounced crosswise rib
Performance Expectations subject to abrasion; absorbent; dimensional stability; poor elongation; flexible; nonpilling; subject to yarn and seam slippage
Drapability Qualities falls into crisp wide cones; fullness retains crisp fall; retains shape of garment

Recommended Construction Aids & Techniques

Hand Needles betweens, 3-4; milliners, 6-8; sharps, 4-8
Machine Needles round/set point, medium 11-14
Threads mercerized cotton, 50; silk, industrial size A-B; cotton-covered polyester core, all purpose
Hems book; double-fold or single-fold bias binding; bound/Hong Kong finish; edge-stitched; faced; overedged; stitched and pinked; seam binding
Seams plain; safety-stitched; top-stitched; welt; double welt
Seam Finishes single-ply bound; Hong Kong bound; single-ply overcast; pinked; pinked and stitched; serging/single-ply overedged; single-ply zigzag
Pressing steam; safe temperature limit 300°F (150.1°C)
Care dry clean
Fabric Resource American Silk Mills Corp.

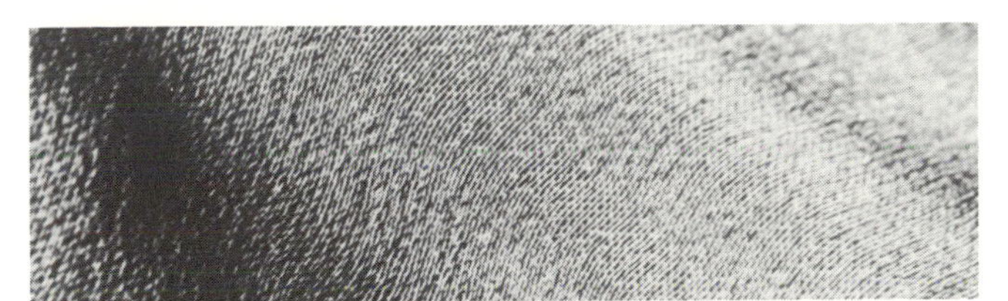

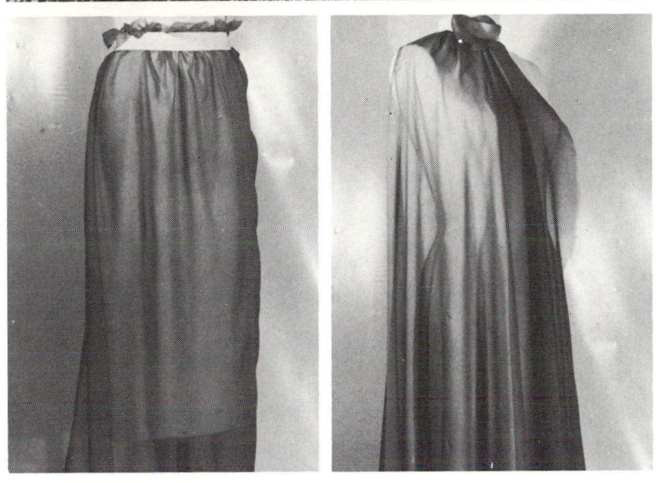

Crepe Georgette

Fiber Content 100% silk
Yarn Type natural filament fiber
Yarn Construction creped; tight twist
Fabric Structure plain weave; alternating left- and right-hand twisted yarns
Finishing Processes singeing; stiffening; stretching
Color Application piece dyed; yarn dyed
Width 47 inches (119.4 cm)
Weight fine to lightweight
Hand dry; harsh; thin
Texture crepey; sandy; transparent
Performance Expectations absorbent; dimensional stability; flexible; nonpilling
Drapability Qualities falls into soft flares; accommodates fullness by pleating, gathering, elasticized shirring; fullness retains soft fall

Recommended Construction Aids & Techniques

Hand Needles beading, 10-13; betweens, 5-7; milliners, 8-10; sharps, 8-10
Machine Needles round/set point, fine 9
Threads mercerized cotton, 50; silk, industrial size A; cotton-covered polyester core, extra fine
Hems bound/Hong Kong finish; double fold; edge-stitched; edge-stitched only; double edge-stitched; horsehair; machine-stitched; hand- or machine-rolled; seam binding
Seams flat-felled; French; mock French; plain; tissue-stitched
Seam Finishes self bound; double-ply bound; double-stitched; double-stitched and overcast; edge-stitched; mock French; double-ply overcast; double-ply zigzag
Pressing steam; safe temperature limit 300°F (150.1°C)
Care dry clean
Fabric Resources American Silk Mills Corp.; Kabat Textile Corp.; Sormani Co. Inc.

Flat Crepe

Fiber Content 100% trilobal polyester
Yarn Type filament fiber
Yarn Construction conventional; hard twist
Fabric Structure plain weave; alternating Z- and S-twist filling yarns
Finishing Processes calendering; softening; stabilizing
Color Application piece dyed; solution dyed; yarn dyed
Width 44-45 inches (111.8-114.3 cm)
Weight medium-light
Hand pliable; soft; springy
Texture flat surface; slippery
Performance Expectations dimensional stability; durable; elongation; heat-set properties; resilient; high tensile strength
Drapability Qualities falls into soft flares; accommodates fullness by pleating, gathering, elasticized shirring; fullness retains soft graceful fall

Recommended Construction Aids & Techniques

Hand Needles beading, 10-13; betweens, 5-7; milliners, 8-10; sharps, 8-10
Machine Needles round/set point, medium-fine 9-11
Threads mercerized cotton, 50; silk, industrial size A-B; cotton-covered polyester core, all purpose
Hems bound/Hong Kong finish; double fold; edge-stitched; horsehair; machine-stitched; hand- or machine-rolled; seam binding
Seams flat-felled; French; false French; mock French; safety-stitched; top-stitched; welt; double welt
Seam Finishes self bound; double-ply bound; double-stitched; double-stitched and overcast; mock French; double-ply zigzag
Pressing steam; safe temperature limit 375°F (192.1°C)
Care launder; dry clean
Fabric Resource Springs Mills Inc.

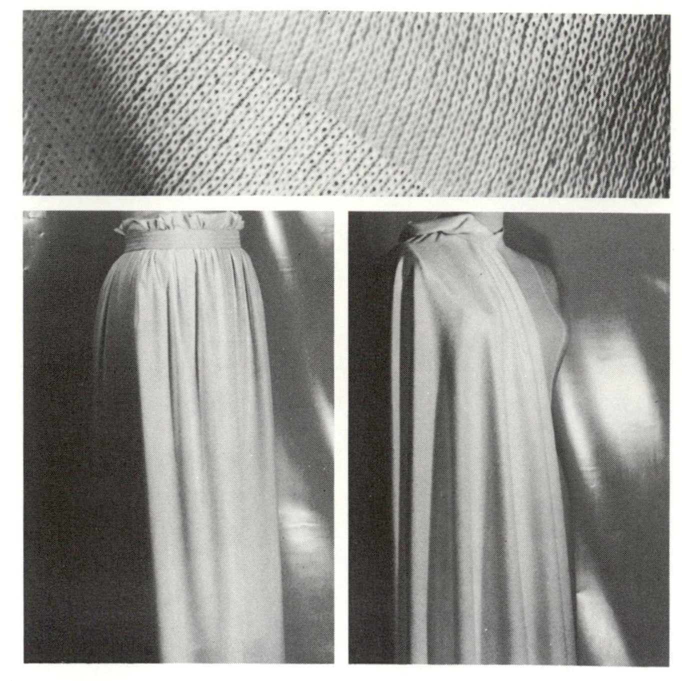

Lingerie/"French" Crepe (lightweight)

Fiber Content 100% nylon
Yarn Type filament fiber
Yarn Construction textured
Fabric Structure tricot knit
Finishing Processes antistatic; crepe-set
Color Application piece dyed; solution dyed; yarn dyed
Width 54 inches (137.2 cm)
Weight light
Hand soft; supple
Texture crepey; pebbly
Performance Expectations abrasion resistant; antistatic; nonclinging; dimensional stability; durable; elongation; heat-set properties; mildew and moth resistant; high tensile strength
Drapability Qualities falls into soft flares and ripples; accommodates fullness by gathering, elasticized shirring; fullness retains soft fall

Recommended Construction Aids & Techniques

Hand Needles beading, 10-13; betweens, 5-7; milliners, 8-10; sharps, 8-10
Machine Needles ball point, 9
Threads poly-core waxed, 60; cotton-polyester blend, fine; cotton-covered polyester core, extra fine; spun polyester, fine
Hems bound/Hong Kong finish; double fold; edge-stitched; horsehair; machine-stitched; hand- or machine-rolled; seam binding
Seams flat-felled; French; false French; mock French; overedged; safety-stitched; top-stitched; welt; double welt
Seam Finishes self bound; single-ply or double-ply bound; double-stitched; double-stitched and overcast; edge-stitched; mock French; double-ply overcast; serging/single-ply overedged; double-ply zigzag
Pressing steam; safe temperature limit 350°F (178.1°C)
Care launder
Fabric Resource Native Textiles

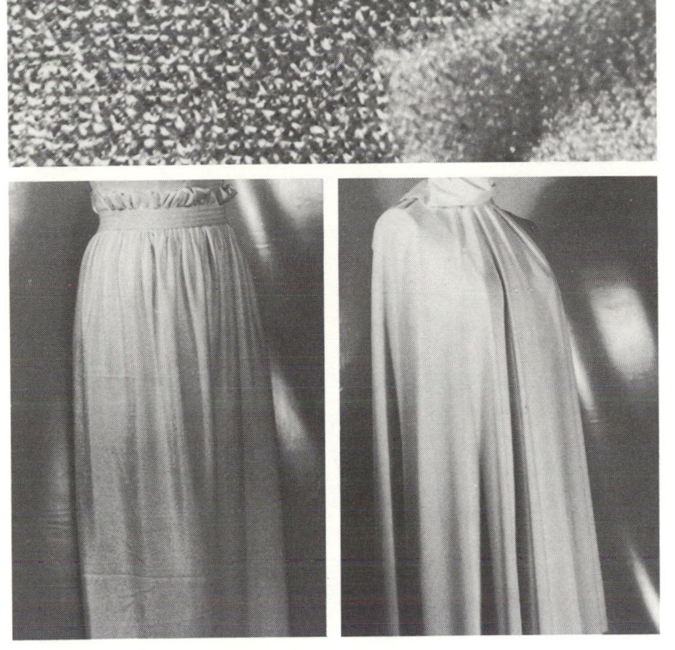

Lingerie/"French" Crepe (heavy texture)

Fiber Content 100% nylon
Yarn Type filament fiber
Yarn Construction textured
Fabric Structure tricot knit
Finishing Processes antistatic; crepe-set
Color Application piece dyed; solution dyed; yarn dyed
Width 54 inches (137.2 cm)
Weight medium
Hand pliable; soft
Texture crepey; pebbly
Performance Expectations abrasion resistant; antistatic; nonclinging; dimensional stability; durable; elongation; heat-set properties; mildew and moth resistant; high tensile strength
Drapability Qualities falls into moderate flares; accommodates fullness by gathering, elasticized shirring; fullness retains moderately soft fall

Recommended Construction Aids & Techniques

Hand Needles beading, 10–13; betweens, 5–7; milliners, 8–10; sharps, 8–10
Machine Needles ball point, 11
Threads poly-core waxed, 50; cotton-polyester blend, all purpose; cotton-covered polyester core, all purpose; spun polyester, all purpose
Hems book; double-fold or single-fold bias binding; bound/Hong Kong finish; double-stitched; stitched and turned flat; machine-stitched; seam binding
Seams flat-felled; French; false French; safety-stitched; top-stitched; welt; double welt
Seam Finishes single-ply or double-ply bound; Hong Kong bound; edge-stitched; single-ply overcast; overedged; pinked and stitched; serging/single-ply overedged; single-ply zigzag
Pressing steam; safe temperature limit 350°F (178.1°C)
Care launder
Fabric Resource **Native Textiles**

Crepe Matelassé

Fiber Content 40% silk/60% wool
Yarn Type natural filament fiber; worsted wool staple
Yarn Construction conventional; fine
Fabric Structure double cloth weave
Finishing Processes calendering; singeing; softening; stabilizing
Color Application piece dyed–cross dyed; yarn dyed
Width 44–45 inches (111.8–114.3 cm)
Weight light
Hand soft; spongy; thin
Texture raised blistered effect
Performance Expectations subject to abrasion; absorbent; flexible; subject to snagging due to weave
Drapability Qualities falls into wide cones; fullness maintains lofty effect; retains shape of garment

Recommended Construction Aids & Techniques

Hand Needles betweens, 3–4; milliners, 6–8; sharps, 4–8
Machine Needles round/set point, medium 11–14
Threads mercerized cotton, 40; silk, industrial size A-B-C; cotton-covered polyester core, heavy duty
Hems book; double-fold or single-fold bias binding; bound/Hong Kong finish; stitched and turned flat; faced; overedged; stitched and pinked; seam binding
Seams plain; safety-stitched; top-stitched; welt; double welt
Seam Finishes single-ply bound; Hong Kong bound; single-ply overcast; pinked; pinked and stitched; serging/single-ply overedged; single-ply zigzag
Pressing steam; safe temperature limit 300°F (150.1°C)
Care dry clean
Fabric Resource **American Silk Mills Corp./Bucol**

Meteor Crepe/Satin-faced Chiffon

Fiber Content 42% silk/58% polyester
Yarn Type natural and man-made filament fibers
Yarn Construction fine; textured
Fabric Structure double cloth weave; satin face/twill back
Finishing Processes calendering; stabilizing; stretching
Color Application piece dyed–cross dyed; yarn dyed
Width 44–48 inches (111.8–121.9 cm)
Weight light
Hand fine; soft; supple
Texture reversible; soft face; crepey back
Performance Expectations subject to abrasion; dimensional stability; flexible; subject to snagging due to weave
Drapability Qualities falls into soft flares; flows and is fluid; fullness retains soft graceful fall

Recommended Construction Aids & Techniques

Hand Needles beading, 10–13; betweens, 5–7; milliners, 8–10; sharps, 8–10
Machine Needles round/set point, medium-fine 9–11
Threads mercerized cotton, 50; silk, industrial size A-B; cotton-covered polyester core, all purpose
Hems bound/ Hong Kong finish; double fold; edge-stitched; horsehair; machine-stitched; hand- or machine-rolled; seam binding
Seams flat-felled; French; false French; mock French; safety-stitched; top-stitched; welt; double welt
Seam Finishes single-ply or double-ply bound; Hong Kong bound; edge-stitched; single-ply overcast; pinked and stitched; serging/single-ply overedged; single-ply zigzag
Pressing steam; safe temperature limit 300°F (150.1°C)
Care dry clean
Fabric Resource **American Silk Mills Corp./Bucol**

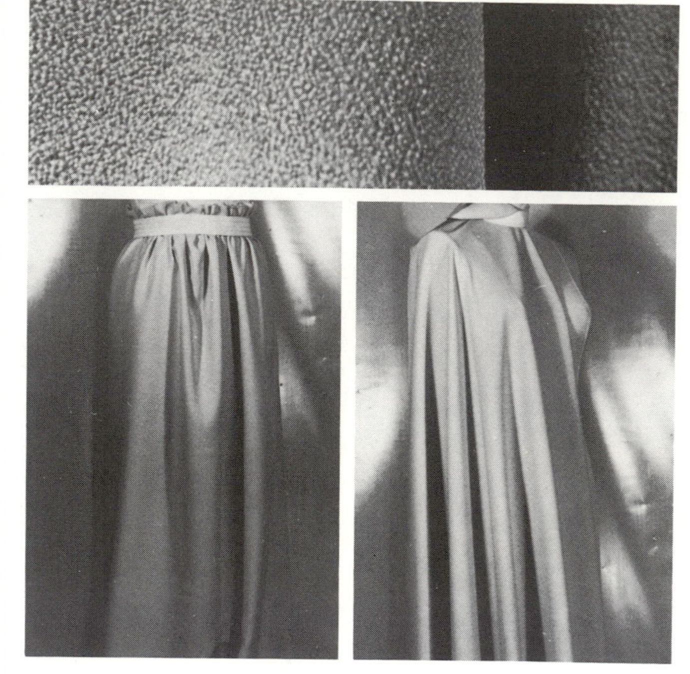

Pebbly/Mossy Crepe/Sand Crepe®

Fiber Content 100% polyester
Yarn Type filament fiber
Yarn Construction creped; ply; textured
Fabric Structure plain weave; alternating one twisted and one creped yarn
Finishing Processes creping; softening; stabilizing
Color Application piece dyed; solution dyed; yarn dyed
Width 44–45 inches (111.8–114.3 cm)
Weight medium
Hand harsh; springy; supple
Texture crepey; pebbly; sandy
Performance Expectations dimensional stability; durable; elongation; heat-set properties; resilient; high tensile strength
Drapability Qualities falls into soft flares; accommodates fullness by pleating, gathering, elasticized shirring; fullness retains soft fall

Recommended Construction Aids & Techniques

Hand Needles beading, 10–13; betweens, 5–7; milliners, 8–10; sharps, 8–10
Machine Needles round/set point, medium-fine 9–11
Threads poly-core waxed, 50; cotton-polyester blend, all purpose; cotton-covered polyester core, all purpose; spun polyester, all purpose
Hems book; double-fold or single-fold bias binding; bound/Hong Kong finish; edge-stitched; machine-stitched; seam binding
Seams flat-felled; French; false French; safety-stitched; top-stitched; welt; double welt
Seam Finishes single-ply or double-ply bound; Hong Kong bound; edge-stitched; single-ply overcast; pinked and stitched; serging/single-ply overedged; single-ply zigzag
Pressing steam; safe temperature limit 325°F (164.1°C)
Care launder; dry clean
Fabric Resources **Bloomsburg Mills; Hargro Fabrics Inc.; Springs Mills Inc.**

6 – Satin/Satin-type Fabrics

Satin is a term applied to a fabric usually characterized by a lustrous, smooth face and a dull back. Satin fabrics may be made of natural or man-made fibers, blends, or in any combination.

Satin and satin-type fabrics may be produced:

- in different types and qualities
- from light to heavy weight
- from fine and sheer to thick opacity
- in a variety of surface luster or shine
- with natural silk or man-made filament fiber face and cotton staple back
- with a highly lustrous face and dull back
- with double face
- with a silky face and pebbly back
- with stretch yarns or be elasticized
- in yard goods or ribbon

The word *satin* is used with equal validity to name a fabric and to designate a weave structure. Satin weave refers to the structure, but many fabrics with a designated satin weave are not called satin. Different names or trade names are used for many fabrics made with a satin weave with regard to different finishes, surface interest, hand, weight, end use, and origin of cloth.

Antique Satin (thick doupioni yarns)

Fiber Content 100% silk
Yarn Type natural filament
Yarn Construction conventional; fine and heavy doupioni yarns
Fabric Structure satin weave; irregular yarn in filling
Finishing Processes stabilizing; stiffening
Color Application piece dyed–cross dyed; yarn dyed; white or natural color warp yarn/different colored filling yarns
Width 48 inches (121.9 cm)
Weight medium-heavy
Hand crisp; thick
Texture rough; uneven yarns in crosswise direction
Performance Expectations subject to abrasion; absorbent; dimensional stability; poor elongation; flexible; nonpilling; subject to snagging due to weave
Drapability Qualities falls into wide cones; fullness maintains crisp bouffant effect; retains shape of garment

Recommended Construction Aids & Techniques

Hand Needles cotton darners, 9–10; milliners, 3–6; sharps, 1–4
Machine Needles round/set point, medium 14
Threads mercerized cotton, 40; silk, industrial size A-B-C; cotton-covered polyester core, heavy duty
Hems book; double-fold or single-fold bias binding; bound/Hong Kong finish; edge-stitched; faced; overedged; stitched and pinked; seam binding
Seams plain; safety-stitched; top-stitched; welt; double welt
Seam Finishes single-ply bound; Hong Kong bound; single-ply overcast; pinked; pinked and stitched; serging/single-ply overedged; single-ply zigzag
Pressing steam; safe temperature limit 300°F (150.1°C)
Care dry clean
Fabric Resource **American Silk Mills Corp.**

Antique Satin (thin doupioni yarns)

Fiber Content 100% silk
Yarn Type natural filament
Yarn Construction conventional; fine doupioni yarns
Fabric Structure satin weave; irregular yarn in filling
Finishing Processes stabilizing; stiffening
Color Application piece dyed–cross dyed; yarn dyed; white or natural color warp yarn/different colored filling yarns
Width 44–45 inches (111.8–114.3 cm)
Weight medium
Hand semi-crisp; firm
Texture irregularly slubbed yarn in crosswise direction
Performance Expectations subject to abrasion; absorbent; dimensional stability; poor elongation; flexible; nonpilling; subject to snagging due to weave
Drapability Qualities falls into moderately crisp cones; accomodates fullness by pleating, gathering; fullness maintains moderately crisp effect

Recommended Construction Aids & Techniques

Hand Needles betweens, 3–4; milliners, 6–8; sharps, 4–8
Machine Needles round/set point, medium 11
Threads poly-core waxed, 50; cotton-polyester blend, all purpose; cotton-covered polyester core, all purpose; spun polyester, all purpose
Hems book; double-fold or single-fold bias binding; bound/Hong Kong finish; edge-stitched; machine-stitched; seam binding
Seams flat-felled; French; false French; safety-stitched; top-stitched; welt; double welt
Seam Finishes single-ply or double-ply bound; Hong Kong bound; edge-stitched; single-ply overcast; pinked and stitched; serging/single-ply overedged; single ply zigzag
Pressing steam; safe temperature limit 300°F (150.1°C)
Care dry clean
Fabric Resource **American Silk Mills Corp.**

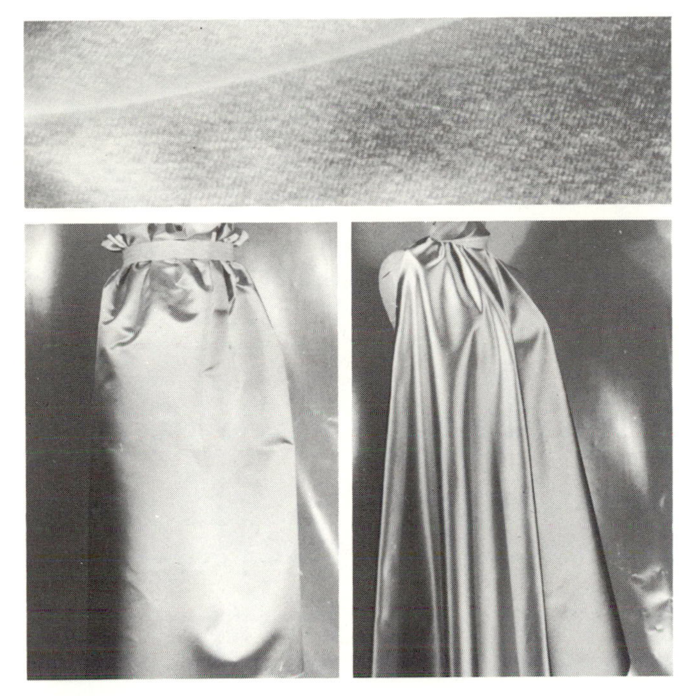

Bridal Satin

Fiber Content 100% silk
Yarn Type natual filament fiber
Yarn Construction conventional; fine
Fabric Structure satin weave
Finishing Processes calendering; singeing; softening
Color Application piece dyed; yarn dyed
Width 50 inches (127 cm)
Weight medium-heavy
Hand compact; rigid; soft
Texture lustrous; smooth
Performance Expectations subject to abrasion; absorbent; dimensional stability; poor elongation; flexible
Drapability Qualities falls into crisp wide cones; fullness maintains crisp bouffant effect; retains shape or silhouette of garment

Recommended Construction Aids & Techniques

Hand Needles cotton darners, 9–10; milliners, 3–6; sharps, 1–4
Machine Needles round/set point, medium 11–14
Threads mercerized cotton, 40; silk, industrial size A-B-C; cotton-covered polyester core, heavy duty
Hems book; double-fold or single-fold bias binding; bound/Hong Kong finish; edge-stitched; faced; overedged; stitched and pinked; seam binding
Seams plain; safety-stitched; top-stitched; welt; double welt
Seam Finishes single-ply bound; Hong Kong bound; single-ply overcast; pinked; pinked and stitched; serging/single-ply overedged; single-ply zigzag
Pressing steam; safe temperature limit 300°F (150.1°C)
Care dry clean
Fabric Resource **Lafitte Inc.**

Satin Brocade

Fiber Content 100% silk
Yarn Type natural filament fiber
Yarn Construction conventional; fine; slack twist
Fabric Structure Jacquard weave
Finishing Processes softening; stabilizing
Color Application piece dyed; yarn dyed
Width 48 inches (121.9 cm)
Weight light
Hand loose; soft; supple
Texture weave creates smooth and textured surface; yarns and weave create light-reflecting qualities
Performance Expectations subject to abrasion; absorbent; flexible; subject to snagging due to weave
Drapability Qualities falls into soft flares; accommodates fullness by pleating, gathering; fullness retains soft fall

Recommended Construction Aids & Techniques

Hand Needles beading, 10–13; betweens, 5–7; milliners, 8–10; sharps, 8–10
Machine Needles round/set point, medium-fine 9–11
Threads mercerized cotton, 50; silk, industrial size A-B; cotton-covered polyester core, all purpose
Hems book; double-fold or single-fold bias binding; bound/Hong Kong finish; edge-stitched; machine-stitched; seam binding
Seams flat-felled; French; false French; plain; safety-stitched; top-stitched; welt; double welt
Seam Finishes single-ply or double-ply bound; Hong Kong bound; edge-stitched; single-ply overcast; pinked and stitched; serging/single-ply overedged; single-ply zigzag
Pressing steam; safe temperature limit 300°F (150.1°C)
Care dry clean
Fabric Resource **American Silk Mills Corp.**

Charmeuse

Fiber Content 100% polyester
Yarn Type filament fiber
Yarn Construction conventional; crepe; hard twist
Fabric Structure satin weave; hard twist warp yarn; crepe or spun filling yarn
Finishing Processes calendering; stabilizing; stretching
Color Application piece dyed; yarn dyed
Width 45 inches (114.3 cm)
Weight light
Hand firm; soft; supple
Texture slightly crepey; high lustrous face; dull back
Performance Expectations subject to color crocking and edge abrasion; dimensional stability; elongation; heat-set properties; resilient; high tensile strength
Drapability Qualities falls into soft flares; falls close to body contour; fullness retains soft graceful fall

Recommended Construction Aids & Techniques

Hand Needles beading, 10–13; betweens, 5–7; milliners, 8–10; sharps, 8–10
Machine Needles round/set point, fine 9
Threads poly-core waxed, 60; cotton-polyester blend, fine; cotton-covered polyester core, extra fine; spun polyester, fine
Hems bound/Hong Kong finish; double fold; edge-stitched; double edge-stitched; horsehair; machine-stitched; hand- or machine-rolled; seam binding
Seams flat-felled; French; mock French; tissue-stitched
Seam Finishes self bound; double-ply bound; double-stitched; double-stitched and overcast; edge-stitched; mock French; double-ply overcast; double-ply zigzag
Pressing steam; safe temperature limit 325°F (164.1°C)
Care launder; dry clean
Fabric Resources Concord Fabrics Inc.; Hargro Fabrics Inc.; Private Collections Fabrics Ltd.

Charmeuse Dobby Stripe

Fiber Content 100% polyester
Yarn Type filament fiber
Yarn Construction conventional; fine
Fabric Structure dobby weave
Finishing Processes calendering; stabilizing; stretching
Color Application piece dyed; solution dyed; yarn dyed
Width 44–45 inches (111.8–114.3 cm)
Weight light
Hand firm; semi-soft; supple
Texture flat; yarns and weave create light-reflecting qualities
Performance Expectations subject to color crocking and edge abrasion; dimensional stability; durable; elongation; heat-set properties; resilient; high tensile strength
Drapability Qualities falls into soft flares; accommodates fullness by pleating, gathering; fullness retains soft fall

Recommended Construction Aids & Techniques

Hand Needles beading, 10–13; betweens, 5–7; milliners, 8–10; sharps, 8–10
Machine Needles round/set point, fine 9
Threads poly-core waxed, 60; cotton-polyester blend, fine; cotton-covered polyester core, extra fine; spun polyester, fine
Hems bound/Hong Kong finish; double fold; edge-stitched; double edge-stitched; horsehair; machine-stitched; hand- or machine-rolled; seam binding
Seams flat-felled; French; false French; tissue-stitched
Seam Finishes single-ply or double-ply bound; Hong Kong bound; edge-stitched; single-ply overcast; pinked and stitched; serging/single-ply overedged; single-ply zigzag
Pressing steam; safe temperature limit 325°F (164.1°C)
Care launder; dry clean
Fabric Resource Hargro Fabrics Inc.

Ciré Satin

Fiber Content 100% silk
Yarn Type natural filament fiber
Yarn Construction conventional; fine; slack twist
Fabric Structure satin weave
Finishing Processes cire calendering; stabilizing; wax resin and polished
Color Application piece dyed; yarn dyed
Width 36 inches (91.4 cm)
Weight light
Hand firm; soft; supple
Texture highly glossed face; dull back; smooth
Performance Expectations dimensional stability; poor elongation; flexible; nonpilling; water-repellent properties due to finish
Drapability Qualities falls into soft flares; accommodates fullness by pleating, gathering; fullness retains soft fall

Recommended Construction Aids & Techniques

Hand Needles beading, 10–13; betweens, 3–4; milliners, 6–8; sharps, 4–8
Machine Needles round/set point, fine 9
Threads mercerized cotton, 50; silk, industrial size A-B; cotton-covered polyester core, all purpose
Hems book; double-fold or single-fold bias binding; bound/Hong Kong finish; edge-stitched; machine-stitched; seam binding
Seams flat-felled; French; false French; plain; safety-stitched; top-stitched; welt; double welt
Seam Finishes single-ply or double-ply bound; Hong Kong bound; edge-stitched; single-ply overcast; pinked and stitched; serging/single-ply overedged; single-ply zigzag
Pressing steam; safe temperature limit 300°F (150.1°C)
Care dry clean
Fabric Resource **Lafitte Inc.**

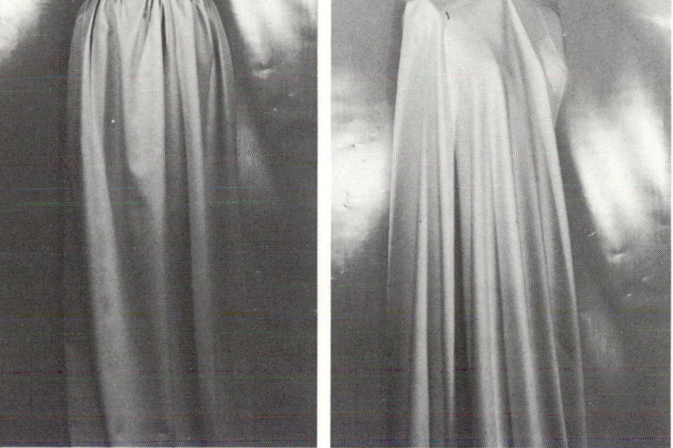

Crepe Satin/Satin Crepe

Fiber Content 100% polyester
Yarn Type trilobal filament fiber
Yarn Construction textured
Fabric Structure satin weave face; creped back
Finishing Processes calendering; stabilizing; stretching
Color Application piece dyed; yarn dyed
Width 45 inches (114.3 cm)
Weight medium-heavy
Hand firm; soft
Texture flat smooth face; slightly creped back
Performance Expectations subject to color crocking and edge abrasion; dimensional stability; durable; elongation; heat-set properties; resilient; high tensile strength
Drapability Qualities falls into soft flares; fullness retains moderately lofty effect; retains shape of garment

Recommended Construction Aids & Techniques

Hand Needles betweens, 3–4; milliners, 6–8; sharps, 4–8
Machine Needles round/set point, medium 11
Threads poly-core waxed, 50; cotton-polyester blend, all purpose; cotton-covered polyester core, all purpose; spun polyester, all purpose
Hems book; double-fold or single-fold bias binding; bound/Hong Kong finish; edge-stitched; machine-stitched; seam binding
Seams flat-felled; French; false French; safety-stitched; top-stitched; welt; double welt
Seam Finishes single-ply or double-ply bound; Hong Kong bound; edge-stitched; single-ply overcast; pinked and stitched; serging/single-ply overedged; single-ply zigzag
Pressing steam; safe temperature limit 325°F (164.1°C)
Care launder; dry clean
Fabric Resource **Springs Mills Inc.**

Creped Satin

Fiber Content 100% acetate
Yarn Type filament fiber
Yarn Construction conventional; fine; slack twist
Fabric Structure satin weave
Finishing Processes heat-creped finish; stabilizing
Color Application piece dyed; yarn dyed
Width 44–45 inches (111.8–114.3 cm)
Weight medium
Hand firm; soft; supple
Texture smooth face; crepey back
Performance Expectations subject to abrasion; dimensional stability; elongation; flexible; heat-set properties; mildew and moth resistant; nonpilling
Drapability Qualities falls into moderately soft flares; accommodates fullness by pleating, gathering; retains shape of garment

Recommended Construction Aids & Techniques

Hand Needles betweens, 3–4; milliners, 6–8; sharps, 4–8
Machine Needles round/set point, medium 11
Threads poly-core waxed, 50; cotton-polyester blend, all purpose; cotton-covered polyester core, all purpose; spun polyester, all purpose
Hems book; double-fold or single-fold bias binding; bound/Hong Kong finish; edge-stitched; machine-stitched; seam binding
Seams flat-felled; French; false French; safety-stitched; top-stitched; welt; double welt
Seam Finishes single-ply or double-ply bound; Hong Kong bound; edge-stitched; single-ply overcast; pinked and stitched; serging/single-ply overedged; single-ply zigzag
Pressing steam; safe temperature limit 250°F (122.1°C)
Care hand washable; dry clean
Fabric Resources **Cameo Fabrics Inc.; Lafitte Inc.**

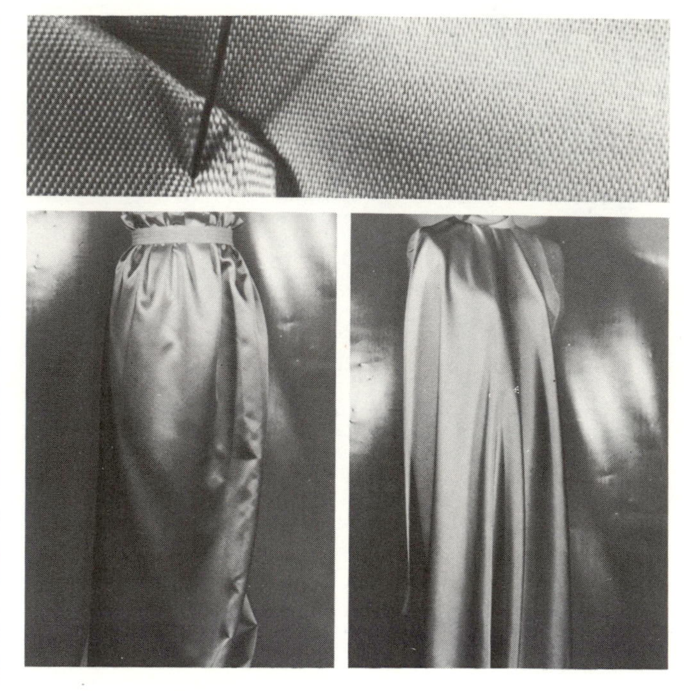

Duchesse Satin

Fiber Content 100% acetate
Yarn Type filament fiber
Yarn Construction conventional
Fabric Structure satin weave; 8–12 end warp
Finishing Processes calendering; stabilizing
Color Application piece dyed; yarn dyed
Width 44–45 inches (111.8–114.3 cm)
Weight medium-heavy
Hand firm; rigid; stiff
Texture smooth face; ribbed back
Performance Expectations subject to abrasion; dimensional stability; flexible; elongation; heat-set properties; mildew and moth resistant; nonpilling
Drapability Qualities falls into stiff wide cones; fullness maintains crisp bouffant effect; retains shape or silhouette of garment

Recommended Construction Aids & Techniques

Hand Needles betweens, 3–4; milliners, 6–8; sharps, 4–8
Machine Needles round/set point, medium 14
Threads poly-core waxed, 50; cotton-polyester blend, all purpose; cotton-covered polyester core, all purpose; spun polyester, all purpose
Hems book; double-fold or single-fold bias binding; bound/Hong Kong finish; edge-stitched; faced; overedged; stitched and pinked; seam binding
Seams plain; safety-stitched; top-stitched; welt; double welt
Seam Finishes single-ply bound; Hong Kong bound; single-ply overcast; pinked; pinked and stitched; serging/single-ply overedged; single-ply zigzag
Pressing steam; safe temperature limit 250°F (122.1°C)
Care dry clean
Fabric Resources **Cameo Fabrics Inc; Springs Mills Inc.**

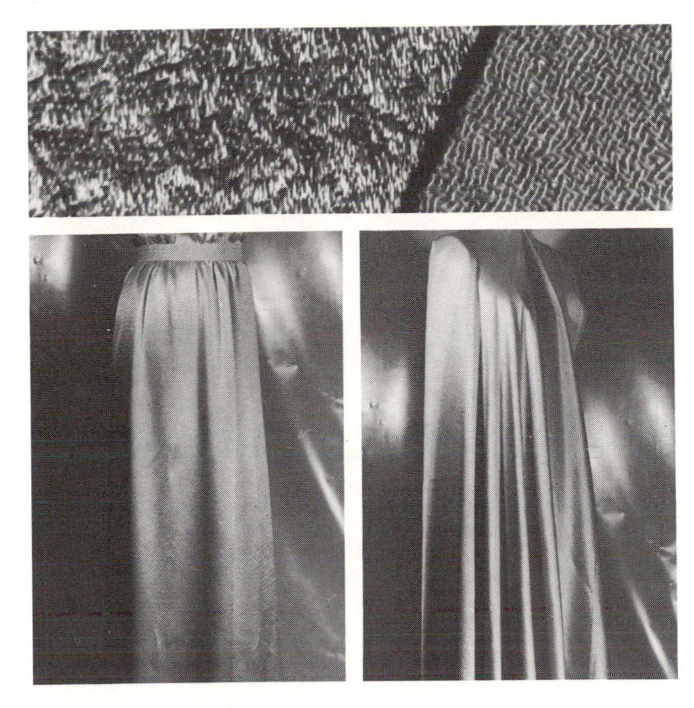

Hammered Satin

Fiber Content 100% silk
Yarn Type natural filament fiber
Yarn Construction textured
Fabric Structure double weave satin face; creped back
Finishing Processes creping; heat-set embossing; stabilizing
Color Application piece dyed; yarn dyed
Width 40–42 inches (101.6–106.7 cm)
Weight[1] medium-light
Hand soft; spongy; supple
Texture uneven and indented surface
Performance Expectations absorbent; flexible; subject to snagging due to weave and finish; wrinkle resistant
Drapability Qualities falls into moderately soft flares; accommodates fullness by gathering; fullness retains moderately soft fall

Recommended Construction Aids & Techniques

Hand Needles betweens, 3–4; milliners, 6–8; sharps, 4–8
Machine Needles round/set point, medium 11
Threads mercerized-cotton, 40; silk, industrial size A-B-C; cotton-covered polyester core, heavy duty
Hems book; double-fold or single-fold bias binding; bound/Hong Kong finish; edge-stitched; faced; overedged; stitched and pinked; seam binding
Seams plain; safety-stitched; top-stitched; welt; double welt
Seam Finishes single-ply bound; Hong Kong bound; single-ply overcast; pinked; pinked and stitched; serging/single-ply overedged; single-ply zigzag
Pressing steam; safe temperature limit 300°F (150.1°C)
Care dry clean
Fabric Resource **American Silk Mills Corp.**

Jacquard Satin

Fiber Content 100% polyester
Yarn Type trilobal filament fiber
Yarn Construction textured
Fabric Structure Jacquard weave
Finishing Processes calendering; softening; stretching
Color Application piece dyed; yarn dyed
Width 44–45 inches (111.8–114.3 cm)
Weight light
Hand firm; smooth; springy
Texture flat; yarns and weave create glossy pattern
Performance Expectations subject to abrasion; dimensional stability; durable; elongation; heat-set properties; resilient; high tensile strength
Drapability Qualities falls into moderately soft flares; accommodates fullness by pleating, gathering; fullness retains moderately soft fall

Recommended Construction Aids & Techniques

Hand Needles beading, 10–13; betweens, 5–7; milliners, 8–10; sharps, 8–10
Machine Needles round/set point, fine 9
Threads poly-core waxed, 60; cotton-polyester blend, fine; cotton-covered polyester core, extra fine; spun polyester, fine
Hems bound/Hong Kong finish; double fold; edge-stitched; horsehair; machine-stitched; hand- or machine-rolled; seam binding
Seams flat-felled; French; mock French; plain; safety-stitched; top-stitched; welt; double welt
Seam Finishes self bound; double-ply bound; double-stitched; double-stitched and overcast; edge-stitched; mock French; double-ply overcast; pinked and stitched; double-ply zigzag
Pressing steam; safe temperature limit 325°F (164.1°C)
Care launder; dry clean
Fabric Resource **Springs Mills Inc.**

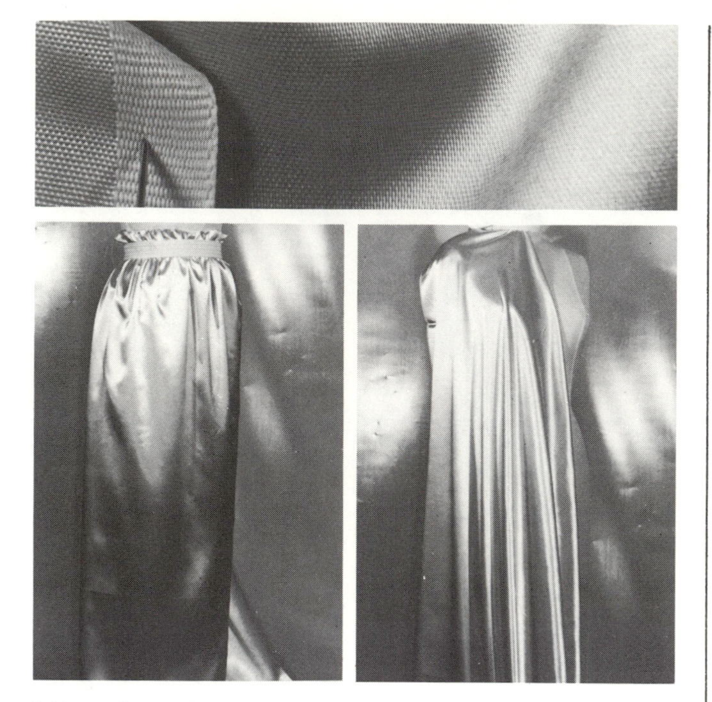

Messaline Satin

Fiber Content 100% acetate
Yarn Type filament fiber
Yarn Construction conventional; fine; low twist
Fabric Structure 5-shaft loosely woven satin weave
Finishing Processes calendering; softening; stabilizing
Color Application piece dyed; yarn dyed
Width 44–45 inches (111.8–114.3 cm)
Weight light
Hand sleazy; slippery; soft
Texture glossy; smooth face; dull back
Performance Expectations subject to abrasion; dimensional stability; elongation; flexible; heat-set properties; mildew and moth resistant; nonpilling; subject to snagging
Drapability Qualities falls into soft flares; accommodates fullness by pleating, gathering; fullness retains moderately soft fall

Recommended Construction Aids & Techniques

Hand Needles beading, 10–13; betweens, 5–7; milliners, 8–10; sharps, 8–10
Machine Needles round/set point, medium 11
Threads poly-core waxed, 60; cotton-polyester blend, fine; cotton-covered polyester core, extra fine; spun polyester, fine
Hems book; double-fold or single-fold bias binding; bound/Hong Kong finish; edge-stitched; machine-stitched; seam binding
Seams flat-felled; French; false French; plain; safety-stitched; top-stitched; welt; double welt
Seam Finishes single-ply or double-ply bound; Hong Kong bound; edge-stitched; single-ply overcast; pinked and stitched; serging/single-ply overedged; single-ply zigzag
Pressing steam; safe temperature limit 250°F (122.1°C)
Care dry clean
Fabric Resources Lawrence Textile Co. Inc.; Springs Mills Inc.

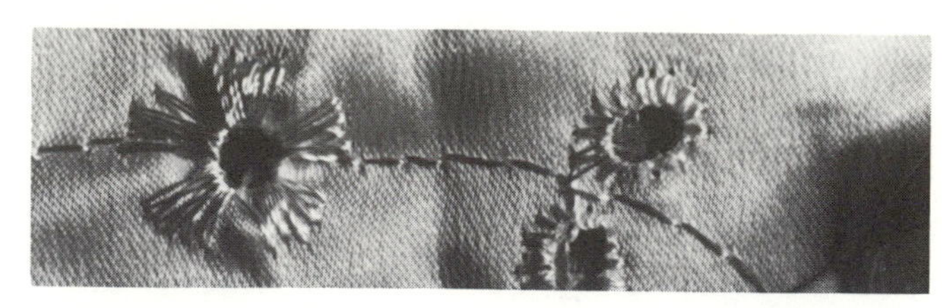

Embroidered Satin

Moiré Satin

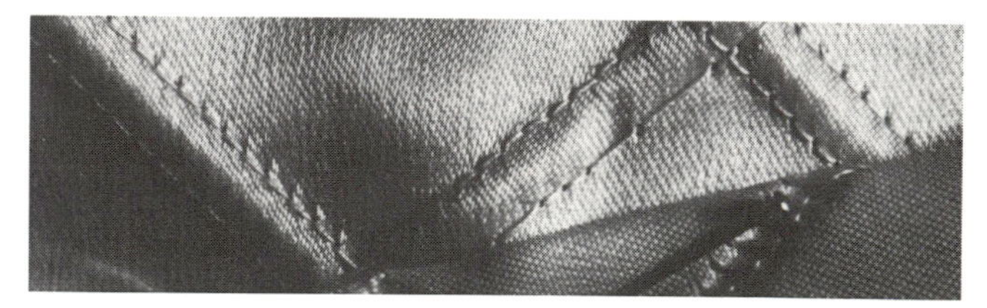

Printed Satin

Trapunto Satin

Fabric Resource Cameo Fabrics Inc.

Sequin-embroidered Satin

Fiber Content 100% acetate; plastic paillettes
Yarn Type filament fiber
Yarn Construction textured; low twist
Fabric Structure Schiffli embroidery on satin weave base
Finishing Processes brushing; calendering; flame retardant
Color Application piece dyed
Width 45 inches (114.3 cm)
Weight medium
Hand base: pliable; paillettes: rough, scratchy
Texture base: silky, smooth, opaque; paillettes form high and low raised pattern on surface
Performance Expectations dimensional stability; durable; elasticity; elongation; resilient; high tensile strength; paillettes subject to abrasion, breaking, snagging
Drapability Qualities falls into wide cones; fullness maintains bouffant effect; retains shape of garment

Recommended Construction Aids & Techniques

Hand Needles beading, 13–16; betweens, 4–6; embroidery, 6–8; milliners, 7–9; sharps, 6–8
Machine Needles round/set point, medium 11–14
Threads silk, industrial size A; poly-core waxed, fine; cotton-covered polyester core, all purpose; spun polyester, all purpose; nylon/Dacron, monocord A
Hems single-fold or double-fold bias binding; bound/Hong Kong finish; faced; overedged; seam binding
Seams plain; safety-stitched; tissue-stitched; overlap and hand-stitch sequins on seamline after stitching
Seam Finishes single-ply bound; Hong Kong bound; edge-stitched; single-ply overcast; pinked and stitched; serging/single-ply overedged; untreated/plain; remove sequins in seam allowance
Pressing use pressing cloth; safe temperature limit 250°F (122.1°C)
Care dry clean only
Fabric Resource **Sequin International Corp.**

Milium®-backed Satin

Fiber Content 100% acetate
Yarn Type filament fiber
Yarn Construction conventional
Fabric Structure satin weave
Finishing Processes Milium®-coated back; stretching; stabilizing
Color Application piece dyed; yarn dyed
Width 45 inches (114.3 cm)
Weight medium
Hand semi-crisp; firm; rigid
Texture smooth; shiny face; dull back
Performance Expectations subject to abrasion; dimensional stability; elongation; flexible; heat-set properties; insulating properties due to Milium® finish; nonpilling
Drapability Qualities falls into moderately crisp cones; fullness maintains crisp effect; retains shape of garment

Recommended Construction Aids & Techniques

Hand Needles betweens, 3–4; milliners, 6–8; sharps, 4–8
Machine Needles round/set point, medium 14
Threads poly-core waxed, 50; cotton-polyester blend, all purpose; cotton-covered polyester core, all purpose; spun polyester, all purpose
Hems book; double-fold or single-fold bias binding; bound/Hong Kong finish; edge-stitched; faced; overedged; stitched and pinked; seam binding
Seams plain; safety-stitched; top-stitched; welt; double welt
Seam Finishes single-ply bound; Hong Kong bound; single-ply overcast; pinked; pinked and stitched; serging/single-ply overedged; single-ply zigzag
Pressing steam; safe temperature limit 250°F (122.1°C)
Care dry clean
Fabric Resource **Springs Mills Inc.**

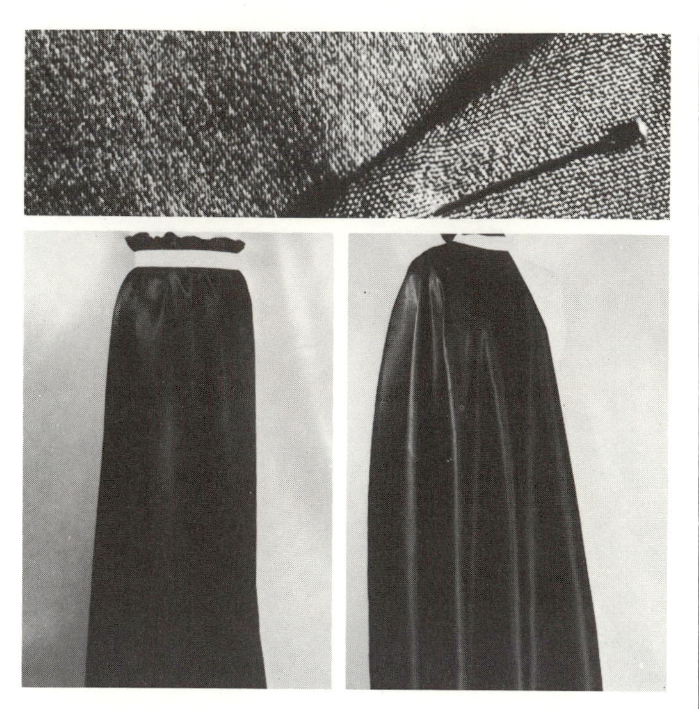

Nap/Fleece-faced Satin

Fiber Content acetate/rayon
Yarn Type filament fiber; filament staple
Yarn Construction conventional; fine yarns low twist; heavy yarns high twist
Fabric Structure satin weave; heavy filling yarn
Finishing Processes brushing; calendering; napping (face)
Color Application piece dyed–cross dyed; yarn dyed
Width 48 inches (121.9 cm)
Weight medium-heavy to heavy
Hand soft; thick; warm
Texture fuzzy; fleece napped face; smooth, lustrous back
Performance Expectations dimensional stability; elongation; flexible; insulating properties due to napping; nap face tends to flatten and abrade with wear
Drapability Qualities falls into moderately soft flares; fullness maintains moderately bouffant effect; retains shape of garment

Recommended Construction Aids & Techniques

Hand Needles betweens, 3–4; milliners, 6–8; sharps, 4–8
Machine Needles round/set point, medium 14
Threads poly-core waxed, 50; cotton-polyester blend, all purpose; cotton-covered polyester core, all purpose; spun polyester, all purpose
Hems book; double-fold or single-fold bias binding; bound/Hong Kong finish; edge-stitched; faced; overedged; stitched and pinked; seam binding
Seams plain; safety-stitched; top-stitched; welt; double welt
Seam Finishes single-ply bound; Hong Kong bound; single-ply overcast; pinked; pinked and stitched; serging/single-ply overedged; single-ply zigzag
Pressing steam; safe temperature limit 250°F (122.1°C)
Care dry clean
Fabric Resource **Truemark Discount Fabrics Inc.**

7 ~ Taffeta/Taffeta-type Fabrics

Antique Taffeta
Faille Taffeta
Jacquard Taffeta
Moiré Taffeta
Paper Taffeta
Tissue Taffeta
Warp-dyed / Warp-printed Taffeta
Yarn-dyed Taffeta

Taffeta is a term applied to a fabric characterized by a fine rib in the crosswise direction. Taffeta is constructed in a firm, closely woven plain weave utilizing heavier or more filling yarns than warp yarns. Taffeta fabrics may be made of natural fiber yarns, man-made thin filament fiber yarns, blends, or in any combination.

Taffeta and taffeta-type fabrics may be produced:

- with a dull, lustrous or shiny surface
- with a fine, smooth surface on one or both sides
- with flat or raised fine ribs
- from soft and pliable to crisp hand
- weighted to produce crispness
- stiffened to produce rustling sound
- in a moiré pattern
- with an iridescent or shimmering appearance

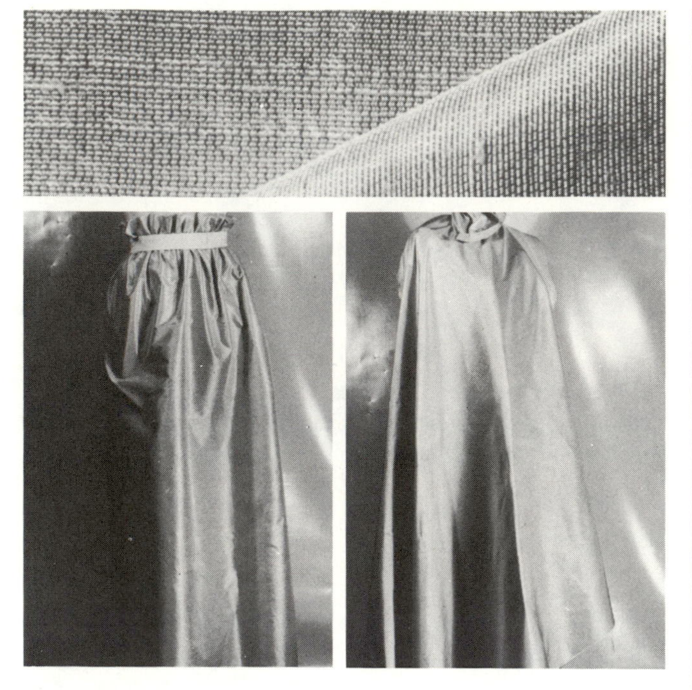

Antique Taffeta

Fiber Content 100% silk
Yarn Type natural filament fiber
Yarn Construction conventional; doupioni yarns
Fabric Structure plain weave; additional filling yarns form fine rib in crosswise direction
Finishing Processes calendering; stabilizing; stiffening
Color Application piece dyed; yarn dyed
Width 50 inches (127 cm)
Weight medium-light
Hand crisp; coarse; firm
Texture rough; iridescent appearance; irregular thick slubbed yarns in crosswise direction
Performance Expectations absorbent; dimensional stability; poor elongation; flexible; nonpilling
Drapability Qualities falls into crisp flares; accommodates fullness by gathering; fullness maintains crisp effect

Recommended Construction Aids & Techniques

Hand Needles betweens, 3–4; milliners, 6–8; sharps, 4–8
Machine Needles round/set point, medium-fine 9–11
Threads mercerized cotton, 40; silk, industrial size A-B; cotton-covered polyester core, all purpose
Hems book; double-fold or single-fold bias binding; bound/Hong Kong finish; edge-stitched; machine-stitched; seam binding
Seams flat-felled; French; false French; plain; tissue-stitched
Seam Finishes single-ply or double-ply bound; Hong Kong bound; edge-stitched; single-ply overcast; pinked and stitched; serging/single-ply overedged; single-ply zigzag
Pressing steam; safe temperature limit 300°F (150.1°C)
Care dry clean
Fabric Resource American Silk Mills Corp./Bucol

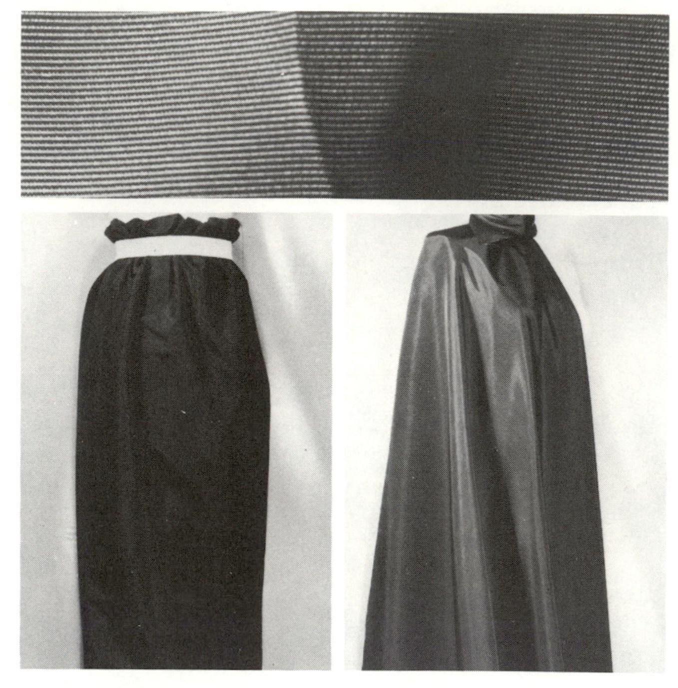

Faille Taffeta

Fiber Content 100% silk
Yarn Type natural filament fiber
Yarn Construction conventional; fine; slack twist
Fabric Structure plain weave; additional filling yarns form rib in crosswise direction
Finishing Processes calendering; stabilizing; stiffening
Color Application piece dyed; yarn dyed
Width 44–45 inches (111.8–114.3 cm)
Weight medium-light
Hand crisp; firm; scroopy
Texture flat with fine noticeable rib in crosswise direction
Performance Expectations absorbent; dimensional stability; poor elongation; flexible; nonpilling
Drapability Qualities falls into crisp flares; accommodates fullness by pleating, gathering; fullness maintains crisp effect

Recommended Construction Aids & Techniques

Hand Needles beading, 10–13; betweens, 5–7; milliners, 8–10; sharps, 8–10
Machine Needles round/set point, medium-fine 9–11
Threads mercerized cotton, 50; silk, industrial size A-B; cotton-covered polyester core, all purpose
Hems bound/Hong Kong finish; double fold; edge-stitched; horsehair; machine-stitched; hand- or machine-rolled; seam binding
Seams flat-felled; French; false French; plain; safety-stitched; top-stitched; welt; double welt
Seam Finishes single-ply or double-ply bound; Hong Kong bound; edge-stitched; single-ply overcast; pinked and stitched; serging/single-ply overedged; single-ply zigzag
Pressing steam; safe temperature limit 300°F (150.1°C)
Care dry clean
Fabric Resources American Silk Mills Corp.; Sormani Co. Inc./Taroni

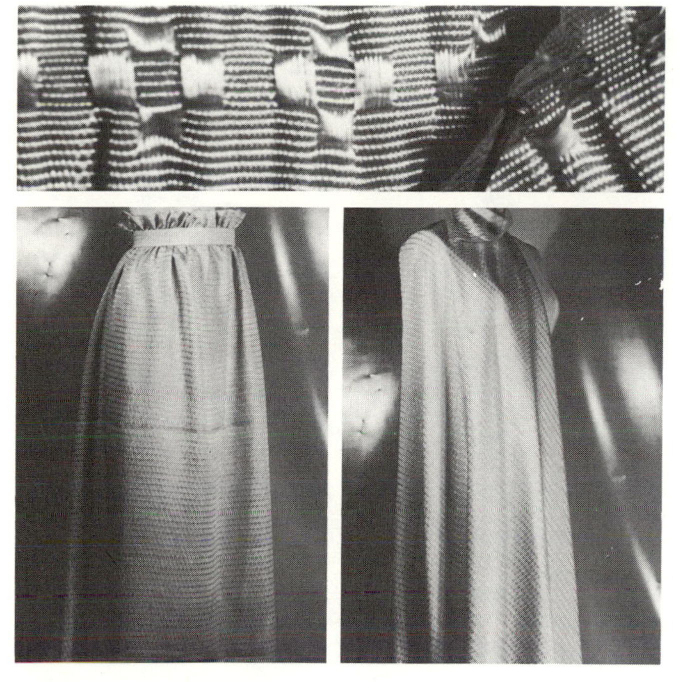

Jacquard Taffeta

Fiber Content 100% silk
Yarn Type natural filament fiber
Yarn Construction conventional; fine; slack twist
Fabric Structure Jacquard weave; rib in crosswise direction
Finishing Processes stabilizing; stiffening; stretching
Color Application piece dyed; yarn dyed
Width 45–46 inches (114.3–116.8 cm)
Weight medium-light
Hand crisp
Texture weave creates puckered uneven surface
Performance Expectations subject to abrasion; absorbent; dimensional stability; flexible; nonpilling; subject to snagging due to weave
Drapability Qualities falls into crisp wide cones; fullness maintains lofty bouffant effect; retains shape of garment

Recommended Construction Aids & Techniques

Hand Needles betweens, 3–4; milliners, 6–8; sharps, 4–8
Machine Needles round/set point, medium 11
Threads mercerized cotton, 40; silk, industrial size A-B-C; cotton-covered polyester core, all purpose
Hems book; double-fold or single-fold bias binding; bound/Hong Kong finish; edge-stitched; faced; overedged; stitched and pinked; seam binding
Seams plain; safety-stitched; top-stitched; welt; double welt
Seam Finishes single-ply bound; Hong Kong bound; single-ply overcast; pinked; pinked and stitched; serging/single-ply overedged; single-ply zigzag
Pressing steam; safe temperature limit 300°F (150.1°C)
Care dry clean
Fabric Resource **American Silk Mills Corp.**

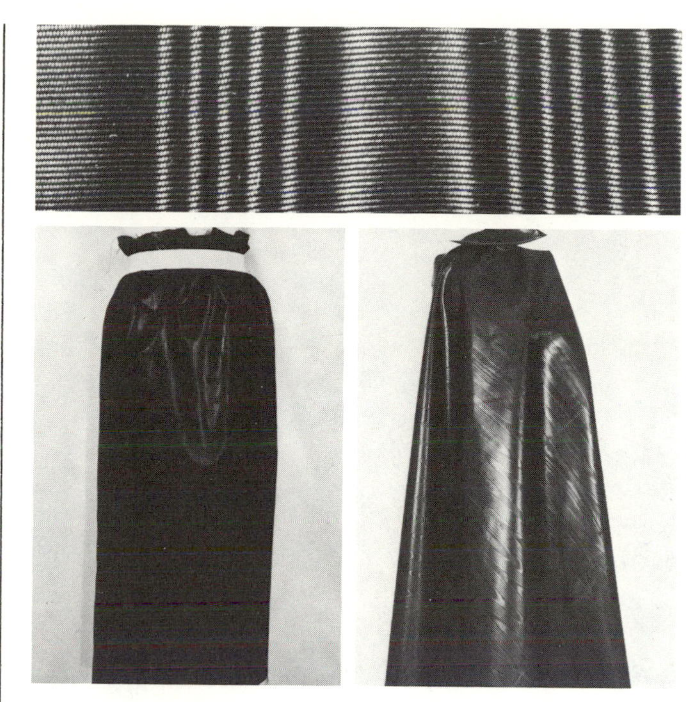

Moiré Taffeta

Fiber Content 100% acetate
Yarn Type filament fiber
Yarn Construction conventional; fine; slack twist
Fabric Structure plain weave; additional filling yarns form rib in crosswise direction
Finishing Processes calendering; moiré finish; stretching
Color Application piece dyed; solution dyed; yarn dyed
Width 44–45 inches (111.8–114.3 cm)
Weight medium-light to medium
Hand crisp; firm; scroopy
Texture flat with fine crosswise rib
Performance Expectations subject to abrasion; dimensional stability; elongation; flexible; heat-set properties; mildew and moth resistant
Drapability Qualities falls into crisp flares; accommodates fullness by pleating, gathering; fullness maintains crisp effect

Recommended Construction Aids & Techniques

Hand Needles beading, 10–13; betweens, 5–7; milliners, 8-10; sharps, 8–10
Machine Needles round/set point, medium-fine 9–11
Threads poly-core waxed, 60; cotton-polyester blend, fine; cotton-covered polyester core, extra fine; spun polyester, fine
Hems book; double-fold or single-fold bias binding; bound/Hong Kong finish; edge-stitched; machine-stitched; seam binding
Seams flat-felled; French; false French; safety-stitched; top-stitched; welt; double welt
Seam Finishes single-ply or double-ply bound; Hong Kong bound; edge-stitched; single-ply overcast; pinked and stitched; serging/single-ply overedged; single-ply zigzag
Pressing steam; safe temperature limit 250°F (122.1°C)
Care dry clean
Fabric Resources **Lawrence Textile Co. Inc.; Sormani Co. Inc./Taroni**

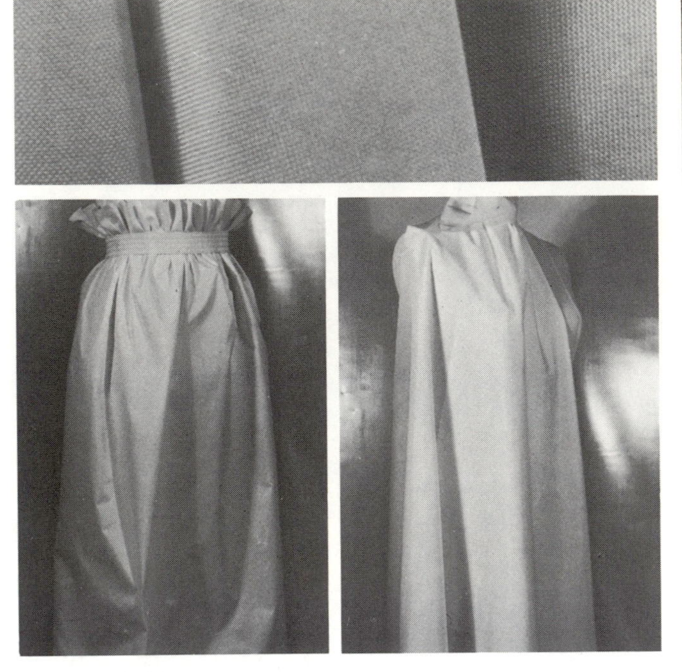

Paper Taffeta

Fiber Content 100% nylon
Yarn Type filament fiber
Yarn Construction conventional; fine; low twist
Fabric Structure closely woven plain weave
Finishing Processes calendering; stiffening; stretching
Color Application piece dyed; solution dyed; yarn dyed
Width 46 inches (116.8 cm)
Weight light
Hand boardy; firm; scroopy; stiff
Texture harsh; papery; smooth
Performance Expectations abrasion resistant; dimensional stability; durable; elongation; heat-set properties; high tensile strength
Drapability Qualities falls into stiff cones; fullness maintains crisp bouffant effect; retains shape or silhouette of garment

Recommended Construction Aids & Techniques

Hand Needles beading, 10–13; betweens, 5–7; milliners, 8–10; sharps, 8–10
Machine Needles round/set point, fine 9
Threads poly-core waxed, 60; cotton-polyester blend, fine; cotton-covered polyester core, extra fine; spun polyester, fine
Hems bound/Hong Kong finish; double fold; edge-stitched; double edge-stitched; horsehair; machine-stitched; hand- or machine-rolled; seam binding
Seams flat-felled; French; mock French; plain; tissue-stitched
Seam Finishes self bound; double-ply bound; double-stitched; double-stitched and overcast; edge-stitched; mock French; double-ply overcast; double-ply zigzag
Pressing steam; safe temperature limit 350°F (178.1°C)
Care dry clean
Fabric Resource **Lawrence Textile Co. Inc.**

Tissue Taffeta

Fiber Content 49% silk/51% polyester
Yarn Type natural and man-made filament fibers
Yarn Construction conventional; fine; slack twist
Fabric Structure plain weave; additional filling yarns form rib in crosswise direction
Finishing Processes calendering; softening; stabilizing
Color Application piece dyed–cross dyed; yarn dyed
Width 36 inches (91.4 cm)
Weight light
Hand firm; papery; soft
Texture harsh; smooth; subtle rib in crosswise direction
Performance Expectations subject to color crocking and edge abrasion; dimensional stability; flexible; heat-set properties; moth resistant
Drapability Qualities falls into moderately soft flares; accommodates fullness by pleating, gathering, elasticized shirring; fullness retains soft fall

Recommended Construction Aids & Techniques

Hand Needles beading, 10–13; betweens, 5–7; milliners, 8–10; sharps, 8–10
Machine Needles round/set point, fine 9
Threads mercerized cotton, 50; silk, industrial size A; cotton-covered polyester core, extra fine
Hems bound/Hong Kong finish; double fold; edge-stitched; double edge-stitched; horsehair; machine-stitched; hand- or machine-rolled; seam binding
Seams flat-felled; French; mock French; plain; tissue-stitched
Seam Finishes self bound; double-ply bound; double-stitched; double-stitched and overcast; edge-stitched; mock French; double-ply overcast; double-ply zigzag
Pressing steam; safe temperature limit 300°F (150.1°C)
Care dry clean
Fabric Resource **American Silk Mills Corp./Bucol**

Warp-dyed/Warp-printed Taffeta

Fiber Content 30% silk/70% polyester
Yarn Type natural and man-made filament fibers
Yarn Construction conventional; fine; slack twist
Fabric Structure plain weave; additional filling yarns form rib in crosswise direction
Finishing Processes calendering; stabilizing; stiffening
Color Application yarn dyed
Width 39–40 inches (99.1–101.6 cm)
Weight light
Hand crisp; firm; scroopy
Texture flat; slight noticeable rib in crosswise direction
Performance Expectations subject to abrasion; dimensional stability; moderately durable; flexible; heat-set properties; moderate tensile strength
Drapability Qualities falls into crisp flares; accommodates fullness by pleating, gathering; fullness maintains crisp effect

Recommended Construction Aids & Techniques

Hand Needles beading, 10–13; betweens, 5–7; milliners, 8–10; sharps, 8–10
Machine Needles round/set point, medium-fine 9–11
Threads silk, industrial size A; cotton-covered polyester core, extra fine; mercerized cotton, 50
Hems book; double-fold or single-fold bias binding; bound/Hong Kong finish; edge-stitched; machine-stitched; seam binding
Seams flat-felled; French; false French; safety-stitched; top-stitched; welt; double welt
Seam Finishes double-ply bound; Hong Kong bound; edge-stitched; single-ply overcast; pinked and stitched; serging/single-ply overedged; single-ply zigzag
Pressing steam; safe temperature limit 300°F (150.1°C)
Care dry clean
Fabric Resource **American Silk Mills Corp./Bucol**

Yarn-dyed Taffeta

Fiber Content 100% acetate
Yarn Type filament fiber
Yarn Construction conventional; fine; low twist
Fabric Structure plain weave; additional filling yarns form rib in crosswise direction
Finishing Processes calendering; stabilizing; stiffening
Color Application yarn dyed; different colored yarns form stripe, check or plaid pattern
Width 48–49 inches (121.9–124.5 cm)
Weight light
Hand crisp, firm; scroopy
Texture flat; fine rib in crosswise direction
Performance Expectations subject to abrasion; dimensional stability; elongation; flexible; heat-set properties; mildew and moth resistant
Drapability Qualities falls into crisp flares; accommodates fullness by pleating, gathering; fullness maintains crisp effect

Recommended Construction Aids & Techniques

Hand Needles beading, 10–13; betweens, 5–7; milliners, 8–10; sharps, 8–10
Machine Needles round/set point, medium-fine 9–11
Threads poly-core waxed, 60; cotton-polyester blend, fine; cotton-covered polyester core, extra fine; spun polyester, fine
Hems book; double-fold or single-fold bias binding; bound/Hong Kong finish; edge-stitched; machine-stitched; seam binding
Seams flat-felled; French; false French; plain; safety-stitched; top-stitched; welt; double welt
Seam Finishes single-ply or double-ply bound; Hong Kong bound; edge-stitched; single-ply overcast; pinked and stitched; serging/single-ply overedged; single-ply zigzag
Pressing steam; safe temperature limit 250°F (122.1°C)
Care dry clean
Fabric Resource **Lawrence Textile Co. Inc.**

8 ~ Wool/Wool-type Fabrics

- Albatross Cloth
  - Albatross Wool
- Astrakhan / Poodle Cloth
- Baize
- Bedford Cord
  - Wool Bedford Cord (medium cord)
  - Wool Bedford Cord (fine cord)
- Bengaline
  - Wool Bengaline
- Bolivia
- Broadcloth
  - Wool Broadcloth (flat clear finish)
  - Wool Broadcloth (napped finish)
- Cassimere
- Cheviot
- Cloqué
- Covert Cloth
  - Wool Covert Suiting
  - Wool Covert Coating
- Crepe
  - Wool Crepe
- Doeskin / Wool Doeskin
- Double Cloth
  - Wool Double Cloth
  - Wool French Back
- Double-faced Wool
  - Double-faced Wool (heavy texture)
  - Double-faced Wool (flat face)
  - Blanket Cloth / Plaid Back
- Duvetyn / Wool Duvetyn / Suede Flannel
- Eiderdown
- Étamine
  - Wool Étamine
- Flannel
  - Wool Flannel
  - Worsted Flannel
  - French Flannel
- Fleece / Wool Fleece
- Gabardine
  - Wool-worsted Gabardine
  - Wool Gabardine (summer weight)
  - Wool Gabardine Coating
- Homespun
  - Wool Homespun
  - Wool Homespun Coating
- Hopsacking
  - Wool Hopsacking
- Iridescent Wool / Wool Iridescent
  - Iridescent Wool / Wool Iridescent (plain weave)
  - Iridescent Wool / Wool Iridescent (twill weave)
- Kasha Cloth
- Kersey
- Loden Cloth
- Mackinaw / Mackinac / Buffalo Plaid
- Melton
  - Melton
  - Melton (face-finished) / Chinchilla Cloth
- Ottoman
  - Wool Ottoman
- Poplin
  - Wool Poplin
- Polo Cloth
- Ratiné / Épongé
  - Ratiné / Épongé Wool (plain weave)
  - Ratiné / Épongé Wool (twill weave)
- Serge
  - Wool Serge
- Sharkskin
  - Wool Sharkskin
- Sheer Wools / Wool Sheers
  - Wool Batiste
  - Wool Challis
  - Wool Voile
- Suede Cloth
  - Wool Suede Cloth
- Tartan Woolens / Clan Plaid Woolens
  - Names of the Various Clans
  - Black Watch Plaid
  - Glen Plaid
  - Hound's-tooth Check
- Tricotine
  - Wool / Tricotine / Cavalry Twill
- Tweeds / Wool Tweeds
  - Bird's Eye Tweed
  - Donegal Tweed / Donegal-type Tweed
  - Heather Tweed / Heather-type Tweed
  - Wool Herringbone Tweed (napped finish)
  - Worsted Herringbone Tweed (clear finish)
  - Oatmeal Tweed
  - Salt-and-Pepper Tweed
  - Heavy Textured Tweed
- Whipcord
  - Wool Whipcord
- Wool Fancies
  - Bouclé / Wool Bouclé
  - Wool Gauze
  - Reverse Twill Weave Wool
  - Waffle Cloth
- Wool Knits
  - Doubleknit Wool / Wool Doubleknit
  - Jersey Wool Knit / Wool Jersey
- Worsted
  - Tropical Worsted
  - Unfinished Worsted
- Listing of Wool Terms
  - Merino Wool

Australian Wool
Botany Wool
Saxony Wool
Shetland Wool
Lambswool
Naked Wool
Chart
Comparison of Woolen & Worsted Fabrics

Wool and wool-type fabrics, as listed in this unit, refer to fabrics generically named and classified as wool fabrics which were originally made of one-hundred percent natural wool fibers. This unit includes wool and wool-type fabrics which may be made of:

- Natural wool fibers
- Man-made fibers simulating wool
- Man-made fiber blends
- Natural and man-made fiber blends
- Mixed or combined yarns of different fiber origins

Regardless of the fiber content, wool and wool-type fabrics maintain or simulate the same outward characteristics of the original generically named fabric. The following characteristics remain the same:

- Appearance
- Surface texture and interest
- Hand or feel
- Yarn type and construction
- Fabric structure
- Drapability qualities
- Weight (usually)

Although the original outward characteristics of the fabrics are maintained, the fiber content used *changes* the performance, thread selection, pressing and care factors of the fabric. Each fiber has its own particular characteristics and properties.

There are variations within the generically named fabric. Variations may be obtained by changing, altering, modifying or combining:

- Yarn type
- Yarn construction
- Yarn count and size
- Yarns in warp and filling
- Usual weave structure
- Finishing processes

By combining or varying the components the textile designer can create new fabrics. When one component is changed, the fabric is changed. The newly designed or produced fabric receives a new trade name. The new fabrics or newly named fabrics presented each season are usually placed into categories which may refer to their hand or feel, texture, surface appearance, structure or weight. Different textile companies have their own trade names or trademarks for the fabrics they produce.

Albatross Wool

Fiber Content 100% polyester
Yarn Type filament fiber
Yarn Construction fine; textured
Fabric Structure 4-yarn double-cloth weave; filling of one set interlaced with another set; weaving process binds the cloth—cannot be separated
Finishing Processes shearing; singeing; stabilizing
Color Application piece dyed; solution dyed; yarn dyed
Width 36 inches (91.4 cm)
Weight light
Hand loose; pliable; stretchy
Texture grainy; harsh; rough; semi-sheer
Performance Expectations dimensional stability; durable; elongation; heat-set properties; mildew and moth resistant; high tensile strength; subject to snagging and pulling due to weave; subject to pilling
Drapability Qualities falls into crisp flares; accommodates fullness by gathering, elasticized shirring; fullness retains soft fall

Recommended Construction Aids & Techniques

Hand Needles betweens, 3–5; cotton darners, 4–8; embroidery, 3–6; milliners, 3–6; sharps, 1–3
Machine Needles round/set point, medium 11–14
Threads cotton-covered polyester core, all purpose; cotton-polyester blend, all purpose; spun polyester, all purpose; poly-core waxed, 60
Hems book; bonded; single-fold bias binding; bound/Hong Kong finish; flat/plain; pinked flat; stitched and pinked flat; interfaced; overedged; seam binding
Seams flat-felled; lapped; plain; top-stitched; welt; double welt
Seam Finishes book; single-ply or double-ply bound; Hong Kong bound; single-ply overcast; pinked; pinked and stitched; serging/single-ply overedged; single-ply zigzag
Pressing steam; safe temperature limit 325°F (164.1°C)
Care launder; dry clean
Fabric Resource **Drexler Associates Inc.**

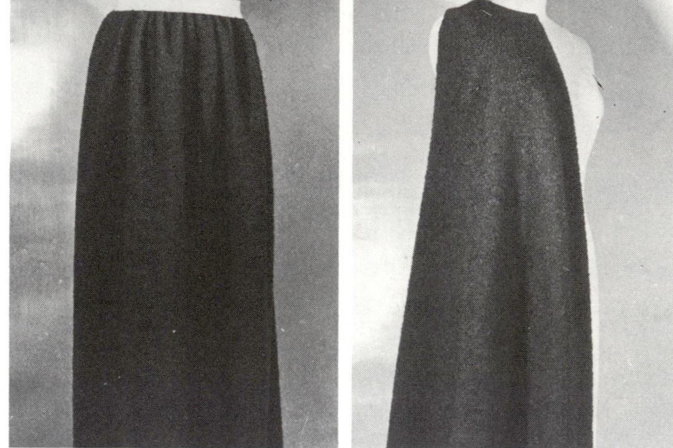

Astrakhan/Poodle Cloth

Fiber Content 88% wool/12% nylon
Yarn Type carded staple; filament fiber
Yarn Construction bulked; novelty curled and looped yarn
Fabric Structure plain weave base; pile structure face
Finishing Processes crabbing; napping; stabilizing
Color Application piece dyed-union dyed; yarn dyed
Width 60 inches (152.4 cm)
Weight heavy
Hand furry; lofty; downy soft; thick
Texture curly loop pile simulates karakul lamb fleece; fuzzy; rough
Performance Expectations elasticity; elongation; insulating properties; tends to shrink and felt; resilient; subject to snagging due to looped yarn and weave; wrinkle resistant
Drapability Qualities falls into wide cones; retains shape or silhouette of garment; better utilized if fitted by seaming and eliminating excess fabric

Recommended Construction Aids & Techniques

Hand Needles chenilles, 18; cotton darners, 1–3; embroidery, 1–4; milliners, 3; tapestry, 18–22
Machine Needles round/set point, coarse 16–18
Threads cotton-covered polyester core, heavy duty; cotton-polyester blend, heavy duty; spun polyester, heavy duty; poly-core waxed, 40
Hems bonded; double-fold or single-fold bias binding; bound/Hong Kong finish; faced; flat/plain; interfaced
Seams plain
Seam Finishes single-ply bound; Hong Kong bound; pinked; pinked and stitched; serging/single-ply overedged; untreated/plain
Pressing *do not* steam; use needle board; safe temperature limit 300°F (150.1°C)
Care dry clean
Fabric Resource **Anglo Fabrics Co. Inc.**

Baize

Fiber Content 100% wool
Yarn Type carded staple
Yarn Construction bulked; conventional
Fabric Structure loosely woven plain weave
Finishing Processes fulling; napping; shearing; processed to resemble felt
Color Application piece dyed; yarn dyed
Width 60 inches (152.4 cm)
Weight medium-heavy
Hand compact; soft; thick
Texture flat; feltlike; smooth
Performance Expectations absorbent; elasticity; elongation; insulating properties; subject to moth damage; resilient
Drapability Qualities falls into wide cones; retains shape or silhouette of garment; better utilized if fitted by seaming and eliminating excess fabric

Recommended Construction Aids & Techniques

Hand Needles chenilles, 20–22; cotton darners, 3–6; embroidery, 1–4; milliners, 3–6
Machine Needles round/set point, medium-coarse, 14–16
Threads silk, industrial size A-B-C; mercerized cotton, heavy duty; six-cord cotton, 40; waxed, 50; cotton-covered polyester core, heavy duty
Hems bonded; double-fold or single-fold bias binding; bound/Hong Kong finish; faced; flat/plain; pinked flat; interfaced; overedged; seam binding
Seams plain; safety-stitched
Seam Finishes single-ply or double-ply bound; Hong Kong bound; single-ply overcast; pinked; pinked and stitched; serging/single-ply overedged; single-ply zigzag
Pressing steam; safe temperature limit 300°F (150.1°C)
Care dry clean
Fabric Resource **Anglo Fabrics Co. Inc.**

Wool Bedford Cord (medium cord)

Fiber Content 100% wool
Yarn Type carded staple
Yarn Construction conventional
Fabric Structure rib weave; rib in lengthwise direction
Finishing Processes crabbing; semidecating; shearing; singeing
Color Application piece dyed; yarn dyed
Width 58-60 inches (147.3-152.4 cm)
Weight medium-heavy
Hand firm; semi-soft
Texture pronounced lengthwise rib on face; high and low surface
Performance Expectations subject to abrasion; elasticity; elongation; subject to moth damage; resilient
Drapability Qualities falls into moderately soft flares; accommodates fullness by pleating, gathering; fullness retains soft fall

Recommended Construction Aids & Techniques

Hand Needles chenilles, 20–22; cotton darners, 1–3; embroidery, 1–4; milliners, 3–6
Machine Needles round/set point, medium 11–14
Threads silk, industrial size A-B-C; mercerized cotton, heavy duty; six-cord cotton, 40; cotton-covered polyester core, heavy duty
Hems book; bonded; single-fold bias binding; bound/Hong Kong finish; flat/plain; pinked flat; stitched and pinked flat; interfaced; overedged; seam binding
Seams plain; safety-stitched; top-stitched; welt; double welt
Seam Finishes book; single-ply or double-ply bound; Hong Kong bound; pinked; pinked and stitched; serging/single-ply overedged; single-ply zigzag
Pressing steam; safe temperature limit 300°F (150.1°C)
Care dry clean
Fabric Resource **Arthur Zeiler Woolens Inc.**

Wool Bedford Cord (fine cord)

Fiber Content 63% wool/23% polyester/9% nylon/5% fur
Yarn Type staple; filament staple; hair fiber
Yarn Construction conventional; textured; tight twist
Fabric Structure rib weave; rib in lengthwise direction
Finishing Processes brushing; napping; stabilizing
Color Application piece dyed–cross dyed, union dyed; yarn dyed
Width 58 inches (147.3 cm)
Weight light
Hand scratchy; soft
Texture fine but pronounced lengthwise rib on face
Performance Expectations subject to abrasion; durable; elongation; heat-set properties; resilient
Drapability Qualities falls into moderately soft flares; accommodates fullness by pleating, gathering, elasticized shirring; maintains a soft fold, unpressed pleat

Recommended Construction Aids & Techniques

Hand Needles chenilles, 20–22; cotton darners, 3–6; embroidery, 1–4; milliners, 3–6
Machine Needles round/set point, medium 11
Threads cotton-covered polyester core, all purpose; cotton-polyester blend, all purpose; spun polyester, all purpose; poly-core waxed, 60
Hems book; bonded; single-fold bias binding; bound/Hong Kong finish; flat/plain; pinked flat; stitched and pinked flat; interfaced; overedged; seam binding
Seams plain; safety-stitched; top-stitched; welt; double welt
Seam Finishes book; single-ply or double-ply bound; Hong Kong bound; single-ply overcast; pinked; pinked and stitched; serging/single-ply overedged; single-ply zigzag
Pressing steam; safe temperature limit 300°F (150.1°C)
Care dry clean
Fabric Resource **Arthur Zeiler Woolens Inc.**

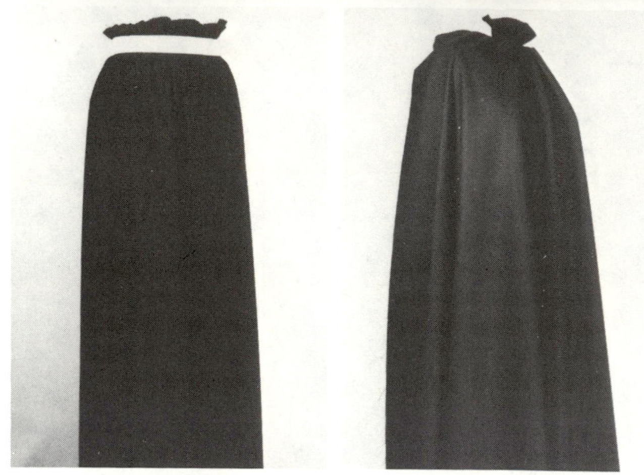

Wool Bengaline

Fiber Content 100% wool
Yarn Type combed staple
Yarn Construction conventional; fine; high twist; worsted spun
Fabric Structure plain weave; heavier filling yarns form fine rib in crosswise direction
Finishing Processes clear finish; decating; shearing; singeing
Color Application piece dyed; yarn dyed
Width 62 inches (157.5 cm)
Weight medium-light
Hand semi-crisp; springy; thin
Texture flat; fine crosswise rib
Performance Expectations subject to edge abrasion; elasticity; elongation; resilient; tends to shine with wear
Drapability Qualities falls into crisp flares; fullness maintains crisp effect; retains shape of garment

Recommended Construction Aids & Techniques

Hand Needles chenilles, 20–22; cotton darners, 3–6; embroidery, 1–4; milliners, 3–6
Machine Needles round/set point, medium 14
Threads silk, industrial size A-B-C; mercerized cotton, heavy duty; six-cord cotton, 50; waxed, 50; cotton-covered polyester core, all purpose
Hems book; bonded; single-fold bias binding; bound/Hong Kong finish; flat/plain; pinked flat; stitched and pinked flat; interfaced; overedged; seam binding
Seams plain; safety-stitched; top-stitched; welt; double welt
Seam Finishes book; single-ply or double-ply bound; Hong Kong bound; single-ply overcast; pinked; pinked and stitched; serging/single-ply overedged; single-ply zigzag
Pressing steam; safe temperature limit 300°F (150.1°C)
Care dry clean
Fabric Resource **Drexler Associates Inc.**

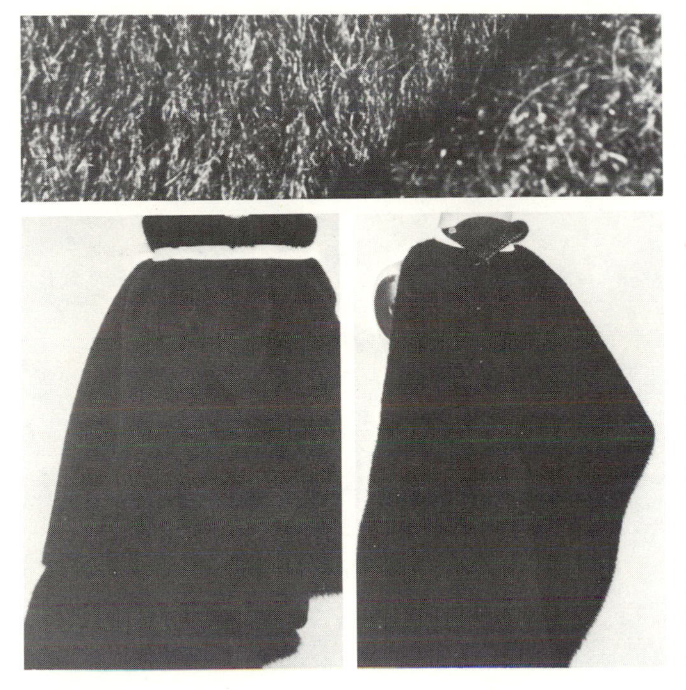

Bolivia

Fiber Content 100% wool
Yarn Type carded staple
Yarn Construction bulked; softly spun
Fabric Structure closely woven twill weave; 3-up and 3-down
Finishing Processes brushing; heavily fulled; napping
Color Application piece dyed; yarn dyed
Width 60 inches (152.4 cm)
Weight heavy
Hand bulky; furry; downy soft; thick
Texture fuzzy; plush pile surface; velvety
Performance Expectations subject to edge abrasion; elasticity; elongation; insulating properties; resilient
Drapability Qualities falls into wide cones; retains shape or silhouette of garment; better utilized if fitted by seaming and eliminating excess fabric

Recommended Construction Aids & Techniques

Hand Needles chenilles, 18; cotton darners, 1–3; embroidery, 1–4; milliners, 3; tapestry, 18–22
Machine Needles round/set point, coarse 16–18
Threads silk, industrial size A-B-C-D; six-cord cotton, 30; waxed, 40 heavy duty; cotton-covered polyester core, heavy duty
Hems bonded; double-fold or single-fold bias binding; bound/Hong Kong finish; faced; flat/plain; interfaced
Seams plain
Seam Finishes single-ply bound; Hong Kong bound; pinked; pinked and stitched; serging/single-ply overedged; untreated/plain
Pressing steam; safe temperature limit 300°F (150.1°C)
Care dry clean
Fabric Resource **Anglo Fabrics Co. Inc.**

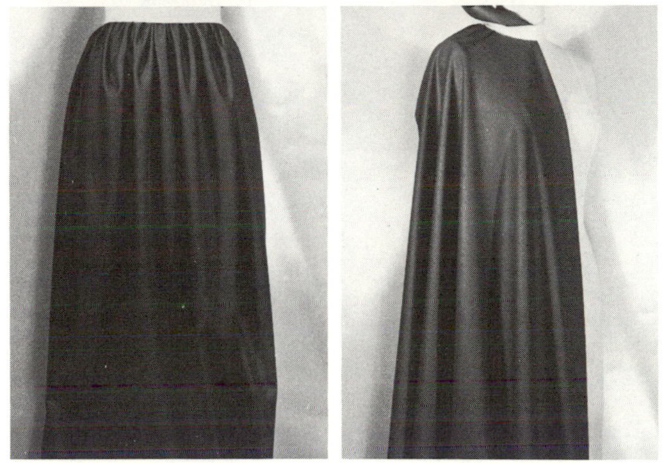

Wool Broadcloth (flat clear finish)

Fiber Content 65% wool/35% nylon
Yarn Type carded staple; filament fiber
Yarn Construction conventional; fine
Fabric Structure loosely woven plain weave
Finishing Processes decating; fulling; shearing; singeing
Color Application piece dyed–union dyed; yarn dyed
Width 60 inches (152.4 cm)
Weight medium-light
Hand light; lofty; soft
Texture flat; smooth
Performance Expectations elasticity; elongation; heat-set properties; insulating properties; resilient; subject to static buildup
Drapability Qualities falls into moderately soft flares; accommodates fullness by pleating, gathering, elasticized shirring; fullness retains soft fall

Recommended Construction Aids & Techniques

Hand Needles chenilles, 20–22; cotton darners, 3–6; embroidery, 1–4; milliners, 3–6
Machine Needles round/set point, medium 14
Threads poly-core waxed, 60; cotton-polyester blend, all purpose; cotton-covered polyester core, all purpose; spun polyester, all purpose
Hems book; bonded; single-fold or double-fold bias binding; flat/plain; pinked flat; stitched and pinked flat; interfaced; overedged; seam binding
Seams plain; safety-stitched; top-stitched; welt; double welt
Seam Finishes book; single-ply or double-ply bound; Hong Kong bound; single-ply overcast; pinked; pinked and stitched; serging/single-ply overedged; single-ply zigzag
Pressing steam; safe temperature limit 300°F (150.1°C)
Care dry clean
Fabric Resource **Twintella Fabrics**

Wool Broadcloth (napped finish)

Fiber Content 100% wool
Yarn Type carded staple
Yarn Construction conventional; fine; low twist
Fabric Structure twill weave; 2-up and 2-down
Finishing Processes fulling; napping; pressed flat; shearing
Color Application piece dyed; yarn dyed
Width 60 inches (152.4 cm)
Weight medium
Hand light; lofty; soft
Texture flat; smooth; velvety
Performance Expectations absorbent; elasticity; elongation; insulating properties; subject to moth damage; resilient; tends to shrink and felt
Drapability Qualities falls into moderately soft flares; fullness maintains moderately soft fall; maintains soft fold, unpressed pleat

Recommended Construction Aids & Techniques

Hand Needles chenilles, 20–22; cotton darners, 3–6; embroidery, 1–4; milliners, 3–6
Machine Needles round/set point, medium 14
Threads mercerized cotton, heavy duty; silk, industrial size A-B-C; six-cord cotton, 50; waxed, 50; cotton-covered polyester core, regular
Hems book; bonded; single-fold bias binding; bound/Hong Kong finish; flat/plain; stitched and pinked flat; interfaced; overedged; seam binding
Seams plain; safety-stitched; top-stitched; welt; double welt
Seam Finishes book; single-ply or double-ply bound; Hong Kong bound; single-ply overcast; pinked; pinked and stitched; serging/single-ply overedged; single-ply zigzag
Pressing steam; safe temperature limit 300°F (150.1°C)
Care dry clean
Fabric Resources Burlington Industries Inc.; Twintella Fabrics

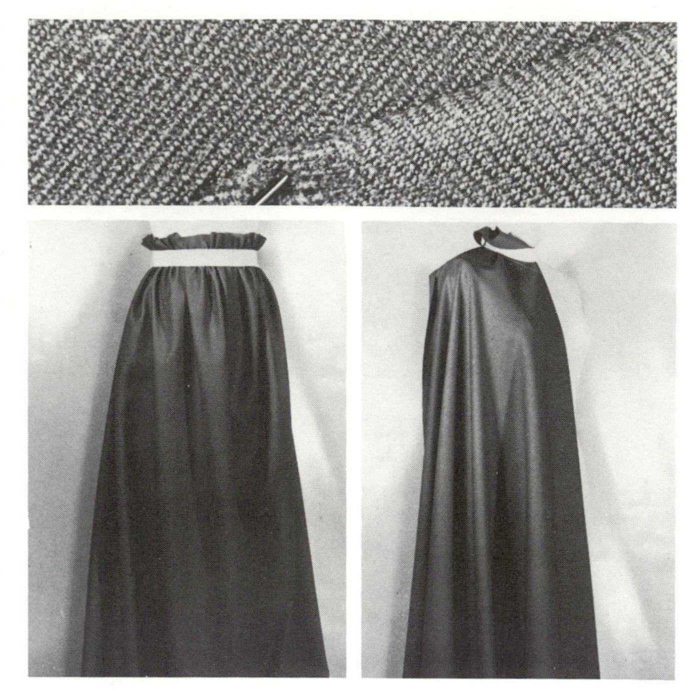

Cassimere[1]

Fiber Content 100% wool
Yarn Type worsted staple
Yarn Construction conventional; high twist
Fabric Structure closely woven right-hand twill weave; 2-up and 2-down
Finishing Processes decating; singeing; sponging
Color Application piece dyed; yarn dyed
Width 60 inches (152.4 cm)
Weight medium
Hand firm; hard; soft
Texture harsh; napless face
Performance Expectations absorbent; elasticity; elongation; insulating properties; subject to moth damage; resilient; tends to shrink and felt; tends to shine with wear
Drapability Qualities falls into moderately soft flares; fullness maintains lofty effect; maintains crisp fold, crease, pleat

Recommended Construction Aids & Techniques

Hand Needles chenilles, 20–22; cotton darners, 3–6; embroidery, 1–4; milliners, 3–6
Machine Needles round/set point, medium 14
Threads mercerized cotton, heavy duty; silk, industrial size A-B-C; six-cord cotton, 50; waxed, 50; cotton-covered polyester core, regular
Hems bonded; double-fold or single-fold bias binding; bound/Hong Kong finish; faced; flat/plain; pinked flat; interfaced; overedged; seam binding
Seams plain; safety-stitched; top-stitched; welt; double welt
Seam Finishes book; single-ply or double-ply bound; Hong Kong bound; single-ply overcast; pinked; pinked and stitched; serging/single-ply overedged; single-ply zigzag
Pressing steam; safe temperature limit 300°F (150.1°C)
Care dry clean
Fabric Resource Burlington Industries Inc.

Cheviot

Fiber Content 100% wool
Yarn Type worsted staple
Yarn Construction conventional
Fabric Structure reverse twill weave
Finishing Processes fulling; singeing; stabilizing
Color Application piece dyed; yarn dyed
Width 60 inches (152.4 cm)
Weight medium-heavy
Hand firm; lofty; soft
Texture slightly napped face; slightly rough; striped appearance created by weave
Performance Expectations absorbent; elasticity; elongation; insulating properties; subject to moth damage; resilient; tends to shrink and felt; tends to shine with wear
Drapability Qualities falls into wide cones; maintains crisp fold, crease, pleat; retains shape of garment

Recommended Construction Aids & Techniques

Hand Needles chenilles, 18; cotton darners, 1–3; embroidery, 1–4; milliners, 3; tapestry, 18–22
Machine Needles round/set point, medium-coarse 14–16
Threads mercerized cotton, heavy duty; silk, industrial size A-B-C; six-cord cotton, 50; waxed, 50; cotton-covered polyester core, regular
Hems bonded; double-fold or single-fold bias binding; bound/Hong Kong finish; faced; flat/plain; pinked flat; interfaced; overedged; seam binding
Seams plain; safety-stitched
Seam Finishes single-ply or double-ply bound; Hong Kong bound; single-ply overcast; pinked; pinked and stitched; serging/single-ply overedged; single-ply zigzag
Pressing steam; safe temperature limit 300°F (150.1°C)
Care dry clean
Fabric Resource **Anglo Fabrics Co. Inc.**

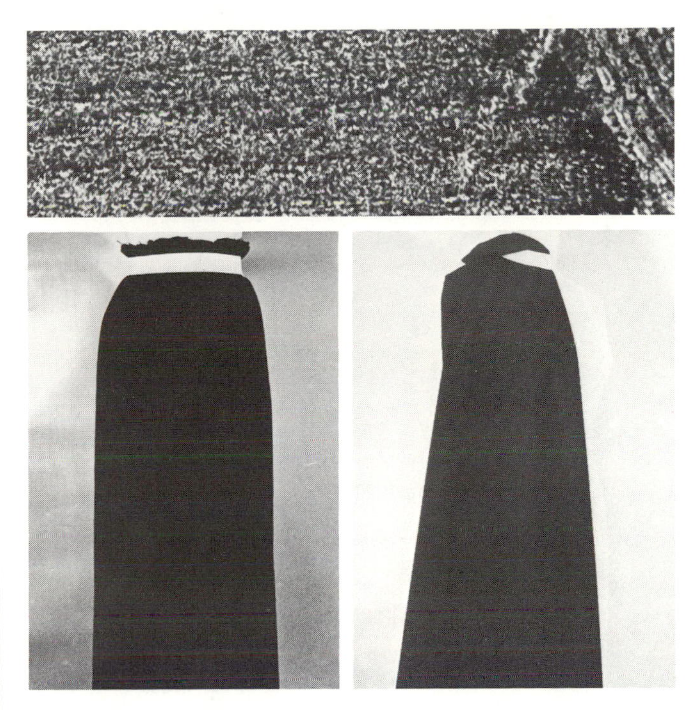

Cloqué

Fiber Content 88% wool/12% nylon
Yarn Type filament fiber; worsted staple
Yarn Construction conventional; fine and thick; high twist
Fabric Structure Jacquard weave; double cloth; fine crisp yarn for back
Finishing Processes decating; pressing; shrinkage control
Color Application piece dyed–union dyed; yarn dyed
Width 60 inches (152.4 cm)
Weight heavy
Hand coarse; firm; hard
Texture blistered effect creates irregularly raised surface
Performance Expectations elasticity; elongation; insulating properties; resilient; subject to abrasion and static buildup; wrinkle resistant
Drapability Qualities falls into wide cones; retains shape or silhouette of garment; better utilized if fitted by seaming and eliminating excess fabric

Recommended Construction Aids & Techniques

Hand Needles chenilles, 18; cotton darners, 1–3; embroidery, 1–4; milliners, 3; tapestry, 18–22
Machine Needles round/set point, medium-coarse 14–16
Threads poly-core waxed, 40; cotton-polyester blend, heavy duty; cotton-covered polyester core, heavy duty; spun polyester, heavy duty
Hems bonded; double-fold or single-fold bias binding; bound/Hong Kong finish; faced; flat/plain; interfaced
Seams plain
Seam Finishes single-ply bound; Hong Kong bound; pinked; pinked and stitched; serging/single-ply overedged; untreated/plain
Pressing steam; safe temperature limit 300°F (150.1°C)
Care dry clean
Fabric Resource **Anglo Fabrics Co. Inc.**

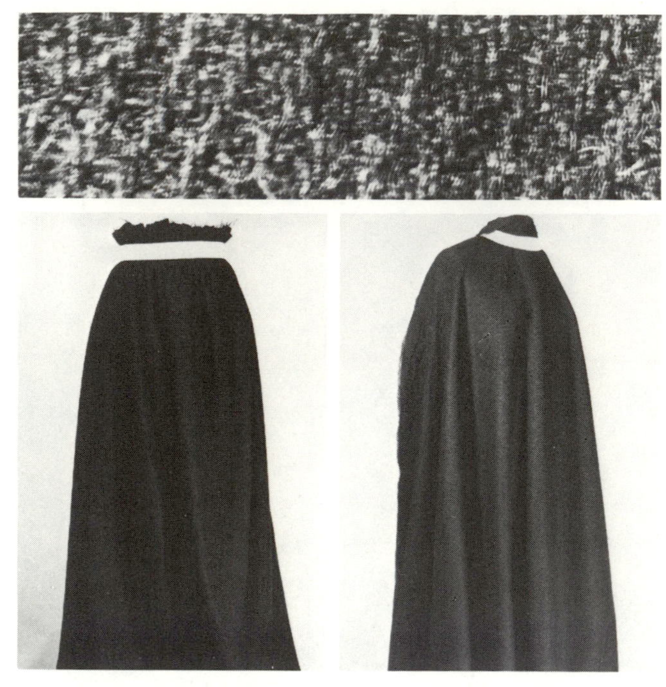

Wool Covert Suiting

Fiber Content 45% acrylic/30% polyester/25% wool
Yarn Type filament fiber; filament staple; worsted staple
Yarn Construction conventional; 2-ply; high twist
Fabric Structure warp-faced twill weave
Finishing Processes clear finish; decating; singeing
Color Application piece dyed–cross dyed; yarn dyed—2 different colored plies
Width 60 inches (152.4 cm)
Weight medium
Hand coarse; firm; semi-soft
Texture diagonal weave surface; flecked appearance
Performance Expectations dimensional stability; elasticity; elongation; resilient; subject to pilling and static buildup
Drapability Qualities falls into moderately soft flares; maintains crisp fold, crease, pleat; retains shape of garment

Recommended Construction Aids & Techniques

Hand Needles chenilles, 20–22; cotton darners, 3–6; embroidery, 1–4; milliners, 3–6
Machine Needles round/set point, medium-coarse 14–16
Threads poly-core waxed, 50; cotton-polyester blend, heavy duty; cotton-covered polyester core, heavy duty; spun polyester, heavy duty
Hems bonded; double-fold or single-fold bias binding; bound/Hong Kong finish; faced; flat/plain; pinked flat; interfaced; overedged; seam binding
Seams plain; safety-stitched
Seam Finishes single-ply or double-ply bound; Hong Kong bound; single-ply overcast; pinked; pinked and stitched; serging/single-ply overedged; single-ply zigzag
Pressing steam; safe temperature limit 300°F (150.1°C)
Care dry clean
Fabric Resource Twintella Fabrics

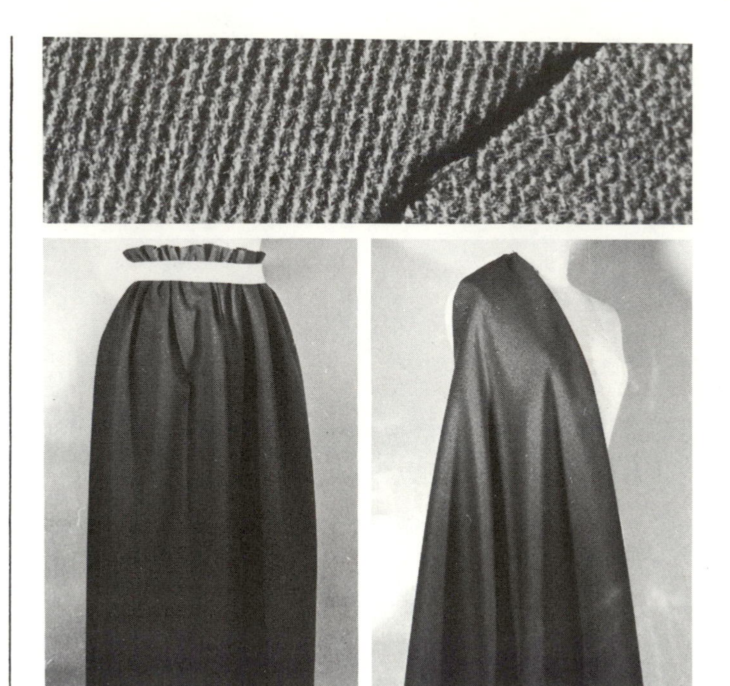

Wool Covert Coating

Fiber Content 100% wool
Yarn Type worsted staple
Yarn Construction conventional; 2-ply; high twist
Fabric Structure warp-faced, right-hand twill weave
Finishing Processes fulling; napping; sponging; water repellent
Color Application piece dyed; yarn dyed
Width 60 inches (152.4 cm)
Weight heavy
Hand compact; soft; thick
Texture pronounced diagonal twill on surface
Performance Expectations durable; elasticity; elongation; insulating properties; subject to pilling; resilient; tends to shrink and felt; water-repellent properties
Drapability Qualities falls into wide cones; retains shape or silhouette of garment; better utilized if fitted by seaming and eliminating excess fabric

Recommended Construction Aids & Techniques

Hand Needles chenilles, 18; cotton darners, 1–3; embroidery, 1–4; milliners, 3; tapestry, 18–22
Machine Needles round/set point, coarse 16–18
Threads silk, industrial size A-B-C-D; six-cord cotton, 30; waxed, 40 heavy duty; cotton-covered polyester core, heavy duty
Hems bonded; double-fold or single-fold bias binding; bound/Hong Kong finish; faced; flat/plain; interfaced
Seams plain
Seam Finishes single-ply bound; Hong Kong bound; pinked; pinked and stitched; serging/single-ply overedged; untreated/plain
Pressing steam; safe temperature limit 300°F (150.1°C)
Care dry clean
Fabric Resource Anglo Fabrics Co. Inc.

Wool Crepe

Fiber Content 100% wool
Yarn Type worsted staple
Yarn Construction creped; alternating S and Z tightly twisted warp yarns
Fabric Structure plain weave
Finishing Processes decating; singeing; stabilizing
Color Application piece dyed; yarn dyed
Width 60 inches (152.4 cm)
Weight light
Hand loose; springy; supple
Texture pebbly; rough
Performance Expectations absorbent; elasticity; elongation; resilient; wrinkle resistant
Drapability Qualities falls into soft flares; accommodates fullness by pleating, gathering, elasticized shirring; fullness retains soft fall

Recommended Construction Aids & Techniques

Hand Needles betweens, 3–5; cotton darners, 4–8; embroidery, 3–6; milliners, 3–6; sharps, 1–3
Machine Needles round/set point, medium 11–14
Threads mercerized cotton, heavy duty; silk, industrial size A; six-cord cotton, 50; waxed, 50; cotton-covered polyester core, all purpose
Hems book; bound/Hong Kong finish; pinked flat; stitched and pinked flat; machine-stitched; overedged; seam binding
Seams flat-felled; lapped; plain; top-stitched; welt; double welt
Seam Finishes single-ply bound; Hong Kong bound; double-stitched; edge-stitched; pinked; pinked and stitched; serging/single-ply overedged; untreated/plain; single-ply zigzag
Pressing steam; safe temperature limit 300°F (150.1°C)
Care dry clean
Fabric Resource **Burlington Industries Inc.**

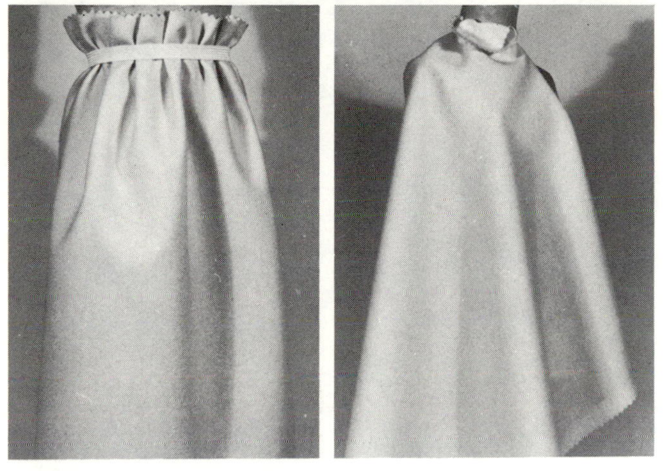

Doeskin/Wool Doeskin

Fiber Content 100% wool
Yarn Type worsted staple
Yarn Construction conventional; thin
Fabric Structure plain weave
Finishing Processes decating; fulling; sanding/sueding; singeing
Color Application piece dyed; yarn dyed
Width 60 inches (152.4 cm)
Weight medium-light
Hand compact; light; downy soft
Texture feltlike; velvety smooth face
Performance Expectations elasticity; elongation; insulating properties; subject to moth damage; resilient
Drapability Qualities falls into moderately soft flares; fullness maintains lofty effect; retains shape of garment

Recommended Construction Aids & Techniques

Hand Needles chenilles, 20–22; cotton darners, 3–6; embroidery, 1–4; milliners, 3–6
Machine Needles round/set point, medium 11–14
Threads mercerized cotton, heavy duty; silk, industrial size A-B; six-cord cotton, 50; waxed, 50; cotton-covered polyester core, all purpose
Hems book; bonded; single-fold bias binding; bound/Hong Kong finish; flat/plain; pinked flat; stitched and pinked flat; interfaced; overedged; seam binding
Seams plain; safety-stitched; top-stitched; welt; double welt
Seam Finishes book; single-ply or double-ply bound; Hong Kong bound; single-ply overcast; pinked; pinked and stitched; serging/single-ply overedged; single-ply zigzag
Pressing steam; safe temperature limit 300°F (150.1°C)
Care dry clean
Fabric Resource **Anglo Fabrics Co. Inc.**

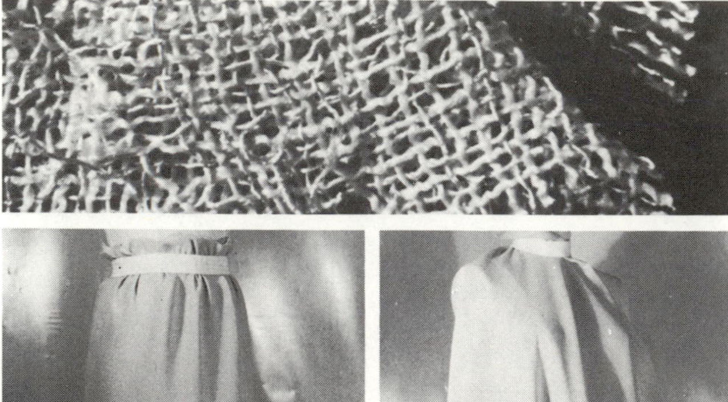

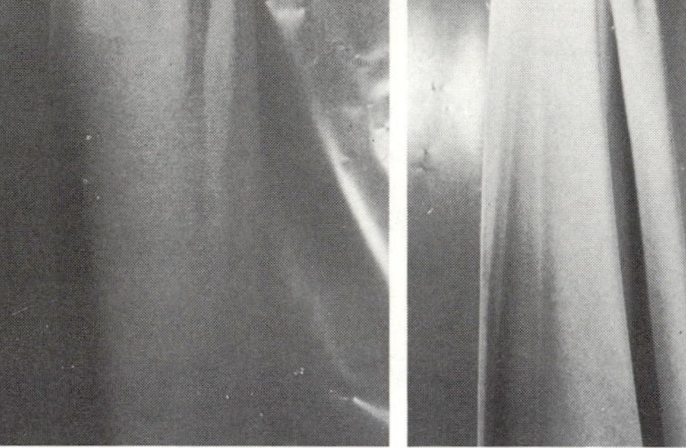

Wool Double Cloth

Fiber Content 100% polyester
Yarn Type filament fiber
Yarn Construction fine; textured
Fabric Structure 5-yarn double-cloth weave; additional filling yarn joins both cloths; weaving process binds the cloth—cannot be separated
Finishing Processes sponging; stabilizing
Color Application piece dyed; yarn dyed
Width 36 inches (91.4 cm)
Weight light
Hand loose; pliable; stretchy
Texture grainy; harsh; rough; semi-sheer
Performance Expectations dimensional stability; durable; elongation; heat-set properties; mildew and moth resistant; subject to pilling; subject to snagging and pulling due to weave; high tensile strength
Drapability Qualities falls into crisp flares; accommodates fullness by gathering, elasticized shirring; fullness retains soft fall

Recommended Construction Aids & Techniques

Hand Needles betweens, 3-5; cotton darners, 4-8; embroidery, 3-6; milliners, 3-6; sharps, 1-3
Machine Needles round/set point, medium 11-14
Threads poly-core waxed, 60; cotton-polyester blend, all purpose; cotton-covered polyester core, all purpose; spun polyester, all purpose
Hems book; bonded; single-fold bias binding; bound/Hong Kong finish; flat/plain; pinked flat; stitched and pinked flat; interfaced; overedged; seam binding
Seams flat-felled; lapped; plain; top-stitched; welt; double welt
Seam Finishes book; single-ply or double-ply bound; Hong Kong bound; single-ply overcast; pinked and stitched; serging/single-ply overedged; single-ply zigzag
Pressing steam; safe temperature limit 325°F (164.1°C)
Care launder; dry clean
Fabric Resource Drexler Associates Inc.

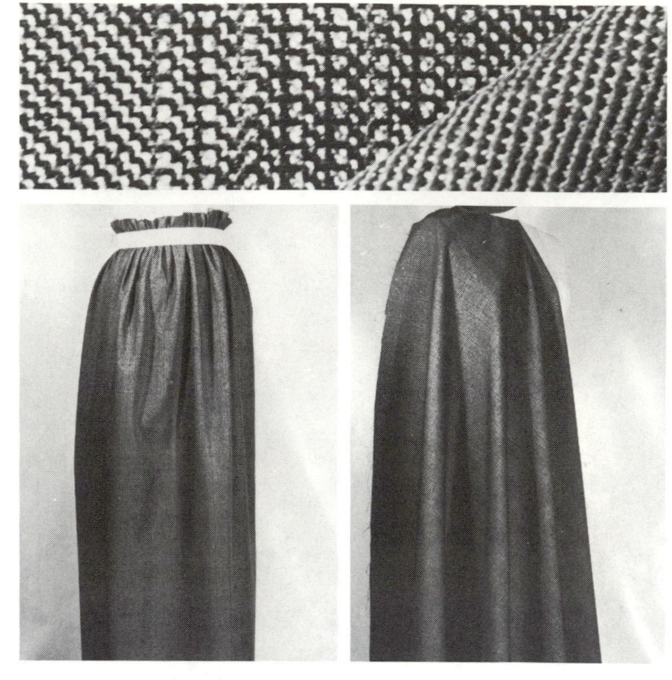

Wool French Back

Fiber Content 100% polyester
Yarn Type filament fiber
Yarn Construction 150-denier; fine; textured
Fabric Structure 3-yarn double-cloth weave; two warp yarns/one filling yarn
Finishing Processes sponging; stabilizing
Color Application solution dyed; yarn dyed
Width 60 inches (152.4 cm)
Weight medium-light
Hand nonelastic; firm; hard
Texture flat; smooth; simulated worsted suiting
Performance Expectations subject to edge abrasion; dimensional stability; durable; elongation; heat-set properties; mildew and moth resistant; high tensile strength
Drapability Qualities falls into crisp wide flares; maintains crisp fold, crease, pleat; retains shape of garment

Recommended Construction Aids & Techniques

Hand Needles chenilles, 20-22; cotton darners, 3-6; embroidery, 1-4; milliners, 3-6
Machine Needles round/set point, medium 14
Threads poly-core waxed, 60; cotton-polyester blend, all purpose; cotton-covered polyester core, all purpose; spun polyester, all purpose
Hems book; bonded; single-fold bias binding; bound/Hong Kong finish; flat/plain; pinked flat; stitched and pinked flat; interfaced; overedged; seam binding
Seams plain; safety-stitched; top-stitched; welt; double welt
Seam Finishes book; single-ply or double-ply bound; Hong Kong bound; single-ply overcast; pinked; pinked and stitched; serging/single-ply overedged; single-ply zigzag
Pressing steam; safe temperature limit 325°F (164.1°C)
Care dry clean
Fabric Resource Drexler Associates Inc.

Double-faced Wool (heavy texture)

Fiber Content 68% wool/19% acrylic/10% nylon/3% polyester
Yarn Type filament fiber; filament staple; staple
Yarn Construction conventional; thick-and-thin; high and slack twist
Fabric Structure 5-yarn double cloth weave; closely woven plain weave (face); loosely woven plain weave (back); separate binder yarn joins face and back—can be separated
Finishing Processes crabbing; fulling; sponging; stabilizing
Color Application yarn dyed
Width 54–56 inches (137.2–142.2 cm)
Weight heavy
Hand bulky; soft; thick
Texture reversible; tweedy; uneven surface
Performance Expectations compressible; elasticity; elongation; insulating properties; resilient; subject to snagging due to weave; subject to pilling and static buildup
Drapability Qualities falls into a wide cone; retains shape or silhouette of garment; better utilized if fitted by seaming and eliminating excess fabric

Recommended Construction Aids & Techniques

Hand Needles chenilles, 18; cotton darners, 1–3; embroidery, 1–4; milliners, 3; tapestry, 18–22
Machine Needles round/set point, coarse 16–18
Threads poly-core waxed, 40; cotton-polyester blend, heavy duty; cotton-covered polyester core, heavy duty; spun polyester, heavy duty
Hems bonded; double-fold or single-fold bias binding; bound/Hong Kong finish; faced; flat/plain; each ply turned to inside for clean-finished face and back; trimmed and self bound
Seams plain
Seam Finishes single-ply bound; Hong Kong bound; pinked; pinked and stitched; serging/single-ply overedged; untreated/plain; each ply turned to inside for clean-finished face and back; trimmed and self bound
Pressing steam; safe temperature limit 300°F (150.1°C)
Care dry clean
Fabric Resource **Auburn Fabrics Inc.**

Double-faced Wool (flat face)

Fiber Content 90% wool/10% nylon
Yarn Type staple; filament staple
Yarn Construction bulked; conventional
Fabric Structure 5-yarn double-cloth weave; twill weave face and back; separate binder yarn joins face and back—can be separated
Finishing Processes fulling; napping; shearing; sponging
Color Application yarn dyed
Width 58 inches (147.3 cm)
Weight heavy
Hand compact; lofty; soft; thick
Texture bulky; flat; napped face and back; reversible
Performance Expectations absorbent; elasticity; elongation; insulating properties; resilient; water-repellent properties
Drapability Qualities falls into wide cones; retains shape or silhouette of garment; better utilized if fitted by seaming and eliminating excess fabric

Recommended Construction Aids & Techniques

Hand Needles chenilles, 18; cotton darners, 1–3; embroidery, 1–4; milliners, 3; tapestry, 18–22
Machine Needles round/set point, coarse 16–18
Threads silk, industrial size A-B-C; six-cord cotton, 30; waxed, 40 heavy duty; cotton-covered polyester core, heavy duty
Hems bonded; double-fold or single-fold bias binding; bound/Hong Kong finish; flat/plain; overedged; each ply turned to inside for clean-finished face and back; trimmed and self bound
Seams plain; safety-stitched and top-stitched; single-ply overedged and top-stitched; welt; double welt
Seam Finishes single-ply bound; Hong Kong bound; pinked; pinked and stitched; serging/single-ply overedged; untreated/plain; each ply turned to inside for clean-finished face and back; trimmed and self bound
Pressing steam; use needle board; safe temperature limit 300°F (150.1°C)
Care dry clean
Fabric Resource **Amical Fabrics**

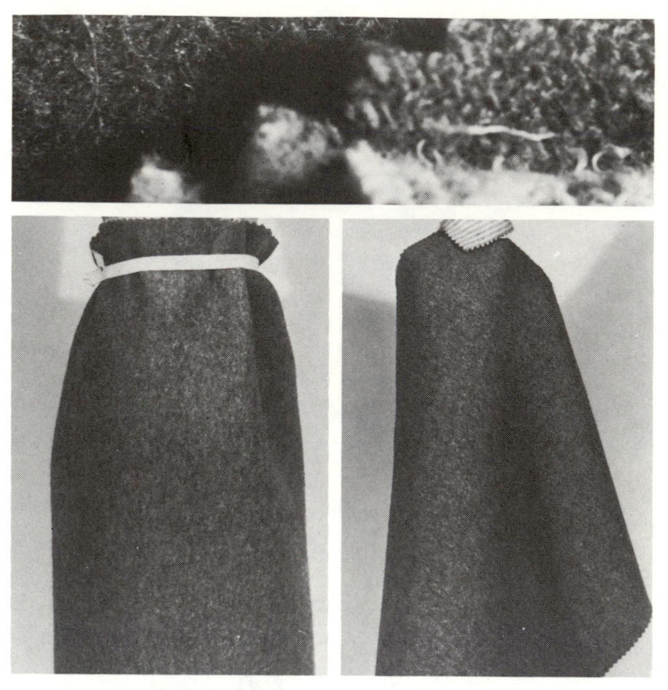

Blanket Cloth/Plaid Back

Fiber Content 99% wool/1% nylon
Yarn Type carded staple; filament binder yarn
Yarn Construction bulked; softly spun
Fabric Structure 5-yarn double-faced cloth; twill weave face and back; separate binder yarn joins face and back—can be separated
Finishing Processes fulling; napping; sponging; stabilizing
Color Application yarn dyed
Width 60 inches (152.4 cm)
Weight heavy
Hand compact; semi-soft; thick
Texture fuzzy; napped; reversible
Performance Expectations elasticity; elongation; insulating properties; subject to pilling; resilient; tends to shrink and felt
Drapability Qualities falls into a wide cone; retains shape or silhouette of garment; better utilized if fitted by seaming and eliminating excess fabric

Recommended Construction Aids & Techniques

Hand Needles chenilles, 18; cotton darners, 1–3; embroidery, 1–4; milliners, 3; tapestry, 18–22
Machine Needles round/set point, coarse 18
Threads silk, industrial size A-B-C-D; six-cord cotton, 30; waxed, 40 heavy duty; cotton-covered polyester core, heavy duty
Hems bonded; double-fold or single-fold bias binding; bound/Hong Kong finish; faced; flat/plain; overedged; each ply turned to inside for clean-finished face and back; trimmed and self bound
Seams plain; top-stitched
Seam Finishes single-ply bound; Hong Kong bound; pinked; pinked and stitched; serging/single-ply overedged; untreated/plain; each ply turned to inside for clean-finished face and back
Pressing steam; safe temperature limit 300°F (150.1°C)
Care dry clean
Fabric Resource Anglo Fabrics Co. Inc.

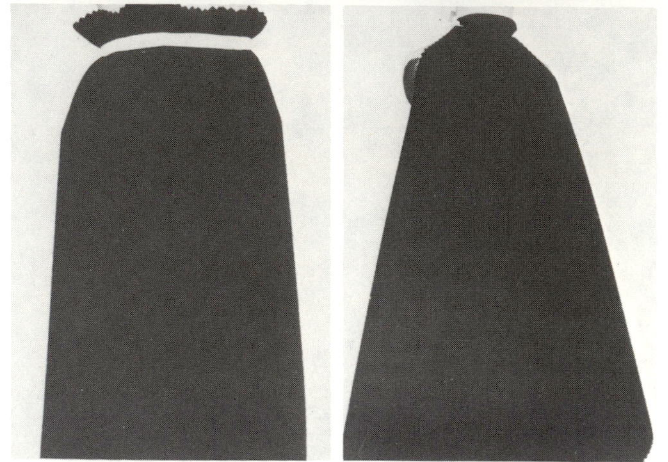

Duvetyn/Wool Duvetyn/Suede Flannel

Fiber Content 100% wool
Yarn Type carded staple
Yarn Construction conventional; softly spun
Fabric Structure filling-faced satin or twill weave
Finishing Processes brushing; fulling; napping; shearing; singeing
Color Application piece dyed; yarn dyed
Width 60 inches (152.4 cm)
Weight medium-heavy
Hand silky; soft; suedelike; velvety
Texture fine, downy soft nap; nap conceals weave structure; brushed and pressed finish creates smooth lustrous face
Performance Expectations elasticity; elongation; insulating properties; subject to pilling; resilient; tends to shine with wear
Drapability Qualities falls into a wide cone; retains shape or silhouette of garment; better utilized if fitted by seaming and eliminating excess fabric

Recommended Construction Aids & Techniques

Hand Needles chenilles, 18; cotton darners, 1–3; embroidery, 1–4; milliners, 3; tapestry, 18–22
Machine Needles round/set point, coarse 16–18
Threads silk, industrial size A-B-C; mercerized cotton, heavy duty; six-cord cotton, 30; waxed, 40 heavy duty; cotton-covered polyester core, heavy duty
Hems bonded; double-fold or single-fold bias binding; bound/Hong Kong finish; faced; flat/plain; pinked flat; interfaced; overedged; seam binding
Seams plain; safety-stitched
Seam Finishes single-ply bound; Hong Kong bound; pinked; pinked and stitched; serging/single-ply overedged; untreated/plain
Pressing steam; safe temperature limit 300°F (150.1°C)
Care dry clean
Fabric Resource Anglo Fabrics Co. Inc.

Eiderdown

Fiber Content 100% wool
Yarn Type carded berber yarn
Yarn Construction conventional; softly spun; thick
Fabric Structure loosely woven twill weave
Finishing Processes brushing; napping on face; sponging
Color Application yarn dyed
Width 56–58 inches (142.2–147.3 cm)
Weight medium-light
Hand elastic; lofty; soft
Texture raised fuzzy hairs on face; heavily napped face
Performance Expectations compressible; elasticity; flexible; subject to pilling; resilient; subject to snagging due to weave; high tear strength
Drapability Qualities falls into a few soft cones; fullness maintains lofty bouffant effect; retains shape or silhouette of garment

Recommended Construction Aids & Techniques

Hand Needles chenilles, 18; cotton darners, 1–3; embroidery, 1–4; milliners, 3; tapestry, 18–22
Machine Needles round/set point, coarse 16–18
Threads silk, industrial size A-B; mercerized cotton, heavy duty; six-cord cotton, 50; waxed, 50; cotton-covered polyester core, regular
Hems bonded; double-fold or single-fold bias binding; bound/Hong Kong finish; faced; pinked flat; interfaced; overedged; seam binding
Seams plain; safety-stitched
Seam Finishes single-ply or double-ply bound; Hong Kong bound; single-ply overcast; serging/single-ply overedged; untreated/plain; single-ply zigzag
Pressing steam; safe temperature limit 300°F (150.1°C)
Care dry clean
Fabric Resource **Auburn Fabrics Inc.**

Wool Étamine

Fiber Content 100% wool
Yarn Type staple
Yarn Construction conventional; thick round yarns; tight twist
Fabric Structure plain weave; openweave
Finishing Processes decating; shearing; singeing; stabilizing
Color Application piece dyed; yarn dyed
Width 60 inches (152.4 cm)
Weight light to medium-light
Hand lofty; loose; semi-soft
Texture coarse; flat
Performance Expectations absorbent; elasticity; elongation; subject to pilling; resilient; tends to shrink and felt
Drapability Qualities falls into firm flares; fullness maintains lofty effect; maintains soft fold, unpressed pleat

Recommended Construction Aids & Techniques

Hand Needles chenilles, 20–22; cotton darners, 3–6; embroidery, 1–4; milliners, 3–6
Machine Needles round/set point, medium 14
Threads silk, industrial size A-B; mercerized cotton, heavy duty; six-cord cotton, 50; waxed, 50; cotton-covered polyester core, regular
Hems book; bonded; single-fold bias binding; bound/Hong Kong finish; flat/plain; pinked flat; stitched and pinked flat; interfaced; overedged; seam binding
Seams plain; safety-stitched; top-stitched; welt; double welt
Seam Finishes single-ply or double-ply bound; Hong Kong bound; single-ply overcast; pinked; pinked and stitched; serging/single-ply overedged; single-ply zigzag
Pressing steam; safe temperature limit 300°F (150.1°C)
Care dry clean
Fabric Resources **Anglo Fabrics Co. Inc.; Twintella Fabrics**

Wool Flannel

Fiber Content 100% wool
Yarn Type carded staple
Yarn Construction conventional; ply
Fabric Structure[2] 2/1 right-hand twill weave
Finishing Processes[2] fulling; napping; sponging
Color Application piece dyed; yarn dyed
Width 60 inches (152.4 cm)
Weight[2] medium
Hand lofty; soft; supple
Texture smooth; slightly napped face
Performance Expectations absorbent; elasticity; elongation; subject to pilling; resilient; tends to shrink and felt
Drapability Qualities falls into firm flares; fullness maintains lofty effect; maintains soft fold, unpressed pleat

Recommended Construction Aids & Techniques

Hand Needles chenilles, 20–22; cotton darners, 3–6; embroidery, 1–4; milliners, 3–6
Machine Needles round/set point, medium-coarse 14–16
Threads silk, industrial size A-B; mercerized cotton, heavy duty; six-cord cotton, 50; waxed, 50; cotton-covered polyester core, all purpose
Hems bonded; double-fold or single-fold bias binding; bound/Hong Kong finish; faced; flat/plain; pinked flat; interfaced; overedged; seam binding
Seams plain; safety-stitched; top-stitched; welt; double welt
Seam Finishes single-ply or double-ply bound; Hong Kong bound; single-ply overcast; pinked; pinked and stitched; serging/single-ply overedged; single-ply zigzag
Pressing steam; safe temperature limit 300°F (150.1°C)
Care dry clean
Fabric Resource **Carleton Woolen Mills Inc.**

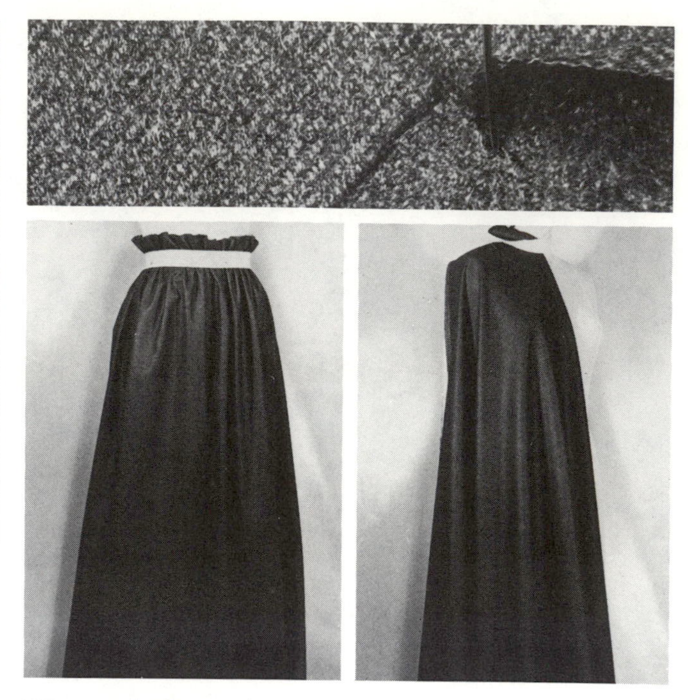

Worsted Flannel

Fiber Content 100% wool
Yarn Type worsted staple
Yarn Construction conventional; tight twist
Fabric Structure[2] twill weave
Finishing Processes[2] decating; slight napping; singeing; stabilizing
Color Application top-dyed yarn; yarn dyed
Width 60 inches (152.4 cm)
Weight[2] medium-light
Hand nonelastic; firm; hard
Texture flat; harsh
Performance Expectations elasticity; elongation; subject to moth damage; tends to shine with wear
Drapability Qualities falls into crisp flares; maintains crisp fold, crease, pleat; retains shape of garment

Recommended Construction Aids & Techniques

Hand Needles chenilles, 20–22; cotton darners, 3–6; embroidery, 1–4; milliners, 3–6
Machine Needles round/set point, medium 14
Threads silk, industrial size A-B; mercerized cotton, heavy duty; six-cord cotton, 50; waxed, 50; cotton-covered polyester core, all purpose
Hems book; bonded; single-fold bias binding; bound/Hong Kong finish; flat/plain; pinked flat; stitched and pinked flat; interfaced; overedged; seam binding
Seams plain; safety-stitched; top-stitched; welt; double welt
Seam Finishes book; single-ply or double-ply bound; Hong Kong bound; single-ply overcast; pinked; pinked and stitched; serging/single-ply overedged; single-ply zigzag
Pressing steam; safe temperature limit 300°F (150.1°C)
Care dry clean
Fabric Resources **Burlington Industries Inc.; Drexler Associates Inc.**

French Flannel

Fiber Content 100% wool
Yarn Type worsted staple
Yarn Construction conventional; fine; high twist
Fabric Structure tightly woven twill weave
Finishing Process fulling; slight napping; shearing; singeing
Color Application piece dyed; yarn dyed
Width 61–62 inches (154.9–157.5 cm)
Weight medium-light
Hand smooth; downy soft; suedelike
Texture compact; flat; velvety napped face and back
Performance Expectations elasticity; elongation; subject to moth damage; tends to shine with wear
Drapability Qualities falls into firm flares; fullness maintains lofty effect; retains shape of garment

Recommended Construction Aids & Techniques

Hand Needles chenilles, 20–22; cotton darners, 3–6; embroidery, 1–4; milliners, 3–6
Machine Needles round/set point, medium 14
Threads silk, industrial size A-B; mercerized cotton, heavy duty; six-cord cotton, 50; waxed, 50; cotton-covered polyester core, all purpose
Hems book; bonded; single-fold bias binding; bound/Hong Kong finish; flat/plain; pinked flat; stitched and pinked flat; interfaced; overedged; seam binding
Seams plain; safety-stitched; top-stitched; welt; double welt
Seam Finishes book; single-ply or double-ply bound; Hong Kong bound; single-ply overcast; pinked; pinked and stitched; serging/single-ply overedged; single-ply zigzag
Pressing steam; safe temperature limit 300°F (150.1°C)
Care dry clean
Fabric Resources **Spring Mills Inc.**

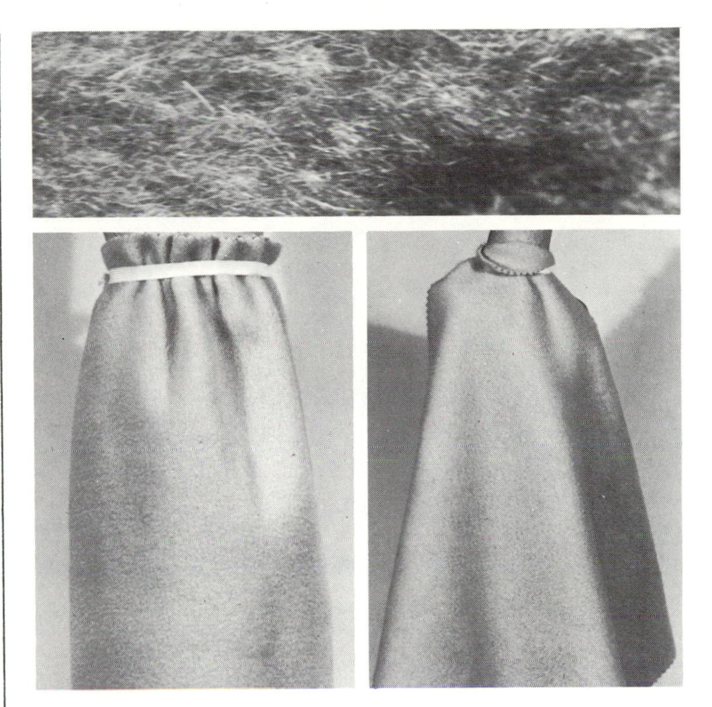

Fleece/Wool Fleece

Fiber Content 100% wool
Yarn Type carded staple
Yarn Construction bulked; softly spun; thick
Fabric Structure loosely woven twill weave
Finishing Processes heavily fulled; napping; shearing
Color Application piece dyed; yarn dyed
Width 60 inches (152.4 cm)
Weight heavy
Hand bulky; soft; thick
Texture fuzzy; long pile
Performance Expectations subject to edge abrasion; elasticity; elongation; insulating properties; subject to pilling; resilient
Drapability Qualities falls into a wide cone; retains shape or silhouette of garment; better utilized if fitted by seaming and eliminating excess fabric

Recommended Construction Aids & Techniques

Hand Needles chenilles, 18; cotton darners, 1–3; embroidery, 1–4; milliners, 3; tapestry, 18–22
Machine Needles round/set point, coarse 18
Threads silk, industrial size A-B-C-D; six-cord cotton, 30; waxed, 40 heavy duty; cotton-covered polyester core, heavy duty
Hems bonded; double-fold or single-fold bias binding; bound/Hong Kong finish; faced; flat/plain; interfaced
Seams plain
Seam Finishes single-ply bound; Hong Kong bound; pinked; pinked and stitched; serging/single-ply overedged; untreated/plain
Pressing steam; use needle board; safe temperature limit 300°F (150.1°C)
Care dry clean
Fabric Resource **Anglo Fabrics Co. Inc.**

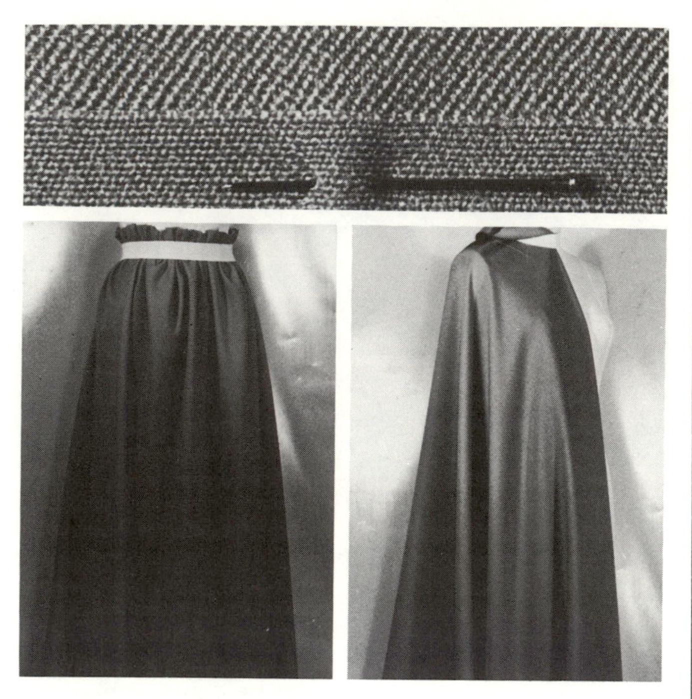

Wool-worsted Gabardine

Fiber Content 100% wool
Yarn Type worsted staple
Yarn Construction conventional; high twist
Fabric Structure tightly woven 2x2 twill weave (59°-63° twill)
Finishing Processes clear finish; decating; pressing; singeing
Color Application piece dyed; yarn dyed
Width 60 inches (152.4 cm)
Weight medium
Hand coarse; firm; hard
Texture harsh; pronounced twill rib on face; flat smooth back without rib
Performance Expectations subject to edge abrasion; durable; elasticity; elongation; resilient; tends to shine with wear
Drapability Qualities falls into firm flares; maintains a crisp fold, crease, pleat; retains shape of garment

Recommended Construction Aids & Techniques

Hand Needles chenilles, 20-22; cotton darners, 3-6; embroidery, 1-4; milliners, 3-6
Machine Needles round/set point, medium-coarse 14-16
Threads silk, industrial size A-B; mercerized cotton, heavy duty; six-cord cotton, 50; waxed, 50; cotton-covered polyester core, all purpose
Hems bonded; double-fold or single-fold bias binding; bound/Hong Kong finish; faced; flat/plain; pinked flat; interfaced; overedged; seam binding
Seams plain; safety-stitched
Seam Finishes single-ply or double-ply bound; Hong Kong bound; single-ply overcast; pinked; pinked and stitched; serging/single-ply overedged; single-ply zigzag
Pressing steam; safe temperature limit 300°F (150.1°C)
Care dry clean
Fabric Resource **Burlington Industries Inc.**

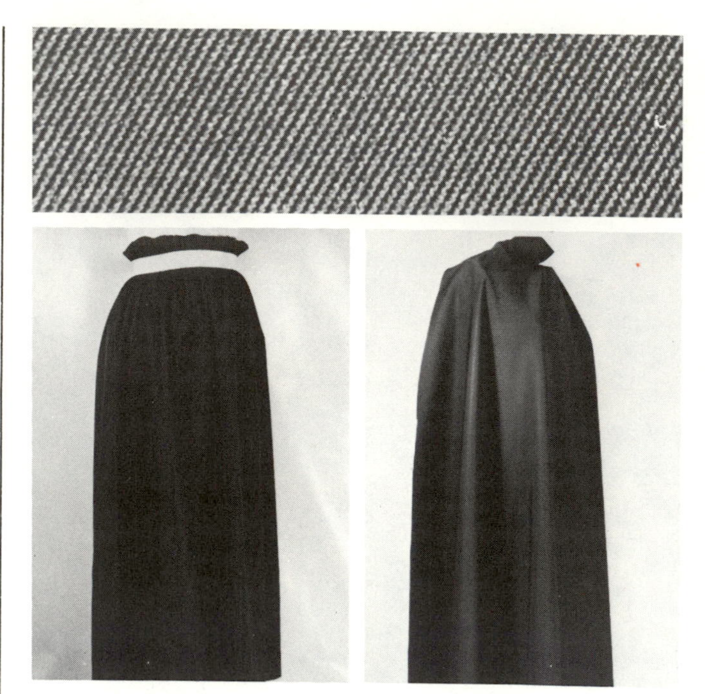

Wool Gabardine (summer weight)

Fiber Content 100% polyester
Yarn Type filament fiber
Yarn Construction fine; textured
Fabric Structure tightly woven 2x2 twill weave
Finishing Processes decating; singeing; stabilizing
Color Application piece dyed; yarn dyed
Width 60 inches (152.4 cm)
Weight light
Hand nonelastic; firm; springy
Texture flat; silky; fine rib on face; smooth back without rib
Performance Expectations subject to color crocking and edge abrasion; dimensional stability; durable; elongation; heat-set properties; mildew and moth resistant; high tensile strength
Drapability Qualities falls into crisp flares; fullness maintains crisp effect; maintains crisp fold, crease, pleat

Recommended Construction Aids & Techniques

Hand Needles betweens, 3-5; cotton darners, 4-8; embroidery, 3-6; milliners, 3-6; sharps, 1-5
Machine Needles round/set point, medium 14
Threads poly-core waxed, 60; cotton-polyester blend, all purpose; cotton-covered polyester core, all purpose; spun polyester, all purpose
Hems book; bonded; single-fold bias binding; bound/Hong Kong finish; flat/plain; pinked flat; stitched and pinked flat; interfaced; overedged; seam binding
Seams plain; safety-stitched; top-stitched; welt; double welt
Seam Finishes book; single-ply or double-ply bound; Hong Kong bound; single-ply overcast; pinked; pinked and stitched; serging/single-ply overedged; single-ply zigzag
Pressing steam; safe temperature limit 325°F (164.1°C)
Care launder; dry clean
Fabric Resource **Drexler Associates Inc.**

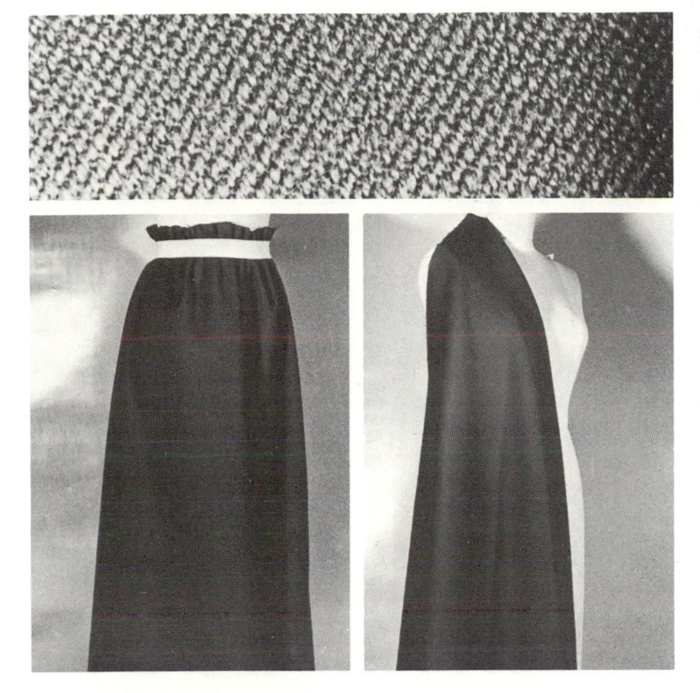

Wool Gabardine Coating

Fiber Content 100% wool
Yarn Type carded staple
Yarn Construction conventional; thick
Fabric Structure 2x2 twill weave
Finishing Processes semidecating; fulling; shearing; stabilizing
Color Application piece dyed; yarn dyed
Width 60 inches (152.4 cm)
Weight heavy
Hand compact; hard; thick
Texture harsh; pronounced rib on face; flat smooth back without rib
Performance Expectations subject to edge abrasion; durable; elasticity; elongation; insulating properties; resilient; tends to shine with wear; water-repellent properties
Drapability Qualities falls into wide cones; retains shape or silhouette of garment; better utilized if fitted by seaming and eliminating excess fabric

Recommended Construction Aids & Techniques

Hand Needles chenilles, 18; cotton darners, 1-3; embroidery, 1-4; milliners, 3; tapestry, 18-22
Machine Needles round/set point, coarse 16-18
Threads silk, industrial size A-B-C-D; six-cord cotton, 30; waxed, 40 heavy duty; cotton-covered polyester core, heavy duty
Hems bonded; double-fold or single-fold bias binding; bound/Hong Kong finish; faced; flat/plain; interfaced
Seams plain
Seam Finishes single-ply bound; Hong Kong bound; pinked; pinked and stitched; serging/single-ply overedged; untreated/plain
Pressing steam; safe temperature limit 300°F (150.1°C)
Care dry clean
Fabric Resource **Anglo Fabrics Co. Inc.**

Wool Homespun

Fiber Content 100% wool
Yarn Type carded berber yarn
Yarn Construction thick irregular yarns; unevenly spun; tight twist
Fabric Structure loosely woven plain weave
Finishing Processes pressing; shearing; stabilizing
Color Application yarn dyed and/or natural wool coloring
Width 58-60 inches (147.3-152.4 cm)
Weight medium
Hand harsh; lofty; loose; rough
Texture coarse; heavy tweedy appearance; loomed by hand or made to simulate handwoven appearance
Performance Expectations absorbent; elasticity; elongation; subject to pilling; resilient; tends to shrink and felt
Drapability Qualities falls into wide cones; fullness maintains lofty bouffant effect; maintains soft fold, unpressed pleat

Recommended Construction Aids & Techniques

Hand Needles chenilles, 20-22; cotton darners, 3-6; embroidery, 1-4; milliners, 3-6
Machine Needles round/set point, medium 14
Threads mercerized cotton, heavy duty; silk, industrial size A-B; six-cord cotton, 50; waxed, 50; cotton-covered polyester core, all purpose
Hems bonded; double-fold or single-fold bias binding; bound/Hong Kong finish; faced; flat/plain; pinked flat; interfaced; overedged; seam binding
Seams plain; safety-stitched; top-stitched; welt; double welt
Seam Finishes single-ply or double-ply bound; Hong Kong bound; single-ply overcast; pinked; pinked and stitched; serging/single-ply overedged; single-ply zigzag
Pressing steam; safe temperature limit 300°F (150.1°C)
Care dry clean
Fabric Resource **Auburn Fabrics Inc.**

Wool Homespun Coating

Fiber Content 100% wool
Yarn Type carded staple
Yarn Construction thick, coarse conventional yarn; loose twist
Fabric Structure plain weave
Finishing Processes fulling; shearing; stabilizing
Color Application yarn dyed and/or natural wool coloring
Width 72 inches (182.9 cm)
Weight heavy
Hand lofty; soft; thick
Texture coarse; rough; tweedy appearance; loomed by hand or made to simulate handwoven appearance
Performance Expectations subject to edge abrasion; elasticity; elongation; insulating properties; resilient; water-repellent properties; wrinkle resistant
Drapability Qualities falls into wide cones; retains shape or silhouette of garment; better utilized if fitted by seaming and eliminating excess fullness

Recommended Construction Aids & Techniques

Hand Needles chenilles, 18; cotton darners, 1-3; embroidery, 1-4; tapestry, 18–22
Machine Needles round/set point, coarse 18
Threads silk, industrial size A-B-C-D; six-cord cotton, 30; waxed, 40 heavy duty; cotton-covered polyester core, heavy duty
Hems bonded; double-fold or single-fold bias binding; bound/Hong Kong finish; faced; flat/plain; interfaced
Seams plain; top-stitched
Seam Finishes single-ply bound; Hong Kong bound; pinked; pinked and stitched; serging/single-ply overedged; untreated/plain
Pressing steam; safe temperature limit 300°F (150.1°C)
Care dry clean
Fabric Resource **Jack Larsen Incorporated**

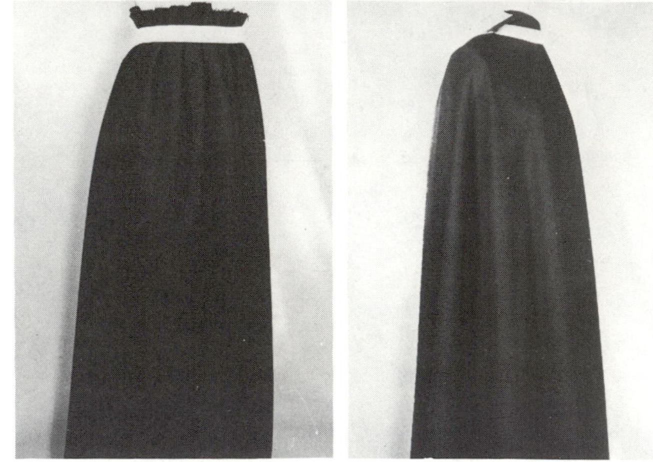

Wool Hopsacking

Fiber Content 100% wool
Yarn Type carded staple
Yarn Construction coarse; conventional; thick
Fabric Structure 2x2 basket weave; openweave
Finishing Processes pressing; shearing; stabilizing
Color Application piece dyed; yarn dyed
Width 72 inches (182.9 cm)
Weight medium-heavy (20 oz.)
Hand harsh; loose
Texture coarse; rough
Performance Expectations elasticity; elongation; insulating properties; resilient; subject to pilling and edge abrasion; subject to snagging due to weave; subject to yarn and seam slippage
Drapability Qualities falls into wide cones; fullness maintains lofty bouffant effect; retains shape of garment

Recommended Construction Aids & Techniques

Hand Needles chenilles, 18; cotton darners, 1–3; embroidery, 1–4; milliners, 3; tapestry, 18–22
Machine Needles round/set point, coarse 16–18
Threads mercerized cotton, heavy duty; silk, industrial size A-B-C; six-cord cotton, 40; cotton-covered polyester core, heavy duty
Hems bonded; double-fold or single-fold bias binding; bound/Hong Kong finish; faced; flat/plain; pinked flat; interfaced; overedged; seam binding
Seams plain; safety-stitched
Seam Finishes single-ply bound; Hong Kong bound; single-ply overcast; pinked; pinked and stitched; serging/single-ply overedged; single-ply zigzag
Pressing steam; safe temperature limit 300°F (150.1°C)
Care dry clean
Fabric Resource **Carleton Woolen Mills Inc.**

Iridescent Wool/Wool Iridescent (plain weave)

Fiber Content 45% acrylic/30% polyester/25% wool
Yarn Type carded staple; filament fiber; filament staple
Yarn Construction conventional; thick-and-thin
Fabric Structure plain weave; fine warp/thick filling
Finishing Processes crabbing; shearing; stabilizing
Color Application piece dyed–cross dyed; yarn dyed—filling and warp yarns different colors
Width 59–60 inches (149.9–152.4 cm)
Weight medium
Hand firm; semi-soft
Texture flat; slightly napped face; three-tone changeable hue and shade
Performance Expectations compressible; dimensional stability; elasticity; elongation; heat-set properties
Drapability Qualities falls into moderately soft flares; fullness maintains lofty effect; retains shape of garment

Recommended Construction Aids & Techniques

Hand Needles chenilles, 20–22; cotton darners, 3–6; embroidery, 1–4; milliners, 3–6
Machine Needles round/set point, medium 14
Threads poly-core waxed, 60; cotton-polyester blend, all purpose; cotton-covered polyester core, all purpose; spun polyester, all purpose
Hems book; bonded; single-fold bias binding; bound/Hong Kong finish; flat/plain; pinked flat; stitched and pinked flat; interfaced; overedged; seam binding
Seams plain; safety-stitched; top-stitched; welt; double welt
Seam Finishes book; single-ply or double-ply bound; Hong Kong bound; single-ply overcast; pinked; pinked and stitched; serging/single-ply overedged; single-ply zigzag
Pressing steam; safe temperature limit 300°F (150.1°C)
Care dry clean
Fabric Resources **Reeves Brothers, Inc.; Twintella Fabrics**

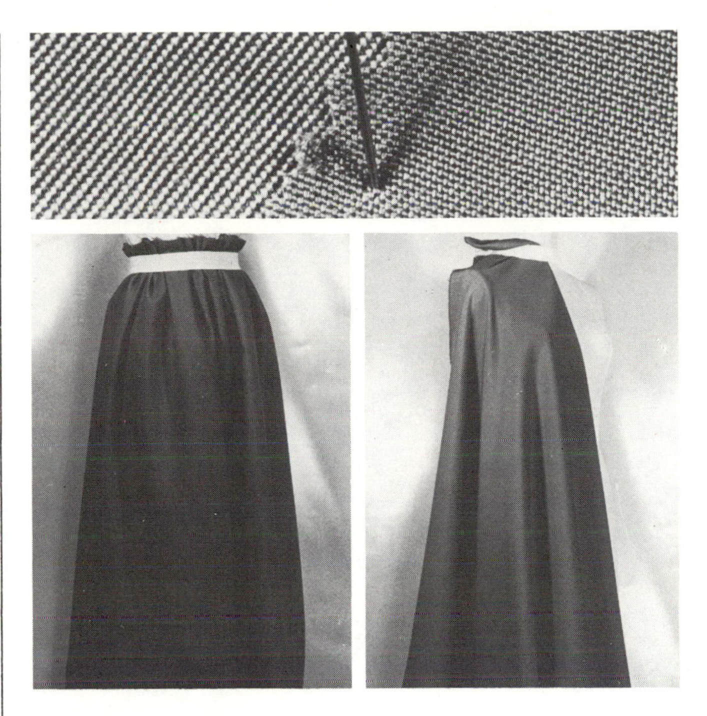

Iridescent Wool/Wool Iridescent (twill weave)

Fiber Content 100% polyester
Yarn Type filament fiber
Yarn Construction textured; thin; slack twist
Fabric Structure 1/2 twill weave
Finishing Processes delustering; stabilizing; texturizing
Color Application piece dyed–cross dyed; yarn dyed—filling and warp yarns different colors
Width 59–60 inches (149.9–152.4 cm)
Weight medium-light
Hand firm; hard; springy
Texture grainy; harsh; three-tone changeable hue and shade
Performance Expectations subject to color crocking and edge abrasion; dimensional stability; durable; elongation; heat-set properties; high tensile strength
Drapability Qualities falls into crisp cones; fullness maintains crisp effect; retains shape of garment

Recommended Construction Aids & Techniques

Hand Needles chenilles, 20–22; cotton darners, 3–6; embroidery, 1–4; milliners, 3–6
Machine Needles round/set point, medium 14
Threads poly-core waxed, 60; cotton-polyester blend, all purpose; cotton-covered polyester core, all purpose; spun polyester, all purpose
Hems book; bonded; single-fold bias binding; bound/Hong Kong finish; flat/plain; pinked flat; stitched and pinked flat; interfaced; overedged; seam binding
Seams plain; safety-stitched; top-stitched; welt; double welt
Seam Finishes book; single-ply or double-ply bound; Hong Kong bound; single-ply overcast; pinked; pinked and stitched; serging/single-ply overedged; single-ply zigzag
Pressing steam; safe temperature limit 325°F (164.1°C)
Care dry clean
Fabric Resource **Reeves Brothers, Inc.**

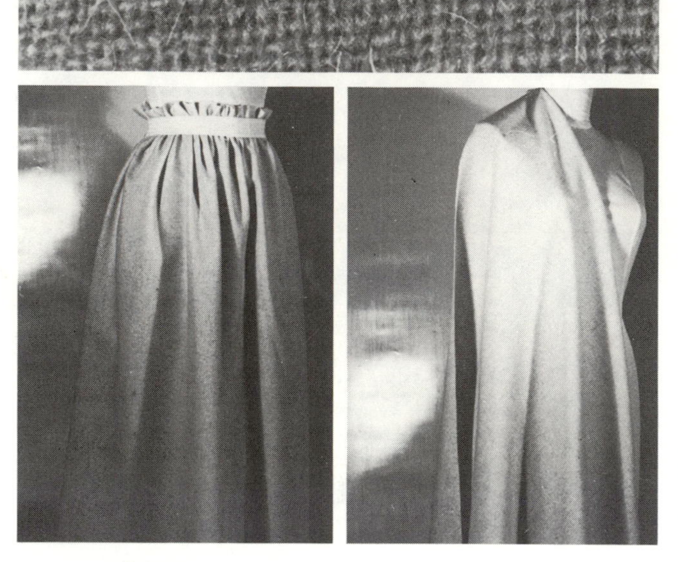

Kasha Cloth[3]

Fiber Content 100% wool
Yarn Type staple fiber
Yarn Construction conventional
Fabric Structure plain weave
Finishing Processes napping; pressing; shearing
Color Application bale dyed; sized warp yarns take dye; natural lanolin filling yarns resist dye
Width 60 inches (152.4 cm)
Weight medium
Hand lofty; smooth; soft
Texture flat; flannel type with mottled appearance (usually tan or brown)
Performance Expectations absorbent; elasticity; elongation; insulating properties; subject to pilling; resilient
Drapability Qualities falls into moderately soft cones; maintains soft falls, unpressed pleats; retains shape of garment

Recommended Construction Aids & Techniques

Hand Needles chenilles, 20–22; cotton darners, 3–6; embroidery, 1–4; milliners, 3–6
Machine Needles round/set point, medium 14
Threads mercerized cotton, heavy duty; silk, industrial size A-B-C; six-cord cotton, 40; waxed, 50; cotton-covered polyester core, heavy duty
Hems book; bonded; single-fold bias binding; bound/Hong Kong finish; flat/plain; pinked flat; stitched and pinked flat; interfaced; overedged; seam binding
Seams plain; safety-stitched; top-stitched; welt; double welt
Seam Finishes book; single-ply or double-ply bound; Hong Kong bound; single-ply overcast; pinked; pinked and stitched; serging/single-ply overedged; single-ply zigzag
Pressing steam; safe temperature limit 300°F (150.1°C)
Care dry clean
Fabric Resource Anglo Fabrics Co. Inc.

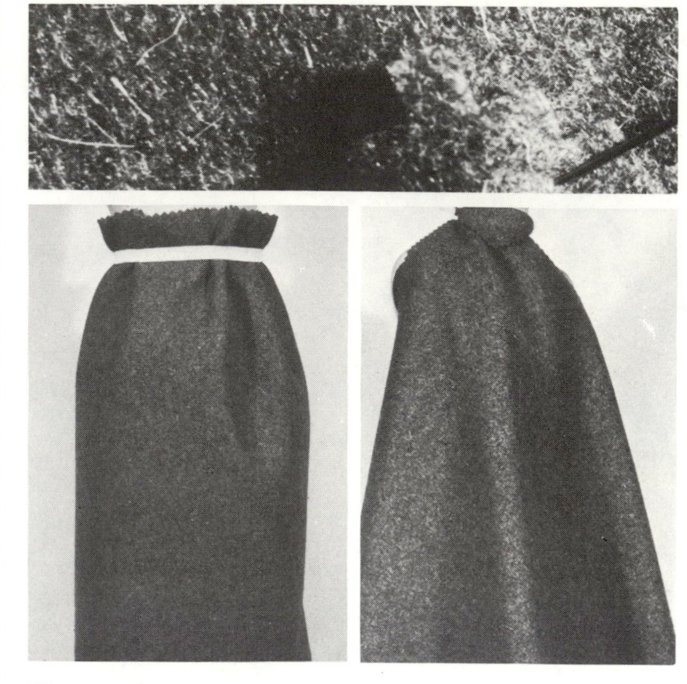

Kersey

Fiber Content 85% wool/10% nylon/5% hair
Yarn Type carded staple; filament staple; hair fiber
Yarn Construction bulked; thick; slack twist
Fabric Structure 3-yarn double-cloth twill weave; 2 filling yarns/1 warp yarn
Finishing Processes felting; fulling; napping; closely sheared face
Color Application piece dyed–cross dyed; yarn dyed
Width 60 inches (152.4 cm)
Weight heavy
Hand boardy; bulky; thick
Texture feltlike; short nap; napped surface obscures weave
Performance Expectations durable; elasticity; elongation; insulating properties; subject to pilling and edge abrasion; resilient; natural water-repellent properties
Drapability Qualities falls into wide cones; retains shape or silhouette of garment; better utilized if fitted by seaming and eliminating excess fabric

Recommended Construction Aids & Techniques

Hand Needles chenilles, 18; cotton darners, 1-3; embroidery, 1–4; milliners, 3; tapestry, 18–22
Machine Needles round/set point, coarse 16–18
Threads silk, industrial size A-B-C-D; six-cord cotton, 30; waxed, 40 heavy duty; cotton-covered polyester core, heavy duty
Hems bonded; double-fold or single-fold bias binding; bound/Hong Kong finish; faced; flat/plain; interfaced
Seams plain
Seam Finishes single-ply bound; Hong Kong bound; pinked; pinked and stitched; serging/single-ply overedged; untreated/plain
Pressing steam; safe temperature limit 300°F (150.1°C)
Care dry clean
Fabric Resource Anglo Fabrics Co. Inc.

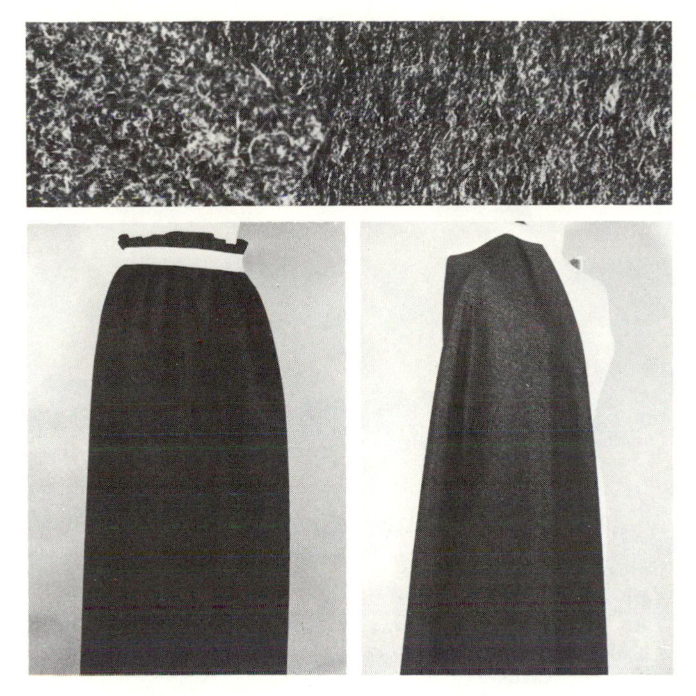

Loden Cloth

Fiber Content 100% wool
Yarn Type carded staple; coarse grade; natural lanolin retained on yarns
Yarn Construction bulked; thick; soft twist
Fabric Structure loosely woven double-cloth effect
Finishing Processes felting; heavily fulled; napping; shearing
Color Application piece dyed; yarn dyed
Width 60 inches (152.4 cm)
Weight heavy
Hand bulky; soft; thick
Texture feltlike; deep fleecy, napped face
Performance Expectations durable; elasticity; elongation; insulating properties; subject to pilling and edge abrasion; resilient; natural water-repellent properties
Drapability Qualities falls into wide cones; retains shape or silhouette of garment; better utilized if fitted by seaming and eliminating excess fabric

Recommended Construction Aids & Techniques

Hand Needles chenilles, 18; cotton darners, 1–3; embroidery, 1–4; milliners, 3; tapestry, 18–22
Machine Needles round/set point, coarse 18
Threads silk, industrial size A-B-C-D; six-cord cotton, 30; waxed, 40, heavy duty; cotton-covered polyester core, heavy duty
Hems bonded; double-fold or single-fold bias binding; bound/Hong Kong finish; faced; flat/plain; interfaced
Seams plain
Seam Finishes single-ply bound; Hong Kong bound; pinked; pinked and stitched; serging/single-ply overedged; untreated/plain
Pressing steam; use needle board; safe temperature limit 300°F (150.1°C)
Care dry clean
Fabric Resource **Anglo Fabrics Co. Inc.**

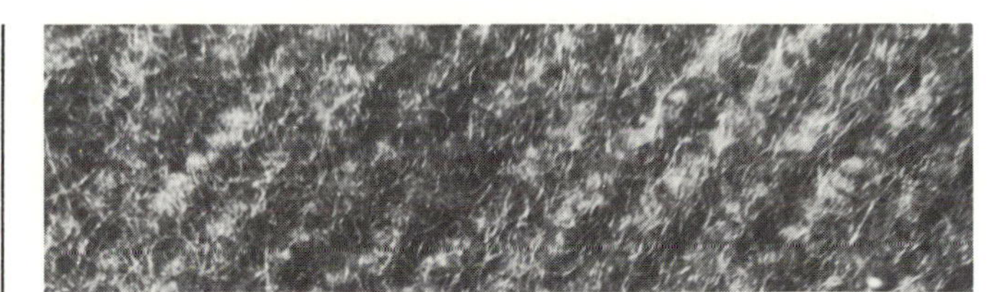

Mackinaw/Mackinac/Buffalo Plaid

Fiber Content 85% wool/15% nylon
Yarn Type staple; filament staple
Yarn Construction conventional; slack twist
Fabric Structure twill weave
Finishing Processes brushing; heavily fulled; napping; sponging
Color Application yarn dyed
Width 58 inches (147.3 cm)
Weight heavy
Hand feltlike; firm; soft; thick
Texture heavily napped surface obscures weave; flat back
Performance Expectations elasticity; elongation; insulating properties; water-repellent properties
Drapability Qualities falls into wide cones; retains shape of garment; better utilized if fitted by seaming and eliminating excess fabric

Recommended Construction Aids & Techniques

Hand Needles chenilles, 18; cotton darners, 1–3; embroidery, 1–4; milliners, 3; tapestry, 18–22
Machine Needles round/set point, coarse 16–18
Threads poly-core waxed, 40; cotton-polyester blend, heavy duty; cotton-covered polyester core, heavy duty; spun polyester, heavy duty
Hems bonded; double-fold or single-fold bias binding; bound/Hong Kong finish; faced; flat/plain; overedged
Seams plain
Seam Finishes single-ply bound; Hong Kong bound; pinked; pinked and stitched; serging/single-ply overedged; untreated/plain
Pressing steam; use needle board; safe temperature limit 300°F (150.1°C)
Care dry clean
Fabric Resource **Arthur Zeiler Woolens Inc.**

Melton

Fiber Content 100% wool
Yarn Type carded staple
Yarn Construction bulked; thick; soft twist
Fabric Structure 3-yarn double-cloth twill weave; 2 filling yarns/1 warp yarn
Finishing Processes felting; heavily fulled; napping; closely sheared face
Color Application piece dyed; yarn dyed
Width 60 inches (152.4 cm)
Weight heavy
Hand bulky; soft; thick
Texture feltlike; short nap; smooth; felting process obscures weave on face; noticeable weave on back
Performance Expectations durable; elasticity; elongation; insulating properties; subject to pilling and edge abrasion; resilient; natural water-repellent properties
Drapability Qualities falls into wide cones; retains shape or silhouette of garment; better utilized if fitted by seaming and eliminating excess fabric

Recommended Construction Aids & Techniques

Hand Needles chenilles, 18; cotton darners, 1–3; embroidery, 1–4; milliners, 3; tapestry, 18–22
Machine Needles round/set point, coarse 18
Threads silk, industrial size A-B-C-D; six-cord cotton, 30; waxed, 40 heavy duty; cotton-covered polyester core, heavy duty
Hems bonded; double-fold or single-fold bias binding; bound/Hong Kong finish; faced; flat/plain; interfaced
Seams plain
Seam Finishes single-ply bound; Hong Kong bound; pinked; pinked and stitched; serging/single-ply overedged; untreated/plain
Pressing steam; safe temperature limit 300°F (150.1°C)
Care dry clean
Fabric Resource **Anglo Fabrics Co. Inc.**

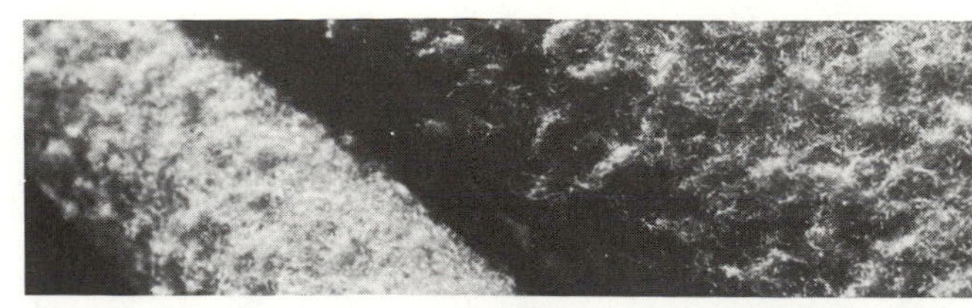

FACE

BACK

Melton (face-finished)/Chinchilla Cloth

Fiber Content 100% wool
Yarn Type carded staple
Yarn Construction conventional; wool and worsted spun
Fabric Structure 3-yarn double-cloth plain or twill weave; 2 filling yarns/1 warp yarn
Finishing Processes brushing; heavily fulled; napping; special novelty finish creates small round tufts or nubs
Color Application yarn dyed
Width 60 inches (152.4 cm)
Weight heavy
Hand compact; firm; heavy
Texture napped; surface rolled into evenly spaced tufts or nubs on face by finishing process; smooth back
Performance Expectations subject to edge abrasion; durable; elasticity; elongation; insulating properties; resilient; natural water-repellent properties
Drapability Qualities falls into wide cones; retains shape or silhouette of garment; better utilized if fitted by seaming and eliminating excess fabric

Recommended Construction Aids & Techniques

Hand Needles chenilles, 18; cotton darners, 1–3; embroidery, 1–4; milliners, 3; tapestry, 18–22
Machine Needles round/set point, coarse 18
Threads silk, industrial size A-B-C-D; six-cord cotton, 30; waxed, 40 heavy duty; cotton-covered polyester core, heavy duty
Hems bonded; double-fold or single-fold bias binding; bound/Hong Kong finish; faced; flat/plain; overedged
Seams plain
Seam Finishes single-ply bound; Hong Kong bound; pinked; serging/single-ply overedged; untreated/plain
Pressing steam; use needle board; safe temperature limit 300°F (150.1°C)
Care dry clean
Fabric Resource **Anglo Fabrics Co. Inc.**

Wool Ottoman

Fiber Content 100% wool
Yarn Type worsted staple
Yarn Construction conventional; high twist
Fabric Structure plain weave; fine warp; heavy filling yarn creates rib effect in crosswise direction
Finishing Processes decating; singeing; tentering
Color Application yarn dyed
Width 60 inches (152.4 cm)
Weight heavy
Hand compact; hard; harsh
Texture wide raised crosswise rib
Performance Expectations durable; elasticity; elongation; insulating properties; subject to pilling and edge abrasion; resilient; natural water-repellent properties
Drapability Qualities falls into wide cones; retains shape or silhouette of garment; better utilized if fitted by seaming and eliminating excess fabric

Recommended Construction Aids & Techniques

Hand Needles chenilles, 18; cotton darners, 1–3; embroidery, 1–4; milliners, 3; tapestry, 18–22
Machine Needles round/set point, coarse 18
Threads silk, industrial size A-B-C-D; six-cord cotton, 30; waxed, 40 heavy duty; cotton-covered polyester core, heavy duty
Hems bonded; double-fold or single-fold bias binding; bound/Hong Kong finish; faced; flat/plain; interfaced
Seams plain; top-stitched
Seam Finishes single-ply bound; Hong Kong bound; pinked; pinked and stitched; serging/single-ply overedged; untreated/plain
Pressing steam; safe temperature limit 300°F (150.1°C)
Care dry clean
Fabric Resource Anglo Fabrics Co. Inc.

Wool Poplin

Fiber Content 100% wool
Yarn Type worsted staple
Yarn Construction conventional; high twist
Fabric Structure plain weave; slight rib effect in crosswise direction
Finishing Processes decating; shearing; singeing
Color Application piece dyed; yarn dyed
Width 60 inches (152.4 cm)
Weight medium
Hand lofty; smooth; soft
Texture flat; harsh
Performance Expectations elasticity; elongation; subject to moth damage; resilient; tends to shine with wear
Drapability Qualities falls into moderately soft flares; fullness maintains lofty effect; retains shape of garment

Recommended Construction Aids & Techniques

Hand Needles chenilles, 20–22; cotton darners, 3–6; embroidery, 1–4; milliners, 3–6
Machine Needles round/set point, medium 14
Threads mercerized cotton, heavy duty; silk, industrial size A-B-C; six-cord cotton, 40; waxed, 50; cotton-covered polyester core, heavy duty
Hems book; bonded; single-fold bias binding; bound/Hong Kong finish; flat/plain; pinked flat; stitched and pinked flat; interfaced; overedged; seam binding
Seams plain; safety-stitched; top-stitched; welt; double welt
Seam Finishes book; single-ply or double-ply bound; Hong Kong bound; single-ply overcast; pinked; pinked and stitched; serging/single-ply overedged; single-ply zigzag
Pressing steam; safe temperature limit 300°F (150.1°C)
Care dry clean
Fabric Resource Anglo Fabrics Co. Inc.

Polo Cloth

Fiber Content 93% wool/7% nylon
Yarn Type carded staple; filament fiber
Yarn Construction bulked; loosely spun
Fabric Structure 3-yarn loosely woven double-cloth twill weave; 2 filling yarns/1 warp yarn
Finishing Processes felting; heavily fulled; napping; shearing
Color Application piece dyed–cross dyed; yarn dyed
Width 60 inches (152.4 cm)
Weight heavy
Hand bulky; soft; thick
Texture boardy; compact; feltlike
Performance Expectations durable; elasticity; elongation; insulating properties; subject to pilling and edge abrasion; resilient; natural water-repellent properties
Drapability Qualities falls into wide cones; retains shape or silhouette of garment; better utilized if fitted by seaming and eliminating excess fabric

Recommended Construction Aids & Techniques

Hand Needles chenilles, 18; cotton darners, 1–3; embroidery, 1–4; milliners, 3; tapestry, 18–22
Machine Needles round/set point, coarse 18
Threads silk, industrial size A-B-C-D; six-cord cotton, 30; waxed, 40 heavy duty; cotton-covered polyester core, heavy duty
Hems bonded; double-fold or single-fold bias binding; bound/Hong Kong finish; faced; flat/plain; interfaced
Seams plain
Seam Finishes single-ply bound; Hong Kong bound; pinked; pinked and stitched; serging/single-ply overedged; untreated/plain
Pressing steam; safe temperature limit 300°F (150.1°C)
Care dry clean
Fabric Resource **Anglo Fabrics Co. Inc.**

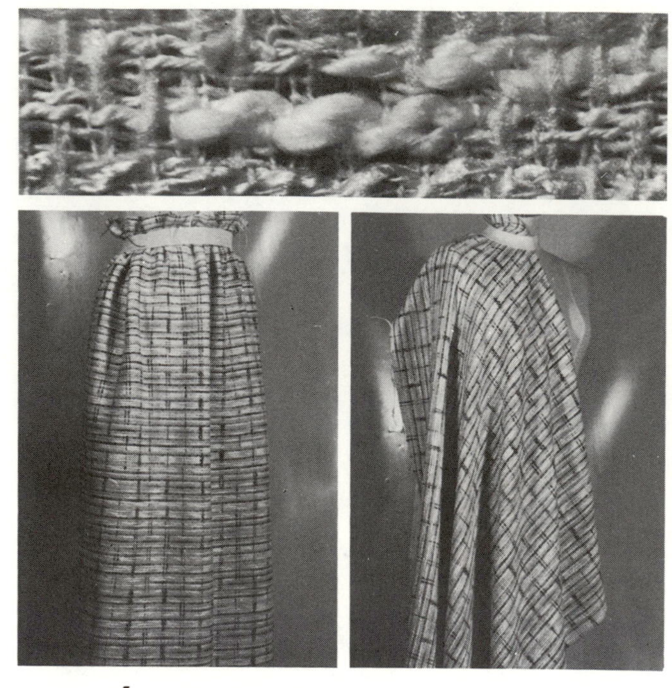

Ratiné/Épongé Wool (plain weave)

Fiber Content 100% wool
Yarn Type carded staple
Yarn Construction conventional; thick-and-thin novelty yarns; slack twist
Fabric Structure plain weave variation; skip denting
Finishing Processes semidecating; singeing; stabilizing
Color Application yarn dyed
Width 56 inches (142.2 cm)
Weight medium-heavy
Hand loose; pliable; semi-soft
Texture openweave; thick-and-thin irregular yarn creates raised uneven surface
Performance Expectations absorbent; elasticity; elongation; flexible; subject to pulling and snagging due to yarn and weave; resilient
Drapability Qualities falls into moderately soft flares; accommodates fullness by pleating, gathering; maintains soft fold, unpressed pleat

Recommended Construction Aids & Techniques

Hand Needles chenilles, 20–22; cotton darners, 3–6; embroidery, 1–4; milliners, 3–6
Machine Needles round/set point, coarse 16–18
Threads mercerized cotton, heavy duty; silk, industrial size A-B-C; six-cord cotton, 30; waxed, 40 heavy duty; cotton-covered polyester core, heavy duty
Hems bonded; double-fold or single-fold bias binding; bound/Hong Kong finish; faced; flat/plain; interfaced; overedged
Seams plain; safety-stitched; top-stitched; welt; double welt
Seam Finishes single-ply bound; Hong Kong bound; pinked; pinked and stitched; serging/single-ply overedged; untreated/plain
Pressing steam; use pressing cloth; safe temperature limit 300°F (150.1°C)
Care dry clean
Fabric Resource **Arthur Zeiler Woolens Inc.**

Ratiné/Épongé Wool (twill weave)

Fiber Content 67% wool/18% Orlon®/15% nylon
Yarn Type staple; filament staple; filament fiber
Yarn Construction bulked; conventional; thick-and-thin
Fabric Structure twill weave
Finishing Processes semidecating; sizing; stabilizing
Color Application yarn dyed
Width 60 inches (152.4 cm)
Weight medium-heavy
Hand semi-crisp; firm; flexible
Texture thick-and-thin irregular yarn creates raised uneven surface
Performance Expectations elasticity; elongation; flexible; subject to pulling and snagging due to yarn and weave; resilient
Drapability Qualities falls into firm wide flares; fullness maintains lofty bouffant effect; retains shape of garment

Recommended Construction Aids & Techniques

Hand Needles chenilles, 14; cotton darners, 1–3; embroidery, 1–4; milliners, 3; tapestry, 18–22
Machine Needles round/set point, coarse 16–18
Threads poly-core waxed, 40; cotton-polyester blend, heavy duty; cotton-covered polyester core, heavy duty; spun polyester, heavy duty
Hems bonded; double-fold or single-fold bias binding; bound/Hong Kong finish; faced; flat/plain; interfaced; overedged
Seams plain; safety-stitched
Seam Finishes single-ply bound; Hong Kong bound; pinked; pinked and stitched; serging/single-ply overedged; untreated/plain
Pressing steam; use pressing cloth; safe temperature limit 300°F (150.1°C)
Care dry clean
Fabric Resource Arthur Zeiler Woolens Inc.

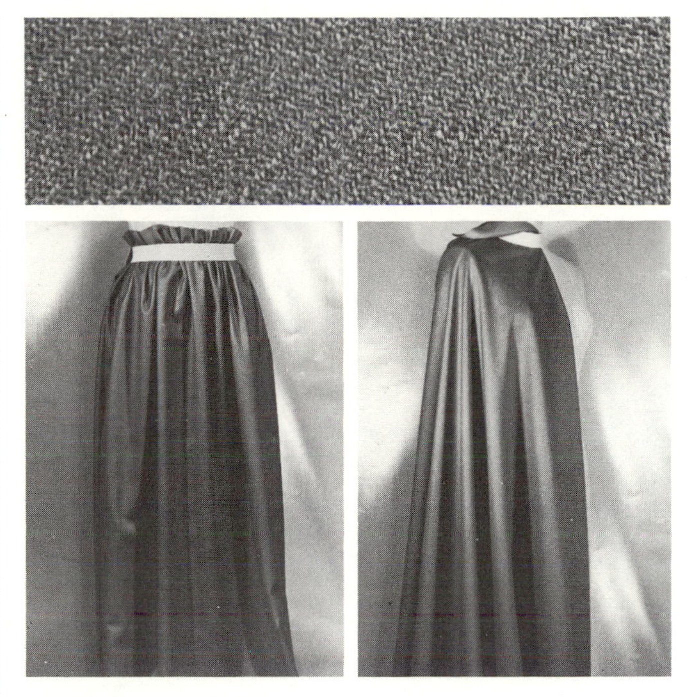

Wool Serge

Fiber Content 100% wool
Yarn Type worsted staple
Yarn Construction conventional; fine; high twist
Fabric Structure left-hand 45° closely woven twill weave
Finishing Processes clear finish; decating; napping; shearing; singeing
Color Application piece dyed; yarn dyed
Width 60 inches (152.4 cm)
Weight medium
Hand firm; soft
Texture nap pressed flat; diagonal of twill visible on face and back
Performance Expectations elasticity; elongation; subject to moth damage; resilient; tends to shine with wear
Drapability Qualities falls into moderately soft flares; fullness maintains lofty effect; retains shape of garment

Recommended Construction Aids & Techniques

Hand Needles chenilles, 20–22; cotton darners, 3–6; embroidery, 1–4; milliners, 3–6
Machine Needles round/set point, medium-coarse 14–16
Threads mercerized cotton, heavy duty; silk, industrial size A-B; six-cord cotton, 50; waxed, 50; cotton-covered polyester core, all purpose
Hems book; bonded; single-fold bias binding; bound/Hong Kong finish; flat/plain; pinked flat; stitched and pinked flat; interfaced; overedged; seam binding
Seams plain; safety-stitched; top-stitched; welt; double welt
Seam Finishes book; single-ply or double-ply bound; Hong Kong bound; single-ply overcast; pinked; pinked and stitched; serging/single-ply overedged; single-ply zigzag
Pressing steam; safe temperature limit 300°F (150.1°C)
Care dry clean
Fabric Resource Burlington Industries Inc.

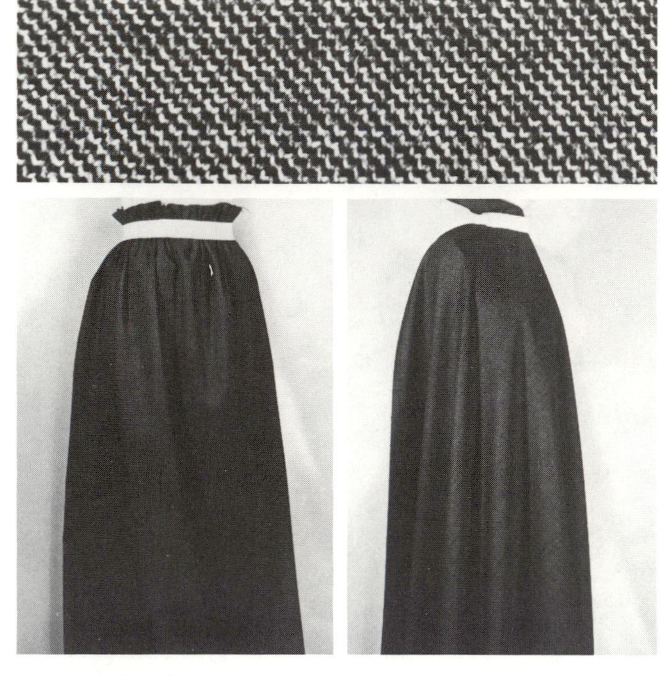

Wool Sharkskin

Fiber Content 55% Dacron®/45% wool
Yarn Type filament fiber; worsted staple
Yarn Construction conventional; fine; tight twist
Fabric Structure 2/2 right-hand twill weave
Finishing Processes clear finish; decating; shearing; singeing
Color Application piece dyed-cross dyed; yarn dyed
Width 60 inches (152.4 cm)
Weight medium-light
Hand semi-crisp; firm; hard
Texture flat; sleek; smooth; color line runs diagonally to the left and opposite to the direction of the twill
Performance Expectations subject to edge abrasion; durable; elasticity; elongation; heat-set properties; resilient
Drapability Qualities falls into firm flares; maintains crisp fold, crease, pleat; retains shape of garment

Recommended Construction Aids & Techniques

Hand Needles chenilles, 20–22; cotton darners, 3–6; embroidery, 1–4; milliners, 3–6
Machine Needles round/set point, medium 14
Threads mercerized cotton, heavy duty; silk, industrial size A-B; six-cord cotton, 50; waxed, 50; cotton-covered polyester core, all purpose
Hems book; bonded; single-fold bias binding; bound/Hong Kong finish; flat/plain; pinked flat; stitched and pinked flat; interfaced; overedged; seam binding
Seams flat-felled; lapped; plain; top-stitched; welt; double welt
Seam Finishes book; single-ply or double-ply bound; Hong Kong bound; single-ply overcast; pinked; pinked and stitched; serging/single-ply overedged; single-ply zigzag
Pressing steam; safe temperature limit 300°F (150.1°C)
Care dry clean
Fabric Resource **Burlington Industries Inc.**

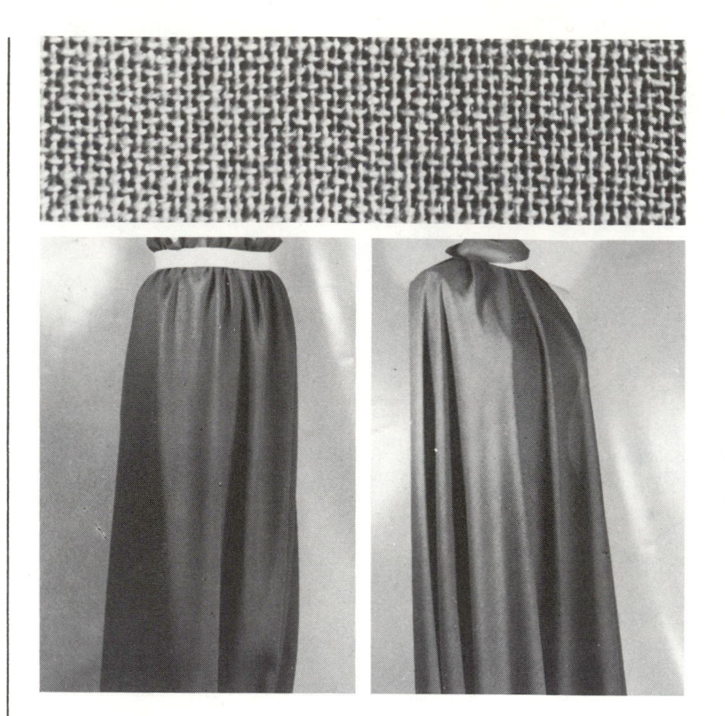

Wool Batiste

Fiber Content 100% wool
Yarn Type carded staple
Yarn Construction conventional; high twist
Fabric Structure loosely woven plain weave
Finishing Processes decating; shearing; singeing; softening
Color Application piece dyed; yarn dyed
Width 60 inches (152.4 cm)
Weight light
Hand pliable; semi-soft
Texture fine; semi-sheer; heavier, thicker yarn and less sheer than wool voile; smooth
Performance Expectations absorbent; air permeable; elasticity; elongation; resilient
Drapability Qualities falls into soft flares; accommodates fullness by pleating, gathering, elasticized shirring; fullness retains soft fall

Recommended Construction Aids & Techniques

Hand Needles betweens, 3–5; cotton darners, 4–8; embroidery, 3–6; milliners, 3–6; sharps, 1–3
Machine Needles round/set point, medium 11–14
Threads mercerized cotton, 50; silk, industrial size A; six-cord cotton, 50; waxed, 60; cotton-covered polyester core, all purpose
Hems book; bound/Hong Kong finish; pinked flat; stitched and pinked flat; machine-stitched; overedged; seam binding
Seams flat-felled; lapped; plain; top-stitched; welt; double welt
Seam Finishes single-ply bound; Hong Kong bound; double-stitched; edge-stitched; pinked; pinked and stitched; serging/single-ply overedged; single-ply zigzag
Pressing steam; safe temperature limit 300°F (150.1°C)
Care dry clean
Fabric Resources **Anglo Fabrics Co. Inc.; Auburn Fabrics Inc.**

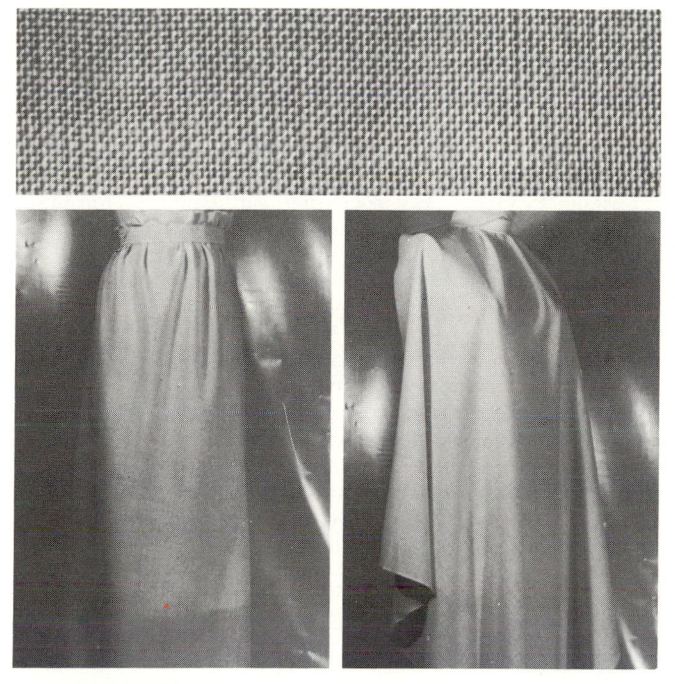

Wool Challis

Fiber Content 100% wool
Yarn Type combed worsted staple
Yarn Construction conventional; fine; thin; tight twist
Fabric Structure closely woven plain weave
Finishing Processes decating; shearing; singeing; softening
Color Application piece dyed; yarn dyed
Width 36 inches (91.4 cm)
Weight light
Hand firm; soft; supple
Texture fine; lacks luster; thin; softest and finest of all wool sheers
Performance Expectations absorbent; air permeable; elasticity; elongation; resilient
Drapability Qualities falls into soft flares; accommodates fullness by pleating, gathering, elasticized shirring; fullness retains soft fall

Recommended Construction Aids & Techniques

Hand Needles betweens, 3–5; cotton darners, 4–8; embroidery, 3–6; milliners, 3–6; sharps, 1–3
Machine Needles round/set point, medium 11–14
Threads mercerized cotton, 50; silk, industrial size A; six-cord cotton, 50; waxed, 60; cotton-covered polyester core, all purpose
Hems book; bound/Hong Kong finish; pinked flat; stitched and pinked flat; machine-stitched; overedged; seam binding
Seams flat-felled; lapped; plain; top-stitched; welt; double welt
Seam Finishes single-ply bound; Hong Kong bound; double-stitched; edge-stitched; pinked; pinked and stitched; serging/single-ply overedged; single-ply zigzag
Pressing steam; safe temperature limit 300°F (150.1°C)
Care dry clean
Fabric Resources **Anglo Fabrics Co. Inc.; Auburn Fabrics Inc.**

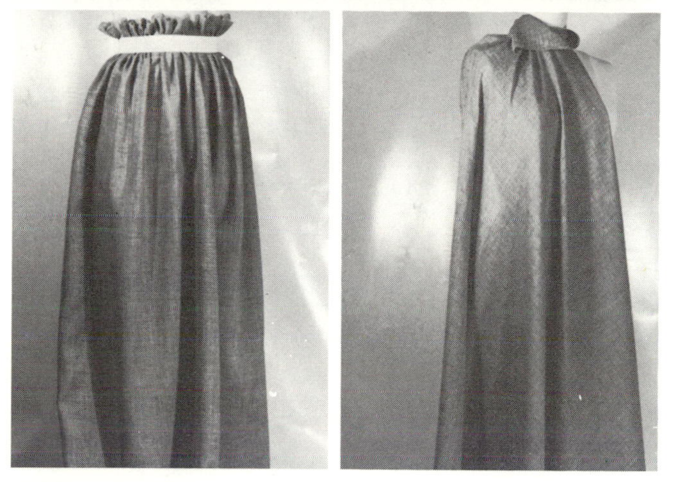

Wool Voile

Fiber Content 100% wool
Yarn Type combed worsted staple
Yarn Construction conventional; thin; tight twist
Fabric Structure loosely woven plain weave
Finishing Processes decating; shearing; singeing; softening
Color Application piece dyed; yarn dyed
Width 44–45 inches (111.8–114.3 cm)
Weight light
Hand pliable; semi-soft
Texture fine; semi-sheer; smooth; heavier than challis, but lighter and more transparent than batiste
Performance Expectations absorbent; air permeable; elasticity; elongation; resilient
Drapability Qualities falls into soft flares; accommodates fullness by pleating, gathering, elasticized shirring; fullness retains soft fall

Recommended Construction Aids & Techniques

Hand Needles betweens, 3–5; cotton darners, 4–8; embroidery, 3–6; milliners, 3–6; sharps, 1–3
Machine Needles round/set point, medium 11–14
Threads mercerized cotton, 50; silk, industrial size A; six-cord cotton, 50; waxed, 60; cotton-covered polyester core, all purpose
Hems book; bound/Hong Kong finish; pinked flat; stitched and pinked flat; machine-stitched; overedged; seam binding
Seams flat-felled; lapped; plain; top-stitched; welt; double welt
Seam Finishes single-ply bound; Hong Kong bound; double-stitched; edge-stitched; pinked; pinked and stitched; serging/single-ply overedged; single-ply zigzag
Pressing steam; safe temperature limit 300°F (150.1°C)
Care dry clean
Fabric Resources **Anglo Fabrics Co. Inc.; Loomskill/Gallery Screen Print**

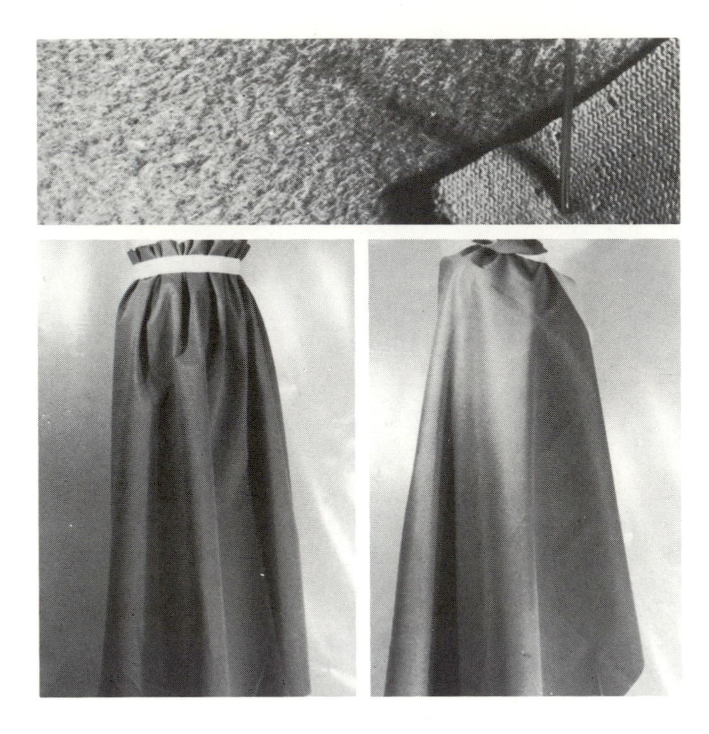

Wool Suede Cloth

Fiber Content 100% polyester
Yarn Type filament fiber; filament staple
Yarn Construction conventional; fine; microfilament flock
Fabric Structure compound fabric: warp-knit ground; flocked face
Finishing Processes electrocoating; stabilizing; tentering; finished and processed to simulate suede hide or Ultrasuede®
Color Application piece dyed; solution dyed; yarn dyed
Width 56 inches (142.2 cm)
Weight medium
Hand compact; nonelastic; soft
Texture napped face; flat back; velvety
Performance Expectations subject to color crocking and edge abrasion; dimensional stability; heat-set properties; resilient
Drapability Qualities falls into wide cones; retains shape of garment; better utilized if fitted by seaming and eliminating excess fabric

Recommended Construction Aids & Techniques

Hand Needles chenilles, 20–22; cotton darners, 3–6; embroidery, 1–4; milliners, 3–6
Machine Needles ball point, 14
Threads poly-core waxed, 60; cotton-polyester blend, all purpose; cotton-covered polyester core, all purpose; spun polyester, all purpose
Hems flat/plain; stitched and pinked flat; double machine-stitched
Seams plain; safety-stitched
Seam Finishes Hong Kong bound; pinked; plain; serging/single-ply overedged; untreated/plain
Pressing steam; use needle board; safe temperature limit 325°F (164.1°C)
Care dry clean
Fabric Resource **Drexler Associates Inc.**

Tartan Woolens/Clan Plaid Woolens

Tartan or clan plaid woolens refers to a group of cross-barred wool and wool-type plaids or checks in the various designs and colors of the Scottish clans. The design of a true tartan or clan plaid consists of a sequence of colored stripes of various widths which start at the edge or selvage of the cloth and run the full width of the cloth. The sequence or arrangement of the stripe is repeated in reverse order, then repeated and reversed again across the width of the cloth.

The sequence, arrangement, proportion and colors of the stripes in the tartan represent and identify the clan.

Names of the Various Clans Include:

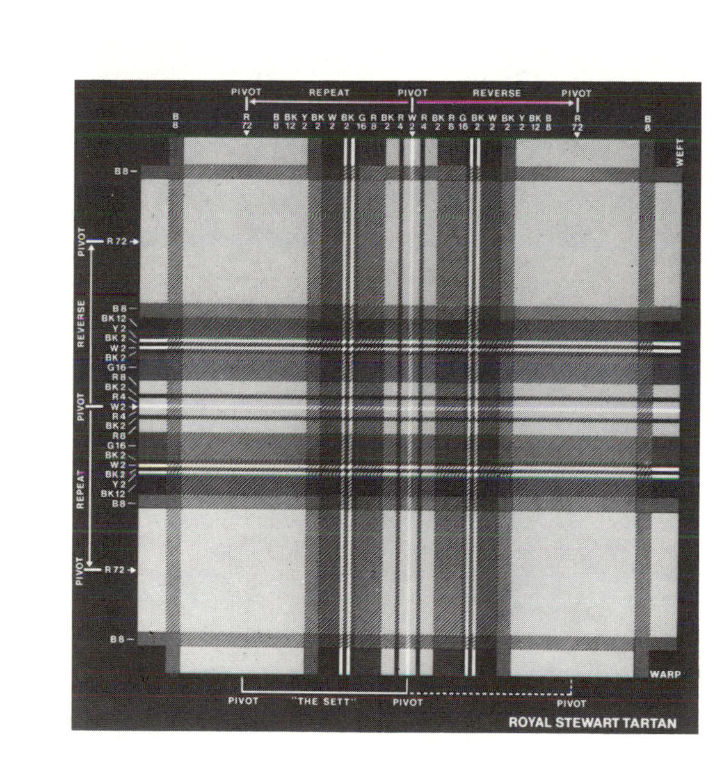

ROYAL STEWART TARTAN

Abercrombie
Anderson
Arbuthnott
Armstrong
Baird
Barclay
Borthwick
Boyd
Brodie
Bruce, Brus
Buchan
Buchanan
Cameron
Campbell of Argyll
Carnegie
Chattan
Chisholm, Chisholme
Cockburn
Colquhoun
Crawford
Cumming
Cunningham
Dalziel
Davidson
Douglas
Drummond
Dunbar
Duncan
Dundas
Elliot
Erskine
Farquharson
Ferguson, Fergusson
Fletcher
Forbes
Fraser
Gordon
Gow
Graeme of Menteith
Grant
Gunn
Hamilton
Hay
Henderson
Home, Hume
Huntly
Innes, Innie
Johnston
Keith
Kennedy
Kerr
Lamont
Leslie
Lindsay
Livingston
Logan
MacAlister
MacAlpine
MacArthur
MacAulay
MacBain
MacBeth
MacCallum
MacColl
MacDonald
MacDonald of Clanranald
MacDonell of Glengarry
MacDonell of Keppoch
MacDougall
MacDuff
MacEwen
MacFarlane
MacFee, Macfie
MacGillivray
MacGregor
MacIan
Macinnes
Macintyre
MacIver
Mackay
MacKenzie
Mackinlay
Mackinnon
Mackintosh
MacLachlan
Maclaine
MacLaren
Maclean
MacLeod of Harris
MacLeod of Lewes
Macmillan
Macnab
Macnaughton
Macneil of Barra
MacNicol
MacPherson
MacQuaire
Macqueen
Macrae
MacTavish
MacThomas
Malcolm
Matheson
Maxwell
Menzies
Montgomerie
Morison, Morrison
Mowat
Munro, Monroe
Murray
Napier
Ogilvie, Ogilvy
Oliphant
Ramsay
Rattray
Robertson
Rose
Ross
Ruthven
Scott
Seton
Shaw
Sinclair
Skene
Steuart
Stewart of Appin
Stewart of Atholl
Stuart of Bute
Sutherland
Urquhart
Wallace
Wemyss

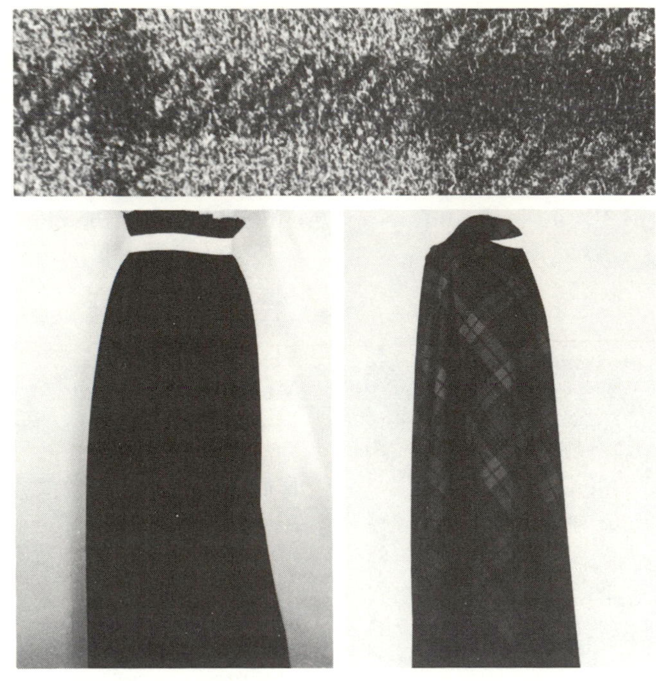

Black Watch Plaid

Fiber Content 100% wool
Yarn Type carded staple
Yarn Construction conventional
Fabric Structure 2/2 twill weave
Finishing Processes fulling; napping; shearing; stabilizing
Color Application yarn dyed
Width 60 inches (152.4 cm)
Weight medium
Hand firm; lofty; semi-soft
Texture slightly napped face
Performance Expectations elasticity; elongation; insulating properties; subject to pilling; resilient; tends to shrink and felt
Drapability Qualities falls into moderately soft flares; maintains soft fold, unpressed pleat; retains shape of garment

Recommended Construction Aids & Techniques

Hand Needles chenilles, 20–22; cotton darners, 3–6; embroidery, 1–4; milliners, 3–6
Machine Needles round/set point, medium-coarse 14–16
Threads mercerized cotton, heavy duty; silk, industrial size A–B; six-cord cotton, 50; waxed, 50; cotton-covered polyester core, all purpose
Hems book; bonded; single-fold bias binding; bound/Hong Kong finish; flat/plain; pinked flat; stitched and pinked flat; interfaced; overedged; seam binding
Seams plain; safety-stitched; top-stitched; welt; double welt
Seam Finishes book; single-ply or double-ply bound; Hong Kong bound; single-ply overcast; pinked; pinked and stitched; serging/single-ply overedged; single-ply zigzag
Pressing steam; safe temperature limit 300°F (150.1°C)
Care dry clean
Fabric Resources Carleton Woolen Mills Inc.; Twintella Fabrics

Glen Plaid

Fiber Content 45% wool/55% polyester
Yarn Type filament fiber; carded staple
Yarn Construction conventional
Fabric Structure twill weave
Finishing Processes fulling; napping; shearing; stabilizing
Color Application piece dyed–cross dyed; yarn dyed
Width 60 inches (152.4 cm)
Weight medium
Hand lofty; semi-soft; springy
Texture slightly napped face
Performance Expectations durable; elasticity; elongation; heat-set properties; resilient; subject to edge abrasion and static buildup
Drapability Qualities falls into moderately soft flares; maintains crisp fold, crease, pleat; retains shape of garment

Recommended Construction Aids & Techniques

Hand Needles chenilles, 20–22; cotton darners, 3–6; embroidery, 1–4; milliners, 3–6
Machine Needles round/set point, medium-coarse 14–16
Threads mercerized cotton, heavy duty; silk, industrial size A–B; six-cord cotton, 50; waxed, 50; cotton-covered polyester core, all purpose
Hems book; bonded; single-fold bias binding; bound/Hong Kong finish; flat/plain; pinked flat; stitched and pinked flat; interfaced; overedged; seam binding
Seams plain; safety-stitched; top-stitched; welt; double welt
Seam Finishes book; single-ply or double-ply bound; Hong Kong bound; single-ply overcast; pinked; pinked and stitched; serging/single-ply overedged; single-ply zigzag
Pressing steam; safe temperature limit 300°F (150.1°C)
Care dry clean
Fabric Resources Carleton Woolens Mills Inc.; Twintella Fabrics

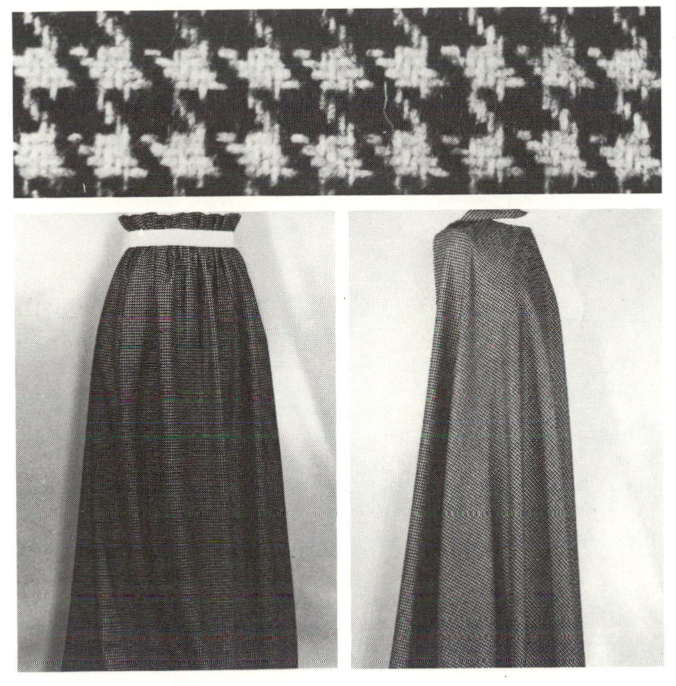

Hound's-tooth Check

Fiber Content 45% wool/55% polyester
Yarn Type carded staple; filament staple
Yarn Construction conventional
Fabric Structure broken twill weave; 2-up and 2-down: 4 ends and 4 picks in a repeat
Finishing Processes fulling; napping; shearing; stabilizing
Color Application piece dyed–cross dyed; yarn dyed
Width 60 inches (152.4 cm)
Weight medium
Hand lofty; semi-soft; springy
Texture slightly napped face
Performance Expectations durable; elasticity; elongation; heat-set properties; resilient; subject to edge abrasion and static buildup
Drapability Qualities falls into moderately soft flares; maintains crisp fold, crease, pleat; retains shape of garment

Recommended Construction Aids & Techniques

Hand Needles chenilles, 20–22; cotton darners, 3–6; embroidery, 1–4; milliners, 3–6
Machine Needles round/set point, medium-coarse 14–16
Threads mercerized cotton, heavy duty; silk, industrial size A-B; six-cord cotton, 50; waxed, 50; cotton-covered polyester core, all purpose
Hems book; bonded; single-fold bias binding; bound/Hong Kong finish; flat/plain; pinked flat; stitched and pinked flat; interfaced; overedged; seam binding
Seams plain; safety-stitched; top-stitched; welt; double welt
Seam Finishes book; single-ply or double-ply bound; Hong Kong bound; single-ply overcast; pinked; pinked and stitched; serging/single-ply overedged; single-ply zigzag
Pressing steam; safe temperature limit 300°F (150.1°C)
Care dry clean
Fabric Resource **Carleton Woolen Mills Inc.**

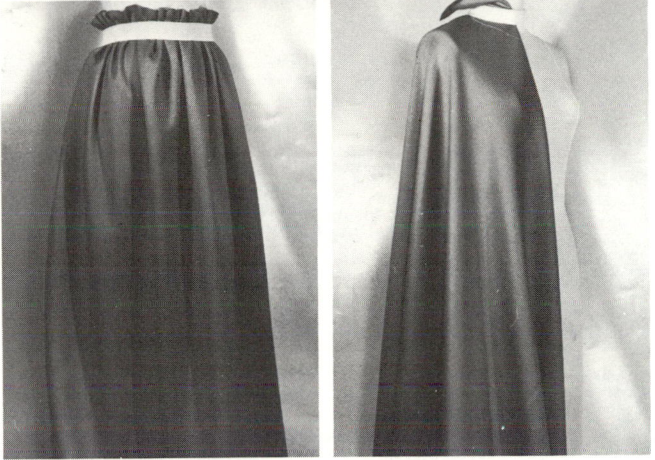

Wool Tricotine/Cavalry Twill

Fiber Content 100% wool
Yarn Type carded staple
Yarn Construction conventional; high twist
Fabric Structure closely woven right-hand twill weave
Finishing Processes clear finish; decating; fulling; singeing; stabilizing
Color Application piece dyed; yarn dyed
Width 60 inches (152.4 cm)
Weight heavy
Hand compact; firm; soft
Texture pronounced steep, diagonal double line on face
Performance Expectations durable; elasticity; elongation; insulating properties; subject to pilling; resilient; tends to shrink and felt
Drapability Qualities falls into wide cones; retains shape or silhouette of garment; better utilized if fitted by seaming and eliminating excess fabric

Recommended Construction Aids & Techniques

Hand Needles chenilles, 18; cotton darners, 1–3; embroidery, 1–4; milliners, 3; tapestry, 18–22
Machine Needles round/set point, coarse 16–18
Threads silk, industrial size A-B-C-D; six-cord cotton, 30; waxed, 40 heavy duty; cotton-covered polyester core, heavy duty
Hems bonded; double-fold or single-fold bias binding; bound/Hong Kong finish; faced; flat/plain; pinked flat; interfaced; overedged; seam binding
Seams plain; safety-stitched
Seam Finishes single-ply bound; Hong Kong bound; single-ply overcast; pinked; pinked and stitched; serging/single-ply overedged; untreated/plain; single-ply zigzag
Pressing steam; safe temperature limit 300°F (150.1°C)
Care dry clean
Fabric Resource **Burlington Industries Inc.**

Tweeds/Wool Tweeds

Tweed is a term used to describe a wide range of fabrics characterized by their mixed color appearance achieved by yarns of two or more colors and/or shades or tones of color. Tweed and tweed-type fabrics may utilize thin, thick or any combination of yarn construction, may be made in any weave structure, and may vary from:

1. light to heavy weight;
2. smooth to rough texture;
3. soft to firm hand;
4. open to closely woven.

Bird's Eye Tweed

Fiber Content 100% wool
Yarn Type carded staple
Yarn Construction conventional
Fabric Structure dobby weave
Finishing Processes semidecating; fulling; shearing; stabilizing
Color Application yarn dyed
Width 58 inches (147.3 cm)
Weight medium
Hand firm; semi-soft
Texture flat; harsh; small diamond pattern produced by weave which suggests bird's eyes
Performance Expectations absorbent; elasticity; elongation; resilient
Drapability Qualities falls into moderately soft flares; fullness maintains lofty effect; maintains soft fold, unpressed pleat

Recommended Construction Aids & Techniques

Hand Needles chenilles, 20–22; cotton darners, 3–6; embroidery, 1–4; milliners, 3–6
Machine Needles round/set point, medium 14
Threads mercerized cotton, heavy duty; silk, industrial size A-B; six-cord cotton, 50; waxed, 50; cotton-covered polyester core, all purpose
Hems book; bonded; single-fold bias binding; bound/Hong Kong finish; flat/plain; pinked flat; stitched and pinked flat; interfaced; overedged; seam binding
Seams plain; safety-stitched; top-stitched; welt; double welt
Seam Finishes single-ply or double-ply bound; Hong Kong bound; single-ply overcast; pinked; pinked and stitched; serging/single-ply overedged; single-ply zigzag
Pressing steam; safe temperature limit 300°F (150.1°C)
Care dry clean
Fabric Resources Twintella Fabrics; Arthur Zeiler Woolens Inc.

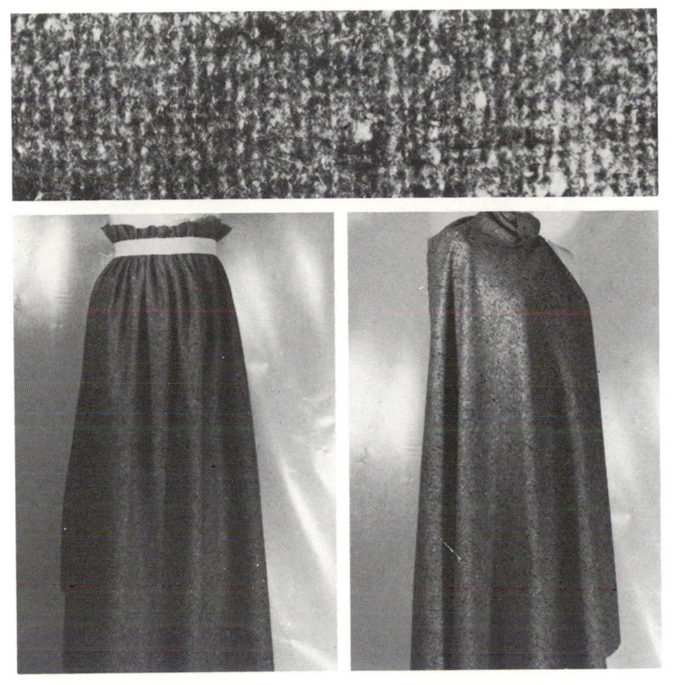

Donegal Tweed/Donegal-type Tweed[4]

Fiber Content 35% acrylic/25% wool/25% polyester/15% nylon
Yarn Type carded staple; filament staple
Yarn Construction bulked; even warp yarns; irregular, thick slub filling yarns
Fabric Structure plain weave
Finishing Processes decating; fulling; shearing
Color Application yarn dyed
Width 60 inches (152.4 cm)
Weight medium-heavy
Hand coarse; harsh; lofty
Texture nubby; rough; thick, multi-colored slubs woven in at random
Performance Expectations compressible; dimensional stability; elasticity; elongation; heat-set properties; resilient
Drapability Qualities falls into wide flares; fullness maintains lofty effect; retains shape of garment

Recommended Construction Aids & Techniques

Hand Needles chenilles, 18; cotton darners, 1–3; embroidery, 1–4; milliners, 3; tapestry, 18–22
Machine Needles round/set point, medium-coarse 14–16
Threads poly-core waxed, 50; cotton-polyester blend, heavy duty; cotton-covered polyester core, heavy duty; spun polyester, heavy duty
Hems bonded; double-fold or single-fold bias binding; bound/Hong Kong finish; faced; flat/plain; pinked flat; interfaced; overedged; seam binding
Seams plain; safety-stitched
Seam Finishes single-ply bound; Hong Kong bound; single-ply overcast; pinked; pinked and stitched; serging/single-ply overedged; single-ply zigzag
Pressing steam; safe temperature limit 300°F (150.1°C)
Care dry clean
Fabric Resources **Drexler Associates Inc.; Twintella Fabrics**

Heather Tweed/Heather-type Tweed

Fiber Content 45% acrylic/30% polyester/25% wool
Yarn Type carded staple; filament staple
Yarn Construction bulked; textured
Fabric Structure twill weave
Finishing Processes fulling; napping; shearing
Color Application mixed stock dyed yarn; piece dyed–cross dyed; yarn dyed
Width 60 inches (152.4 cm)
Weight medium-heavy
Hand compact; lofty; semi-soft
Texture fuzzy nap; harsh; multi-colored yarns combined to produce a blended color appearance
Performance Expectations compressible; dimensional stability; elasticity; elongation; heat-set properties; resilient
Drapability Qualities falls into moderately firm flares; maintains soft folds, unpressed pleats; retains shape of garment

Recommended Construction Aids & Techniques

Hand Needles chenilles, 18; cotton darners, 1–3; embroidery, 1–4; milliners, 3; tapestry, 18–22
Machine Needles round/set point, medium-coarse 14–16
Threads poly-core waxed, 50; cotton-polyester blend, heavy duty; cotton-covered polyester core, heavy duty; spun polyester, heavy duty
Hems bonded; double-fold or single-fold bias binding; bound/Hong Kong finish; faced; flat/plain; pinked flat; interfaced; overedged; seam binding
Seams plain; safety-stitched
Seam Finishes single-ply bound; Hong Kong bound; single-ply overcast; pinked; pinked and stitched; serging/single-ply overedged; single-ply zigzag
Pressing steam; safe temperature limit 300°F (150.1°C)
Care dry clean
Fabric Resource **Twintella Fabrics**

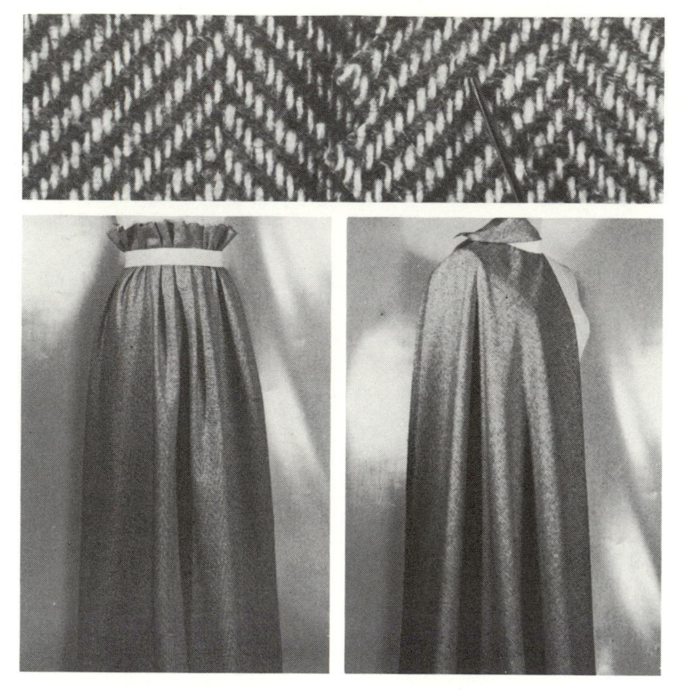

Wool Herringbone Tweed (napped finish)

Fiber Content 55% polyester/45% wool
Yarn Type carded staple; filament staple
Yarn Construction bulked
Fabric Structure 2/2 herringbone twill weave
Finishing Processes fulling; napping; shearing; stabilizing
Color Application piece dyed–cross dyed; yarn dyed
Width 60 inches (152.4 cm)
Weight medium
Hand lofty; scratchy; soft
Texture slightly napped face; broken twill weave pattern with a balanced zigzag design resembles skeletal structure of the herring
Performance Expectations dimensional stability; durable; elasticity; elongation; heat-set properties; resilient; subject to edge abrasion and static buildup
Drapability Qualities falls into crisp flares; maintains crisp fold, crease, pleat; retains shape of garment

Recommended Construction Aids & Techniques

Hand Needles chenilles, 20–22; cotton darners, 3–6; embroidery, 1–4; milliners, 3–6
Machine Needles round/set point, medium-coarse 14-16
Threads poly-core waxed, 60; cotton-polyester blend, all purpose; cotton-covered polyester core, all purpose; spun polyester, all purpose
Hems bonded; double-fold or single-fold bias binding; bound/Hong Kong finish; faced; flat/plain; pinked flat; interfaced; overedged; seam binding
Seams plain; safety-stitched
Seam Finishes single-ply or double-ply bound; Hong Kong bound; single-ply overcast; pinked; pinked and stitched; serging/single-ply overedged; single-ply zigzag
Pressing steam; safe temperature limit 300°F (150.1°C)
Care dry clean
Fabric Resource **Carleton Wool Mills Inc.**

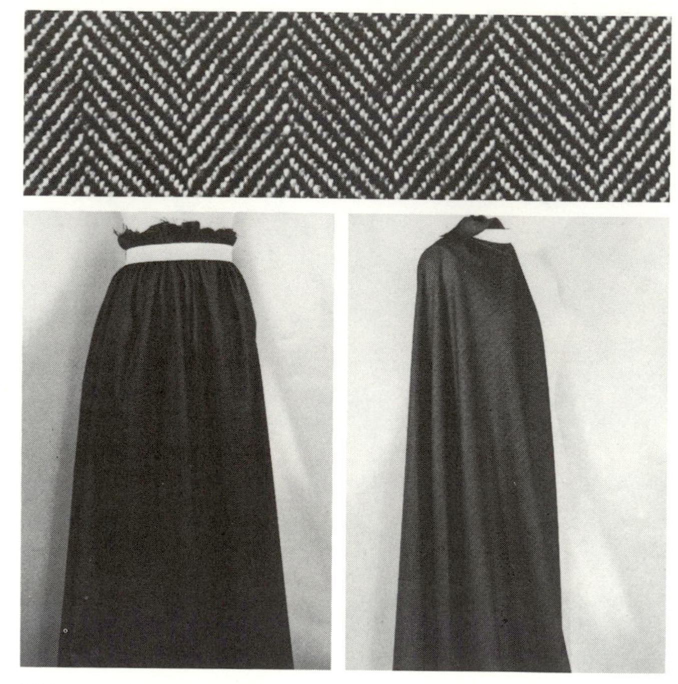

Worsted Herringbone Tweed (clear finish)

Fiber Content 55% polyester/45% wool
Yarn Type combed worsted staple; filament fiber
Yarn Construction conventional; textured; thin filament yarns; high twist
Fabric Structure 2/2 herringbone twill weave
Finishing Processes clear finish; decating; singeing; stabilizing
Color Application piece dyed–cross dyed; yarn dyed
Width 60 inches (152.4 cm)
Weight medium-light
Hand nonelastic; firm; hard
Texture flat; harsh; weave pattern resembles chevrons
Performance Expectations dimensional stability; durable; elasticity; elongation; heat-set properties; resilient; subject to edge abrasion and static buildup
Drapability Qualities falls into crisp flares; maintains crisp fold, crease, pleat; retains shape of garment

Recommended Construction Aids & Techniques

Hand Needles chenilles, 20–22; cotton darners, 3–6; embroidery, 1–4; milliners, 3–6
Machine Needles round/set point, medium 14
Threads poly-core waxed, 60; cotton-polyester blend, all purpose; cotton-covered polyester core, all purpose; spun polyester, all purpose
Hems book; bonded; single-fold bias binding; bound/Hong Kong finish; flat/plain; pinked flat; stitched and pinked flat; interfaced; overedged; seam binding
Seams plain; safety-stitched; top-stitched; welt; double welt
Seam Finishes book; single-ply or double-ply bound; Hong Kong bound; single-ply overcast; pinked; pinked and stitched; serging/single-ply overedged; single-ply zigzag
Pressing steam; safe temperature limit 300°F (150.1°C)
Care dry clean
Fabric Resource **Burlington Industries Inc.**

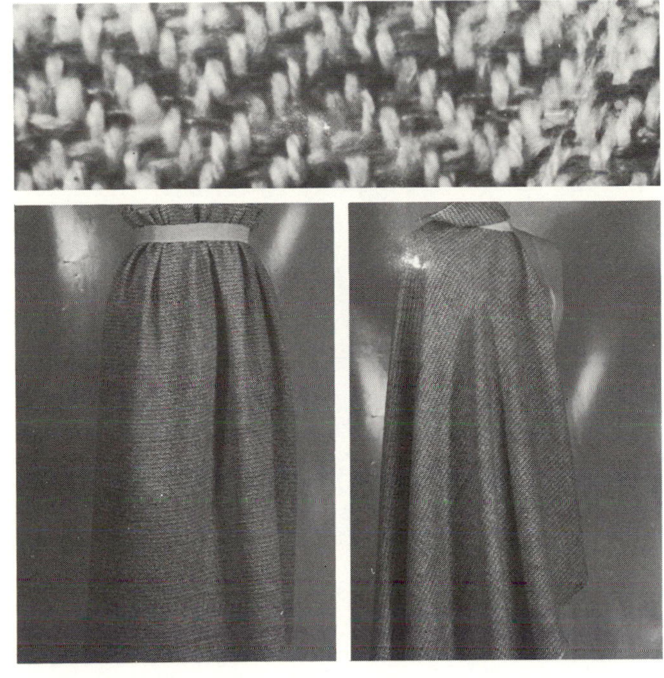

Oatmeal Tweed

Fiber Content 53% mohair/22% cotton/19% silk/6% acrylic
Yarn Type filament fiber; hair fiber; staple
Yarn Construction bulked; conventional; thick-and-thin
Fabric Structure herringbone twill weave
Finishing Processes brushing; singeing; stabilizing
Color Application yarn dyed
Width 58–60 inches (147.3–152.4 cm)
Weight medium-heavy
Hand lofty; loose; soft
Texture coarse; grainy; pattern and color resemble oatmeal
Performance Expectations elasticity; flexible; resilient; subject to snagging and pulling due to loose weave
Drapability Qualities falls into soft flares; accommodates fullness by pleating, gathering; fullness maintains lofty effect

Recommended Construction Aids & Techniques

Hand Needles chenilles, 18; cotton darners, 1–3; embroidery, 1–4; milliners, 3; tapestry, 18–22
Machine Needles round/set point, medium-coarse 14–16
Threads poly-core waxed, 50; cotton-polyester blend, heavy duty; cotton-covered polyester core, heavy duty; spun polyester, heavy duty
Hems bonded; double-fold or single-fold bias binding; bound/Hong Kong finish; faced; flat/plain; interfaced; overedged
Seams plain; safety-stitched
Seam Finishes single-ply bound; Hong Kong bound; pinked; pinked and stitched; serging/single-ply overedged; untreated/plain
Pressing *do not* steam; use needle board; protect with pressing cloth; safe temperature limit 300°F (150.1°C)
Care dry clean
Fabric Resource **Arthur Zeiler Woolens Inc.**

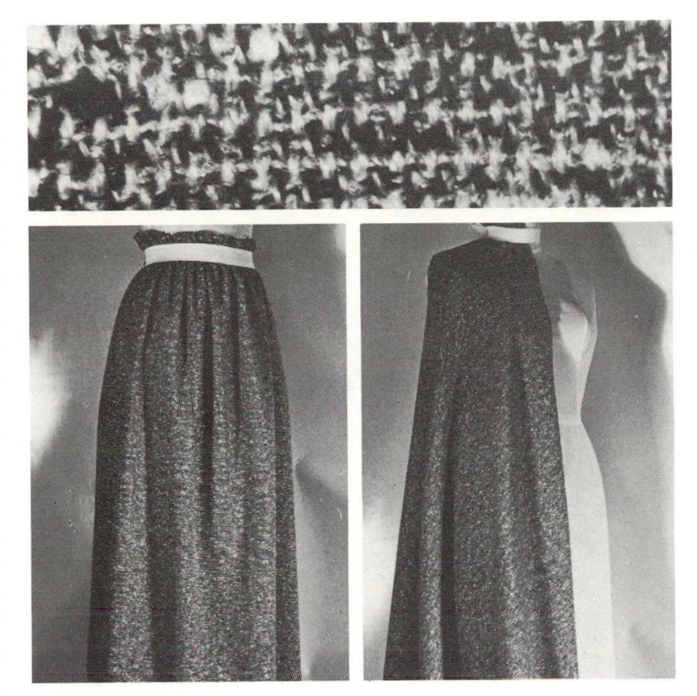

Salt-and-Pepper Tweed

Fiber Content 49% wool/22% acetate/18% polyester/ 11% nylon
Yarn Type combed worsted staple; filament fiber
Yarn Construction novelty slubbed yarn; loose twist
Fabric Structure plain weave
Finishing Processes decating; pressing; stabilizing
Color Application stock dyed yarn; solution dyed; yarn dyed
Width 58-60 inches (147.3–152.4 cm)
Weight medium
Hand coarse; harsh; lofty
Texture nubby tweed; slubbed; alternating and mixed dark and light effect
Performance Expectations elasticity; elongation; flexible; heat-set properties; resilient; wrinkle resistant
Drapability Qualities falls into crisp flares; maintains crisp fold, crease, pleat; retains shape of garment

Recommended Construction Aids & Techniques

Hand Needles chenilles, 20–22; cotton darners, 3–6; embroidery, 1–4; milliners, 3–6
Machine Needles round/set point, medium-coarse 14–16
Threads poly-core waxed, 60; cotton-polyester blend, all purpose; cotton-covered polyester core, all purpose; spun polyester, all purpose
Hems book; bonded; single-fold bias binding; bound/ Hong Kong finish; flat/plain; pinked flat; stitched and pinked flat; interfaced; overedged; seam binding
Seams plain; safety-stitched; top-stitched; welt; double welt
Seam Finishes book; single-ply or double-ply bound; Hong Kong bound; single-ply overcast; pinked; pinked and stitched; serging/single-ply overedged; single-ply zigzag
Pressing steam; safe temperature limit 250°F (122.1°C)
Care dry clean
Fabric Resource **Auburn Fabrics Inc.**

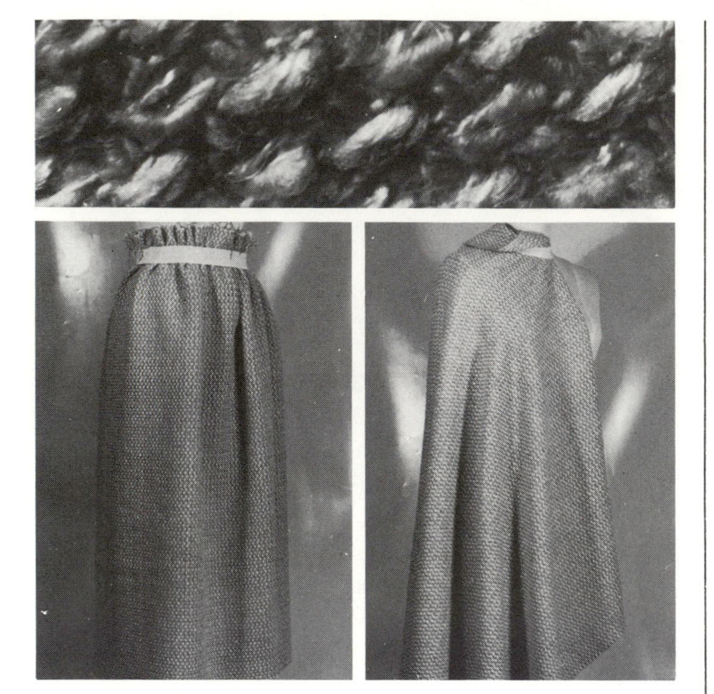

Heavy Textured Tweed

Fiber Content 65% wool/13% rayon/11% cotton/8% acrylic
Yarn Type filament fiber; filament staple; staple
Yarn Construction conventional; novelty bouclé yarn; thick-and-thin brushed yarn
Fabric Structure basket weave
Finishing Processes brushing; shearing; stabilizing
Color Application yarn dyed
Width 58 inches (147.3 cm)
Weight heavy
Hand loose; semi-soft; thick
Texture thick-and-thin/high and low; uneven surface
Performance Expectations subject to abrasion; absorbent; compressible; elasticity; elongation; resilient; subject to snagging and pulling due to loose weave
Drapability Qualities falls into wide cones; retains shape or silhouette of garment; better utilized if fitted by seaming and eliminating excess fabric

Recommended Construction Aids & Techniques

Hand Needles chenilles, 18; cotton darners, 1–3; embroidery, 1-4; milliners, 3; tapestry, 18–22
Machine Needles round/set point, coarse 16–18
Threads cotton-covered polyester core, heavy duty; cotton-polyester blend, heavy duty; spun polyester, heavy duty; poly-core waxed, 40
Hems bonded; double-fold or single-fold bias binding; bound/Hong Kong finish; faced; flat/plain; overedged
Seams plain
Seam Finishes single-ply bound; Hong Kong bound; pinked; pinked and stitched; serging/single-ply overedged; untreated/plain
Pressing steam; use needle board; protect with pressing cloth; safe temperature limit 300°F (150.1°C)
Care dry clean
Fabric Resource **Arthur Zeiler Woolens Inc.**

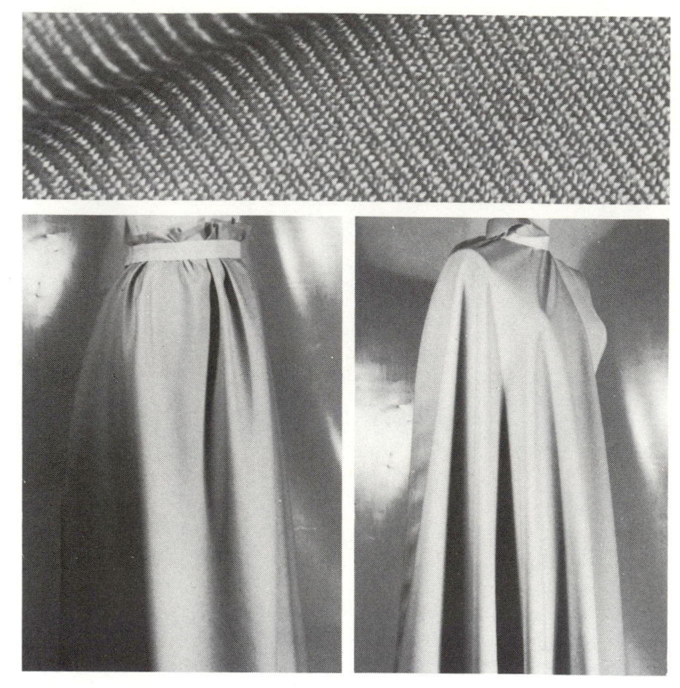

Wool Whipcord

Fiber Content 65% polyester/35% cotton
Yarn Type combed staple; filament fiber
Yarn Construction conventional; textured; high twist
Fabric Structure steep twill weave
Finishing Processes decating; pressing; singeing; Zepel® water/stain repellent
Color Application piece dyed–cross dyed; yarn dyed
Width 59–60 inches (149.9–152.4 cm)
Weight medium-heavy
Hand crisp; nonelastic; firm
Texture pronounced diagonal twill on surface; harsh
Performance Expectations dimensional stability; durable; elasticity; heat-set properties; water and stain resistant due to finish
Drapability Qualities falls into crisp flares; fullness maintains crisp bouffant effect; retains shape of garment

Recommended Construction Aids & Techniques

Hand Needles chenilles, 20–22; cotton darners, 3–6; embroidery, 1–4; milliners, 3–6
Machine Needles round/set point, coarse 16–18
Threads poly-core waxed, 60; cotton-polyester blend; all purpose; cotton-covered polyester core, regular; spun polyester, all purpose
Hems bonded; double-fold or single-fold bias binding; bound/Hong Kong finish; faced; flat/plain; pinked flat; interfaced; overedged; seam binding
Seams plain; safety-stitched; top-stitched; welt; double welt
Seam Finishes single-ply bound; Hong Kong bound; single-ply overcast; pinked; pinked and stitched; serging/single-ply overedged; single-ply zigzag
Pressing steam; safe temperature limit 325°F (164.1°C)
Care dry clean
Fabric Resource **Reeves Brothers Inc.**

Wool Fancies

Wool fancies is a term used to describe a large variety of new or unusual fabrications not listed in original wool generic classifications. Wool fancies may be composed of one or more shades, tones or colors; may utilize any novelty or combination of yarn; may be made in any combination or variety of weave structure; and may vary in:

1. unusual surface texture;
2. internal or external finish;
3. soft to stiff hand.

Some manufacturers may refer to their line of plaids, tweeds, surface patterns or finishes or decorative woolens as fancies.

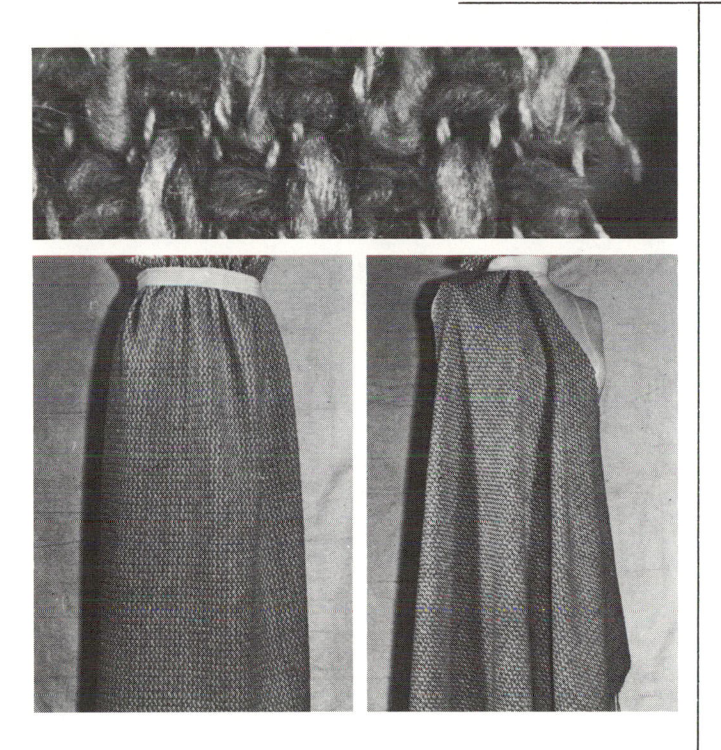

Bouclé/Wool Bouclé

Fiber Content 65% wool/13% rayon/11% cotton/8% acrylic
Yarn Type filament fiber; filament staple; staple
Yarn Construction novelty bouclé yarn; brushed; thick-and-thin
Fabric Structure basket weave
Finishing Processes brushing; shearing; stabilizing
Color Application yarn dyed
Width 58 inches (147.3 cm)
Weight heavy
Hand loose; semi-soft; thick
Texture thick-and-thin areas; yarns create uneven surface
Performance Expectations subject to abrasion; absorbent; compressible; elasticity; elongation; resilient; subject to snagging and pulling due to loose weave
Drapability Qualities falls into wide cones; retains shape of garment; better utilized if fitted by seaming and eliminating excess fabric

Recommended Construction Aids & Techniques

Hand Needles chenilles, 18; cotton darners, 1–3; embroidery, 1–4; milliners, 3; tapestry, 18–22
Machine Needles round/set point, coarse 16–18
Threads poly-core waxed, 40; cotton-polyester blend, heavy duty; cotton-covered polyester core, heavy duty; spun polyester, heavy duty
Hems bonded; double-fold or single-fold bias binding; bound/Hong Kong finish; faced; flat/plain; overedged
Seams plain
Seam Finishes single-ply bound; Hong Kong bound; pinked; pinked and stitched; serging/single-ply overedged; untreated/plain
Pressing steam; use needle board; protect with pressing cloth; safe temperature limit 300°F (150.1°C)
Care dry clean
Fabric Resource **Arthur Zeiler Woolens Inc.**

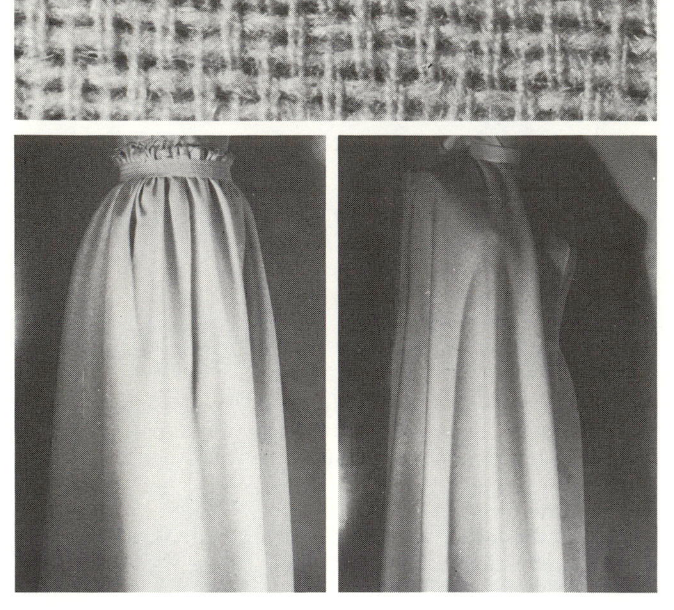

Wool Gauze

Fiber Content 47% acrylic/30% wool/16% mohair/7% nylon
Yarn Type filament fiber; filament staple; hair fiber; combed staple
Yarn Construction bulked; conventional
Fabric Structure loosely woven plain weave
Finishing Processes fulling; napping; stabilizing
Color Application solution dyed yarn; stock dyed yarn; piece dyed–cross dyed
Width 58-60 inches (147.3-152.4 cm)
Weight light
Hand lofty; downy soft; spongy
Texture fluffy; rough; uneven surface
Performance Expectations compressible; elasticity; elongation; heat-set properties; insulating properties; subject to snagging due to weave; wrinkle resistant
Drapability Qualities falls into moderately soft flares; accommodates fullness by gathering; fullness retains moderately soft fall

Recommended Construction Aids & Techniques

Hand Needles chenilles, 20–22; cotton darners, 3–6; embroidery, 1–4; milliners, 3–6
Machine Needles round/set point, medium 14
Threads poly-core waxed, 60; cotton-polyester blend, all purpose; cotton-covered polyester core, all purpose; spun polyester, all purpose
Hems book; bound/Hong Kong finish; flat/plain; stitched and turned flat; machine-stitched; overedged; seam binding
Seams plain; safety-stitched; taped; top-stitched; welt; double welt
Seam Finishes single-ply bound; Hong Kong bound; double-stitched; edge-stitched; serging/single-ply overedged; single-ply zigzag
Pressing *do not* steam; safe temperature limit 300°F (150.1°C)
Care dry clean
Fabric Resource **Auburn Fabrics Inc.**

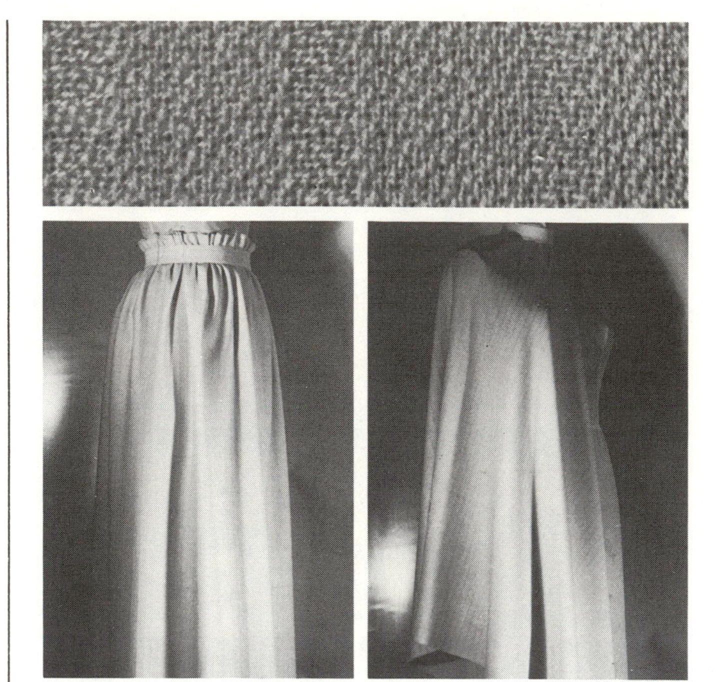

Reverse Twill Weave Wool

Fiber Content 100% wool
Yarn Type worsted staple
Yarn Construction conventional; thin; high twist
Fabric Structure novelty reverse twill weave
Finishing Processes clear finish; crabbing; decating; singeing
Color Application piece dyed; yarn dyed
Width 58-60 inches (147.3-152.4 cm)
Weight medium-light
Hand firm; hard
Texture harsh; striped effect created by weave
Performance Expectations subject to edge abrasion; elasticity; elongation; resilient; tends to shine with wear
Drapability Qualities falls into firm flares; maintains crisp fold, crease, pleat; retains shape of garment

Recommended Construction Aids & Techniques

Hand Needles chenilles, 20–22; cotton darners, 3–6; embroidery, 1–4; milliners, 3–6
Machine Needles round/set point, medium 14
Threads mercerized cotton, heavy duty; silk, industrial size A-B; six-cord cotton, 50; waxed, 50; cotton-covered polyester core, all purpose
Hems book; bonded; single-fold bias binding; bound/Hong Kong finish; flat/plain; pinked flat; stitched and pinked flat; interfaced; overedged; seam binding
Seams flat-felled; lapped; plain; top-stitched; welt; double welt
Seam Finishes single-ply bound; Hong Kong bound; double-stitched; edge-stitched; pinked; pinked and stitched; serging/single-ply overedged; single-ply zigzag
Pressing steam; safe temperature limit 300°F (150.1°C)
Care dry clean
Fabric Resource **Auburn Fabrics Inc.**

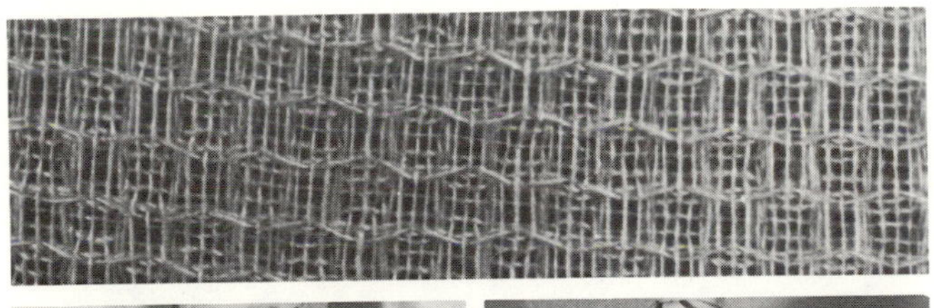

Waffle Cloth

Fiber Content 85% polyester/15% wool
Yarn Type carded staple; filament fiber
Yarn Construction textured; loose twist
Fabric Structure piqué weave
Finishing Processes crabbing; pressing; stabilizing
Color Application piece dyed–cross dyed; yarn dyed
Width 45 inches (114.3 cm)
Weight light
Hand lofty; loose; spongy
Texture high and low areas; uneven surface
Performance Expectations dimensional stability; durable; elasticity; elongation; heat-set properties; resilient; high tensile strength; subject to abrasion and static buildup
Drapability Qualities falls into moderately soft flares; accommodates fullness by pleating, gathering; fullness retains moderately soft fall

Recommended Construction Aids & Techniques

Hand Needles chenilles, 20–22; cotton darners, 3–6; embroidery, 1–4; milliners, 3–6
Machine Needles round/set point, medium-coarse 14–16
Threads poly-core waxed, 60; cotton-polyester blend, all purpose; cotton-covered polyester core, all purpose; spun polyester, all purpose
Hems book; bound/Hong Kong finish; flat/plain; pinked flat; stitched and pinked flat; machine-stitched; overedged; seam binding
Seams plain; safety-stitched; top-stitched; welt; double welt
Seam Finishes book; single-ply or double-ply bound; Hong Kong bound; single-ply overcast; pinked; pinked and stitched; serging/single-ply overedged; untreated/plain; single-ply zigzag
Pressing steam; safe temperature limit 300°F (150.1°C)
Care dry clean
Fabric Resource **J. P. Fabrics (sample courtesy of College Town)**

Doubleknit Wool/Wool Doubleknit

Fiber Content 88% Dacron®/12% wool
Yarn Type combed worsted staple; filament fiber
Yarn Construction fine spun; textured
Fabric Structure double knit
Finishing Processes decating; pressing; stabilizing
Color Application piece dyed–cross dyed; yarn dyed
Width 60 inches (152.4 cm)
Weight medium
Hand hard; scratchy; thick
Texture flat; grainy
Performance Expectations subject to edge abrasion; dimensional stability; durable; elasticity; elongation; high recovery; resilient; wrinkle resistant
Drapability Qualities falls into moderately firm flares; fullness maintains lofty effect; retains shape of garment

Recommended Construction Aids & Techniques

Hand Needles chenilles, 20–22; cotton darners, 3–6; embroidery, 1–4; milliners, 3–6
Machine Needles ball point, 14
Threads poly-core waxed, 60; cotton-polyester blend, all purpose; cotton-covered polyester core, all purpose; spun polyester, all purpose
Hems flat/plain; stitched and turned flat; double machine-stitched
Seams overedged; plain; safety-stitched; zigzagged
Seam Finishes pinked and stitched; serging/single-ply overedged; untreated/plain
Pressing steam; safe temperature limit 300°F (150.1°C)
Care dry clean
Fabric Resource **Burlington Industries Inc.**

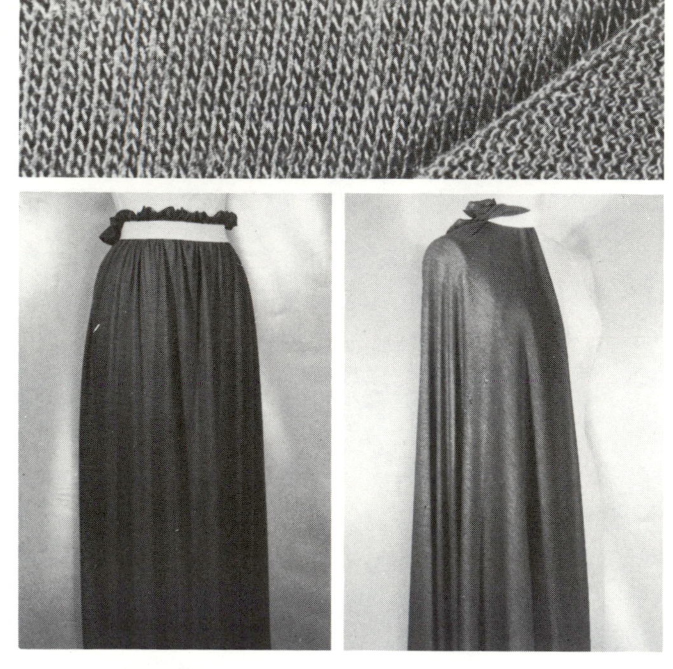

Jersey Wool Knit/Wool Jersey

Fiber Content 80% Namelle® acrylic/20% wool
Yarn Type combed worsted staple; filament fiber
Yarn Construction fine; textured; tight twist
Fabric Structure jersey single-knit structure
Finishing Processes decating; pressing; stabilizing
Color Application piece dyed–cross dyed; yarn dyed
Width 64–65 inches (162.7–165.1 cm)
Weight medium-light
Hand pliable; soft; thin
Texture flat; knit-stitch face; purl-stitch back
Performance Expectations abrasion resistant; compressible; dimensional stability; elasticity; elongation; heat-set properties; resilient
Drapability Qualities falls into soft flares; accommodates fullness by gathering, elasticized shirring; fullness retains soft fall

Recommended Construction Aids & Techniques

Hand Needles chenilles, 20–22; cotton darners, 3–6; embroidery, 1–4; milliners, 3–6
Machine Needles ball point, 11
Threads poly-core waxed, 60; cotton-polyester blend, all purpose; cotton-covered polyester core, all purpose; spun polyester, all purpose
Hems flat/plain; stitched and turned flat; double machine-stitched
Seams overedged; plain; safety-stitched; zigzagged
Seam Finishes double-stitched; plain; serging/single-ply overedged; untreated/plain
Pressing steam; safe temperature limit 300°F (150.1°C)
Care dry clean
Fabric Resource **Gloversville Mills Inc.**

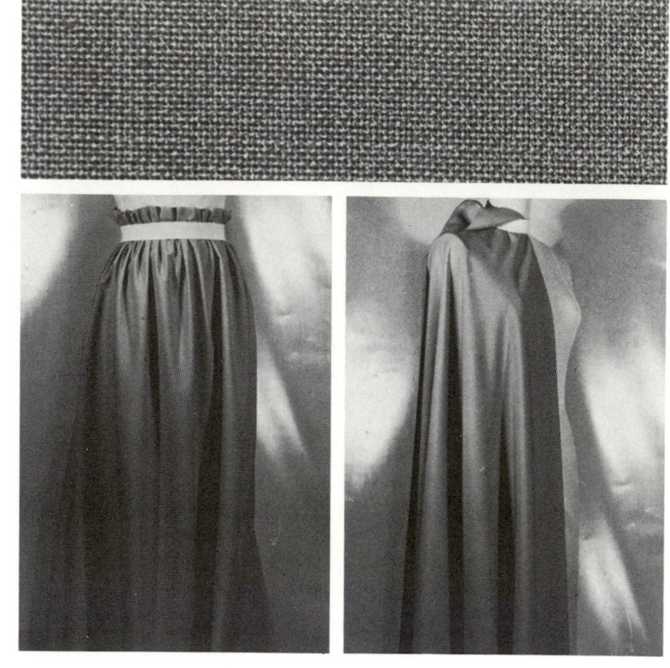

Tropical Worsted

Fiber Content 55% Dacron®/45% wool
Yarn Type combed worsted staple; filament fiber
Yarn Construction conventional; fine; textured; high twist
Fabric Structure tightly woven plain weave
Finishing Processes clear finish; decating; singeing; stabilizing
Color Application piece dyed–cross dyed; yarn dyed
Width 60 inches (152.4 cm)
Weight light
Hand fine; firm; hard
Texture flat; harsh
Performance Expectations dimensional stability; durable; elasticity; elongation; heat-set properties; resilient; subject to edge abrasion and static buildup
Drapability Qualities falls into firm flares; fullness maintains crisp effect; maintains crisp fold, crease, pleat

Recommended Construction Aids & Techniques

Hand Needles chenilles, 20–22; cotton darners, 3–6; embroidery, 1–4; milliners, 3–6
Machine Needles round/set point, medium 14
Threads poly-core waxed, 60; cotton-polyester blend, all purpose; cotton-covered polyester core, all purpose; spun polyester, all purpose
Hems book; bound/Hong Kong finish; pinked flat; stitched and pinked flat; machine-stitched; overedged; seam binding
Seams flat-felled; lapped; plain; top-stitched; welt; double welt
Seam Finishes single-ply bound; Hong Kong bound; double-stitched; edge-stitched; pinked; pinked and stitched; serging/single-ply overedged; single-ply zigzag
Pressing steam; safe temperature limit 300°F (150.1°C)
Care dry clean
Fabric Resource **Burlington Industries Inc.**

Unfinished Worsted[5]

Fiber Content 100% wool
Yarn Type combed worsted staple
Yarn Construction conventional; slack twist
Fabric Structure closely woven plain weave
Finishing Processes fulling; slightly napped; shearing
Color Application stock yarn dyed; piece dyed
Width 60 inches (152.4 cm)
Weight medium-light
Hand firm; hard; scratchy
Texture flat; harsh; napped finish obscures weave
Performance Expectations subject to edge abrasion; absorbent; elasticity; elongation; resilient; tends to shine with wear
Drapability Qualities falls into moderately soft flares; fullness maintains lofty effect; retains shape of garment

Recommended Construction Aids & Techniques

Hand Needles chenilles, 20–22; cotton darners, 3–6; embroidery, 1–4; milliners, 3–6
Machine Needles round/set point, medium 14
Threads mercerized cotton, heavy duty; silk, industrial size A-B; six-cord cotton, 50; waxed, 50; cotton-covered polyester core, all purpose
Hems book; bonded; single-fold bias binding; bound/Hong Kong finish; flat/plain; pinked flat; stitched and pinked flat; interfaced; overedged; seam binding
Seams plain; safety-stitched; top-stitched; welt; double welt
Seam Finishes book; single-ply or double-ply bound; Hong Kong bound; single-ply overcast; pinked; pinked and stitched; serging/single-ply overedged; single-ply zigzag
Pressing steam; safe temperature limit 300°F (150.1°C)
Care dry clean
Fabric Resource **Burlington Industries Inc.**

Listing of Wool Terms

The following woolens are named for the origin or type of sheep from which the fleece is shorn and do not relate to yarn, structure, finish or properties of the fabric.

Merino Wool A fine soft wool from the Merino sheep.
Australian Wool Wool from Spanish Merino stock raised in Australia.
Botany Wool Wool from the Merino sheep raised in the Botany Bay area of Australia.
Saxony Wool Originally referred to a wool fabric made from sheep raised in Saxony Germany. Today, the term refers to fine woolens similar to Saxony woolens.
Shetland Wool The term Shetland should be used only when describing wool or wool products from the Shetland Islands in Scotland.
Lambswool The first or virgin wool fiber clippings obtained from seven- to eight-month old lambs.
Naked Wool A general term used to describe lightweight, sheer wool fabric constructed into garments without backing or linings.

Comparison of Woolen & Worsted Fabrics

Worsted woolens are made from a select choice of fine long staple fibers. The worsted process incorporates a carding and combing procedure that removes foreign matter and short fibers from the shorn wool.

Worsted yarns are:

- Composed of long parallel fibers
- Composed of fibers which are uniform in length
- Spun evenly, producing a fine yarn
- Tightly or firmly twisted
- Compactly woven into cloth

The worsted process produces a fabric with a smooth, fine surface texture having a crisp hand and firm appearance. The finish of worsted fabrics is hardy and more durable than other woolen fiber fabrics.

Comparison of Woolen & Worsted Fabrics Chart

	Woolen Fabrics	Worsted Fabrics
Fiber	short curly	long straight
Yarn	carded only weak bulky uneven twist slack twist —	carded and combed great tensile strength fine, smooth, even even twist tight twist generally yarn dyed
Weave	indistinct pattern usually plain weave sometimes twill weave low thread count loosely woven	distinct pattern usually twill weave sometimes plain weave high thread count closely woven
Finish	soft fulling, napping steaming napping can conceal quality of construction	hard singeing steaming unfinished worsteds are napped
Appearance	soft fuzzy thick	harsh rough flat
Hand	soft lofty	hard crisp
Characteristics	warmer not as durable nap reduces shine soft surface holds dirt stains easily removed does not hold crease well	less insulatory more durable becomes shiny with use resistant to dust and dirt shows stains quickly holds creases and shape
Cost	generally less expensive to produce	costlier yarns, more expensive to produce

9 ~ Specialty Hair/Specialty Hair Blended Fabrics

Alpaca Hair
- Alpaca Hair (lightweight)
- Alpaca Hair Cloth (medium weight)
- Alpaca Hair Cloth (heavyweight)
- Alpaca Hair Cloth Coating

Camel's Hair
- Camel's Hair (lightweight)
- Camel's Hair Cloth (medium weight)
- Camel's Hair Coating

Cashmere
- Cashmere (lightweight)
- Cashmere (medium weight)
- Cashmere (heavyweight)
- Cashmere Cloth Coating

Mohair
- Gauze Mohair
- Looped-faced Mohair / Bouclé Mohair
- Clear-face Mohair / Hard-finished Mohair
- Single-faced Napped Mohair
- Double-faced Napped Mohair
- Mohair Coating / Zebaline®

Rabbit Hair Cloth
- Rabbit Hair Cloth (lightweight)
- Rabbit Hair Cloth (medium weight)
- Rabbit Hair Cloth Coating

Hair or hair-blended fabrics contain the specialty hair taken from a variety of animals. The type and amount of specialty hair fiber contained in the fabric must be specified and listed in accordance with the Textile Fiber Products Identification Act.

Specialty hair fibers may be used alone to produce a cloth of one-hundred percent specialty hair fibers. The specialty hair fibers, available in limited quantities, are usually blended with wool or other natural or man-made fibers. When blended, the specialty hair fibers add softness which affects the hand and drapability qualities of the fabric. The fabric retains the properties, characteristics and care factors of the natural or man-made fibers added to the specialty hair fibers.

Specialty hair fibers used alone or in any combination with other fibers may be produced:

- With fine conventional yarns or coarse novelty yarns
- In woven or knitted structure
- From loose and open to firm and dense structure
- From bulky, lofty effect to firm, flat or fine surface texture

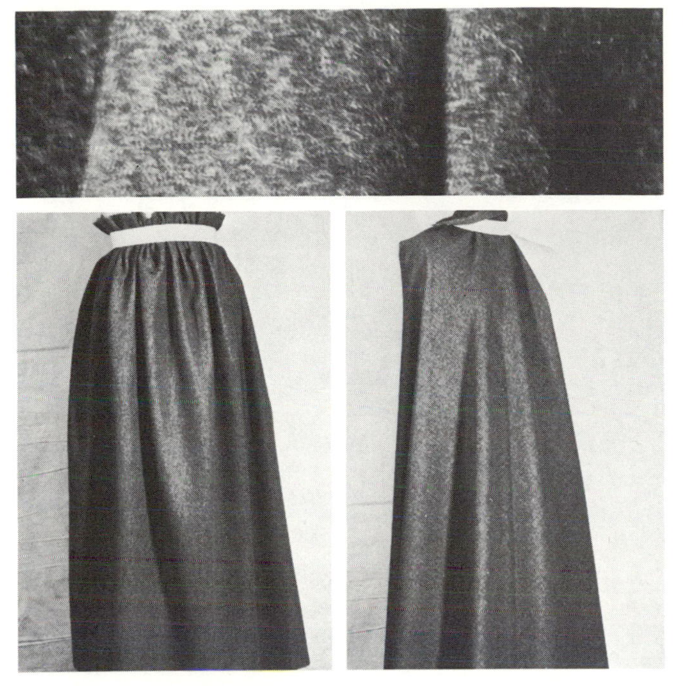

Alpaca Hair (lightweight)

Fiber Content 100% alpaca hair
Yarn Type hair fiber; staple; woolen spun
Yarn Construction conventional; loose twist
Fabric Structure loosely woven plain weave
Finishing Processes brushing; fulling; napping; sponging
Color Application yarn dyed
Width 63 inches (160 cm)
Weight light
Hand fine; downy soft; supple
Texture furry; napped face obliterates weave
Performance Expectations absorbent; elasticity; elongation; insulating properties; resilient; wrinkle resistant
Drapability Qualities falls into soft flares; accommodates fullness by pleating, gathering, elasticized shirring; fullness retains soft graceful fall

Recommended Construction Aids & Techniques

Hand Needles betweens, 3–5; cotton darners, 4–8; embroidery, 3–6; milliners, 3–6; sharps, 1–3
Machine Needles round/set point, medium 11–14
Threads mercerized cotton, 50; silk, industrial size A; six-cord cotton, 50; cotton-covered polyester core, all purpose
Hems book; bound/Hong Kong finish; pinked flat; stitched and pinked flat; machine-stitched; overedged; seam binding
Seams flat-felled; lapped; plain; top-stitched; welt; double welt
Seam Finishes single-ply or double-ply bound; Hong Kong bound; double-stitched; edge-stitched; pinked; pinked and stitched; serging/single-ply overedged; single-ply zigzag
Pressing *do not* steam; use needle board; safe temperature limit 300°F (150.1°C)
Care dry clean
Fabric Resource **Amical Fabrics**

Alpaca Hair Cloth (medium weight)

Fiber Content 85% wool/15% alpaca hair
Yarn Type hair fiber; staple; worsted spun
Yarn Construction bulked; conventional; alpaca hair in filling
Fabric Structure plain weave
Finishing Processes napping; shearing; singeing
Color Application piece dyed; yarn dyed
Width 60 inches (152.4 cm)
Weight medium-light
Hand lofty; soft; springy
Texture flat; lustrous; smooth
Performance Expectations absorbent; elasticity; elongation; insulating properties; subject to moth damage and pilling; resilient; tends to shrink and felt; wrinkle resistant
Drapability Qualities falls into moderately soft flares; fullness maintains soft lofty effect; retains shape of garment

Recommended Construction Aids & Techniques

Hand Needles chenilles, 20–22; cotton darners, 3–6; embroidery, 1–4; milliners, 3–6
Machine Needles round/set point, medium 11–14
Threads mercerized cotton, heavy duty; silk, industrial size A-B; six-cord cotton, 40; waxed, 50; cotton-covered polyester core, heavy duty
Hems bonded; double-fold or single-fold bias binding; bound/Hong Kong finish; faced; flat/plain; pinked flat; interfaced; overedged; seam binding
Seams plain; safety-stitched; top-stitched; welt; double welt
Seam Finishes single-ply or double-ply bound; Hong Kong bound; single-ply overcast; pinked; pinked and stitched; serging/single-ply overedged; single-ply zigzag
Pressing *do not* steam; safe temperature limit 300°F (150.1°C)
Care dry clean
Fabric Resource **Burlington Industries Inc.**

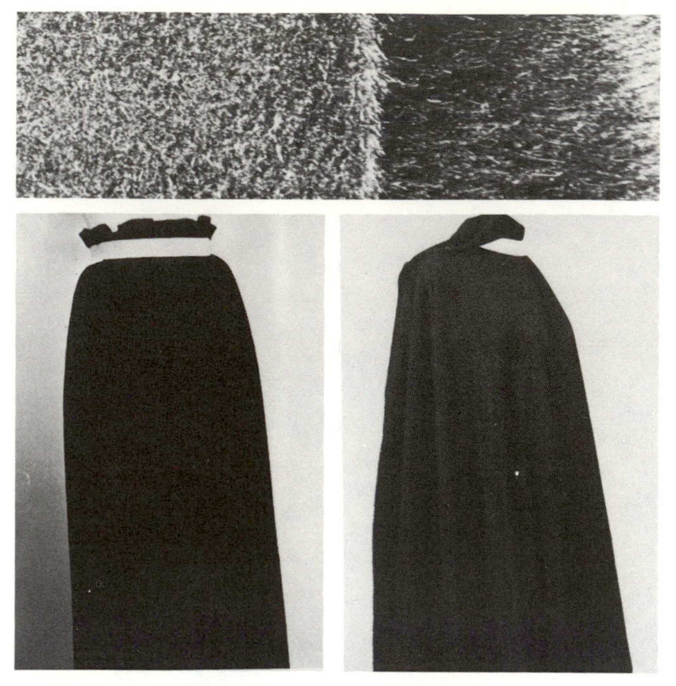

Alpaca Hair Cloth (heavyweight)

Fiber Content 52% alpaca hair/48% wool
Yarn Type hair fiber; staple; woolen and worsted spun
Yarn Construction conventional; loose twist
Fabric Structure double cloth
Finishing Processes brushing; heavily fulled and felted; napping; sponging
Color Application piece dyed; yarn dyed
Width 62 inches (157.5 cm)
Weight heavy
Hand compact; soft; thick
Texture furry; napped face obliterates weave; nap pressed flat in one direction
Performance Expectations elongation; thermo-insulating properties; resilient; water-repellent properties due to long nap and structure; wrinkle resistant
Drapability Qualities falls into wide cones; retains shape or silhouette of garment; better utilized if fitted by seaming and eliminating excess fabric

Recommended Construction Aids & Techniques

Hand Needles chenilles, 18; cotton darners, 1–3; embroidery, 1–4; milliners, 3; tapestry, 18–22
Machine Needles round/set point, coarse 16–18
Threads silk, industrial size A-B-C-D; six-cord cotton, 30; waxed, 40 heavy duty; cotton-covered polyester core, heavy duty
Hems bonded; double-fold or single-fold bias binding; bound/Hong Kong finish; faced; flat/plain; interfaced; overedged
Seams plain
Seam Finishes single-ply bound; Hong Kong bound; pinked; pinked and stitched; serging/single-ply overedged; untreated/plain
Pressing *do not* steam; use needle board; safe temperature limit 300°F (150.1°C)
Care dry clean
Fabric Resource Amical Fabrics

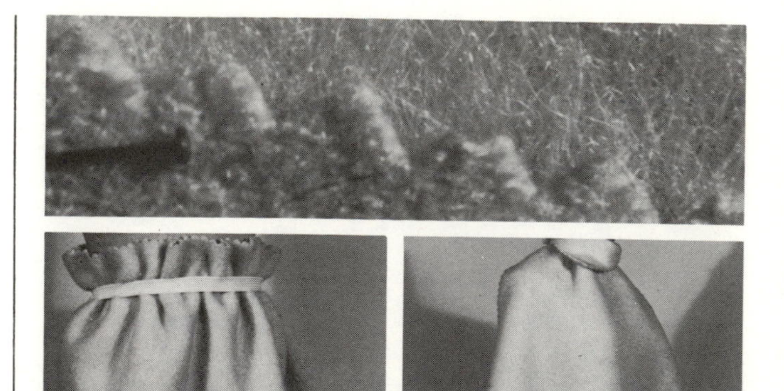

Alpaca Hair Cloth Coating

Fiber Content 63% wool/37% alpaca hair
Yarn Type hair fiber; staple
Yarn Construction conventional; thick-and-thin; alpaca hair in filling
Fabric Structure plain weave
Finishing Processes brushing; heavily fulled; napping
Color Application piece dyed; yarn dyed
Width 60 inches (152.4 cm)
Weight heavy
Hand compact; thick; warm
Texture fuzzy; napped; long nap obscures weave
Performance Expectations elongation; thermo-insulating properties; subject to moth damage; resilient; tends to shrink and felt; water-repellent properties due to long nap and structure; wrinkle resistant
Drapability Qualities falls into wide cones; retains shape or silhouette of garment; better utilized if fitted by seaming and eliminating excess fabric

Recommended Construction Aids & Techniques

Hand Needles chenilles, 18; cotton darners, 1–3; embroidery, 1–4; milliners, 3; tapestry, 18–22
Machine Needles round/set point, coarse 16–18
Threads silk, industrial size A-B-C-D; six-cord cotton, 30; waxed, 40 heavy duty; cotton-covered polyester core, heavy duty
Hems bonded; double-fold or single-fold bias binding; bound/Hong Kong finish; faced; flat/plain; interfaced
Seams plain
Seam Finishes single-ply bound; Hong Kong bound; pinked; pinked and stitched; serging/single-ply overedged; untreated/plain
Pressing *do not* steam; use needle board; safe temperature limit 300°F (150.1°C)
Care dry clean
Fabric Resource Anglo Fabrics Co. Inc.

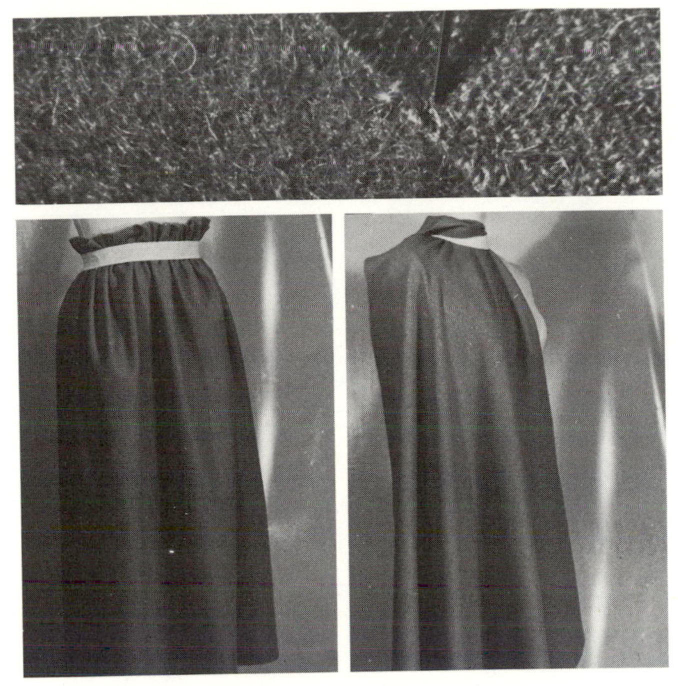

Camel's Hair (lightweight)

Fiber Content 100% camel's hair
Yarn Type hair fiber; staple; woolen and worsted spun
Yarn Construction conventional; average twist
Fabric Structure loosely woven twill weave
Finishing Processes brushing; fulling; napping; pressing; sponging
Color Application undyed/natural color hair; yarn dyed
Width 60 inches (152.4 cm)
Weight light
Hand fine; soft; woolly
Texture short nap obscures weave; napped face; clear back
Performance Expectations elongation; insulating properties; resilient; wrinkle resistant
Drapability Qualities falls into soft cones; accommodates fullness by pleating, gathering; maintains soft fold, unpressed pleat

Recommended Construction Aids & Techniques

Hand Needles betweens, 3–5; cotton darners, 4–8; embroidery, 3–6; milliners, 3–6; sharps, 1–3
Machine Needles round/set point, medium 11–14
Threads mercerized cotton, 50; silk, industrial size A; six-cord cotton, 50; waxed, 60; cotton-covered polyester core, all purpose
Hems book; bound/Hong Kong finish; pinked flat; stitched and pinked flat; machine-stitched; overedged; seam binding
Seams flat-felled; lapped; plain; top-stitched; welt; double welt
Seam Finishes single-ply bound; Hong Kong bound; double-stitched; edge-stitched; pinked; pinked and stitched; serging/single-ply overedged; single-ply zigzag
Pressing *do not* steam; use needle board; brush surface; safe temperature limit 300°F (150.1°C)
Care dry clean
Fabric Resource **Amical Fabrics**

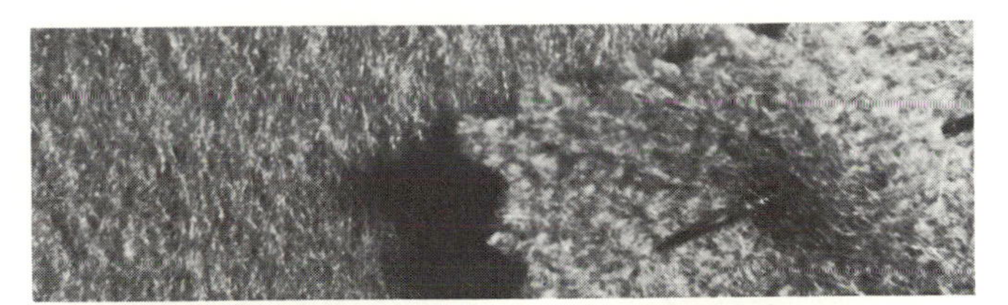

Camel's Hair Cloth (medium weight)

Fiber Content 85% wool/15% camel's hair
Yarne Type hair fiber; staple
Yarn Construction conventional; thick-and-thin
Fabric Structure plain weave
Finishing Processes brushing; fulling; napping
Color Application piece dyed; undyed/natural color hair; yarn dyed
Width 60 inches (152.4 cm)
Weight medium
Hand soft; thick
Texture napped face; nap obscures weave; nap brushed flat in one direction
Performance Expectations elongation; thermo-insulating properties; subject to moth damage; resilient; tends to shrink and felt; natural water-repellent properties; wrinkle resistant
Drapability Qualities falls into wide cones; retains shape or silhouette of garment; better utilized if fitted by seaming and eliminating excess fabric

Recommended Construction Aids & Techniques

Hand Needles chenilles, 18; cotton darners, 1–3; embroidery; 1–4; milliners, 3; tapestry, 18–22
Machine Needles round/set point, coarse 16–18
Threads mercerized cotton, heavy duty; silk, industrial size A-B; six-cord cotton, 50; waxed, 50; cotton-covered polyester core, all purpose
Hems bonded; double-fold or single-fold bias binding; bound/Hong Kong finish; faced; flat/plain; interfaced
Seams plain;
Seam Finishes single-ply bound; Hong Kong bound; pinked; pinked and stitched; serging/single-ply overedged; untreated/plain
Pressing *do not* steam; use needle board; safe temperature limit 300°F (150.1°C)
Care dry clean
Fabric Resource **Anglo Fabrics Co. Inc.**

Camel's Hair Coating

Fiber Content 100% camel's hair
Yarn Type hair fiber; staple; woolen and worsted spun
Yarn Construction conventional; heavy warp yarns; thick filling yarns
Fabric Structure twill weave
Finishing Processes brushing; heavily fulled; napping; shearing; sponging
Color Application undyed/natural color hair; yarn dyed
Width 60 inches (152.4 cm)
Weight heavy
Hand lofty; soft; thick
Texture fuzzy; heavily napped face obliterates weave; opaque
Performance Expectations elongation; thermo-insulating properties; resilient; wrinkle resistant
Drapability Qualities falls into moderately soft flares; retains shape or silhouette of garment; better utilized if fitted by seaming and eliminating excess fabric

Recommended Construction Aids & Techniques

Hand Needles chenilles, 18; cotton darners, 1–3; embroidery, 1–4; milliners, 3; tapestry, 18–22
Machine Needles round/set point, coarse 16–18
Threads silk, industrial size A-B-C-D; six-cord cotton, 30; waxed, 40 heavy duty; cotton-covered polyester core, 60
Hems bonded; double-fold or single-fold bias binding; bound/Hong Kong finish; faced; flat/plain; interfaced; overedged
Seams plain; safety-stitched
Seam Finishes single-ply bound; Hong Kong bound; pinked; pinked and stitched; serging/single-ply overedged; untreated/plain
Pressing *do not* steam; use needle board; safe temperature limit 300°F (150.1°C)
Care dry clean
Fabric Resource **Amical Fabrics**

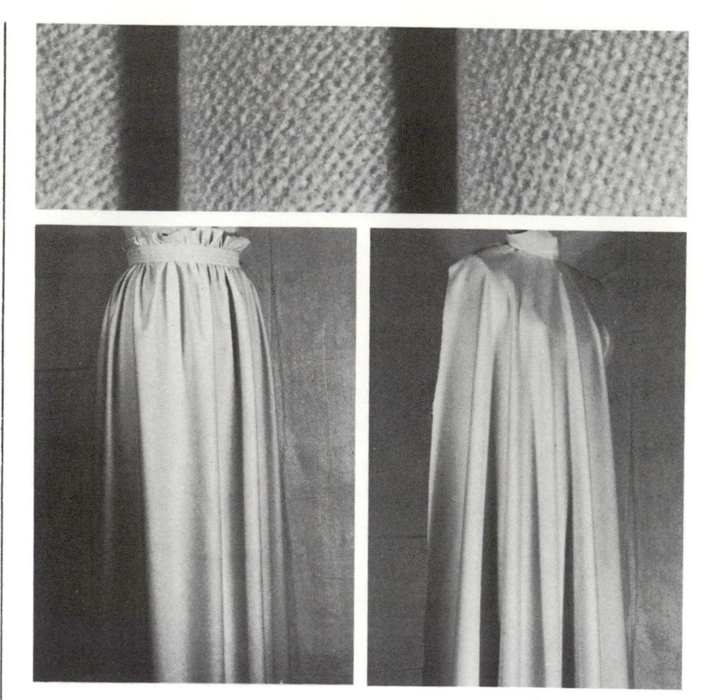

Cashmere (lightweight)

Fiber Content 100% cashmere goat hair
Yarn Type hair fiber; staple; woolen and worsted spun
Yarn Construction conventional; average twist
Fabric Structure plain weave
Finishing Processes brushing; fulling; shearing; singeing; sponging
Color Application yarn dyed
Width 60 inches (152.4 cm)
Weight light
Hand fine; pliable, downy soft
Texture flat; slightly napped face
Performance Expectations flexible; elasticity; elongation; resilient; wrinkle resistant
Drapability Qualities falls into soft flares; accommodates fullness by pleating, gathering, elasticized shirring; fullness retains soft fall

Recommended Construction Aids & Techniques

Hand Needles betweens, 3–5; cotton darners, 4–8; embroidery, 3–6; milliners, 3–6; sharps, 1–3
Machine Needles round/set point, medium 11
Threads mercerized cotton, 50; silk, industrial size A; six-cord cotton, 50; waxed, 60; cotton-covered polyester core, all purpose
Hems book; bound/Hong Kong finish; pinked flat; stitched and pinked flat; machine-stitched; overedged; seam binding
Seams lapped; plain; safety-stitched; top-stitched; welt; double welt
Seam Finishes book; single-ply bound; Hong Kong bound; double-stitched; edge-stitched; pinked; pinked and stitched; serging/single-ply overedged; single-ply zigzag
Pressing *do not* steam; use pressing cloth; safe temperature limit 300°F (150.1°C)
Care dry clean
Fabric Resource **Amical Fabrics**

Cashmere (medium weight)

Fiber Content 100% cashmere goat hair
Yarn Type hair fiber; staple; woolen spun
Yarn Construction conventional
Fabric Structure twill weave
Finishing Processes clear finish; decating; singeing; sponging
Color Application yarn dyed
Width 59 inches (149.9 cm)
Weight medium
Hand firm; lofty; downy soft
Texture visible pattern of twill weave; flat; smooth
Performance Expectations flexible; elasticity; elongation; resilient; wrinkle resistant
Drapability Qualities falls into soft cones; accommodates fullness by pleating, gathering; maintains soft folds, unpressed pleats

Recommended Construction Aids & Techniques

Hand Needles chenilles, 20–22; cotton darners, 3–6; embroidery, 1–4; milliners, 3–6
Machine Needles round/set point, medium 14
Threads mercerized cotton, heavy duty; silk, industrial size A-B; six-cord cotton, 50; waxed, 50; cotton-covered polyester core, all purpose
Hems book; bonded; single-fold bias binding; bound/Hong Kong finish; flat/plain; pinked; stitched and pinked flat; interfaced; overedged; seam binding
Seams plain; safety-stitched; top-stitched; welt; double welt
Seam Finishes book; single-ply or double-ply bound; Hong Kong bound; single-ply overcast; pinked; pinked and stitched; serging/single-ply overedged; single-ply zigzag
Pressing *do not* steam; use pressing cloth; safe temperature limit 300°F (150.1°C)
Care dry clean
Fabric Resource Amical Fabrics

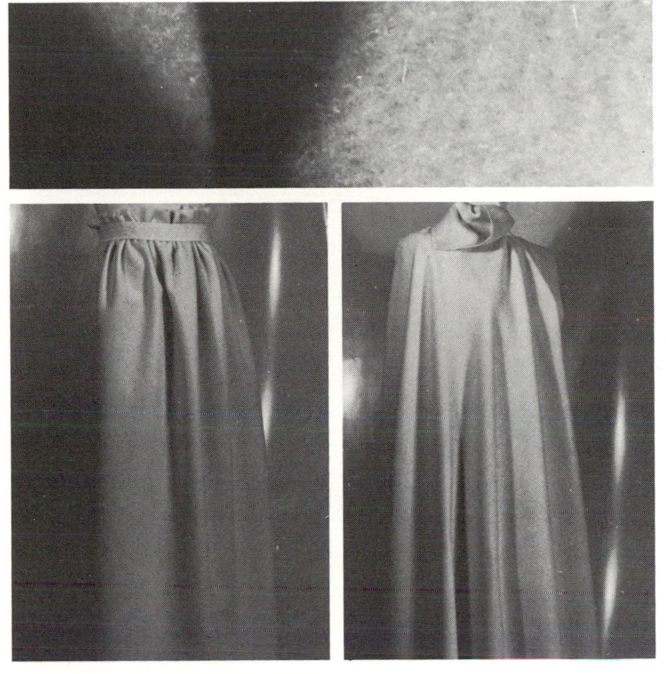

Cashmere (heavyweight)

Fiber Content 100% cashmere goat hair
Yarn Type hair fiber; staple; woolen and worsted spun
Yarn Construction conventional; loosely twisted woolen yarn; tightly twisted worsted yarn
Fabric Structure twill weave
Finishing Processes brushing; heavily fulled; napping; shearing; sponging
Color Application yarn dyed
Width 60 inches (152.4 cm)
Weight heavy
Hand compact; lofty; soft
Texture heavily napped face obscures weave; feltlike
Performance Expectations absorbent; compressible; flexible; insulating properties; water-repellent properties due to nap and weave; wrinkle resistant
Drapability Qualities falls into wide cones; retains shape or silhouette of garment; better utilized if fitted by seaming and eliminating excess fabric

Recommended Construction Aids & Techniques

Hand Needles chenilles, 18; cotton darners, 1–3; embroidery, 1–4; milliners, 3; tapestry, 18–22
Machine Needles round/set point, coarse 16–18
Threads silk, industrial size A-B-C-D; six-cord cotton, 30; waxed, 40 heavy duty; cotton-covered polyester core, heavy duty
Hems bonded; double-fold or single-fold bias binding; bound/Hong Kong finish; flat/plain; pinked flat; interfaced; overedged
Seams plain; safety-stitched
Seam Finishes single-ply bound; Hong Kong bound; pinked; pinked and stitched; serging/single-ply overedged; untreated/plain
Pressing *do not* steam; use needle board and pressing cloth; safe temperature limit 300°F (150.1°C)
Care dry clean
Fabric Resource Amical Fabrics

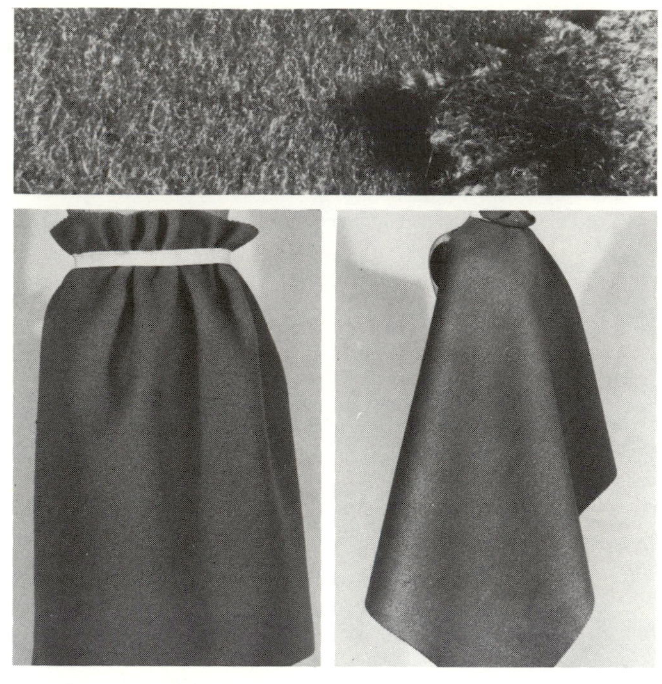

Cashmere Cloth Coating

Fiber Content 85% wool/15% cashmere goat hair
Yarn Type hair fiber; staple
Yarn Construction conventional; thick and fine yarns
Fabric Structure closely woven plain or twill weave
Finishing Processes brushing; fulling; napping
Color Application piece dyed; yarn dyed
Width 60 inches (152.4 cm)
Weight heavy
Hand silky; soft; thick
Texture napped; nap obscures weave
Performance Expectations elongation; thermo-insulating properties; subject to moth damage; resilient; tends to shrink and felt; natural water-repellent properties; wrinkle resistant
Drapability Qualities falls into wide cones; retains shape or silhouette of garment; better utilized if fitted by seaming and eliminating excess fabric

Recommended Construction Aids & Techniques

Hand Needles chenilles, 18; cotton darners, 1–3; embroidery, 1–4; milliners, 3; tapestry, 18–22
Machine Needles round/set point, coarse 16–18
Threads silk, industrial size A-B-C-D; six-cord cotton, 30; waxed, 40 heavy duty; cotton-covered polyester core, heavy duty
Hems bonded; double-fold or single-fold bias binding; bound/Hong Kong finish; faced; flat/plain; interfaced
Seams plain
Seam Finishes single-ply bound; Hong Kong bound; pinked; pinked and stitched; serging/single-ply overedged; untreated/plain
Pressing *do not* steam; use needle board; safe temperature limit 300°F (150.1°C)
Care dry clean
Fabric Resource Anglo Fabrics Co. Inc.

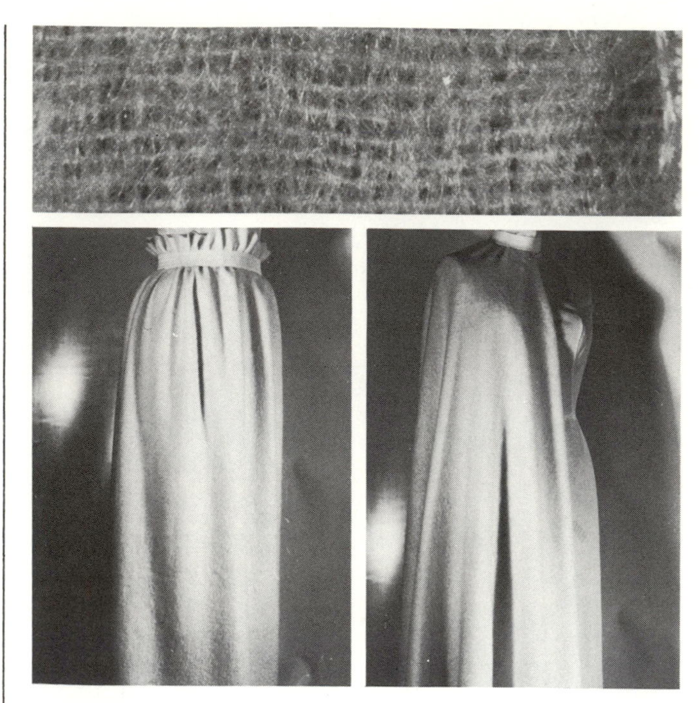

Gauze Mohair

Fiber Content 41% wool/23% nylon/23% acrylic/13% mohair
Yarn Type combed staple; filament fiber; hair fiber; worsted spun
Yarn Construction blended yarns; conventional; low twist
Fabric Structure loosely woven plain weave
Finishing Processes brushing; napping; stabilizing
Color Application piece dyed–cross dyed; yarn dyed
Width 52–54 inches (132.1–137.2 cm)
Weight light
Hand pliable; silky; soft
Texture fluffy; napped; loose open weave structure
Performance Expectations abrasion resistant; elasticity; flexible; insulating properties; resilient; tends to shrink and felt; subject to snagging due to weave and pilling
Drapability Qualities falls into moderately soft flares; accommodates fullness by gathering; fullness maintains soft lofty effect

Recommended Construction Aids & Techniques

Hand Needles chenilles, 20–22; cotton darners, 3–6; embroidery, 1–4; milliners, 3–6
Machine Needles round/set point, medium 11–14
Threads mercerized cotton, heavy duty; silk, industrial size A; six-cord cotton, 50; waxed, 50; cotton-covered polyester core, all purpose
Hems bonded; single-fold bias binding; bound/Hong Kong finish; flat/plain; pinked flat; stitched and pinked flat; interfaced; overedged; seam binding
Seams plain; safety-stitched; tissue-stitched
Seam Finishes book; single-ply or double-ply bound; Hong Kong bound; single-ply overcast; pinked; pinked and stitched; serging/single-ply overedged; single-ply zigzag
Pressing *do not* steam; use needle board and pressing cloth; safe temperature limit 300°F (150.1°C)
Care dry clean
Fabric Resource Auburn Fabrics Inc.

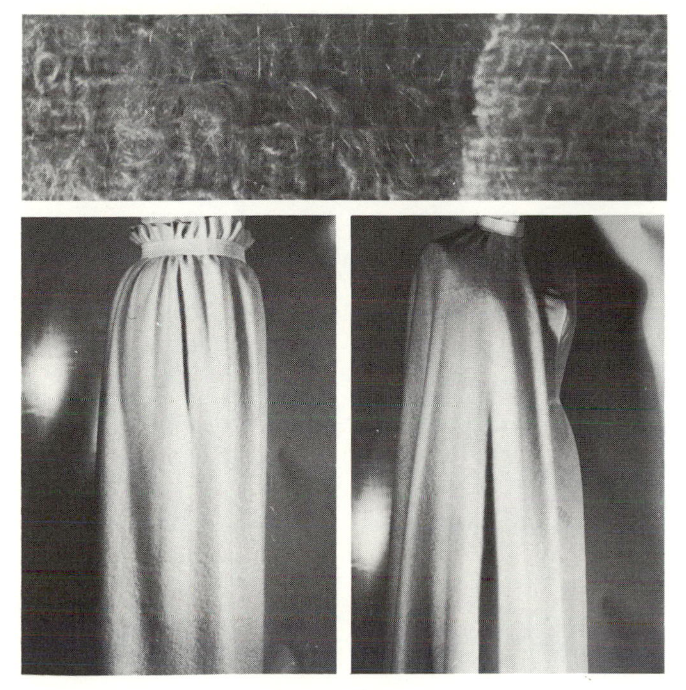

Looped-faced Mohair/Bouclé Mohair

Fiber Content 63% wool/37% mohair
Yarn Type staple; hair fiber
Yarn Construction novelty looped yarn; slack twist
Fabric Structure loosely woven plain weave
Finishing Processes brushing; stabilizing
Color Application piece dyed–union dyed; yarn dyed
Width 60 inches (152.4 cm)
Weight light
Hand silky; soft; springy
Texture napped; loose bouclé yarn creates looped surface on face; open structure
Performance Expectations abrasion resistant; elasticity; flexible; insulating properties; subject to pilling; resilient; tends to shrink and felt; subject to snagging due to yarn and weave
Drapability Qualities falls into soft flares; fullness maintains soft lofty effect; maintains soft fold, unpressed pleat

Recommended Construction Aids & Techniques

Hand Needles chenilles, 20–22; cotton darners, 3–6; embroidery, 1–4; milliners, 3–6
Machine Needles round/set point, medium 11–14
Threads mercerized cotton, heavy duty; silk, industrial size A-B; six-cord cotton, 50; waxed, 50; cotton-covered polyester core, heavy duty
Hems bonded; single-fold bias binding; bound/Hong Kong finish; flat/plain; pinked flat; stitched and pinked flat; interfaced; overedged; seam binding
Seams plain; safety-stitched; tissue-stitched
Seam Finishes single-ply or double-ply bound; Hong Kong bound; single-ply overcast; pinked; pinked and stitched; serging/single-ply overedged; single-ply zigzag
Pressing *do not* steam; use needle board and pressing cloth; safe temperature limit 300°F (150.1°C)
Care dry clean
Fabric Resource **Anglo Fabrics Co. Inc.**

Hard-faced Mohair/Clear Mohair

Fiber Content 75% wool/15% mohair/10% nylon
Yarn Type carded staple; filament fiber; hair fiber
Yarn Construction blended; conventional; low twist
Fabric Structure twill weave
Finishing Processes clear finish; decating; shearing; singeing; stabilizing
Color Application yarn dyed
Width 56–58 inches (142.2–147.3 cm)
Weight medium-light
Hand soft; wiry
Texture flat; smooth
Performance Expectations abrasion resistant; elasticity; flexible; insulating properties; subject to pilling; resilient; tends to shrink and felt
Drapability Qualities falls into moderately firm cones; fullness maintains lofty effect; retains shape of garment

Recommended Construction Aids & Techniques

Hand Needles chenilles, 20–22; cotton darners, 3–6; embroidery, 1–4; milliners, 3–6
Machine Needles round/set point, medium 11–14
Threads mercerized cotton, heavy duty; silk, industrial size A-B-C; six-cord cotton, 40; waxed, 50; cotton-covered polyester core, heavy duty
Hems bonded; double-fold or single-fold bias binding; bound/Hong Kong finish; faced; flat/plain; pinked flat; interfaced; overedged; seam binding
Seams plain; safety-stitched
Seam Finishes book; single-ply bound; Hong Kong bound; single-ply overcast; pinked; pinked and stitched; serging/single-ply overedged; single-ply zigzag
Pressing *do not* steam; use pressing cloth; safe temperature limit 300°F (150.1°C)
Care dry clean
Fabric Resource **Auburn Fabrics Inc.**

Single-faced Napped Mohair

Fiber Content 85% mohair/15% nylon
Yarn Type filament fiber; hair fiber; staple
Yarn Construction novelty spiral yarn; conventional tightly twisted yarn
Fabric Structure loosely woven plain weave with laid-in yarns
Finishing Processes brushing; napping; pressing; stabilizing
Color Application yarn dyed
Width 60 inches (152.4 cm)
Weight medium
Hand lofty; downy soft; springy
Texture heavily napped face; clear back; hair fibers lay flat in one direction
Performance Expectations compressible; flexible; elasticity; hair fibers tend to shed and abrade with wear; wrinkle resistant
Drapability Qualities falls into wide cones; fullness maintains lofty bouffant effect; retains shape or silhouette of garment

Recommended Construction Aids & Techniques

Hand Needles chenilles, 20–22; cotton darners, 3–6; embroidery, 1–4; milliners, 3–6
Machine Needles round/set point, medium 14
Threads poly-core waxed, 60; cotton-polyester blend, all purpose; cotton-covered polyester core, all purpose; spun polyester, all purpose
Hems double-fold or single-fold bias binding; bound/Hong Kong finish; flat/plain; overedged
Seams plain; safety-stitched
Seam Finishes single-ply or double-ply bound; Hong Kong bound; serging/single-ply overedged; untreated plain
Pressing *do not* press flat; steam over needle board; safe temperature limit 300°F (150.1°C)
Care dry clean
Fabric Resource **Arthur Zeiler Woolens Inc.**

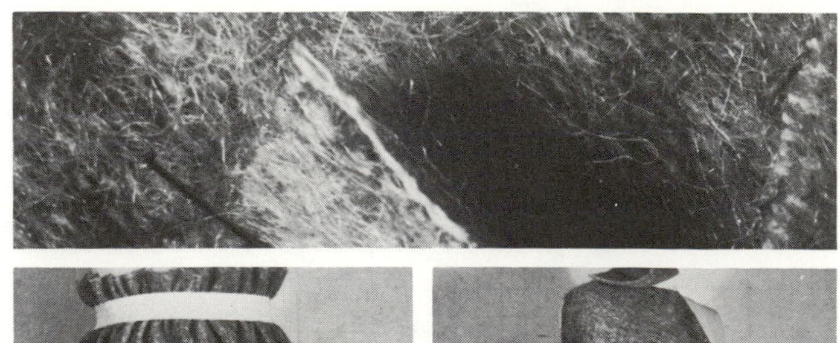

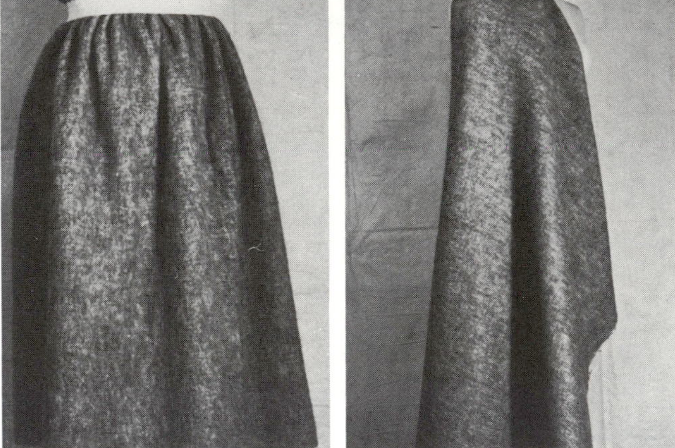

Double-faced Napped Mohair

Fiber Content 87% mohair/13% nylon
Yarn Type filament fiber; hair fiber; staple
Yarn Construction bulked; conventional
Fabric Structure leno weave with laid-in yarns
Finishing Processes brushing; napping; pressing; stabilizing
Color Application yarn dyed
Width 58–60 inches (147.3–152.4 cm)
Weight medium
Hand downy soft; springy
Texture hairy; fuzzy napped face and back; loose weave visible through transparent nap
Performance Expectations compressible; flexible; elasticity; hair fibers tend to shed and abrade with wear; wrinkle resistant
Drapability Qualities falls into wide cones; fullness maintains lofty effect; retains shape of garment

Recommended Construction Aids & Techniques

Hand Needles chenilles, 20–22; cotton darners, 3–6; embroidery, 1–4; milliners, 3–6
Machine Needles round/set point, medium 14
Threads poly-core waxed, 60; cotton-polyester blend, all purpose; cotton-covered polyester core, all purpose; spun polyester, all purpose
Hems double-fold or single-fold bias binding; bound/Hong Kong finish; flat/plain; overedged
Seams plain; safety-stitched
Seam Finishes single-ply or double-ply bound; Hong Kong bound; serging/single-ply overedged; untreated/plain
Pressing *do not* press flat; steam over needle board; safe temperature limit 300°F (150.1°C)
Care dry clean
Fabric Resource **Arthur Zeiler Woolens Inc.**

Mohair Coating/Zebaline®

Fiber Content 69% wool/31% mohair
Yarn Type hair fiber; staple
Yarn Construction conventional; thick-and-thin
Fabric Structure satin weave; hair fibers in filling
Finishing Processes brushing; fulling; napping; stabilizing
Color Application piece dyed; yarn dyed
Width 60 inches (152.4 cm)
Weight heavy
Hand compact; thick; warm
Texture highly raised surface; long sleek nap; nap obscures weave
Performance Expectations elongation; thermo-insulating properties; subject to moth damage; resilient; tends to shrink and felt; water-repellent properties due to long nap and structure; wrinkle resistant
Drapability Qualities falls into wide cones; retains shape or silhouette of garment; better utilized if fitted by seaming and eliminating excess fabric

Recommended Construction Aids & Techniques

Hand Needles chenilles, 18; cotton darners, 1–3; embroidery, 1–4; milliners, 3; tapestry, 18–22
Machine Needles round/set point, coarse 16–18
Threads silk, industrial size A-B-C; six-cord cotton, 30; waxed, 40 heavy duty; cotton-covered polyester core, heavy duty
Hems bonded; double-fold or single-fold bias binding; bound/Hong Kong finish; faced; flat/plain; interfaced
Seams plain
Seam Finishes single-ply bound; Hong Kong bound; pinked; pinked and stitched; serging/single-ply overedged; untreated/plain
Pressing *do not* steam; use needle board; safe temperature limit 300°F (150.1°C)
Care dry clean
Fabric Resource **Anglo Fabrics Co. Inc.**

Rabbit Hair Cloth (lightweight)

Fiber Content 70% wool/20% rabbit hair/10% nylon
Yarn Type filament staple; hair fiber; staple; woolen and worsted spun
Yarn Construction conventional; average twist
Fabric Structure plain weave
Finishing Processes brushing; semidecating; singeing; sponging
Color Application piece dyed–union dyed; yarn dyed
Width 62 inches (157.5 cm)
Weight light
Hand pliable; soft; thin
Texture flat; hair fibers protrude from surface of cloth
Performance Expectations flexible; elasticity; elongation; resilient; hair fibers tend to abrade with wear; wrinkle resistant
Drapability Qualities falls into moderately soft flares; accommodates fullness by pleating, gathering, elasticized shirring; fullness maintains springy effect

Recommended Construction Aids & Techniques

Hand Needles betweens, 3–5; cotton darners, 4–8; embroidery, 3–6; milliners, 3–6; sharps, 1–3
Machine Needles round/set point, medium 11
Threads mercerized cotton, 50; silk, industrial size A; six-cord cotton, 50; waxed, 60; cotton-covered polyester core, all purpose
Hems book; bound/Hong Kong finish; pinked flat; stitched and pinked flat; machine-stitched; overedged; seam binding
Seams flat-felled; lapped; plain; top-stitched; welt; double welt
Seam Finishes single-ply bound; Hong Kong bound; double-stitched; edge-stitched; pinked; pinked and stitched; serging/single-ply overedged; single-ply zigzag
Pressing *do not* steam; use pressing cloth; safe temperature limit 300°F (150.1°C)
Care dry clean
Fabric Resource **Amical Fabrics**

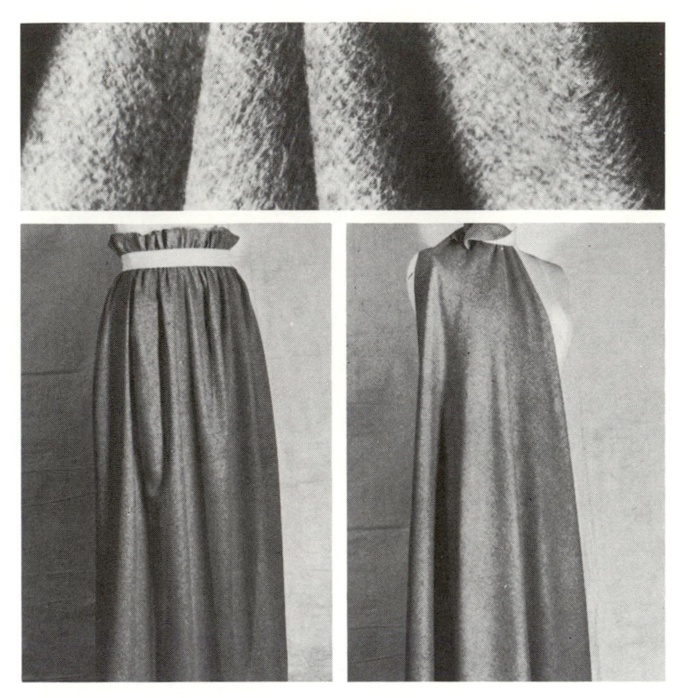

Rabbit Hair Cloth (medium weight)

Fiber Content 50% rabbit hair/50% wool
Yarn Type hair fiber; staple; worsted spun
Yarn Construction conventional
Fabric Structure plain weave
Finishing Processes brushing; fulling; napping; shearing; sponging
Color Application yarn dyed
Width 58 inches (147.3 cm)
Weight medium
Hand fine; pliable; downy soft
Texture napped face; clear flat back
Performance Expectations flexible; elasticity; elongation; resilient; hair fibers tend to abrade with wear; wrinkle resistant
Drapability Qualities falls into soft flares; accommodates fullness by pleating, gathering; fullness retains soft fall

Recommended Construction Aids & Techniques

Hand Needles betweens, 3-5; cotton darners, 4-8; embroidery, 3-6; milliners, 3-6; sharps, 1-3
Machine Needles round/set point, medium 11-14
Threads mercerized cotton, heavy duty; silk, industrial size A-B; six-cord cotton, 50; waxed, 50; cotton-covered polyester core, all purpose
Hems book; bound/Hong Kong finish; pinked flat; stitched and pinked flat; machine-stitched; overedged; seam binding
Seams flat-felled; lapped; plain; top-stitched; welt; double welt
Seam Finishes single-ply bound; Hong Kong bound; double-stitched; edge-stitched; pinked; pinked and stitched; serging/single-ply overedged; single-ply zigzag
Pressing *do not* steam; use pressing cloth; safe temperature limit 300°F (150.1°C)
Care dry clean
Fabric Resource Amical Fabrics

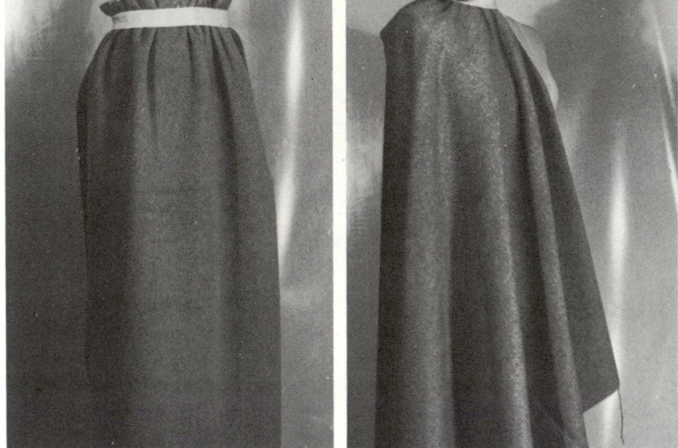

Rabbit Hair Cloth Coating

Fiber Content 60% wool/40% rabbit hair
Yarn Type hair fiber; staple; worsted spun
Yarn Construction conventional; thin warp yarn; thick filling yarn blended with hair
Fabric Structure twill weave with laid-in yarns
Finishing Processes brushing; fulling; napping; pressing; sponging
Color Application yarn dyed
Width 60 inches (152.4 cm)
Weight heavy
Hand furry; soft; thick
Texture heavily napped face; clear back; nap pressed flat in one direction
Performance Expectations elongation; thermo-insulating properties; water-repellent properties due to long overlapping nap and structure; wrinkle resistant
Drapability Qualities falls into wide cones; retains shape or silhouette of garment; better utilized if fitted by seaming and eliminating excess fabric

Recommended Construction Aids & Techniques

Hand Needles chenilles, 18; cotton darners, 1-3; embroidery, 1-4; milliners, 3; tapestry; 18-22
Machine Needles round/set point, coarse 18
Threads silk, industrial size A-B-C-D; six-cord cotton, 30; waxed, 40 heavy duty; cotton-covered polyester core, heavy duty
Hems bonded; double-fold or single-fold bias binding; bound/Hong Kong finish; faced; flat/plain; interfaced; overedged
Seams plain; safety-stitched
Seam Finishes single-ply bound; Hong Kong bound; pinked; pinked and stitched; serging/single-ply overedged; untreated/plain
Pressing *do not* steam; use needle board; safe temperature limit 300°F (150.1°C)
Care dry clean
Fabric Resource Amical Fabrics

10 ~ Felt/Felt-type Fabrics

Felt (heavyweight)
Felt (medium-heavyweight)
Felt (medium weight)
Felt (medium-lightweight)

Felt is a term applied to fabrics characterized by a matted compressed appearance with no apparent system of threads. Felt fabrics may be constructed of wool, fur, cotton or man-made fibers alone or blended in any combination. Felt fabrics are made in a variety of thicknesses and weights depending on their end use.

The performance and drapability qualities of felt fabrics depend on:

- Fiber content
- Method or technique of construction
- Thickness
- Weight
- Hand
- Finishing processes

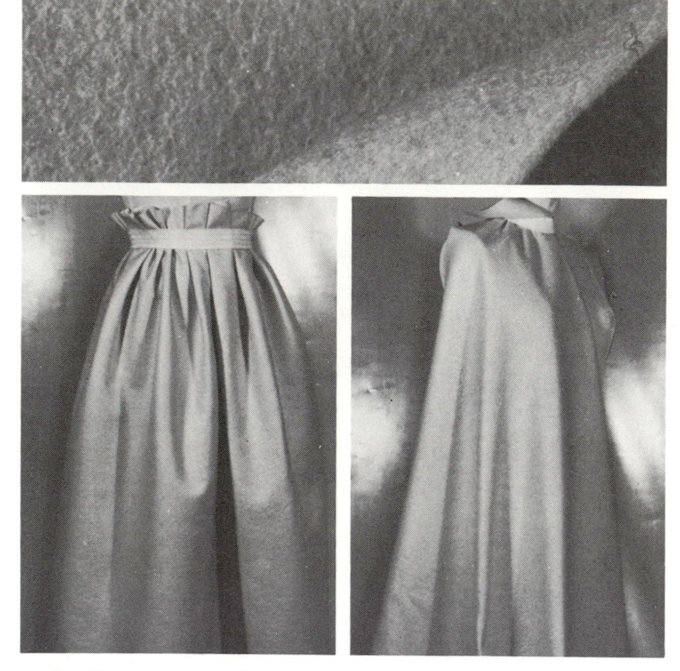

Felt (heavyweight)

Fiber Content 100% wool
Yarn Type staple
Yarn Construction fiber state
Fabric Structure felting
Finishing Processes fulling; napping; stretching
Color Application piece dyed
Width 72 inches (182.9 cm)
Weight heavy
Hand hard; thick; warm
Texture napped; woolly
Performance Expectations insulating properties; does not fray or tear; lacks flexibility and elasticity; resilient
Drapability Qualities falls into wide cones; fullness maintains bouffant effect; can be molded or shaped by moisture and heat

Recommended Construction Aids & Techniques

Hand Needles cotton darners, 1-6; embroidery, 1-4; milliners, 3-6; sharps, 1-3
Machine Needles round/set point, medium-coarse 16
Threads mercerized cotton, heavy duty; silk, industrial size A-B-C-D; six-cord cotton, 30; waxed, 40; cotton-covered polyester core, heavy duty
Hems none needed; decoratively pinked or scalloped edge; double-stitched; flat/plain; machine-stitched
Seams lapped with raw edges; plain
Seam Finishes untreated/plain
Pressing steam; safe temperature limit 300°F (150.1°C)
Care dry clean
Fabric Resource Continental Felt Co.

Felt (medium-heavyweight)

Fiber Content 70% wool/30% rayon
Yarn Type staple; filament staple
Yarn Construction fiber state
Fabric Structure felting
Finishing Processes fulling; napping; stretching
Color Application piece dyed
Width 72 inches (182.9 cm)
Weight medium-heavy
Hand hard; thick; warm
Texture napped; woolly
Performance Expectations insulating properties; does not fray or tear; lacks flexibility and elasticity; resilient
Drapability Qualities falls into few wide cones; fullness maintains bouffant effect; can be shaped or molded by moisture and heat

Recommended Construction Aids & Techniques

Hand Needles cotton darners, 1-6; embroidery, 1-4; milliners, 3-6; sharps, 1-3
Machine Needles round/set point, medium-coarse 16
Threads mercerized cotton, heavy duty; silk, industrial size A-B-C; six-cord cotton, 40; waxed, 50; cotton-covered polyester core, heavy duty
Hems none needed; decoratively pinked or scalloped edge; double-stitched; flat/plain; machine-stitched
Seams lapped with raw edges; plain
Seam Finishes untreated/plain
Pressing steam; safe temperature limit 300°F (150.1°C)
Care dry clean
Fabric Resource Continental Felt Co.

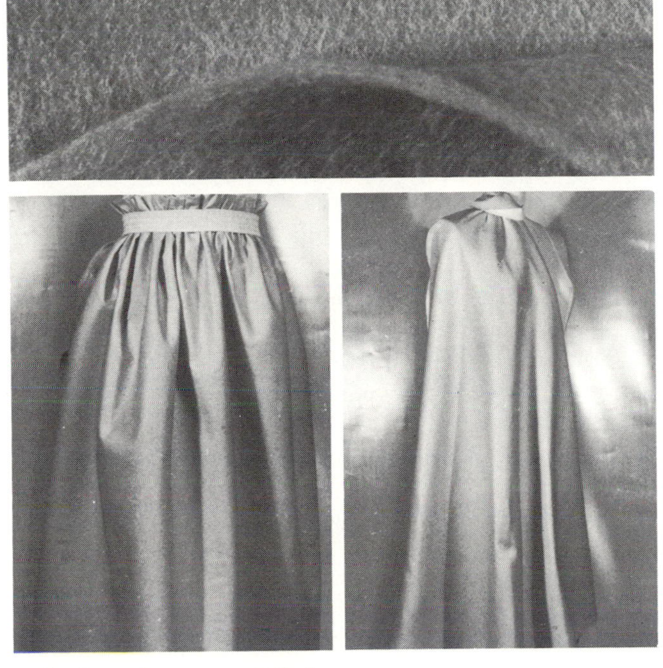

Felt (medium weight)

Fiber Content 50% wool/50% rayon
Yarn Type staple; filament staple
Yarn Construction fiber state
Fabric Structure felting
Finishing Processes fulling; napping; stretching
Color Application piece dyed
Width 72 inches (182.9 cm)
Weight medium
Hand semi-soft; thick; warm
Texture napped; woolly
Performance Expectations insulating properties; does not fray or tear; lacks flexibility and elasticity; resilient
Drapability Qualities falls into wide cones; fullness maintains bouffant effect; can be shaped or molded by moisture and heat

Recommended Construction Aids & Techniques

Hand Needles cotton darners, 1-6; embroidery, 1-4; milliners, 3-6; sharps, 1-3
Machine Needles round/set point, medium-coarse 16
Threads mercerized cotton, heavy duty; silk, industrial size A-B; six-cord cotton, 50; waxed, 50; cotton-covered polyester core, all purpose
Hems none needed; decoratively pinked or scalloped edge; double-stitched; flat/plain; machine-stitched
Seams lapped with raw edges; plain
Seam Finishes untreated/plain
Pressing steam; safe temperature limit 300°F (150.1°C)
Care dry clean
Fabric Resource **Continental Felt Co.**

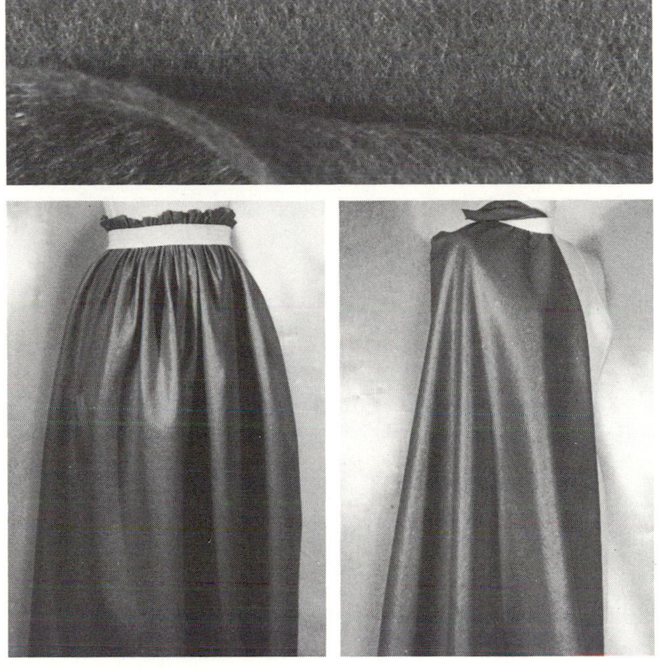

Felt (medium-lightweight)

Fiber Content 40% wool/60% rayon
Yarn Type staple; filament staple
Yarn Construction fiber state
Fabric Structure felting
Finishing Processes fulling; napping; stretching
Color Application piece dyed
Width 72 inches (182.9 cm)
Weight medium-light
Hand soft; comparatively thin; warm
Texture napped; woolly
Performance Expectations insulating properties; does not fray or tear; lacks flexibility and elasticity; resilient
Drapability Qualities falls into wide cones; fullness maintains bouffant effect; can be shaped or molded by moisture and heat

Recommended Construction Aids & Techniques

Hand Needles cotton darners, 1-6; embroidery, 1-4; milliners, 3-6; sharps, 1-3
Machine Needles round/set point, medium-coarse 16
Threads mercerized cotton, heavy duty; silk, industrial size A-B; six-cord cotton, 50; waxed, 50; cotton-covered polyester core, all purpose
Hems none needed; decoratively pinked or scalloped edge; double-stitched; flat/plain; machine-stitched
Seams lapped with raw edges; plain
Seam Finishes untreated/plain
Pressing steam; safe temperature limit 300°F (150.1°C)
Care dry clean
Fabric Resource **Continental Felt Co.**

11 ~ Knit/Knit-type Fabrics

Double Knit
Double-faced Knit
Interlock Knit
- Interlock Knit (fine weight)
- Interlock Knit (lightweight)
- Interlock Knit (medium weight)
- Interlock Knit (medium-heavyweight)
- Interlock Knit (cotton fiber)

Jacquard Knit
- Flat Jacquard Knit
- Blistered Jacquard Knit
- Jersey Jacquard Knit
- Jersey Jacquard Knit Variation
- Rib Jacquard Knit (close up only)
- Smooth-faced, Bird's Eye-backed Jacquard Knit
- Napped-faced, Bird's Eye-backed Jacquard Knit

Jersey Knit
- Jersey Knit / Jersey Sheer
- Jersey Knit (lightweight)
- Jersey Knit (striped)

Jersey Knit Variations
- Jersey Knit Variation (novelty yarn)
- Jersey Knit Variation (tweed appearance)
- Jersey Knit Variation (Lacoste® Knit)
- Jersey Knit Variation (openweave effect)
- Jersey Knit Variation (pleated effect)

Kettenraschel Knits (close up only)
Milanese Knit
Purl Knit
Raschel Knit
- Raschel Knit (crochet effect)
- Raschel Knit (dishcloth effect)
- Raschel Knit (lace effect)
- Raschel Knit (loop-pile fringe)
- Raschel Knit (double cloth effect)
- Raschel Knit (heavy tweed appearance)

Rib Knit
- 1 x 1 Rib Knit
- 2 x 2 Rib Knit
- 4 x 4 Rib Knit

Rib Knit Variations
- Rib Knit Variation (piqué effect)
- Rib Knit Variation (open-work effect)
- Thermo Knit/Brynes Cloth®

Tricot Knit
- Sheer Tricot Knit / Tricot Sheer
- Tricot Knit (lightweight)
- Tricot Knit (medium weight)
- Tricot Knit (heavyweight)

Tricot Knit Variations
- Tricot Knit Variation (soft finish)
- Tricot Knit Variation (stiff finish)
- Tricot Knit Variation (polished)
- Ciré Tricot
- Tricot Knit Variation (patterned)
- Tricot Knit Variation (printed)

Knit Stitch Variations
- Dropped / Missed Stitch Knit
- Transfer Stitch Knit
- Transfer Stitch Knit (close up only)

Tuck Stitch
- Rib Knit Effect (close up only)
- Popcorn Effect (close up only)
- Full Cable Stitch (close up only)
- Half Cable Stitch (close up only)
- Intarsia Knit (close up only)

Sweater Knit
- Sweater Knit
- Sweater Knit (close up only)

Laid-in Yarn Variations
- Single Knit (bouclé yarn)
- Single Knit (looped yarn)
- Single Knit (seed yarn)

General Properties of Knit Fabrics
General Characteristics of Knit Fabrics

A knit fabric is characterized by preceding or succeeding rows of interlocking loops which may vary in appearance from an open, loose construction to a close, compact formation. Knits may be constructed from natural, man-made, or blended fiber yarns alone or in any combination and from conventional, textured or novelty yarns.

The two main classifications of knit fabrics are:

1. Warp knit classification
2. Weft knit classification

Each classification uses different types of machinery to produce a particular knit fabric. Within each classification, different methods and techniques are employed to produce knitted fabrics of various types. Each type of knit has its own particular appearance, properties, characteristics, hand and drapability qualities.

Knit fabrics are produced:

- As plain, pleated or patterned goods
- With flat, textured or raised surfaces
- With degrees of stability

Knit fabrics vary from:

- Fine to thick
- Light to heavy weight
- Sheer to opaque

Knit fabric selection depends on:

- Type and method of construction
- Particular properties of individual knit
- Desired degree of stretch or stability
- Type and design of garment
- End use of garment

Double Knit

Fiber Content 100% polyester
Yarn Type filament fiber
Yarn Construction textured; 135-denier yarn
Fabric Structure double knit; weft knit classification
Finishing Processes brushing; heat-set; stabilizing
Color Application piece dyed; solution dyed; yarn dyed
Width 60 inches (152.4 cm)
Weight medium
Hand firm; semi-soft; controlled stretch in crosswise direction
Texture flat; smooth
Performance Expectations dimensional stability; durable; elongation; resilient; high tensile strength
Drapability Qualities falls into moderately soft flares; retains shape of garment; better utilized if fitted by seaming and eliminating excess fabric

Recommended Construction Aids & Techniques

Hand Needles ball point, 5–10; cotton darners, 1–5; embroidery, 1–5; milliners, 3–5
Machine Needles ball point, 14
Threads poly-core waxed, 50; cotton-covered polyester core, all purpose; spun polyester, all purpose; nylon/Dacron, monocord A
Hems double-stitched; flat/plain; overedged
Seams overedged; plain; safety-stitched; taped; zigzagged
Seam Finishes double-stitched; serging/single-ply overedged; untreated/plain; single-ply or double-ply zigzag
Pressing steam; safe temperature limit 325°F (164.1°C)
Care dry clean
Fabric Resources **Applause Fabrics Ltd.; Klopman Mills; Russell Corp.**

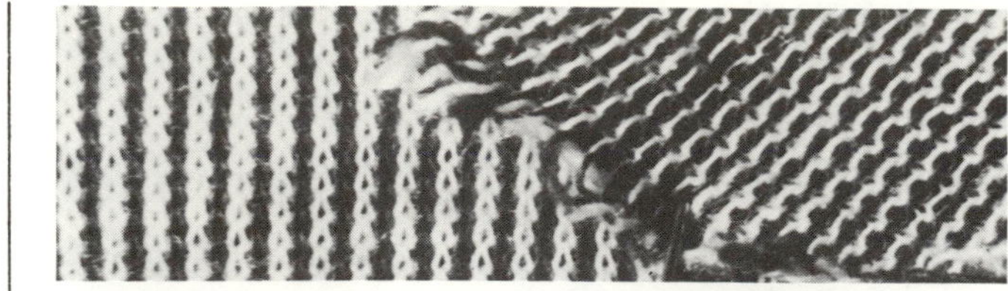

Double-faced Knit

Fiber Content 100% cotton
Yarn Type combed staple
Yarn Construction conventional; fine; tight twist
Fabric Structure double cloth: jersey knit (face); single knit and tuck stitch (back); separate binder yarn joins face and back; weft knit classification
Finishing Processes napping; stabilizing
Color Application yarn dyed
Width 50 inches (127 cm)
Weight medium-heavy
Hand lofty; soft; stretch in lengthwise and crosswise directions; thick
Texture flat; differently colored face and back
Performance Expectations absorbent; antistatic; subject to curling at edges; durable; insulating properties; high tensile strength
Drapability Qualities falls into moderately soft flares; fullness maintains lofty effect; retains shape of garment

Recommended Construction Aids & Techniques

Hand Needles ball point, 5–10; cotton darners, 1–5; embroidery, 1–5; milliners, 3–5
Machine Needles ball point, 14–16
Threads mercerized cotton, heavy duty; six-cord cotton, 50; cotton-covered polyester core, 50
Hems double-stitched; flat/plain; overedged
Seams overedged; plain; safety-stitched; taped; zigzagged
Seam Finishes double-stitched; serging/single-ply overedged; untreated/plain; single-ply or double-ply zigzag
Pressing steam; safe temperature limit 400°F (206.1°C)
Care launder; dry clean
Fabric Resource **Walden Textiles/Barry M. Richards Inc.**

Interlock Knit (fine weight)

Fiber Content 100% Qiana® nylon
Yarn Type filament fiber
Yarn Construction textured
Fabric Structure interlock knit; tight and stable construction; weft knit classification
Finishing Processes heat-set; stabilizing; tentering
Color Application solution dyed; yarn dyed
Width 60–62 inches (152.4–157.5 cm)
Weight light
Hand fine; silky; soft; stretch in crosswise direction
Texture flat; slippery smooth; same appearance on face and back
Performance Expectations abrasion resistant; does not curl at edges; dimensional stability; durable; elasticity; elongation; heat-set properties; resilient; high tensile strength
Drapability Qualities falls into soft languid flares and ripples; accommodates fullness by gathering, elasticized shirring; fullness retains soft graceful fall

Recommended Construction Aids & Techniques

Hand Needles ball point, 5–10
Machine Needles ball point, 9
Threads poly-core waxed, extra fine; spun polyester, fine; nylon/Dacron, monocord A
Hems double-stitched; flat/plain; machine-stitched; overedged
Seams hairline; overedged; plain; safety-stitched; taped; tissue-stitched; zigzagged
Seam Finishes double-stitched; double-stitched and trimmed; serging/single-ply overedged; untreated/plain; single-ply or double-ply zigzag
Pressing steam; safe temperature limit 350°F (178.1°C)
Care launder; dry clean
Fabric Resource Klopman Mills

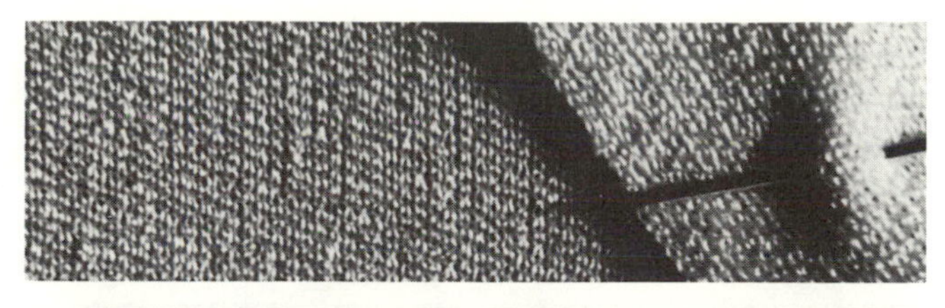

Interlock Knit (lightweight)

Fiber Content 100% Silesta® polyester
Yarn Type filament fiber
Yarn Construction textured
Fabric Structure interlock knit; tight and stable construction; weft knit classification
Finishing Processes heat-set; stabilizing; tentering
Color Application solution dyed; yarn dyed
Width 64 inches (162.7 cm)
Weight medium-light
Hand silky; soft; stretch in crosswise direction; supple
Texture flat; smooth; same appearance on face and back
Performance Expectations does not curl at edges; dimensional stability; durable; elasticity; elongation; heat-set properties; resilient; high tensile strength
Drapability Qualities falls into soft flares; accommodates fullness by gathering, elasticized shirring; fullness retains soft graceful fall

Recommended Construction Aids & Techniques

Hand Needles ball point, 5–10
Machine Needles ball point, 11
Threads poly-core waxed, fine; spun polyester, all purpose; nylon/Dacron, monocord A
Hems double-stitched; flat/plain; machine-stitched; overedged
Seams hairline; overedged; plain; safety-stitched; taped; tissue-stitched; zigzagged
Seam Finishes double-stitched; serging/single-ply overedged; untreated/plain; single-ply or double-ply zigzag
Pressing steam; safe temperature limit 325°F (164.1°C)
Care launder; dry clean
Fabric Resources **Applause Fabrics Ltd.; Gloversville Mills Inc.**

Interlock Knit (medium weight)

Fiber Content 100% nylon
Yarn Type filament fiber
Yarn Construction textured
Fabric Structure interlock knit; tight and stable construction; weft knit classification
Finishing Processes heat-set; stabilizing; tentering
Color Application solution dyed; yarn dyed
Width 56-60 inches (142.2-152.4 cm)
Weight medium
Hand silky; soft; stretch in crosswise direction; supple
Texture flat; smooth; same appearance on face and back
Performance Expectations abrasion resistant; does not curl at edges; dimensional stability; durable; elasticity; elongation; heat-set properties; resilient; high tensile strength
Drapability Qualities falls into soft flares; accommodates fullness by gathering, elasticized shirring; fullness retains soft graceful fall

Recommended Construction Aids & Techniques

Hand Needles ball point, 5-10
Machine Needles ball point, 11
Threads poly-core waxed, fine; spun polyester, all purpose; nylon/Dacron, monocord A
Hems double-stitched; flat/plain; machine-stitched
Seams overedged; plain; safety-stitched; taped; zigzagged
Seam Finishes double-stitched; serging/single-ply overedged; untreated/plain; single-ply or double-ply zigzag
Pressing steam; safe temperature limit 350°F (178.1°C)
Care launder; dry clean
Fabric Resource **Klopman Mills**

Interlock Knit (medium-heavyweight)

Fiber Content 100% polyester
Yarn Type filament fiber
Yarn Construction textured
Fabric Structure interlock knit; tight and stable construction; weft knit classification
Finishing Processes heat-set; stabilizing; tentering
Color Application solution dyed; yarn dyed
Width 60-64 inches (152.4-162.7 cm)
Weight medium-heavy
Hand lofty; silky; stretch in crosswise direction; supple
Texture grainy; same appearance on face and back
Performance Expectations does not curl at edges; dimensional stability; durable; elasticity; elongation; heat-set properties; resilient; high tensile strength
Drapability Qualities falls into soft flares; accommodates fullness by gathering, elasticized shirring; fullness retains soft graceful fall

Recommended Construction Aids & Techniques

Hand Needles ball point, 5-10
Machine Needles ball point, 11-14
Threads poly-core waxed, 50; cotton-covered polyester core, all purpose; spun polyester, all purpose; nylon/Dacron, monocord A
Hems double-stitched; flat/plain; overedged
Seams overedged; plain; safety-stitched; taped; zigzagged
Seam Finishes double-stitched; serging/single-ply overedged; untreated/plain; single-ply or double-ply zigzag
Pressing steam; safe temperature limit 325°F (164.1°C)
Care launder; dry clean
Fabric Resources **Klopman Mills; Milan Textile Machines Inc.; Springs Mills Inc.**

Interlock Knit (cotton fiber)

Fiber Content 100% cotton
Yarn Type combed staple
Yarn Construction conventional; mercerized; hard twist
Fabric Structure interlock knit; tight and stable construction; weft knit classification
Finishing Processes brushing; shearing; stabilizing; tentering
Color Application piece dyed; yarn dyed
Width 56–66 inches (142.2–167.6 cm)
Weight medium
Hand elastic; soft; stretch in crosswise direction; supple
Texture flat; smooth; same appearance on face and back
Performance Expectations absorbent; antistatic; does not curl at edges; durable; elongation; high tensile strength
Drapability Qualities falls into soft flares; can be shaped and molded by stretching; fullness retains soft graceful fall

Recommended Construction Aids & Techniques

Hand Needles ball point, 5–10;
Machine Needles ball point, 11
Threads mercerized cotton, 50; six-cord cotton, 80–100; cotton-covered polyester core, extra fine
Hems double-stitched; flat/plain; machine-stitched
Seams overedged; plain; safety-stitched; taped; zigzagged
Seam Finishes double-stitched; serging/single-ply overedged; untreated/plain; single-ply or double-ply zigzag
Pressing steam; safe temperature limit 400°F (206.1°C)
Care launder
Fabric Resources **Applause Fabrics Ltd.; Gloversville Mills Inc.; Klopman Mills**

Flat Jacquard Knit

Fiber Content 100% polyester
Yarn Type filament fiber
Yarn Construction textured; 135-denier yarn
Fabric Structure 2-color Jacquard knit; weft knit classification
Finishing Processes heat-set; stabilizing; tentering
Color Application solution dyed; yarn dyed
Width 60 inches (152.4 cm)
Weight medium
Hand pliable; semi-soft; controlled stretch in crosswise direction
Texture flat; harsh; patterned face
Performance Expectations dimensional stability; durable; elongation; resilient; high tensile strength
Drapability Qualities falls into wide cones; fullness maintains springy effect; retains shape of garment

Recommended Construction Aids & Techniques

Hand Needles ball point, 5–10; cotton darners, 1–5; embroidery, 1–5; milliners, 3–5
Machine Needles ball point, 14
Threads poly-core waxed, 50; cotton-covered polyester core, all purpose; spun polyester, all purpose; nylon/Dacron, monocord A
Hems double-stitched; flat/plain; overedged
Seams overedged; plain; safety-stitched; taped; zigzagged
Seam Finishes double-stitched; serging/single-ply overedged; untreated/plain; single-ply or double-ply zigzag
Pressing steam; safe temperature limit 325°F (164.1°C)
Care launder; dry clean
Fabric Resource **Milan Textile Machines Inc.**

Blistered Jacquard Knit

Fiber Content 100% polyester
Yarn Type filament fiber
Yarn Construction textured; 135-denier yarn
Fabric Structure double knit; rib gated; weft knit classification
Finishing Processes heat-set; stabilizing; tentering
Color Application solution dyed; yarn dyed
Width 72 inches (182.9 cm)
Weight medium
Hand hard; spongy; springy; controlled stretch in crosswise direction
Texture flat, smooth patterned face; rough back
Performance Expectations dimensional stability; durable; elongation; resilient; high tensile strength
Drapability Qualities falls into moderately soft flares; retains shape or silhouette of garment; better utilized if fitted by seaming and eliminating excess fabric

Recommended Construction Aids & Techniques

Hand Needles ball point, 5–10; cotton darners, 1–5; embroidery, 1–5; milliners, 3–5
Machine Needles ball point, 14
Threads poly-core waxed, 50; cotton-covered polyester core, all purpose; spun polyester, all purpose; nylon/Dacron, monocord A
Hems double-stitched; flat/plain; overedged
Seams overedged; plain; safety-stitched; taped; zigzagged
Seam Finishes double-stitched; serging/single-ply overedged; untreated/plain; single-ply or double-ply zigzag
Pressing steam; safe temperature limit 325°F (164.1°C)
Care launder; dry clean
Fabric Resources Applause Fabrics Ltd.; Milan Textile Machines Inc.

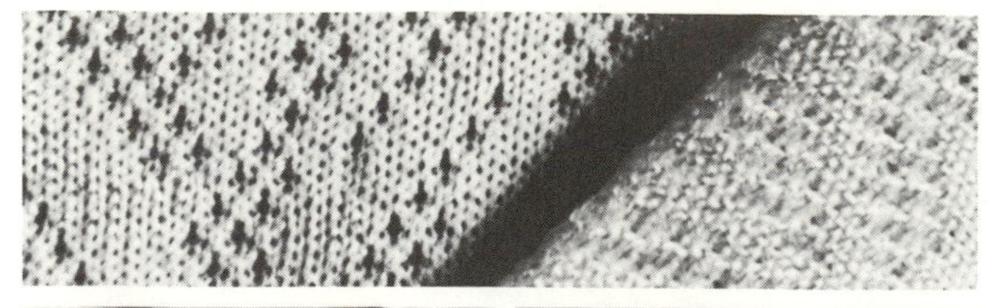

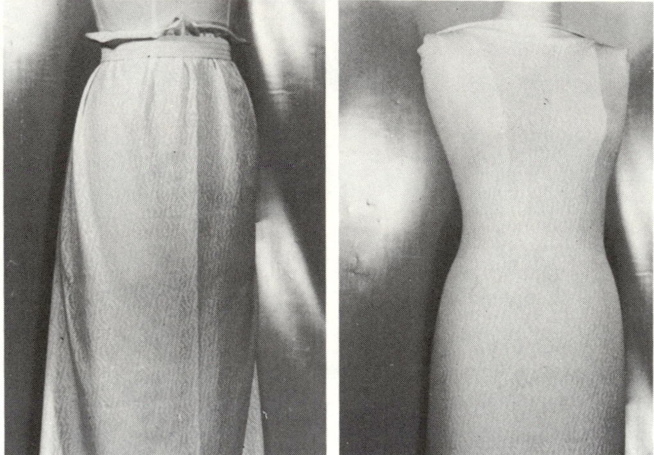

Jersey Jacquard Knit

Fiber Content Helenca® nylon/polyester blend
Yarn Type filament fiber
Yarn Construction textured; 240-denier yarn
Fabric Structure Jacquard knit; weft knit classification
Finishing Processes stabilizing; tentering
Color Application solution dyed; yarn dyed
Width 32 inches (81.3 cm)
Weight medium
Hand soft; stretch in lengthwise and crosswise directions
Texture open pattern; uneven surface
Performance Expectations dimensional stability; durable; elasticity; elongation; heat-set properties; high tensile strength
Drapability Qualities falls into wide cones; can be shaped and molded by stretching; falls close to body contour

Recommended Construction Aids & Techniques

Hand Needles ball point, 5–10; cotton darners, 1–5; embroidery, 1–5; milliners, 3–5
Machine Needles ball point, 11
Threads poly-core waxed, 50; cotton-covered polyester core, all purpose; spun polyester, all purpose; nylon/Dacron, monocord A
Hems double-stitched; flat/plain; overedged
Seams overedged; plain; safety-stitched; taped; zigzagged
Seam Finishes double-stitched; serging/single-ply overedged; untreated/plain; single-ply or double-ply zigzag
Pressing steam; safe temperature limit 325°F (164.1°C)
Care launder; dry clean
Fabric Resources Applause Fabrics Ltd.; Milan Textile Machines Inc.

Jersey Jacquard Knit Variation

Fiber Content 100% polyester
Yarn Type filament fiber
Yarn Construction textured; 135-denier yarn
Fabric Structure Jacquard knit; weft knit classification
Finishing Processes brushing; stabilizing; tentering
Color Application solution dyed; yarn dyed
Width 72 inches (182.9 cm)
Weight medium-heavy
Hand furry; lofty; spongy; controlled stretch in crosswise direction
Texture fuzzy; looped; high and low patterned areas
Performance Expectations chemical resistant; dimensional stability; durable; elongation; mildew and moth resistant; resilient; high tensile strength
Drapability Qualities falls into moderately soft flares; fullness retains moderately soft fall; retains shape of garment

Recommended Construction Aids & Techniques

Hand Needles ball point, 5–10; cotton darners, 1–5; embroidery, 1–5; milliners, 3–5
Machine Needles ball point, 11–14
Threads poly-core waxed, heavy duty; cotton-covered polyester blend, heavy duty; cotton-covered polyester core, heavy duty; spun polyester, heavy duty; nylon/Dacron, monocord A
Hems double-stitched; flat/plain; overedged
Seams overedged; plain; safety-stitched; taped; zigzagged
Seam Finishes serging/single-ply overedged; untreated/plain; single-ply zigzag
Pressing steam; safe temperature limit 325°F (164.1°C)
Care launder
Fabric Resources Applause Fabrics Ltd.; Milan Textile Machines Inc.

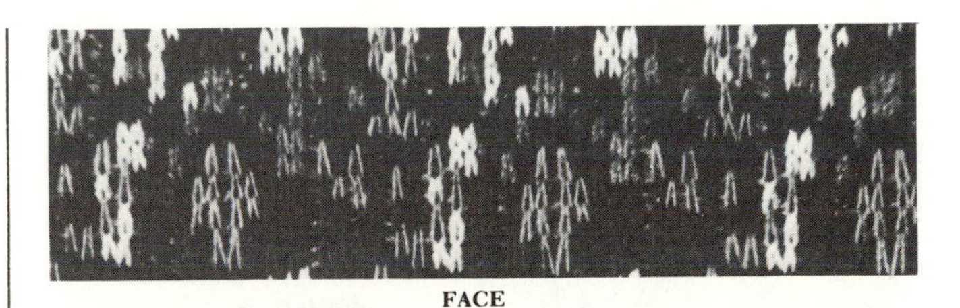

FACE

BACK

Rib Jacquard Knit

Fiber Content 100% polyester
Yarn Type filament fiber
Yarn Construction textured set
Fabric Structure Jacquard knit; weft knit classification
Finishing Process brushing; stabilizing; tentering
Color Application solution dyed; yarn dyed
Width 72 inches (182.9 cm)
Weight medium-heavy
Hand furry; lofty; spongy; controlled stretch in crosswise direction
Texture fuzzy; looped; high and low patterned areas
Performance Expectations chemical resistant; dimensional stability; durable; elongation; mildew and moth resistant; resilient; high tensile strength
Drapability Qualities falls into moderately soft flares; fullness retains moderately soft fall; retains shape of garment

Recommended Construction Aids & Techniques

Hand Needles ball point, 5–10; cotton darners, 1–5; embroidery, 1–5; milliners, 3–5
Machine Needles ball point, 11–14
Threads poly-core waxed, heavy duty; cotton-covered polyester blend, heavy duty; cotton-covered polyester core, heavy duty; spun polyester, heavy duty; nylon/Dacron, monocord A
Hems double-stitched; flat/plain; overedged
Seams overedged; plain; safety-stitched; taped; zigzagged
Seam Finishes serging/single-ply overedged; untreated/plain; single-ply zigzag
Pressing steam; safe temperature limit 325°F (164.1°C)
Care launder; dry clean
Fabric Resources American Cyanamid Co./Fabrics Development Dept.; Applause Fabrics Ltd.

Smooth-faced, Bird's Eye-backed Jacquard Knit

Fiber Content 100% acrylic
Yarn Type filament fiber
Yarn Construction bulked; textured; 240-denier yarn
Fabric Structure 2-color Jacquard knit; weft knit classification
Finishing Processes heat-set; stabilizing; tentering
Color Application yarn dyed
Width 80 inches (203.2 cm)
Weight medium
Hand lofty; springy; stretch in crosswise direction; woolly
Texture flat, smooth face; bird's eye patterned back
Performance Expectations compressible; dimensional stability; elasticity; elongation; resilient
Drapability Qualities falls into wide flares; fullness maintains lofty effect; retains shape of garment

Recommended Construction Aids & Techniques

Hand Needles ball point, 5–10; cotton darners, 1–5; embroidery, 1–5; milliners, 3–5
Machine Needles ball point, 14
Threads poly-core waxed, 50; cotton-covered polyester core, all purpose; spun polyester, all purpose; nylon/Dacron, monocord A
Hems double-stitched; flat/plain; overedged
Seams overedged; plain; safety-stitched; taped; zigzagged
Seam Finishes serging/single-ply overedged; untreated/plain; single-ply zigzag
Pressing steam; safe temperature limit 300°F (150.1°C)
Care dry clean
Fabric Resource **Milan Textile Machines Inc.**

Napped-faced, Bird's Eye-backed Jacquard Knit

Fiber Content 100% acrylic
Yarn Type filament fiber
Yarn Construction textured
Fabric Structure Jacquard knit; weft knit classification
Finishing Processes brushing; napping; shearing; stabilizing
Color Application yarn dyed
Width 64 inches (162.7 cm)
Weight medium
Hand soft; stretch in crosswise direction; thick; woolly
Texture flat napped face; bird's eye patterned back
Performance Expectations compressible; dimensional stability; elasticity; elongation; resilient
Drapability Qualities falls into wide flares; fullness maintains lofty effect; retains shape of garment

Recommended Construction Aids & Techniques

Hand Needles ball point, 5–10; cotton darners, 1–5; embroidery, 1–5; milliners, 3–5
Machine Needles ball point, 14
Threads poly-core waxed, 50; cotton-covered polyester core, all purpose; spun polyester, all purpose; nylon/Dacron, monocord A
Hems double-stitched; flat/plain; overedged
Seams overedged; plain; safety-stitched; taped; zigzagged
Seam Finishes serging/single-ply overedged; untreated/plain; single-ply zigzag
Pressing steam; safe temperature limit 300°F (150.1°C)
Care dry clean
Fabric Resource **Applause Fabrics Ltd.**

Jersey Knit/Jersey Sheer

Fiber Content 100% polyester
Yarn Type filament fiber
Yarn Construction textured
Fabric Structure single knit; weft knit classification
Finishing Processes heat-set; stabilizing; tentering
Color Application solution dyed; yarn dyed
Width 60 inches (152.4 cm)
Weight fine to lightweight
Hand delicate; silky; stretch in crosswise direction; controlled stretch in lengthwise direction
Texture flat; semi-harsh; porous; semi-sheer
Performance Expectations subject to curling at edges; dimensional stability; durable; elongation; heat-set properties; resilient; high tensile strength
Drapability Qualities falls into soft languid flares and ripples; accommodates fullness by gathering, elasticized shirring; fullness retains soft graceful fall

Recommended Construction Aids & Techniques

Hand Needles ball point, 5–10
Machine Needles ball point, 9
Threads poly-core waxed, extra fine; spun polyester, fine; nylon/Dacron, monocord A
Hems double fold; double-stitched; flat/plain; machine-stitched; overedged; hand- or machine-rolled; wired
Seams overedged; plain; hairline; safety-stitched; taped; tissue-stitched; zigzagged
Seam Finishes double-stitched; double-stitched and trimmed; serging/single-ply overedged; untreated/plain; single-ply or double-ply zigzag
Pressing steam; safe temperature limit 325°F (164.1°C)
Care launder; dry clean
Fabric Resource Applause Fabrics Ltd.

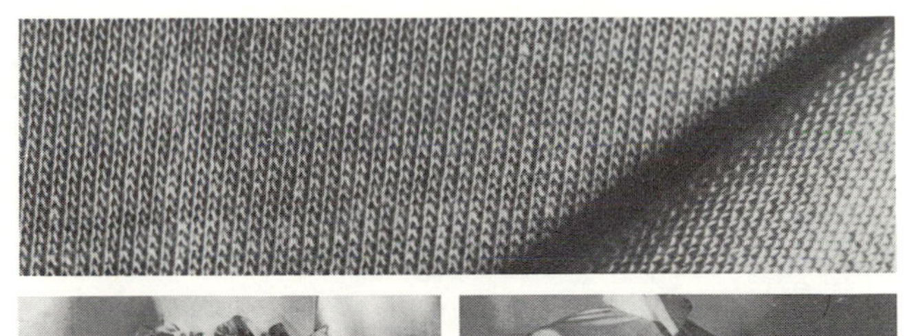

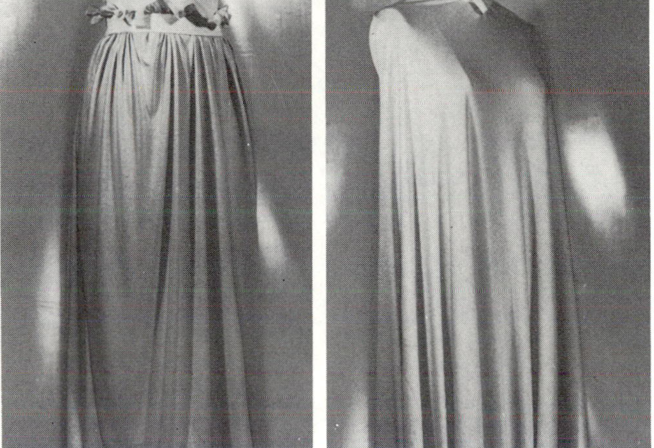

Jersey Knit (lightweight)

Fiber Content 100% cotton
Yarn Type combed staple
Yarn Construction conventional; fine; mercerized; high twist
Fabric Structure single knit; weft knit classification
Finishing Processes mercerizing; stabilizing; tentering
Color Application piece dyed; yarn dyed
Width 36 inches (91.4 cm)
Weight light
Hand light; limp; soft; stretch in crosswise direction; controlled stretch in lengthwise direction
Texture flat; smooth; knitstitch face; purlstitch back
Performance Expectations absorbent; antistatic; subject to curling at edges; durable; elongation; high tensile strength
Drapability Qualities falls into soft flares; can be shaped or molded by stretching; fullness retains soft fall

Recommended Construction Aids & Techniques

Hand Needles ball point, 5–10;
Machine Needles ball point, 11
Threads mercerized cotton, 50; six-cord cotton, 80–100; cotton-covered polyester core, extra fine
Hems double-stitched; flat/plain; machine-stitched; overedged
Seams hairline; overedged; plain; safety-stitched; taped; tissue-stitched; zigzagged
Seam Finishes double-stitched; double-stitched and trimmed; serging/single-ply overedged; untreated/plain; single-ply or double-ply zigzag
Pressing steam; safe temperature limit 400°F (206.1°C)
Care launder
Fabric Resources Cinderella Knitting Mills; Ge-Ray Fabrics (sample courtesy of Jonan Inc.)

Jersey Knit (striped)

Fiber Content 100% cotton
Yarn Type combed staple
Yarn Construction conventional; mercerized; tight twist
Fabric Structure single knit; weft knit classification
Finishing Processes mercerizing; stabilizing; tentering
Color Application yarn dyed
Width 62 inches (157.5 cm)
Weight medium-light
Hand soft; stretch in crosswise direction; controlled stretch in lengthwise direction; supple
Texture flat; smooth; knitstitch face; purlstitch back
Performance Expectations absorbent; antistatic; subject to curling at edges; durable; elongation; high tensile strength
Drapability Qualities falls into soft flares; falls close to body contour; fullness retains soft fall

Recommended Construction Aids & Techniques

Hand Needles ball point, 5-10
Machine Needles ball point, 11
Threads mercerized cotton, 50; six-cord cotton, 80-100; cotton-covered polyester core, extra fine
Hems double fold; double-stitched; flat/plain; machine-stitched; overedged; hand- or machine-rolled; wired
Seams overedged; plain; safety-stitched; taped; zigzagged
Seam Finishes double-stitched; double-stitched and trimmed; serging/single-ply overedged; untreated/plain; single-ply or double-ply zigzag
Pressing steam; safe temperature limit 400°F (206.1°C)
Care launder
Fabric Resources **Applause Fabrics Ltd.; Milan Textile Machines Inc.**

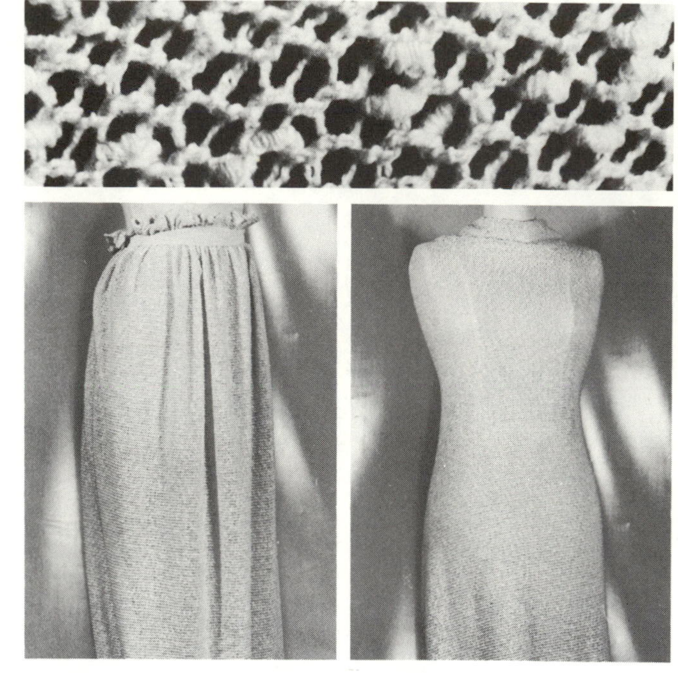

Jersey Knit Variation (novelty yarn)

Fiber Content 100% cotton
Yarn Type staple
Yarn Construction novelty seed yarn
Fabric Structure single knit; weft knit classification
Finishing Processes mercerizing; stabilizing; tentering
Color Application piece dyed; yarn dyed
Width 60 inches (152.4 cm)
Weight heavy
Hand coarse; spongy; stretch in crosswise direction; controlled stretch in lengthwise direction
Texture nubby; openweave; rough
Performance Expectations absorbent; air permeable; antistatic; subject to curling at edges; durable; high tensile strength; subject to snagging and catching due to loose open structure
Drapability Qualities falls into soft flares; can be shaped or molded by stretching; fullness retains moderately soft fall

Recommended Construction Aids & Techniques

Hand Needles ball point, 5-10; cotton darners, 1-5; embroidery, 1-5; milliners, 3-5
Machine Needles ball point, 14
Threads mercerized cotton, heavy duty; six-cord cotton, 50; cotton-covered polyester core, 50
Hems double-stitched; flat/plain; overedged
Seams overedged; plain; safety-stitched; taped; zigzagged
Seam Finishes serging/single-ply overedged; untreated/plain; single-ply zigzag
Pressing steam; use pressing cloth; safe temperature limit 400°F (206.1°C)
Care launder
Fabric Resource **Gloversville Mills Inc.**

Jersey Knit Variation (tweed appearance)

Fiber Content 100% cotton
Yarn Type combed staple
Yarn Construction conventional; mercerized
Fabric Structure single knit with different loop sizes; weft knit classification
Finishing Processes mercerizing; stabilizing; tentering
Color Application yarn dyed
Width 72 inches (182.9 cm)
Weight medium-light
Hand limp; soft; stretch in crosswise direction; controlled stretch in lengthwise direction; supple
Texture flat; smooth; tweedy appearance created by different loop sizes and color of yarns
Performance Expectations absorbent; antistatic; subject to curling at edges; durable; elongation; high tensile strength
Drapability Qualities falls into soft flares; falls close to body contour; fullness retains soft graceful fall

Recommended Construction Aids & Techniques

Hand Needles ball point, 5–10
Machine Needles ball point, 11
Threads mercerized cotton, 50; six-cord cotton, 80–100; cotton-covered polyester core, extra fine
Hems double-stitched; flat/plain; machine-stitched; overedged
Seams hairline; overedged; plain; safety-stitched; taped; tissue-stitched; zigzagged
Seam Finishes double-stitched; double-stitched and trimmed; serging/single-ply overedged; untreated/plain; single-ply or double-ply zigzag
Pressing steam; safe temperature limit 400°F (206.1°C)
Care launder; dry clean
Fabric Resource **Applause Fabrics Ltd.**

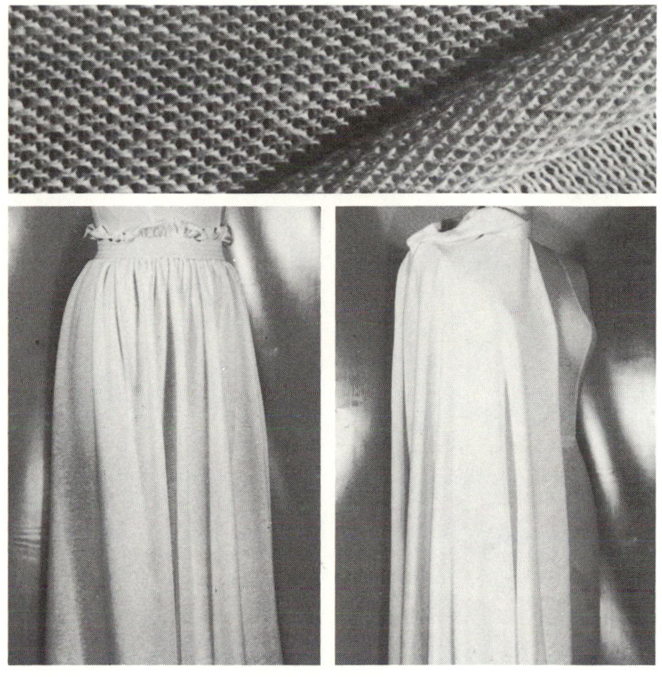

Jersey Knit Variation (Lacoste® knit)

Fiber Content cotton/polyester blend
Yarn Type combed staple; filament fiber
Yarn Construction conventional; textured
Fabric Structure single knit; tuck stitch creates design effect; weft knit classification
Finishing Processes heat-set; stabilizing; tentering
Color Application piece dyed–union dyed; yarn dyed
Width 60 inches (152.4 cm)
Weight medium-light
Hand fine; soft; stretch in crosswise direction; controlled stretch in lengthwise direction; supple
Texture flat face; pebbly, honeycombed-effect back
Performance Expectations dimensional stability; durable; elongation; resilient; high tensile strength
Drapability Qualities falls into soft flares; falls close to body contour; fullness retains moderately soft fall

Recommended Construction Aids & Techniques

Hand Needles ball point, 5–10
Machine Needles ball point, 11
Threads poly-core waxed, 50; cotton-covered polyester core, all purpose; spun polyester, all purpose; nylon/Dacron, monocord A
Hems double-stitched; flat/plain; machine-stitched; overedged
Seams hairline; overedged; plain; safety-stitched; taped; tissue-stitched; zigzagged
Seam Finishes double-stitched; serging/single-ply overedged; untreated/plain; single-ply or double-ply zigzag
Pressing steam; safe temperature limit 325°F (164.1°C)
Care launder
Fabric Resource **Stēvcoknit, Inc.**

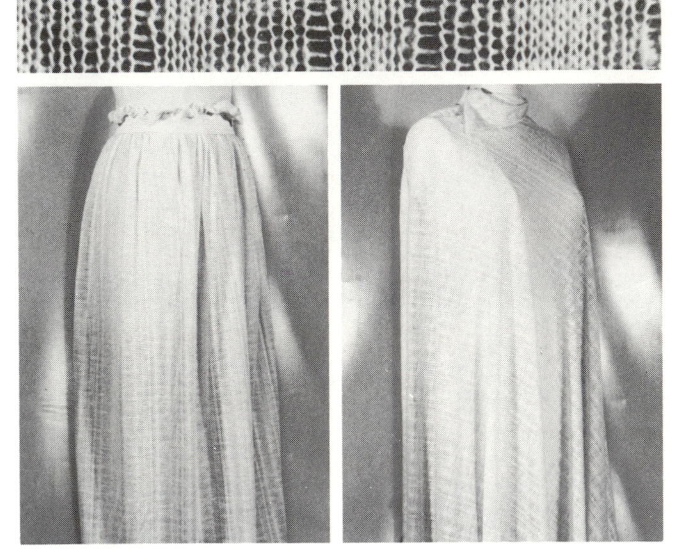

Jersey Knit Variation (openweave effect)

Fiber Content 100% cotton
Yarn Type combed staple
Yarn Construction conventional; mercerized
Fabric Structure single knit; needle deactivated to produce openweave effect; weft knit classification
Finishing Processes mercerizing; stabilizing; stretching
Color Application piece dyed; yarn dyed
Width 72 inches (182.9 cm)
Weight light
Hand fine; porous; soft; stretch in crosswise direction; controlled stretch in lengthwise direction
Texture flat; smooth; open areas of design
Performance Expectations absorbent; antistatic; subject to curling at edges; durable; elongation; high tensile strength
Drapability Qualities falls into soft flares; falls close to body contour; fullness retains soft graceful fall

Recommended Construction Aids & Techniques

Hand Needles ball point, 5–10
Machine Needles ball point, 9
Threads mercerized cotton, 50; six-cord cotton, 80–100; cotton-covered polyester core, extra fine
Hems double-stitched; flat/plain; machine-stitched; overedged
Seams hairline; overedged; plain; safety-stitched; taped; tissue-stitched; zigzagged
Seam Finishes double-stitched; double-stitched and trimmed; serging/single-ply overedged; untreated/plain; single-ply or double-ply zigzag
Pressing steam; safe temperature limit 400°F (206.1°C)
Care launder
Fabric Resource **Applause Fabrics Ltd.**

Jersey Knit Variation (pleated effect)

Fiber Content 100% cotton
Yarn Type combed staple
Yarn Construction conventional; mercerized
Fabric Structure single knit; needle deactivated to produce pleated effect; weft knit classification
Finishing Processes mercerizing; stabilizing; tentering
Color Application yarn dyed
Width 72 inches (182.9 cm)
Weight medium
Hand bulky; semi-soft; spongy; stretch in crosswise direction; controlled stretch in lengthwise direction
Texture uneven surface created by structure; raised lengthwise rib
Performance Expectations absorbent; antistatic; subject to curling at edges; durable; elongation; high tensile strength
Drapability Qualities falls into wide cones; fullness maintains lofty effect; retains shape of garment

Recommended Construction Aids & Techniques

Hand Needles ball point, 5–10; cotton darners, 1–5; embroidery, 1–5; milliners, 3–5
Machine Needles ball point, 14
Threads mercerized cotton, heavy duty; six-cord cotton, 50; cotton-covered polyester core, 50
Hems double-stitched; flat/plain; overedged
Seams overedged; plain; safety-stitched; taped; zigzagged
Seam Finishes double-stitched; serging/single-ply overedged; untreated/plain; single-ply or double-ply zigzag
Pressing steam; safe temperature limit 400°F (206.1°C)
Care launder; dry clean
Fabric Resource **Applause Fabrics Ltd.**

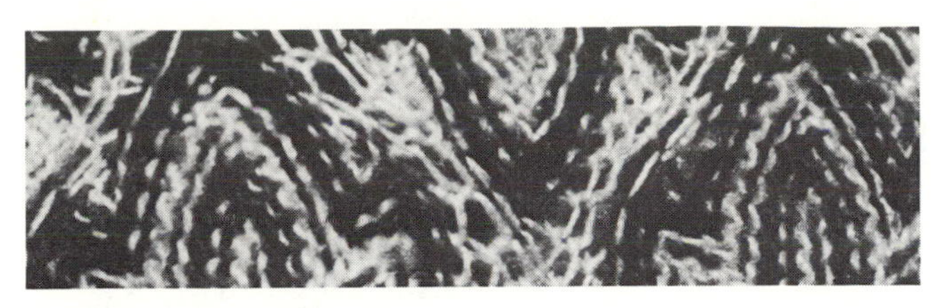

Kettenraschel Knit

Fiber Content cotton/polyester
Yarn Type staple; filament staple
Yarn Construction conventional; textured
Fabric Structure Raschel knit using bearded needle machine with gauges from 9–16 needles and with up to 4 guide bars; warp knit classification
Finishing Processes stabilizing; tentering
Color Application yarn dyed
Width 72 inches (182.9 cm)
Weight medium-light to medium
Hand semi-soft; spongy; controlled stretch in lengthwise and crosswise directions
Texture knitting process creates uneven, three-dimensional effect
Performance Expectations air permeable; dimensional stability; durable; elongation; high tensile strength; subject to snagging and catching due to open knit
Drapability Qualities fullness maintains lofty effect; retains shape of garment; better utilized if fitted by seaming and eliminating excess fabric

Recommended Construction Aids & Techniques

Hand Needles ball point, 5–10; cotton darners, 1–5; embroidery, 1–5; milliners, 3–5
Machine Needles ball point, 14
Threads mercerized cotton, heavy duty; six-cord cotton, 50; cotton-covered polyester core, 50
Hems double-stitched; flat/plain; overedged
Seams overedged; plain; safety-stitched; taped; zigzagged
Seam Finishes double-stitched; serging/single-ply overedged; untreated/plain; single-ply or double-ply zigzag
Pressing steam; safe temperature limit 400°F (206.1°C)
Care launder; dry clean
Fabric Resource **Applause Fabrics Ltd.**

Milanese Knit

Fiber Content 100% nylon
Yarn Type filament fiber
Yarn Construction textured
Fabric Structure 2-needle bar warp knit
Finishing Processes heat-set; stabilizing; tentering
Color Application solution dyed; yarn dyed
Width 48–52 inches (121.9–132.1 cm)
Weight medium-light
Hand silky; slippery; soft; controlled stretch in crosswise direction; no stretch in lengthwise direction
Texture satiny smooth; knitstitch face and back
Performance Expectations abrasion resistant; does not curl at edges; dimensional stability; durable; elasticity; elongation; heat-set properties; resilient; high tensile strength
Drapability Qualities falls into soft flares; accommodates fullness by gathering, elasticized shirring; fullness retains soft fall

Recommended Construction Aids & Techniques

Hand Needles ball point, 5–10
Machine Needles ball point, 11–14
Threads poly-core waxed, fine; spun polyester, all purpose; nylon/Dacron, monocord A
Hems double-stitched; flat/plain; machine-stitched
Seams overedged; plain; safety-stitched; taped; zigzagged
Seam Finishes double-stitched; serging/single-ply overedged; untreated/plain; single-ply or double-ply zigzag
Pressing steam; safe temperature limit 350°F (178.1°C)
Care launder; dry clean
Fabric Resources **Gloversville Mills Inc.; Native Textiles**

Purl Knit

Fiber Content 100% polyester
Yarn Type filament fiber
Yarn Construction textured
Fabric Structure purl knit; weft knit classification
Finishing Processes stabilizing; tentering
Color Application solution dyed; yarn dyed
Width 64 inches (162.7 cm)
Weight medium
Hand elastic; harsh; springy; stretch in crosswise and lengthwise directions
Texture rough; purlstitch face and back; reversible
Performance Expectations does not curl at edges; dimensional stability; durable; elongation; high stretch properties; high tensile strength; subject to snagging and pulling due to loop formation
Drapability Qualities falls into soft wide flares; can be shaped or molded by stretching; fullness maintains springy effect

Recommended Construction Aids & Techniques

Hand Needles ball point, 5–10; cotton darners, 1–5; embroidery, 1–18; milliners, 3–5
Machine Needles ball point, 14
Threads poly-core waxed, 50; cotton-covered polyester core, all purpose; spun polyester, all purpose; nylon/Dacron, monocord A
Hems double-stitched; flat/plain; overedged
Seams overedged; plain; safety-stitched; taped; zigzagged
Seam Finishes double-stitched; serging/single-ply overedged; untreated/plain; single-ply or double-ply zigzag
Pressing steam; safe temperature limit 325°F (164.1°C)
Care launder; dry clean
Fabric Resource **Applause Fabrics Ltd.**

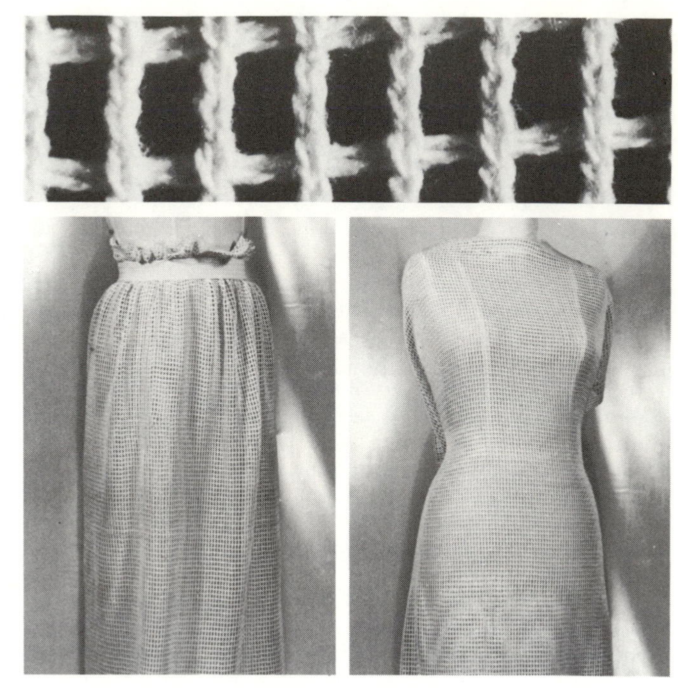

Raschel Knit (crochet effect)

Fiber Content 50% cotton/50% polyester
Yarn Type staple; filament fiber
Yarn Construction conventional; thick; slack twist
Fabric Structure raschel knit; warp knit classification
Finishing Processes stabilizing; tentering
Color Application piece dyed–union dyed; yarn dyed
Width 38 inches (96.5 cm)
Weight medium-heavy
Hand coarse; cottony; thick
Texture fishnet/open mesh simulates handmade crochet
Performance Expectations air permeable; dimensional stability; durable; elongation; no stretch properties; high tensile strength; subject to snagging and catching due to open knit structure
Drapability Qualities falls into moderately soft flares; can be shaped or molded by stretching; fullness retains moderately soft fall

Recommended Construction Aids & Techniques

Hand Needles ball point, 5–10; cotton darners, 1–5; embroidery, 1–5; milliners, 3–5
Machine Needles ball point, 14
Threads mercerized cotton, heavy duty; six-cord cotton, 50; cotton-covered polyester core, 50
Hems double-stitched; flat/plain; overedged
Seams hairline; overedged; plain; safety-stitched; taped; tissue-stitched; zigzagged
Seam Finishes serging/single-ply overedged; untreated/plain; single-ply zigzag
Pressing steam; use pressing cloth; safe temperature limit 325°F (164.1°C)
Care launder
Fabric Resource **Applause Fabrics Ltd.**

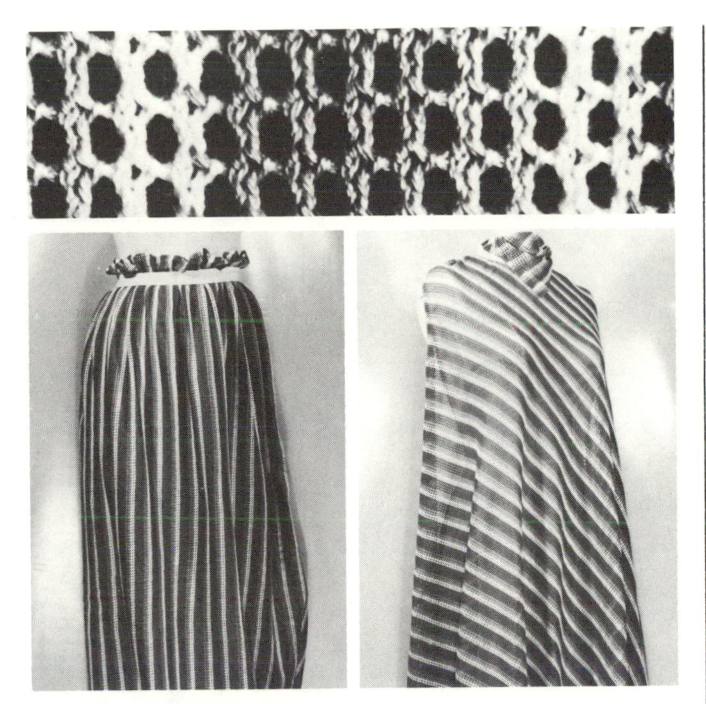

Raschel Knit (dishcloth effect)

Fiber Content 100% cotton
Yarn Type staple
Yarn Construction conventional; thick; slack twist
Fabric Structure raschel knit; warp knit classification
Finishing Processes stabilizing; tentering
Color Application yarn dyed
Width 60 inches (152.4 cm)
Weight medium
Hand coarse; open; thick
Texture open mesh simulates braided or plaited effect
Performance Expectations absorbent; air permeable; anti-static; durable; elongation; minimal stretch properties; high tensile strength
Drapability Qualities falls into moderately soft flares; falls close to body contour; fulness retains moderately soft fall

Recommended Construction Aids & Techniques

Hand Needles ball point, 5-10; cotton darners, 1-5; embroidery, 1-5; milliners, 3-5
Machine Needles ball point, 14
Threads mercerized cotton, heavy duty; six-cord cotton, 50; cotton-covered polyester core, 50
Hems double-stitched; flat/plain; overedged
Seams hairline; overedged; plain; safety-stitched; taped; tissue-stitched; zigzagged
Seam Finishes serging/single-ply overedged; untreated/plain; single-ply zigzag
Pressing steam; use pressing cloth; safe temperature limit 400°F (206.1°C)
Care launder
Fabric Resource **Successful Creations/Shartex Inc.**

Raschel Knit (lace effect)

Fiber Content polyester; Mylar® metallic
Yarn Type filament fiber; extruded metallic film
Yarn Construction laminated metallic monofilament; textured
Fabric Structure raschel knit; warp knit classification
Finishing Processes heat-set; stabilizing; tentering
Color Application solution dyed; yarn dyed
Width 72 inches (182.9 cm)
Weight medium-heavy
Hand coarse; crisp; spongy; controlled stretch in crosswise and lengthwise directions
Texture harsh; uneven surface; raised lace design effect
Performance Expectations resistant to most chemicals; dimensional stability; durable; elongation; mildew and moth resistant; high tensile strength
Drapability Qualities falls into wide cones; fullness maintains lofty effect; retains shape or silhouette of garment

Recommended Construction Aids & Techniques

Hand Needles ball point, 5-10; cotton darners, 1-5; embroidery, 1-5; milliners, 3-5
Machine Needles ball point, 14
Threads poly-core waxed, heavy duty; cotton-polyester blend, heavy duty; cotton-covered polyester core, heavy duty; spun polyester, heavy duty; nylon/Dacron, monocord A
Hems double-stitched; flat/plain; overedged
Seams overedged; plain; safety-stitched; taped; zigzagged
Seam Finishes serging/single-ply overedged; untreated/plain; single-ply zigzag
Pressing steam; safe temperature limit 325°F (164.1°C)
Care dry clean
Fabric Resource **Applause Fabrics Ltd.**

Raschel Knit (loop-pile fringe)

Fiber Content 100% acrylic
Yarn Type filament fiber
Yarn Construction bulked; textured
Fabric Structure raschel knit; warp knit classification
Finishing Processes brushing; pressing; stabilizing
Color Application yarn dyed
Width 50 inches (127 cm)
Weight medium
Hand soft; springy; woolly
Texture fluffy; raised loop-pile fringe
Performance Expectations compressible; dimensional stability; elongation; resilient; no stretch properties; subject to snagging and catching due to large loop-pile fringe
Drapability Qualities falls into moderately soft wide flares; fullness maintains lofty effect; retains shape of garment

Recommended Construction Aids & Techniques

Hand Needles ball point, 5–10; cotton darners, 1–5; embroidery, 1–5; milliners, 3–5
Machine Needles ball point, 14
Threads poly-core waxed, heavy duty; cotton-polyester blend, heavy duty; cotton-covered polyester core, heavy duty; spun polyester, heavy duty; nylon/Dacron, monocord A
Hems double-stitched; flat/plain; overedged
Seams overedged; plain; safety-stitched; taped; zigzagged
Seam Finishes serging/single-ply overedged; untreated/plain; single-ply zigzag
Pressing steam; safe temperature limit 300°F (150.1°C)
Care dry clean
Fabric Resource **Applause Fabrics Ltd.**

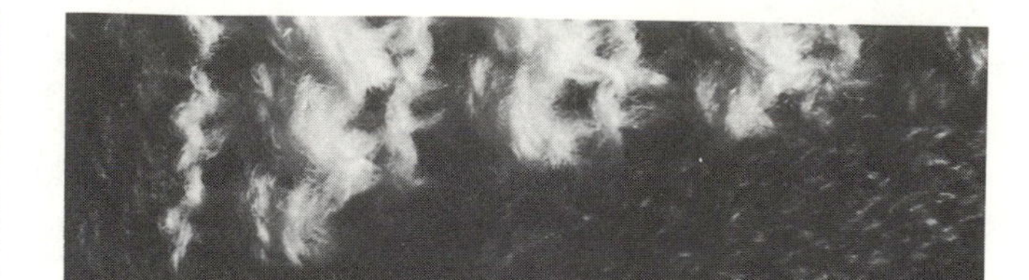

Raschel Knit (double cloth effect)

Fiber Content 100% acrylic
Yarn Type filament fiber
Yarn Construction bulked; textured
Fabric Structure double cloth; raschel knit; warp knit classification
Finishing Processes brushing; pressing; stabilizing
Color Application yarn dyed
Width 59 inches (149.9 cm)
Weight heavy
Hand lofty; semi-soft; thick
Texture uneven surface on face; purlstitch face; flat knit-stitch back; reversible
Performance Expectations compressible; dimensional stability; elongation; resilient; no stretch properties; subject to snagging due to raised loop structure
Drapability Qualities falls into wide cones; retains shape of garment; better utilized if fitted by seaming and eliminating excess fabric

Recommended Construction Aids & Techniques

Hand Needles ball point, 5–10; cotton darners, 1–5; embroidery, 1–5; milliners, 3–5
Machine Needles ball point, 14
Threads poly-core waxed, heavy duty; cotton-polyester blend, heavy duty; cotton-covered polyester core, heavy duty; spun polyester, heavy duty; nylon/Dacron, monocord A
Hems double-stitched; flat/plain; overedged
Seams overedged; plain; safety-stitched; taped; zigzagged
Seam Finishes serging/single-ply overedged; untreated/plain; single-ply zigzag
Pressing steam; safe temperature limit 300°F (150.1°C)
Care dry clean
Fabric Resource **Successful Creations/Shartex Inc.**

Raschel Knit (heavy tweed appearance)

Fiber Content 100% wool
Yarn Type staple
Yarn Construction bulked; conventional; thick
Fabric Structure raschel knit with laid-in yarn; weft knit classification
Finishing Processes brushing; pressing; stabilizing
Color Application yarn dyed
Width 50 inches (127 cm)
Weight heavy
Hand lofty; soft; thick
Texture high and low surface; raised yarn face; flat back
Performance Expectations absorbent; elasticity; elongation; insulating properties; subject to moth damage; resilient; no stretch properties
Drapability Qualities falls into wide cones; retains shape of garment; better utilized if fitted by seaming and eliminating excess fabric

Recommended Construction Aids & Techniques

Hand Needles ball point, 5–10; cotton darners, 1–5; embroidery, 1–5; milliners, 3–5
Machine Needles ball point, medium 14
Threads mercerized cotton, heavy duty; six-cord cotton, 50; cotton-covered polyester core, 50
Hems double-stitched; flat/plain; overedged
Seams overedged; plain; safety-stitched; taped; zigzagged
Seam Finishes serging/single-ply overedged; untreated/plain; single-ply zigzag
Pressing steam; use needle board and pressing cloth; safe temperature limit 300°F (150.1°C)
Care dry clean
Fabric Resource **Arthur Zeiler Woolens Inc.**

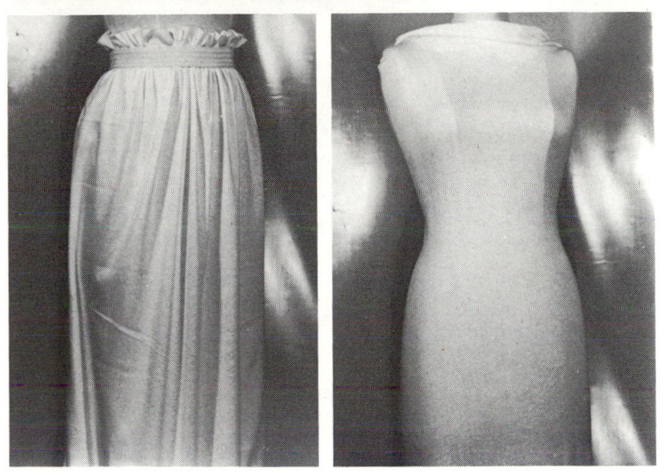

1 x 1 Rib Knit

Fiber Content 100% cotton
Yarn Type combed staple
Yarn Construction conventional; fine; mercerized; tight twist
Fabric Structure rib knit; weft knit classification
Finishing Processes mercerizing; stabilizing; tentering
Color Application piece dyed; yarn dyed
Width 44 inches (111.8 cm)
Weight medium
Hand elastic; semi-soft; high stretch and recovery properties in crosswise direction; thin
Texture flat; vertical rib on face and back; reversible
Performance Expectations absorbent; antistatic; durable; elasticity; elongation; high tensile strength
Drapability Qualities falls into moderately soft flares; can be shaped or molded by stretching; falls close to body contour

Recommended Construction Aids & Techniques

Hand Needles ball point, 5–10
Machine Needles ball point, 11
Threads mercerized cotton, 50; six-cord cotton, 80–100; cotton-covered polyester core, extra fine
Hems double-stitched; flat/plain; machine-stitched; overedged
Seams hairline; overedged; plain; safety-stitched; taped; tissue-stitched; zigzagged
Seam Finishes double-stitched; double-stitched and trimmed; serging/single-ply overedged; untreated/plain; single-ply or double-ply zigzag
Pressing steam; safe temperature limit 400°F (206.1°C)
Care launder
Fabric Resources **Cinderella Knitting Mills Division, Reeves Brothers Inc.; Milan Textile Machines Inc.**

2 x 2 Rib Knit

Fiber Content 100% polyester
Yarn Type filament fiber
Yarn Construction conventional; fine; slack twist
Fabric Structure rib knit; weft knit classification
Finishing Processes heat-set; stabilizing; tentering
Color Application solution dyed; yarn dyed
Width 64 inches (162.7 cm)
Weight medium
Hand elastic; spongy; springy; high stretch and recovery properties in crosswise direction
Texture uneven surface; vertical rib on face and back; reversible
Performance Expectations resistant to most chemicals; dimensional stability; durable; elasticity; elongation; mildew and moth resistant; resilient
Drapability Qualities falls into soft wide flares; can be shaped or molded by stretching; fullness retains moderately soft fall

Recommended Construction Aids & Techniques

Hand Needles ball point, 5-10
Machine Needles ball point, 11-14
Threads poly-core waxed, 50; cotton-covered polyester core, all purpose; spun polyester, all purpose; nylon/Dacron, monocord A
Hems double-stitched; flat/plain; overedged
Seams overedged; plain; safety-stitched; taped; zigzagged
Seam Finishes double-stitched; serging/single-ply overedged; untreated/plain; single-ply or double-ply zigzag
Pressing steam; safe temperature limit 325° (164.1°C)
Care launder; dry clean
Fabric Resource **Applause Fabrics Ltd.**

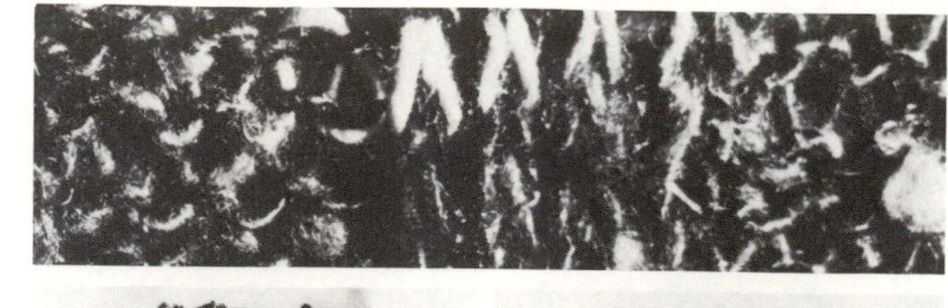

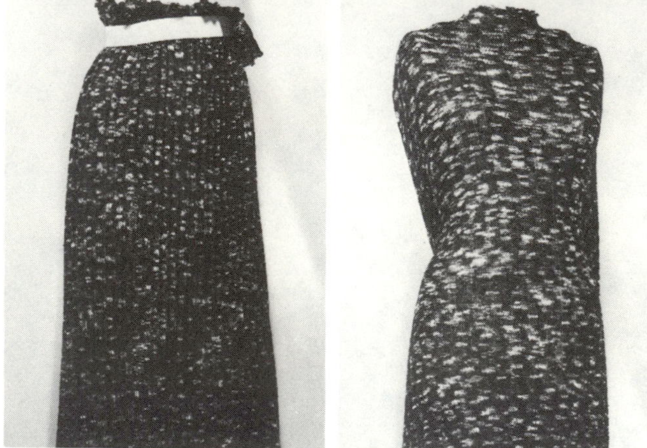

4 x 4 Rib Knit

Fiber Content 100% acrylic
Yarn Type filament fiber
Yarn Construction bulked; textured
Fabric Structure rib knit; weft knit classification
Finishing Processes brushing; stabilizing; tentering
Color Application yarn dyed
Width 56 inches (142.2 cm)
Weight heavy
Hand lofty; soft; high stretch and recovery propeties in crosswise direction; woolly
Texture high and low surface; raised vertical rib on face and back; reversible
Performance Expectations compressible; dimensional stability; elasticity; elongation; resilient
Drapability Qualities can be shaped or molded by stretching; fullness maintains springy effect; retains shape of garment

Recommended Construction Aids & Techniques

Hand Needles ball point, 5-10; cotton darners, 1-5; embroidery, 1-5; milliners, 3-5
Machine Needles ball point, 14
Threads poly-core waxed, heavy duty; cotton-polyester blend, heavy duty; cotton-covered polyester core, heavy duty; spun polyester, heavy duty; nylon/Dacron, monocord A
Hems double-stitched; flat/plain; overedged
Seams overedged; plain; safety-stitched; taped; zigzagged
Seam Finishes serging/single-ply overedged; untreated/plain; single-ply zigzag
Pressing steam; safe temperature limit 300°F (150.1°C)
Care launder; dry clean
Fabric Resource **Applause Fabrics Ltd.**

Rib Knit Variation (piqué effect)

Fiber Content 100% polyester
Yarn Type filament fiber
Yarn Construction textured
Fabric Structure rib knit; weft knit classification
Finishing Processes heat-set; stabilizing; tentering
Color Application solution dyed; yarn dyed
Width 64 inches (162.7 cm)
Weight medium-heavy
Hand coarse; spongy; moderate stretch in crosswise direction; no stretch in lengthwise direction; thick
Texture harsh; pebbly
Performance Expectations dimensional stability; durable; elongation; mildew and moth resistant; resilient
Drapability Qualities falls into soft wide cones; can be shaped or molded by stretching; fullness maintains springy to moderately lofty effect

Recommended Construction Aids & Techniques

Hand Needles ball point, 5–10; cotton darners, 1–5; embroidery, 1–5; milliners, 3–5
Machine Needles ball point, 14
Threads poly-core waxed, 50; cotton-covered polyester core, all purpose; spun polyester, all purpose; nylon/Dacron, monocord A
Hems double-stitched; flat/plain; overedged
Seams overedged; plain; safety-stitched; taped; zigzagged
Seam Finishes serging/single-ply overedged; untreated/plain; single-ply zigzag
Pressing steam; safe temperature limit 325°F (164.1°C)
Care launder; dry clean
Fabric Resource **Applause Fabrics Ltd.**

Rib Knit Variation (open-work effect)

Fiber Content 100% polyester
Yarn Type filament fiber
Yarn Construction textured
Fabric Structure rib knit; needle deactivated to produce open-work effect; weft knit classification
Finishing Processes stabilizing; stiffening; stretching
Color Application solution dyed; yarn dyed
Width 60 inches (152.4 cm)
Weight light
Hand fine; soft; springy; moderate stretch in crosswise direction; no stretch in lengthwise direction
Texture semi-coarse; open-work effect
Performance Expectations dimensional stability; durable; elongation; mildew and moth resistant; subject to snagging due to open-work structure
Drapability Qualities falls into moderately soft flares; accommodates fullness by gathering, shirring; fullness maintains springy effect

Recommended Construction Aids & Techniques

Hand Needles ball point, 5–10; cotton darners, 1–5; embroidery, 1–5; milliners, 3–5
Machine Needles ball point, 11
Threads poly-core waxed, fine; spun polyester, all purpose; nylon/Dacron, monocord A
Hems double-stitched; flat/plain; machine-stitched; overedged
Seams hairline; overedged; plain; safety-stitched; taped; tissue-stitched; zigzagged
Seam Finishes double-stitched; double-stitched and trimmed; serging/single-ply overedged; untreated/plain; single-ply or double-ply zigzag
Pressing steam; safe temperature limit 325°F (164.1°C)
Care launder
Fabric Resource **Successful Creations/Shartex Inc.**

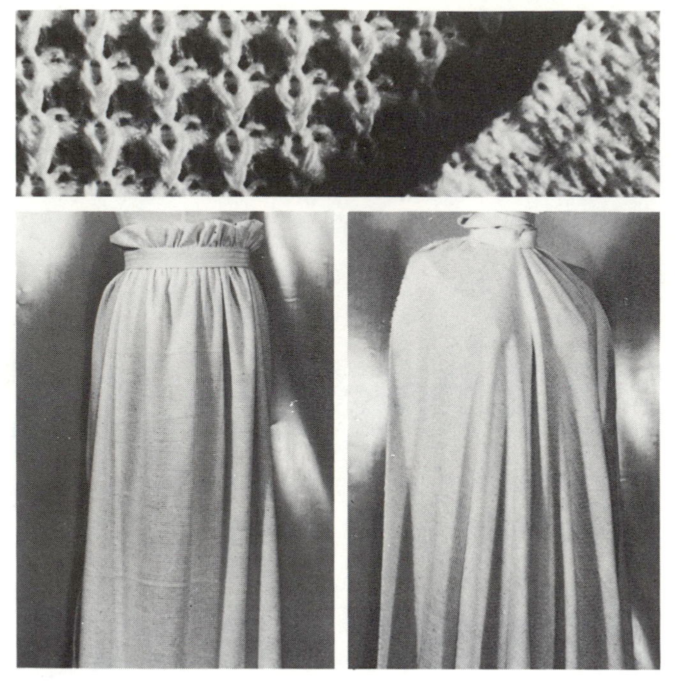

Thermo Knit/Byrnes Cloth®[1]

Fiber Content 77% polyester/23% cotton
Yarn Type combed staple; filament fiber
Yarn Construction bulked; textured
Fabric Structure rib knit with tuck stitch; weft knit classification
Finishing Processes mercerizing; stabilizing; tentering
Color Application yarn dyed
Width 66 inches (167.6 cm)
Weight medium
Hand coarse; lofty; stretch in crosswise direction; no stretch in lengthwise direction; thick
Texture rough; uneven surface; high and low patterned effect
Performance Expectations durable; elasticity; elongation; high tensile strength
Drapability Qualities falls into moderately wide cones; can be shaped or molded by stretching; retains shape of garment

Recommended Construction Aids & Techniques

Hand Needles ball point, 5–10; cotton darners, 1–5; embroidery, 1–5; milliners, 3–5
Machine Needles ball point, 11–14
Threads poly-core waxed, 50; cotton-covered polyester core, all purpose; spun polyester, all purpose; nylon/Dacron, monocord A
Hems double-stitched; flat/plain; overedged
Seams overedged; plain; safety-stitched; taped; zigzagged
Seam Finishes double-stitched; serging/single-ply overedged; untreated/plain; single-ply or double-ply zigzag
Pressing steam; safe temperature limit 325°F (164.1°C)
Care launder
Fabric Resources Applause Fabrics Ltd.; Cinderella Knitting Mills; Successful Creations/Shartex Inc.

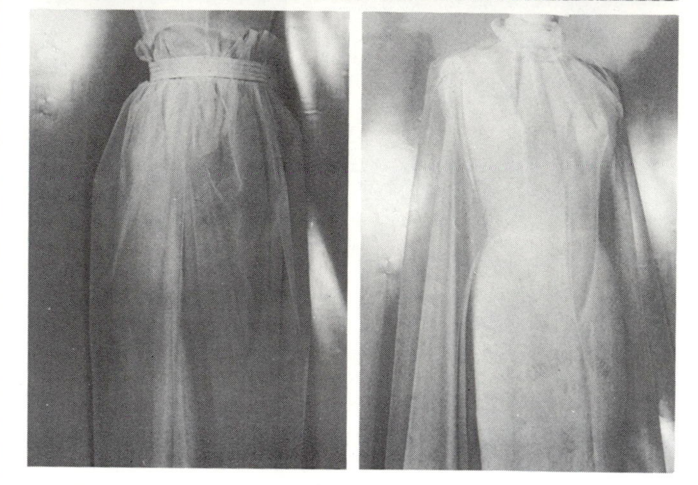

Sheer Tricot Knit/Tricot Sheer

Fiber Content 100% nylon
Yarn Type filament fiber
Yarn Construction fine; textured; thin
Fabric Structure tricot jersey knit; warp knit classification
Finishing Processes heat-set; stabilizing; tentering
Color Application solution dyed; yarn dyed
Width 108 inches (274.3 cm)
Weight fine to lightweight
Hand semi-crisp; silky; moderate stretch in crosswise direction; no stretch in lengthwise direction; thin
Texture flat; sheer; smooth
Performance Expectations abrasion resistant; dimensional stability; durable; elasticity; elongation; heat-set properties; resilient; high tensile strength
Drapability Qualities falls into soft flares; accommodates fullness by gathering, elasticized shirring; fullness maintains springy effect

Recommended Construction Aids & Techniques

Hand Needles ball point, 5–10
Machine Needles ball point, 9
Threads poly-core waxed, extra fine; spun polyester, fine; nylon/Dacron, monocord A
Hems double fold; double-stitched; flat/plain; machine-stitched; overedged; hand- or machine-rolled; wired
Seams hairline; overedged; plain; safety-stitched; taped; tissue-stitched; zigzagged
Seam Finishes double-stitched; double-stitched and trimmed; serging/single-ply overedged; untreated/plain; single-ply or double-ply zigzag
Pressing steam; safe temperature limit 350°F (178.1°C)
Care launder
Fabric Resource Native Textiles

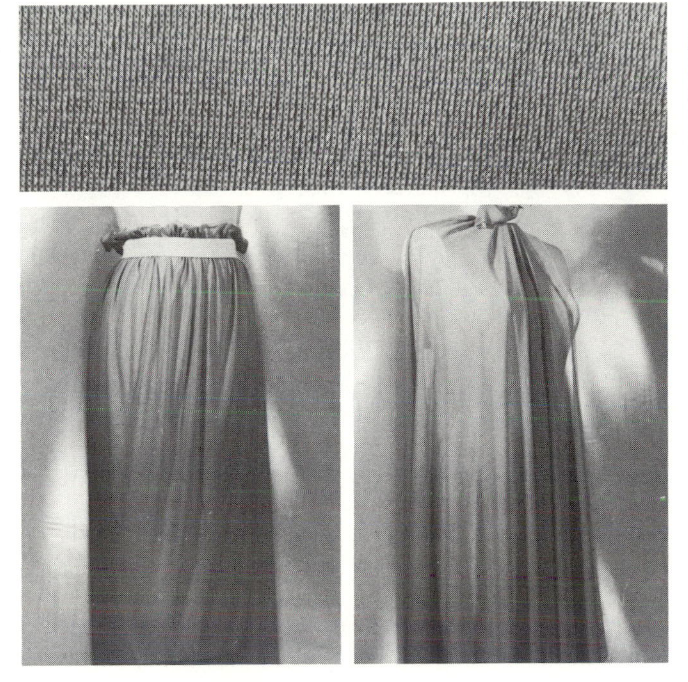

Tricot Knit (lightweight)

Fiber Content 100% Enkalure® nylon
Yarn Type filament fiber
Yarn Construction creped; textured
Fabric Structure tricot knit; warp knit classification
Finishing Processes crepe-set; stabilizing
Color Application solution dyed; yarn dyed
Width 46 inches (116.8 cm)
Weight light
Hand delicate; silky; moderate stretch in crosswise direction; no stretch in lengthwise direction; thin
Texture even surface; flat
Performance Expectations abrasion resistant; dimensional stability; durable; elasticity; elongation; heat-set properties; resilient; high tensile strength
Drapability Qualities falls into soft languid flares and ripples; accommodates fullness by gathering, elasticized shirring, smocking; fullness retains soft graceful fall

Recommended Construction Aids & Techniques

Hand Needles ball point, 5–10
Machine Needles ball point, 9
Threads poly-core waxed, extra fine; spun polyester, fine; nylon/Dacron, monocord A
Hems double fold; double-stitched; flat/plain; machine-stitched; overedged; hand- or machine-rolled; wired
Seams hairline; overedged; plain; safety-stitched; taped; tissue-stitched; zigzagged
Seam Finishes double-stitched; double-stitched and trimmed; serging/single-ply overedged; untreated/plain; single-ply or double-ply zigzag
Pressing steam; safe temperature limit 350°F (178.1°C)
Care launder
Fabric Resource **Native Textiles**

Tricot Knit (medium weight)

Fiber Content 100% Qiana® nylon
Yarn Type filament fiber
Yarn Construction textured
Fabric Structure tricot knit; warp knit classification
Finishing Processes heat-set; stabilizing; tentering
Color Application solution dyed; yarn dyed
Width 65–67 inches (165.1–170.2 cm)
Weight medium-light
Hand silky; soft; springy; moderate stretch in crosswise direction; no stretch in lengthwise direction
Texture smooth
Performance Expectations abrasion resistant; dimensional stability; durable; elasticity; elongation; heat-set properties; resilient; high tensile strength
Drapability Qualities falls into soft flares; accommodates fullness by gathering, elasticized shirring, smocking; fullness retains soft fall

Recommended Construction Aids & Techniques

Hand Needles ball point, 5–10
Machine Needles ball point, 11
Threads poly-core waxed, fine; spun polyester, all purpose; nylon/Dacron, monocord A
Hems double-stitched; flat/plain; machine-stitched; overedged
Seams hairline; overedged; plain; safety-stitched; taped; tissue-stitched; zigzagged
Seam Finishes double-stitched; serging/single-ply overedged; untreated/plain; single-ply or double-ply zigzag
Pressing steam; safe temperature limit 350°F (178.1°C)
Care launder; dry clean
Fabric Resource **Klopman Mills**

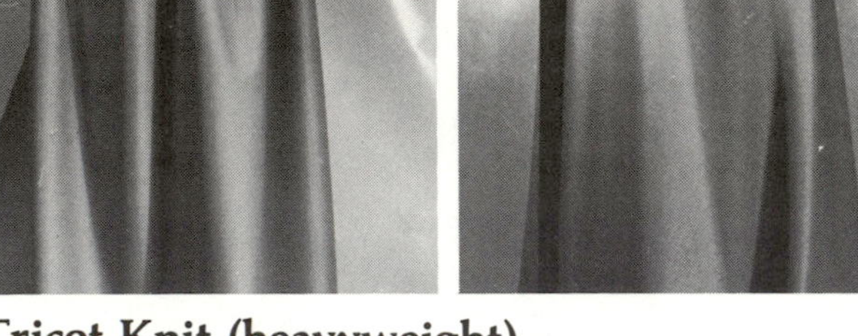

Tricot Knit (heavyweight)

Fiber Content 100% polyester
Yarn Type filament fiber
Yarn Construction textured
Fabric Structure 22-gauge tricot knit; warp knit classification
Finishing Processes stabilizing; stiffening; tentering
Color Application solution dyed; yarn dyed
Width 54 inches (137.2 cm)
Weight medium-heavy
Hand firm; hard
Texture boardy; harsh; opaque
Performance Expectations dimensional stability; durable; elongation; heat-set properties; resilient; no stretch properties; high tensile strength
Drapability Qualities falls into crisp wide cones; retains shape of garment; better utilized if fitted by seaming and eliminating excess fabric

Recommended Construction Aids & Techniques

Hand Needles ball point, 5–10
Machine Needles ball point, 14
Threads poly-core waxed, 50; cotton-covered polyester core, all purpose; spun polyester, all purpose; nylon/Dacron, monocord A
Hems double-stitched; flat/plain; overedged
Seams overedged; plain; safety-stitched; taped; zigzagged
Seam Finishes double-stitched; serging/single-ply overedged; untreated/plain; single-ply or double-ply zigzag
Pressing steam; safe temperature limit 325°F (164.1°C)
Care launder; dry clean
Fabric Resource **Alba-Waldensian Knits**

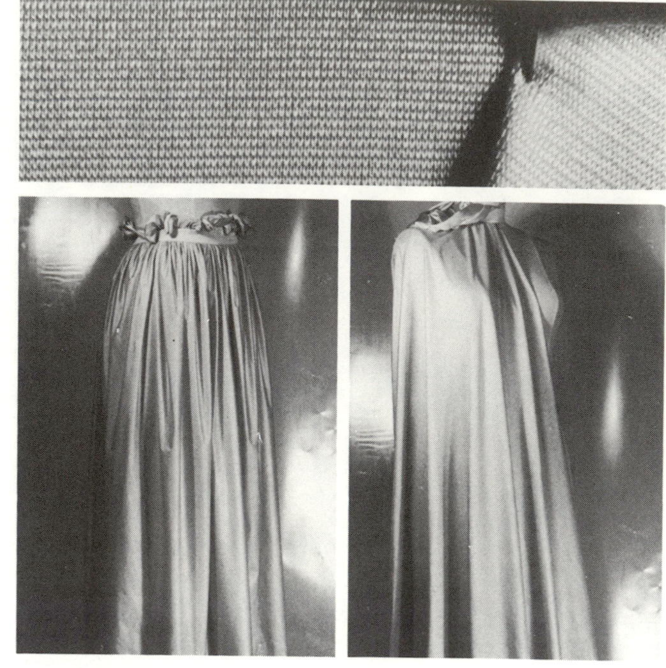

Tricot Knit Variation (soft finish)

Fiber Content 100% Antron® nylon
Yarn Type filament fiber
Yarn Construction textured
Fabric Structure 36-gauge tricot knit; warp knit classification
Finishing Processes softening; stabilizing; tentering
Color Application solution dyed; yarn dyed
Width 92 inches (233.7 cm)
Weight medium-light
Hand fine; firm; soft; moderate stretch in crosswise direction; no stretch in lengthwise direction
Texture silky; smooth
Performance Expectations abrasion resistant; antistatic due to yarn finish; dimensional stability; durable; elasticity; elongation; heat-set properties; resilient; high tensile strength
Drapability Qualities falls into soft flares; accommodates fullness by gathering, elasticized shirring; fullness retains soft fall

Recommended Construction Aids & Techniques

Hand Needles ball point, 5–10
Machine Needles ball point, 11–14
Threads poly-core waxed, extra fine; spun polyester, fine; nylon/Dacron, monocord A
Hems double-stitched; flat/plain; machine-stitched; overedged
Seams hairline; overedged; plain; safety-stitched; taped; tissue-stitched; zigzagged
Seam Finishes double-stitched; double-stitched and trimmed; serging/single-ply overedged; untreated/plain; single-ply or double-ply zigzag
Pressing steam; safe temperature limit 350°F (178.1°C)
Care launder
Fabric Resource **Native Textiles**

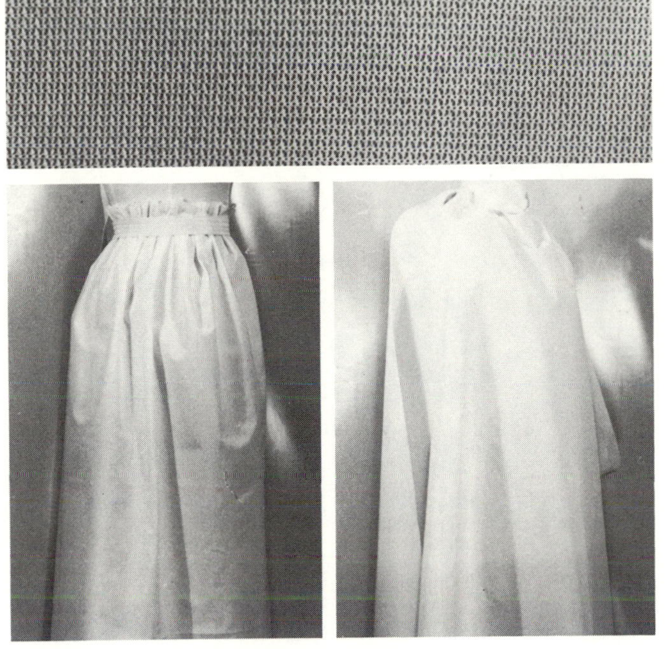

Tricot Knit Variation (stiff finish)

Fiber Content 100% nylon
Yarn Type filament fiber
Yarn Construction delustered; 40-denier yarn; textured
Fabric Structure tricot knit; warp knit classification
Finishing Processes heat-set; stabilizing; stiffening
Color Application solution dyed; yarn dyed
Width 92 inches (233.7 cm)
Weight light
Hand fine; rigid; stiff
Texture flat; harsh
Performance Expectations abrasion resistant; antistatic; dimensional stability; durable; elasticity; elongation; heat-set properties; resilient; no stretch properties; high tensile strength
Drapability Qualities falls into crisp wide cones; accommodates fullness by gathering; fullness maintains crisp bouffant effect

Recommended Construction Aids & Techniques

Hand Needles ball point, 5-10
Machine Needles ball point, 11-14
Threads poly-core waxed, extra fine; spun polyester, fine; nylon/Dacron, monocord A
Hems double-stitched; flat/plain; machine-stitched; overedged
Seams hairline; overedged; plain; safety-stitched; taped; tissue-stitched; zigzagged
Seam Finishes double-stitched; double-stitched and trimmed; serging/single-ply overedged; untreated/plain; single-ply or double-ply zigzag
Pressing steam; safe temperature limit 350°F (178.1°C)
Care launder; dry clean
Fabric Resource Native Textiles

Tricot Knit Variation (polished)

Fiber Content 100% Enkalure® nylon
Yarn Type multilobal filament fiber
Yarn Construction textured
Fabric Structure tricot knit; warp knit classification
Finishing Processes calendering; stabilizing; tentering
Color Application solution dyed; yarn dyed
Width 108 inches (274.3 cm)
Weight medium-light
Hand fine; silky; soft; moderate stretch in crosswise direction; no stretch in lengthwise direction
Texture high luster; satiny; smooth
Performance Expectations abrasion resistant; antistatic due to yarn finish; dimensional stability; durable; elasticity; elongation; heat-set properties; resilient; high tensile strength
Drapability Qualities falls into soft flares; accommodates fullness by gathering, elasticized shirring; fullness retains soft fall

Recommended Construction Aids & Techniques

Hand Needles ball point, 5-10
Machine Needles ball point, 11-14
Threads poly-core waxed, fine; spun polyester, all purpose; nylon/Dacron, monocord A
Hems double-stitched; flat/plain; machine-stitched; overedged
Seams hairline; overedged; plain; safety-stitched; taped; tissue-stitched; zigzagged
Seam Finishes double-stitched; serging/single-ply overedged; untreated/plain; single-ply or double-ply zigzag
Pressing steam; safe temperature limit 350°F (178.1°C)
Care launder
Fabric Resource Native Textiles

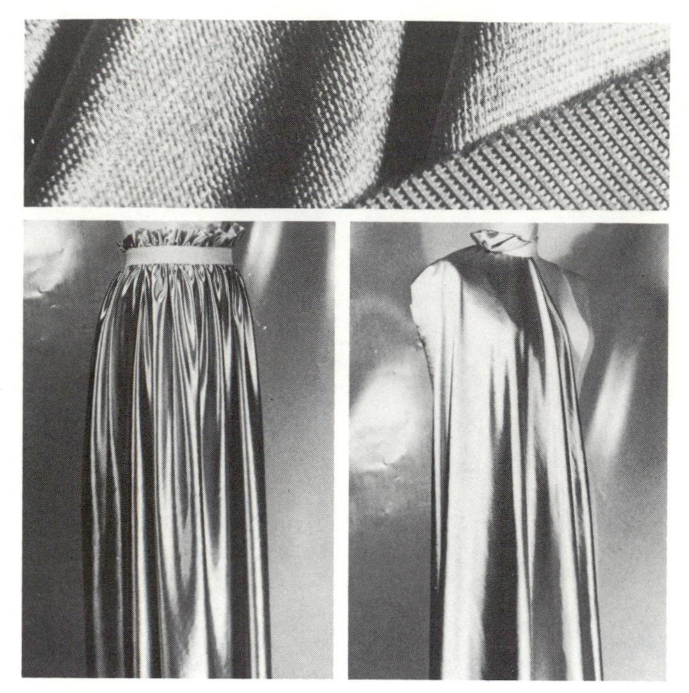

Ciré Tricot

Fiber Content 100% Enkalure® nylon
Yarn Type multilobal filament fiber
Yarn Construction conventional; high luster; low twist
Fabric Structure tricot knit; warp knit classification
Finishing Processes calendering; stabilizing; tentering
Color Application solution dyed; yarn dyed
Width 60 inches (152.4 cm)
Weight medium-light
Hand rigid; silky; soft; moderate stretch in crosswise direction; no stretch in lengthwise direction
Texture high-luster shiny face; dull back; satiny; smooth
Performance Expectations abrasion resistant; antistatic due to yarn finish; dimensional stability; durable; elasticity; elongation; heat-set properties; resilient; high tensile strength
Drapability Qualities falls into soft flares; accommodates fullness by gathering, elasticized shirring; fullness retains soft fall

Recommended Construction Aids & Techniques

Hand Needles ball point, 5–10
Machine Needles ball point, 11–14
Threads poly-core waxed, fine; spun polyester, all purpose; nylon/Dacron, monocord A
Hems double-stitched; flat/plain; machine-stitched; overedged
Seams hairline; overedged; plain; safety-stitched; taped; tissue-stitched; zigzagged
Seam Finishes double-stitched; serging/single-ply overedged; untreated/plain; single-ply or double-ply zigzag
Pressing steam; safe temperature limit 350°F (178.1°C)
Care launder
Fabric Resource **Native Textiles**

Tricot Knit Variation (patterned)

Fiber Content 100% Arnel® triacetate
Yarn Type filament fiber
Yarn Construction textured
Fabric Structure tricot knit; warp knit classification
Finishing Processes heat-set; stabilizing; tentering
Color Application solution dyed; yarn dyed
Width 54 inches (137.2 cm)
Weight medium-light
Hand firm; springy
Texture raised pattern on face creates uneven surface
Performance Expectations dimensional stability; elongation; flexible; heat-set properties; mildew and moth resistant; nonpilling; no stretch properties
Drapability Qualities falls into firm wide cones; accommodates fullness by gathering, elasticized shirring; fullness retains springy effect

Recommended Construction Aids & Techniques

Hand Needles ball point, 5–10
Machine Needles ball point, 11
Threads poly-core waxed, fine; spun polyester, all purpose; nylon/Dacron, monocord A
Hems double-stitched; flat/plain; machine-stitched; overedged
Seams hairline; overedged; plain; safety-stitched; taped; tissue-stitched; zigzagged
Seam Finishes double-stitched; double-stitched and trimmed; serging/single-ply overedged; untreated/plain; single-ply or double-ply zigzag
Pressing steam; safe temperature limit 375°F (192.1°C)
Care launder
Fabric Resource **Native Textiles**

Tricot Knit Variation (printed)

Fiber Content 100% Fortrel® polyester
Yarn Type filament fiber
Yarn Construction textured
Fabric Structure tricot knit; warp knit classification
Finishing Processes crepe-set; stiffening; tentering
Color Application screen printed
Width 60 inches (152.4 cm)
Weight light
Hand fine; soft; controlled stretch in crosswise direction; no stretch in lengthwise direction; supple
Texture crepey
Performance Expectations antistatic due to finish; dimensional stability; durable; elongation; heat-set properties; resilient; high tensile strength
Drapability Qualities falls into soft flares; accommodates fullness by gathering, elasticized shirring; fullness retains soft fall

Recommended Construction Aids & Techniques

Hand Needles ball point, 5–10
Machine Needles ball point, 11
Threads poly-core waxed, extra fine; spun polyester, fine; nylon/Dacron, monocord A
Hems double-stitched; flat/plain; machine-stitched; overedged
Seams hairline; overedged; plain; safety-stitched; taped; tissue-stitched; zigzagged
Seam Finishes double-stitched; double-stitched and trimmed; serging/single-ply overedged; untreated/plain; single-ply or double-ply zigzag
Pressing steam; safe temperature limit 325°F (164.1°C)
Care launder; dry clean
Fabric Resource **Loomskill/Gallery Screen Print**

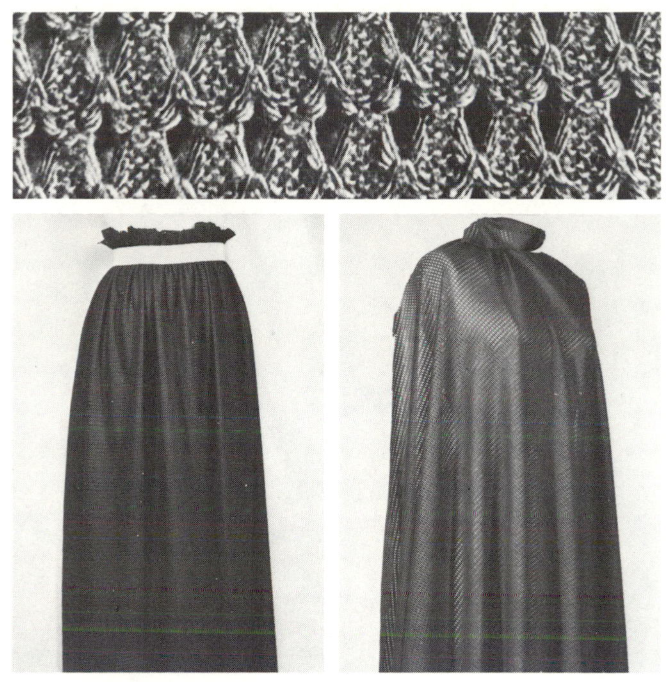

Dropped/Missed Stitch Knit

Fiber Content 100% cotton
Yarn Type combed staple
Yarn Construction conventional; mercerized
Fabric Structure tricot knit; dropped stitch produces design; weft knit classification
Finishing Processes mercerizing; stabilizing; tentering
Color Application piece dyed; yarn dyed
Width 66 inches (167.6 cm)
Weight medium-heavy
Hand semi-soft; spongy; stretch in crosswise direction; no stretch in lengthwise direction
Texture coarse; open-mesh pattern; uneven surface
Performance Expectations absorbent; antistatic; durable; elongation; high tensile strength; subject to catching and pulling due to open structure
Drapability Qualities falls into soft cones; accommodates fullness by gathering, elasticized shirring; fullness retains moderately soft fall

Recommended Construction Aids & Techniques

Hand Needles ball point, 5–10; cotton darners, 1–5; embroidery, 1–5; milliners, 3–5
Machine Needles ball point, 14
Threads mercerized cotton, heavy duty; six-cord cotton, 50; cotton-covered polyester core, 50
Hems double-stitched; flat/plain; overedged
Seams overedged; plain; safety-stitched; taped; zigzagged
Seam Finishes double-stitched; serging/single-ply overedged; untreated/plain; single-ply or double-ply zigzag
Pressing steam; safe temperature limit 400°F (206.1°C)
Care launder
Fabric Resource **Stēvcoknit, Inc.**

Transfer Stitch Knit

Fiber Content 75% polyester/25% metallic
Yarn Type filament fiber; extruded and cut metallic film
Yarn Construction laminated metallic monofilament; textured
Fabric Structure tricot knit; transfer stitch produces design; weft knit classification
Finishing Processes stabilizing; stiffening; stretching
Color Application yarn dyed
Width 76 inches (193 cm)
Weight medium-heavy
Hand semi-crisp; rough; stretch in crosswise direction; no stretch in lengthwise direction
Texture harsh; open-mesh pattern; uneven surface
Performance Expectations dimensional stability; durable; elongation; mildew and moth resistant; resilient; high tensile strength; subject to catching and pulling due to open structure
Drapability Qualities falls into wide flares; falls close to body contour; fullness maintains springy metallic effect

Recommended Construction Aids & Techniques

Hand Needles ball point, 5-10; cotton darners, 1-5; embroidery, 1-5; milliners, 3-5
Machine Needles ball point, 14
Threads poly-core waxed, heavy duty; cotton-polyester blend, heavy duty; cotton-covered polyester core, heavy duty; spun polyester, heavy duty; nylon/Dacron, monocord A
Hems double-stitched; flat/plain; overedged
Seams overedged; plain; safety-stitched; taped; zigzagged
Seam Finishes double-stitched; serging/single-ply overedged; untreated/plain; single-ply or double-ply zigzag
Pressing steam; safe temperature limit 325°F (164.1°C)
Care launder; dry clean
Fabric Resource **Applause Fabrics Ltd.**

Transfer Stitch Knit

Fiber Content 100% Nyesta® nylon
Yarn Type filament fiber
Yarn Construction textured
Fabric Structure tricot knit; transfer stitch produces design; weft knit classification
Finishing Processes heat-set; stabilizing; stretching
Color Application solution dyed; yarn dyed
Width 64 inches (162.7 cm)
Weight medium
Hand silky; soft; stretch in crosswise direction; no stretch in lengthwise direction
Texture crepey; loose; uneven surface
Performance Expectations abrasion resistant; dimensional stability; durable; elasticity; elongation; resilient; high tensile strength
Drapability Qualities falls into moderately soft flares; can be shaped or molded by stretching; falls close to body contour

Recommended Construction Aids & Techniques

Hand Needles ball point, 5-10; cotton darners, 1-5; embroidery, 1-5; milliners, 3-5
Machine Needles ball point, 14
Threads poly-core waxed, 50; cotton-covered polyester core, all purpose; spun polyester, all purpose; nylon/Dacron, monocord A
Hems double-stitched; flat/plain; overedged
Seams overedged; plain; safety-stitched; taped; zigzagged
Seam Finishes double-stitched; serging/single-ply overedged; untreated/plain; single-ply or double-ply zigzag
Pressing steam; safe temperature limit 350°F (178.1°C)
Care launder; dry clean
Fabric Resource **Applause Fabrics Ltd.**

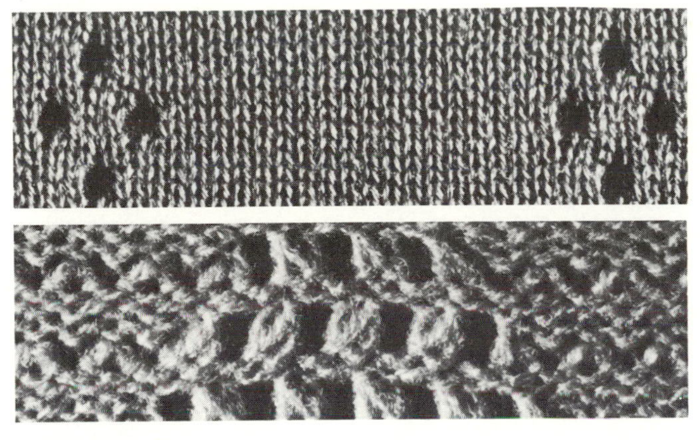

Transfer Stitch

Fabric Resource Universal Knitting Machines Corp.

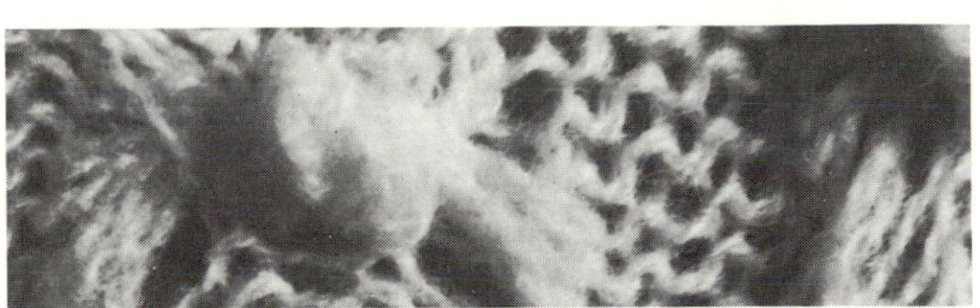

FACE

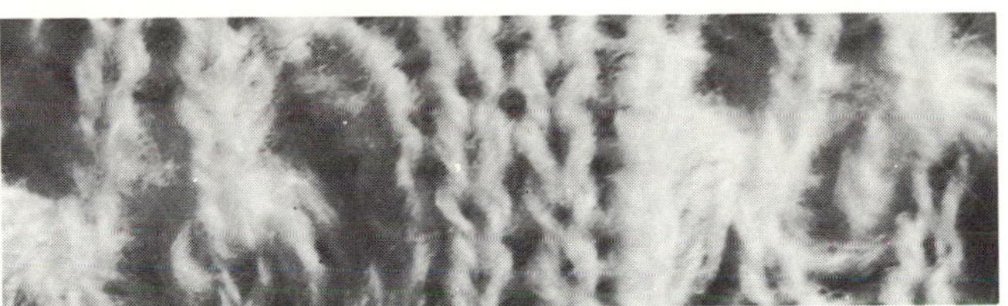

BACK

Popcorn Effect

Fabric Resource Applause Fabrics Ltd.

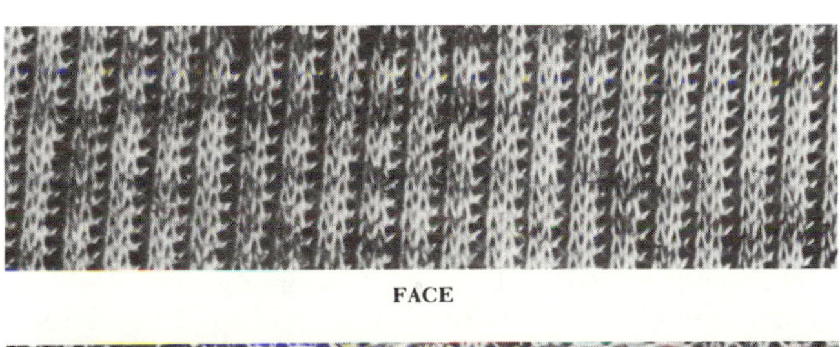

FACE

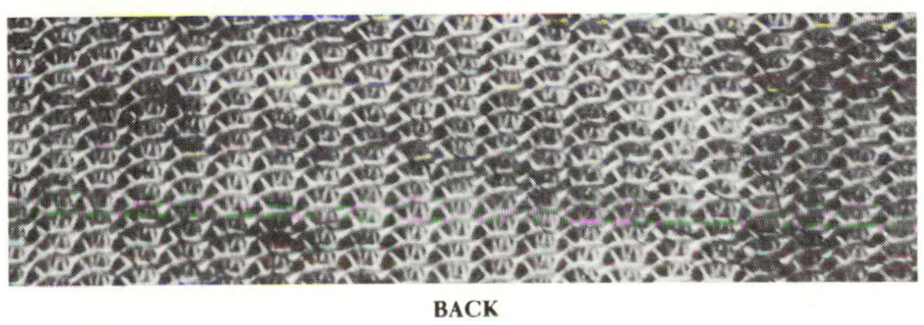

BACK

Rib Knit Effect

Fabric Resource American Cyanamid Co./Fabrics Development Dept.

Full Cable Stitch

Fabric Resource Universal Knitting Machines Corp.

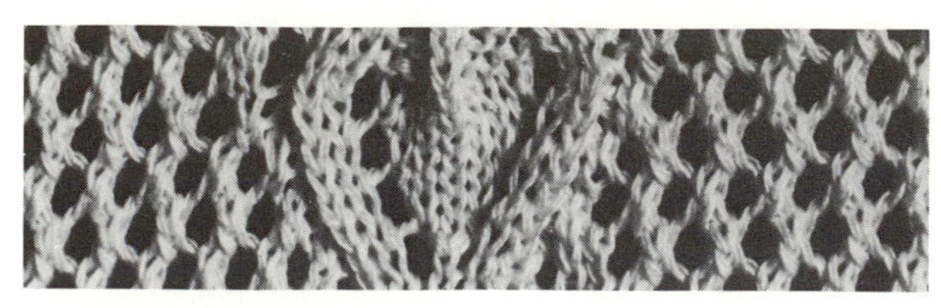

Half Cable Stitch

Fabric Resource Applause Fabrics Ltd.

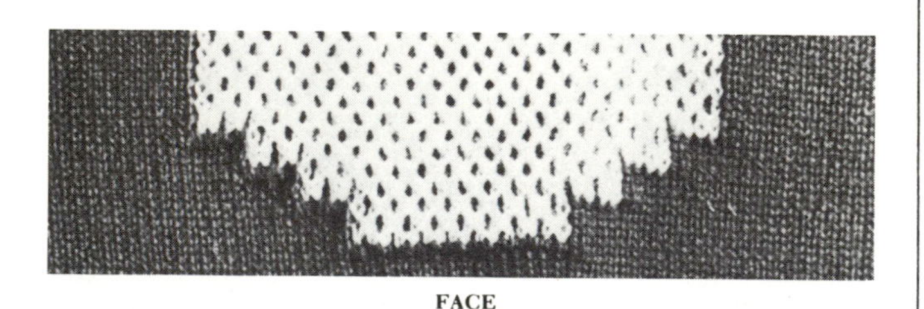

FACE

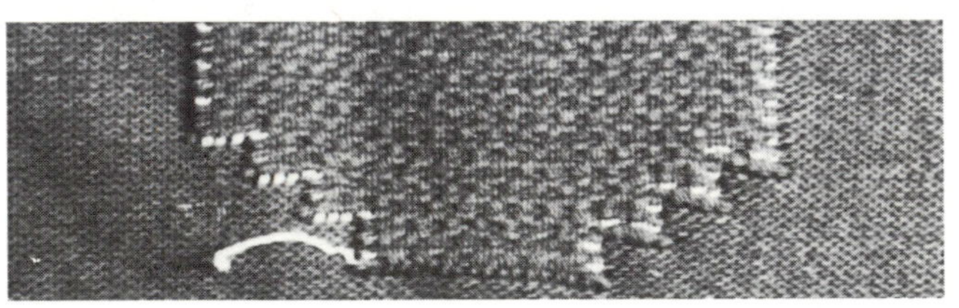

BACK

Intarsia Knit

Fabric Resource Universal Knitting Machines Corp.

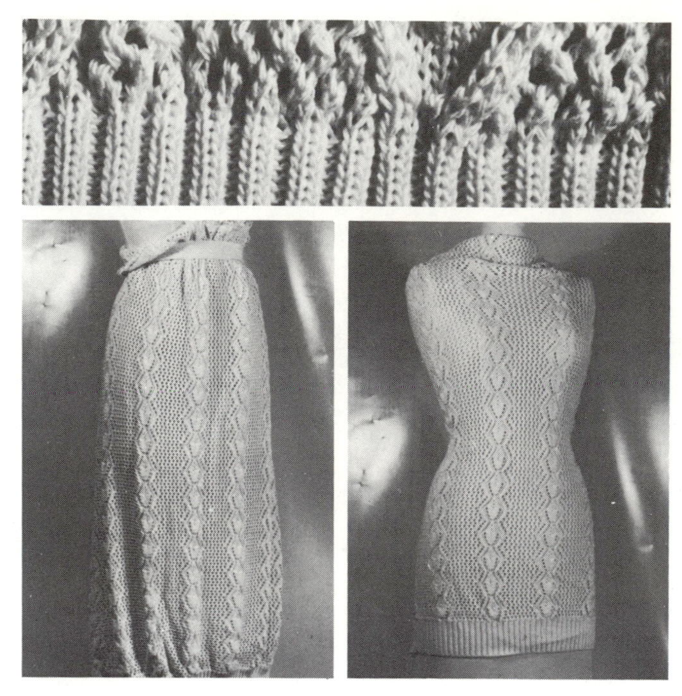

Sweater Knit[2]

Fiber Content 100% polyester
Yarn Type filament fiber
Yarn Construction bulked; textured
Fabric Structure transfer stitch knit; rib-knit finished edge; weft knit classification
Finishing Processes stabilizing; stretching; tentering
Color Application piece dyed; solution dyed; yarn dyed
Width 36–72 inches (91.4–182.9 cm)
Weight heavy
Hand lofty; soft; stretch in lengthwise and crosswise directions; rib knit banding–high stretch and recovery in crosswise direction
Texture loose; open design creates uneven surface
Performance Expectations dimensional stability; durable; elasticity; elongation; high tensile strength
Drapability Qualities can be shaped or molded by stretching; falls close to body contour; better utilized if fitted by seaming and eliminating excess fabric

Recommended Construction Aids & Techniques

Hand Needles ball point, 5–10; cotton darners, 1–5; embroidery, 1–5; milliners, 3–5
Machine Needles ball point, 14
Threads poly-core waxed, heavy duty; cotton-polyester blend, heavy duty; cotton-covered polyester core, heavy duty; spun polyester, heavy duty; nylon/Dacron, monocord A
Hems none needed—knit band is finished edge
Seams overedged; plain; safety-stitched; taped; zigzagged
Seam Finishes serging/single-ply overedged; untreated/plain; single-ply zigzag
Pressing steam; safe temperature limit 325°F (164.1°C)
Care launder; dry clean
Fabric Resource Applause Fabrics Ltd.

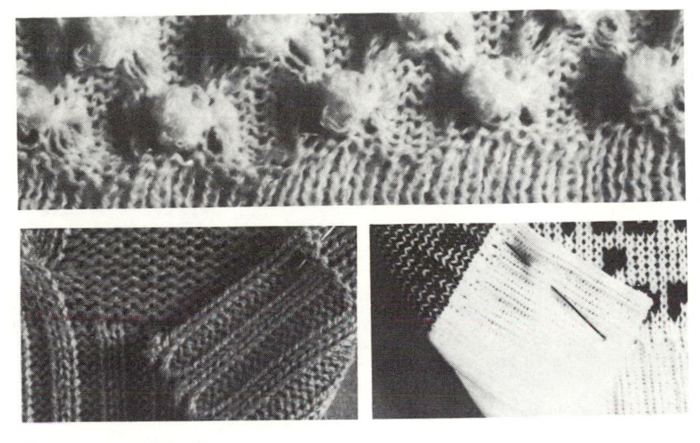

Sweater Knit

Fiber Content 100% polyester
Yarn Type filament fiber
Yarn Construction bulked; textured
Fabric Structure transfer stitch knit; rib-knit finished edge; weft knit classification
Finishing Processes stabilizing; stretching; tentering
Color Application piece dyed; solution dyed; yarn dyed
Width 36-72 inches (91.4-182.9 cm)
Weight heavy
Hand lofty; soft; stretch in lengthwise and crosswise directions; rib knit banding-high stretch and recovery in crosswise direction
Texture loose; open design creates uneven surface
Performance Expectations dimensional stability; durable; elasticity; elongation; high tensile strength
Drapability Qualities can be shaped or molded by stretching; falls close to body contour; better utilized if fitted by seaming and eliminating excess fabric

Recommended Construction Aids & Techniques

Hand Needles ball point, 5-10; cotton darners, 1-5; embroidery, 1-5; milliners, 3-5
Machine Needles ball point, 14
Threads poly-core waxed, heavy duty; cotton-polyester blend, heavy duty; cotton-covered polyester core, heavy duty; spun polyester, heavy duty; nylon/Dacron, monocord A
Hems none needed—knit band is finished edge
Seams overedged; plain; safety-stitched; taped; zigzagged
Seam Finishes serging/single-ply overedged; untreated/plain; single-ply zigzag
Pressing steam; safe temperature limit 325°F (164.1°C)
Care launder; dry clean
Fabric Resource **Universal Knitting Machines Corp.**

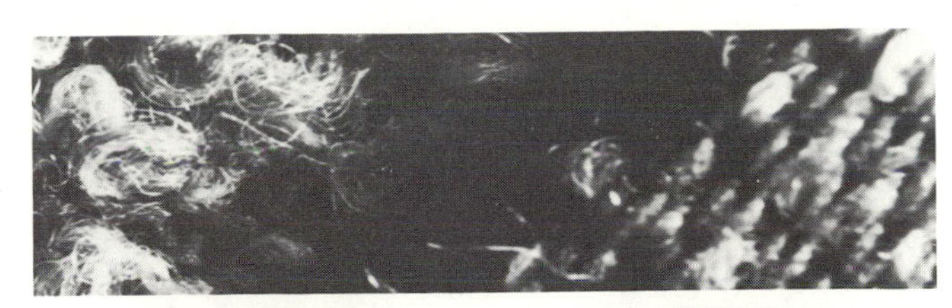

Single Knit (bouclé yarn)

Fiber Content 65% rayon/35% acrylic
Yarn Type filament fiber
Yarn Construction novelty bouclé yarn; bulked; textured
Fabric Structure single knit; laid-in bouclé yarn; weft knit classification
Finishing Processes brushing; napping; stabilizing
Color Application yarn dyed
Width 66 inches (167.6 cm)
Weight medium
Hand lofty; soft; spongy; controlled stretch in lengthwise and crosswise directions
Texture fluffy; fuzzy; loopy
Performance Expectations compressible and resilient due to yarn and structure; subject to snagging due to looped yarn and structure
Drapability Qualities falls into moderately soft flares; fullness maintains lofty bouffant effect; retains shape of garment

Recommended Construction Aids & Techniques

Hand Needles ball point, 5-10; cotton darners, 1-5; embroidery, 1-5; milliners, 3-5
Machine Needles ball point, 14
Threads poly-core waxed, heavy duty; cotton-polyester blend, heavy duty; cotton-covered polyester core, heavy duty; spun polyester, heavy duty; nylon/Dacron, monocord A
Hems double-stitched; flat/plain; overedged
Seams overedged; plain; safety-stitched; taped; zigzagged
Seam Finishes serging/single-ply overedged; untreated/plain; single-ply zigzag
Pressing steam; safe temperature limit 300°F (150.1°C)
Care dry clean
Fabric Resource **Applause Fabrics Ltd.**

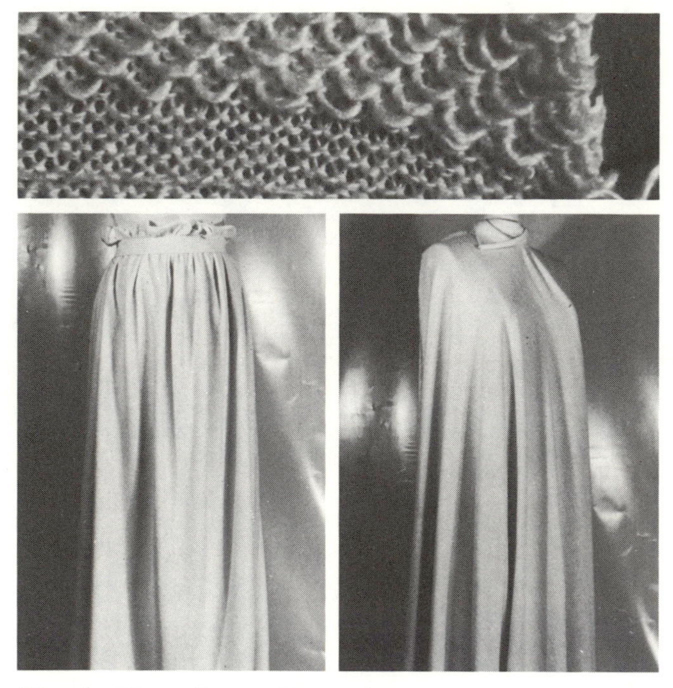

Single Knit (looped yarn)

Fiber Content 50% cotton (face); 50% polyester (back)
Yarn Type combed staple; filament fiber
Yarn Construction conventional; textured
Fabric Structure single knit; laid-in yarn manipulated to form loop; weft knit classification
Finishing Processes napping; pressing; stabilizing
Color Application piece dyed–union dyed; yarn dyed
Width 72 inches (182.9 cm)
Weight medium-heavy
Hand soft; spongy; stretch in lengthwise and crosswise directions; thick
Texture loop pile flattened in one direction; tweedy appearance
Performance Expectations durable; elongation; high tensile strength; subject to pulling and snagging due to looped structure
Drapability Qualities falls into moderately soft flares; fullness maintains lofty effect; retains shape of garment

Recommended Construction Aids & Techniques

Hand Needles ball point, 5–10
Machine Needles ball point, 14
Threads mercerized cotton, heavy duty; six-cord cotton, 50; cotton-covered polyester core, 50
Hems double-stitched; flat/plain; overedged
Seams overedged; plain; safety-stitched; taped; zigzagged
Seam Finishes serging/single-ply overedged; untreated/plain; single-ply zigzag
Pressing steam; safe temperature limit 325°F (164.1°C)
Care launder
Fabric Resource **Gloversville Mills Inc.**

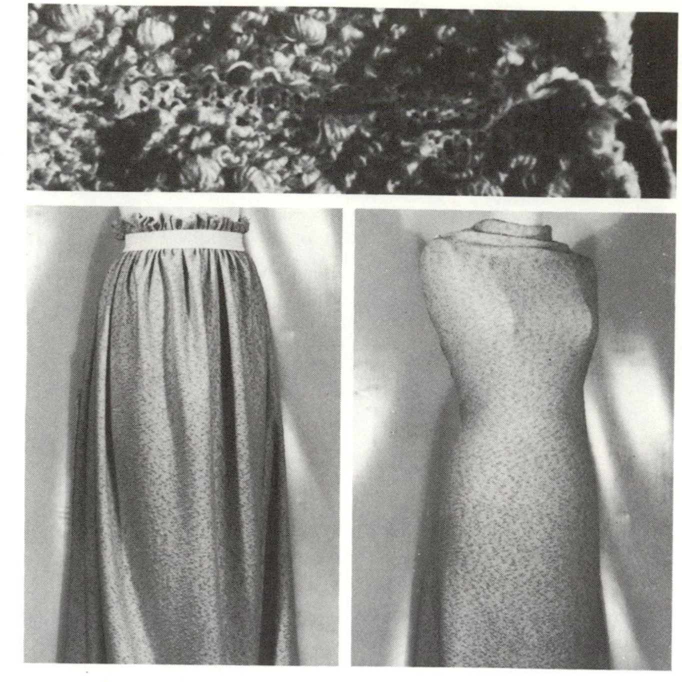

Single Knit (seed yarn)

Fiber Content 100% cotton (face); 100% cotton (back)
Yarn Type filament fiber; staple
Yarn Construction novelty seed yarn; textured
Fabric Structure single knit; laid-in seed yarn; weft knit classification
Finishing Processes mercerizing; stabilizing; tentering
Color Application piece dyed–union dyed; yarn dyed
Width 60 inches (152.4 cm)
Weight medium-heavy
Hand coarse; spongy; stretch in crosswise direction; controlled stretch in lengthwise direction
Texture nubby; rough; seed noil
Performance Expectations durable; elongation; high tensile strength; subject to pulling due to raised yarn structure
Drapability Qualities falls into soft flares; can be shaped or molded by stretching; fullness retains moderately soft fall

Recommended Construction Aids & Techniques

Hand Needles ball point, 5–10
Machine Needles ball point, 14
Threads mercerized cotton, heavy duty; six-cord cotton, 50; cotton-covered polyester core, 50
Hems double-stitched; flat/plain; overedged
Seams overedged; plain; safety-stitched; taped; zigzagged
Seam Finishes double-stitched; serging/single-ply overedged; untreated/plain; single-ply or double-ply zigzag
Pressing steam; safe temperature limit 325°F (164.1°C)
Care launder
Fabric Resource **Gloversville Mills Inc.**

General Properties of Knit Fabrics

Property	Dependent on
absorbency	fiber, yarn formation, finishes, wales and courses per inch
hand and drapability	fiber, yarn type, knit type, stitches, wales and courses per inch
effect of heat/insulatory factors	fiber content, yarn size and amount of twist in yarn, fabric structure
pilling	fiber content, yarn type, fabric structure

General Characteristics of Knit Fabrics

inherent elasticity
considerable give and stretch
returns to original shape after flexing and stretching
freedom of movement in constructed garment
provides comfort
resiliency
good wrinkle and crush resistance
undesirable wrinkles hang out
air permeability
depending on fibers and construction knit fabrics may:
- shrink
- stretch
- be subject to pulls and snags
- not hold crease due to loop formation of construction (thermoplastic fiber may be heat set)

12 ~ Lace/Lace-type Fabrics

- Alençon Lace
  - Alençon-type Designed Lace
- Binche Lace
  - Binche-type Designed Lace
- Blonde Lace
  - Blonde-type Designed Lace
- Brussels Lace
  - Brussels-type Designed Lace
- Chantilly Lace
  - Chantilly-type Designed Lace
  - Re-embroidered Lace / Embroidered Braid on Net Ground Lace
  - Re-embroidered Chantilly-type Designed Lace
- Cluny Lace
  - Cluny-type Designed Lace
- Crochet Lace
  - Crochet-type Designed Lace
- Cut Work Lace
  - Cut Work-type Designed Lace / Eyelet-styled Openwork Lace
  - Cut Work / Openwork
  - Burnt-out Method Openwork (close up only)
- Duchesse lace
  - Duchesse-type Designed Lace
- Embroidered Lace
  - Honiton Bobbin Lace / Darned Netting
  - Embroidered / Appliqué-type Designed Lace (close up only)
- Filet Lace
  - Filet / Filet Crochet Lace
  - Filet-type Designed Lace / Crochet Styling
  - Filet / Darned Lace
  - Filet-type Designed Lace / Darned Lace Styling
- Honiton Lace
  - Honiton-type Designed Lace
- Macramé
  - Knotted Lace
  - Macramé Lace
- Nottingham Lace
  - Machine / Nottingham Lace
  - Nottingham-type Designed Lace
- Plauen / Saxony Lace
  - Plauen / Saxony Lace
  - Plauen-type Designed / Burnt-out Lace
- Point d'Angleterre
  - Point d'Angleterre-type Designed Lace
- Point d'Esprit
  - Normandy Lace
  - Point d'Esprit-type Designed Lace
- Point de France
  - Point de France-type Designed Lace
- Point de Venice
  - Point de Venice / Punto Avorio
  - Point de Venice-type Designed Lace
- Point Plat de Venice
  - Point Plat de Venice (bobbin lace)
  - Point Plat de Venice-type Designed Lace (simulates bobbin lace)
  - Point Plat de Venice (needle point lace)
  - Point Plat de Venice-type Designed Lace (simulates needle point lace)
- Rose Point
  - Rose Point (large motif)
  - Rose Point-type Designed Lace (large motif)
  - Rose Point (small motif)
  - Rose Point-type Designed Lace (small motif)
- Shadow Lace
  - Point de Venice
  - Shadow-type Designed Lace
- Tatting
  - Tatting / Tatted Lace
  - Tatting-type Designed Lace / Simulated Tatted Lace
- Val / Valenciennes Lace
  - Val / Valenciennes-type Designed Lace (close up only)
- Venice Lace
  - Gros Point de Venice
  - Venice-type / Gros Point-type Designed Lace
- Performance Expectations
- Recommended Construction Techniques & Aids

A lace fabric is characterized by a pattern or integral design that forms the solid, closely worked areas on a netlike or openwork ground or filling that holds or joins the pattern together. Lace is also defined as a meshlike, openwork, porous constructed fabric with a pattern and base achieved with threads or yarns that are part of the structure of the cloth.

Lace fabrics may be made using:

- Natural fiber yarns of cotton, linen, silk or wool
- Man-made fiber yarns
- Blends of man-made or natural and man-made fibers
- Mixed or combined fiber yarns
- Textured or novelty yarns
- Fine, tightly twisted yarns

Lace design or pattern and its heavy or fine appearance depends on the degree in which the close and open mesh areas are interchanged.

The traditional name or description of lace usually reflected the geographical or historical development of its association, historical personage, sociological events, technical differences as well as type or impression of design. Specifically named laces and their variations are associated with particular characteristics, arrangements of open and closed areas, pattern or design feature, color and weight. Various laces are also named, grouped or typed by the techniques or methods employed in their construction or by structure.

Traditional laces were made by hand using the pillow and bobbin method or the needlepoint method. Contemporary machine-made laces incorporate a mixture of techniques. Machine methods can copy, duplicate or simulate the patterns and intricate work of the originally designed lace.

Lace fabrics may be made by:

- Hand utilizing hooks, needles or bobbins
- Machine incorporating a wide range of techniques
- Utilizing one or more yarns
- Incorporating different stitch formations

Lace fabrics are produced:

- In a variety of patterns by selecting and combining stitches
- From simple to complex designs
- In open, irregular to close regular patterned designs
- As plaited, braided or knitted base fabrics

Handmade lace construction methods are categorized as:

- Bobbin / Pillow laces
- Crochet laces
- Knitted laces
- Needlepoint laces including cut work and drawn work

Machine-made laces are categorized as:

- Bobbin laces
- Leaver laces
- Nottingham laces
- Raschel laces
- Schiffli laces
- Burnt-out laces

Quality and durability of lace depends on:

- Fiber content
- Yarn type and construction
- Methods and techniques of construction used
- Finishes applied
- End use

Due to the complexity and difficulties in identifying lace fabrics with regard to their name, type and design-inspired origin, this unit will follow a format different from the other units in the text. The following pages are presented in an order that will relate today's machine-made laces to the original handmade laces.

To allow for comparison, close-up views of the original handmade laces are shown on the same page as today's machine-made laces. A description of the pattern and method of construction of the original lace is included.

Photographs of machine-made laces include a close-up view of the pattern and structure and the lace fabric draped on the model form. Information regarding the specifications and resources are listed with each term. Performance expectations and recommended construction techniques and aids are presented as a general guide for all laces discussed in this unit and are listed at the end of the unit.

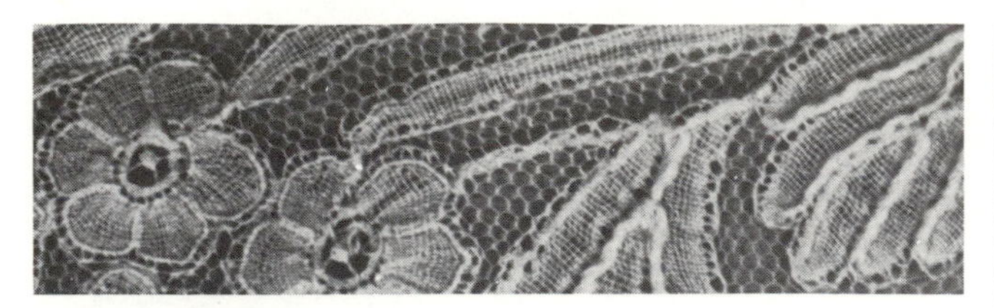

Alençon Lace

Description of Original Lace A delicate lace with a ground of fine net, made of double-twisted threads, and a solid floral motif design. Alençon lace is characterized by a defined cordonnet of firm heavy cord which outlines the motif.
Original Lace-making Method needlepoint
Origin Alençon, France, 1665
Origin of Sample unknown
Sample Courtesy of Design Laboratories, Fashion Institute of Technology

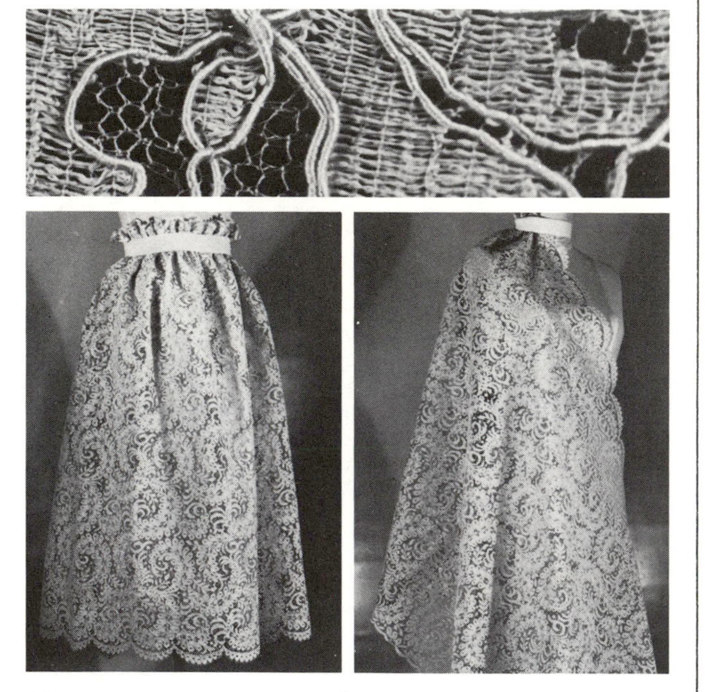

Alençon-type Designed Lace

Fiber Content 70% rayon/20% cotton/10% nylon
Fabric Structure Leavers machine; 9 points; finished with scalloped edge
Width 33 inches (83.9 cm)
Weight medium
Hand crisp
Fabric Resources Lace of France Inc.; Whelan Lace

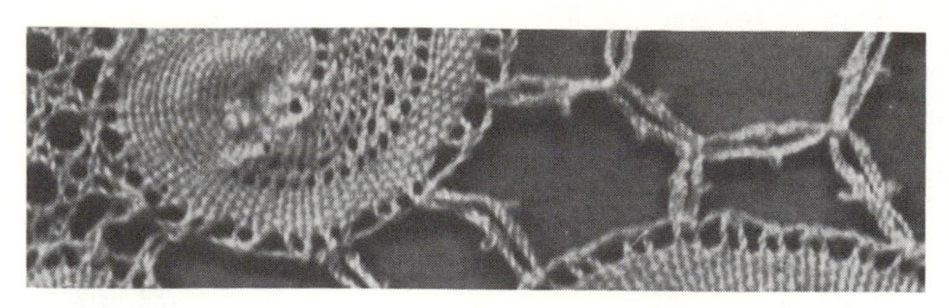

Binche Lace

Description of Original Lace A fine, narrow lace identified by a *cane seat* effect and/or a six-pointed star or snowflake in the ground. The flat, closely woven design may be a geometric or scroll floral motif. No cordonnet is used. However, a thread accentuates the outline of the motif. The edge of this lace has a narrow scallop with picots.
Original Lace-making Method needlepoint
Origin Flanders, 17th century
Origin of Sample Holland, 19th century
Sample Courtesy of Design Laboratories, Fashion Institute of Technology

Binche-type Designed Lace

Fiber Content 100% silk
Fabric Structure Leavers machine
Width 56 inches (142.2 cm)
Weight light
Hand soft
Fabric Resource American Silk Mills Corp./Bucol

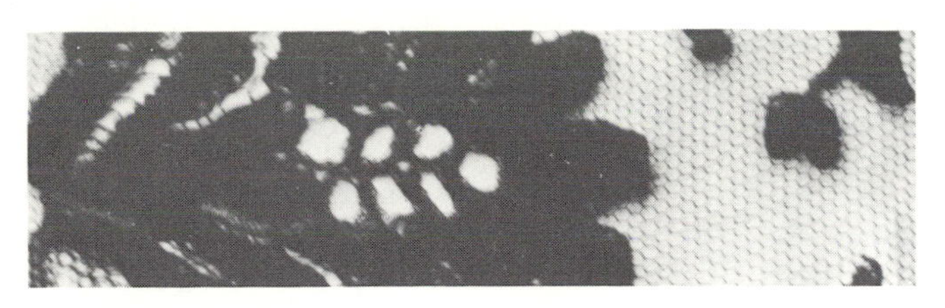

Blonde Lace

Description of Original Lace A combination lace characterized by individual motifs which are worked separately and applied to a bobbin-net ground. Large bold floral motif is closely worked in a heavy, soft flat thread and applied to a delicate mesh ground. A fine, loosely twisted thread accentuates the motif.
Original Lace-making Method mixed lace: bobbin and needlepoint
Origin France
Origin of Sample France, 19th century
Sample Courtesy of Design Laboratories, Fashion Institute of Technology

Blonde-type Designed Lace

Fiber Content 100% nylon
Fabric Structure Raschel machine
Width 56 inches (142.2 cm)
Weight light
Hand semi-soft
Fabric Resource Native Textiles

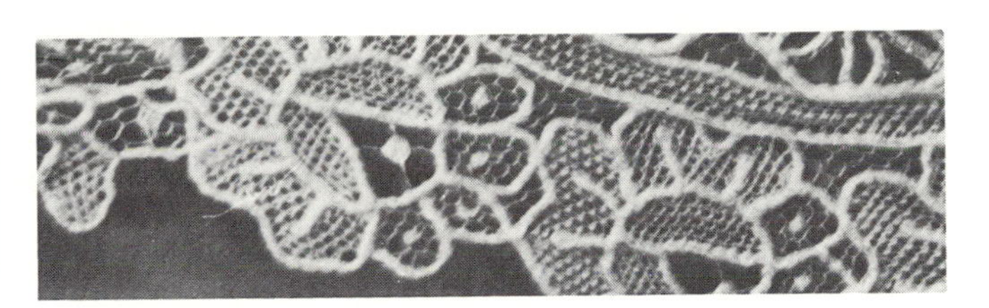

Brussels Lace

Description of Original Lace A combination lace characterized by a needlepoint design which is worked separately and then applied to or joined by a bobbin-net ground. The coarsely worked designs have a needlepoint cordonnet. The cordonnet edging, when applied to the mesh ground, is tacked down flat using a felling stitch instead of a buttonhole stitch.
Original Lace-making Method mixed lace: needlepoint lace motif appliquéd to bobbin-made net ground
Origin Belgium
Origin of Sample England, 19th century
Sample Courtesy of Design Laboratories, Fashion Institute of Technology

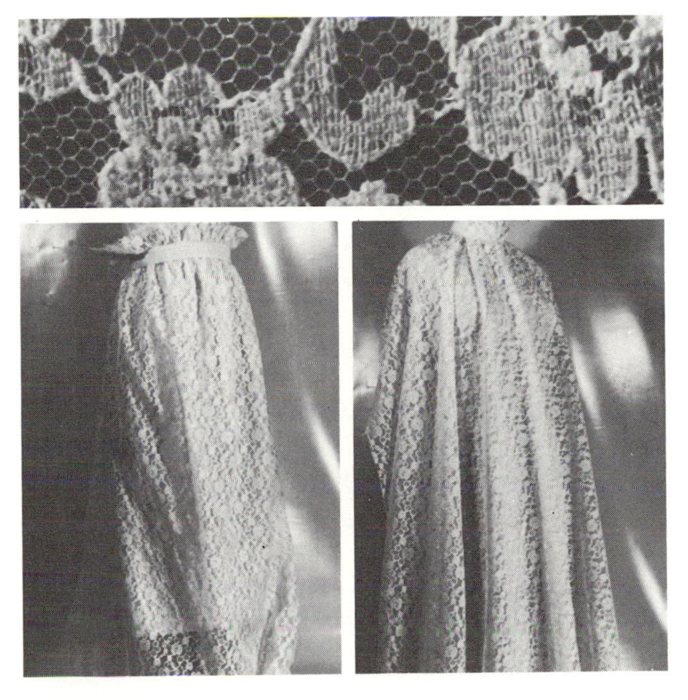

Brussels-type Designed Lace

Fiber Content 100% nylon
Fabric Structure Raschel machine
Width 56 inches (142.2 cm)
Weight light
Hand soft
Fabric Resource North American Lace Co.

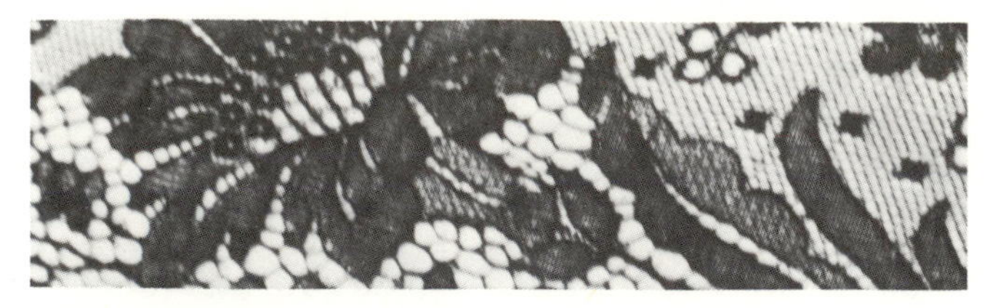

Chantilly Lace

Description of Original Lace A lace characterized by a delicate pattern of floral sprays, flowers or scrolls with a fine, fragile mesh ground. A fine, thin, untwisted thread outlines the flat design.
Original Lace-making Method bobbin
Origin France
Origin of Sample France, 1850
Sample Courtesy of Design Laboratories, Fashion Institute of Technology

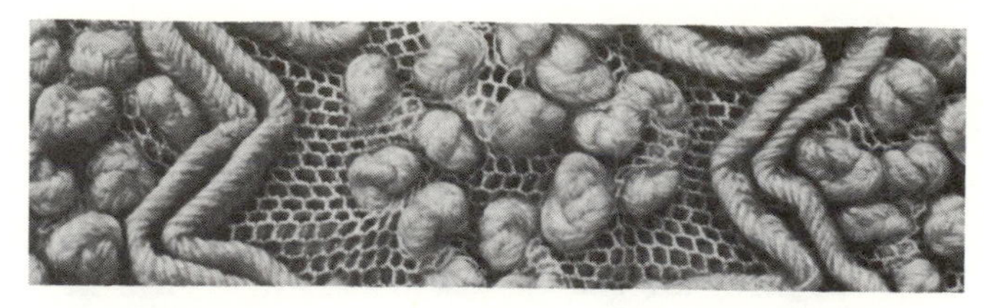

Re-embroidered Lace/Embroidered Braid on Net Ground Lace

Description of Original Lace A bobbin or needlepoint lace characterized by an outlined motif. Heavy cordonnet of thick yarn, heavy cord or gimp braid is used to create a raised re-embroidered effect. Ribbon, narrow lace, beads, sequins and other stones may be used to re-embroider pattern.
Original Lace-making Method mixed lace: embroidery/re-embroidery on lace or net ground
Origin France
Origin of Sample unknown
Sample Courtesy of Design Laboratories, Fashion Institute of Technology

Chantilly-type Designed Lace

Fiber Content 40% cotton/30% nylon/30% silk
Fabric Structure Leavers machine; 12 points; finished with scalloped edge
Width 33 inches (83.9 cm)
Weight light
Hand crisp
Fabric Resources **Lace of France Inc.; Whelan Lace**

Re-embroidered Chantilly-type Designed Lace

Fiber Content 63% cotton/21% rayon/16% nylon
Fabric Structure Leavers machine; 12 points
Width 34 inches (86.4 cm)
Weight heavy
Hand stiff
Fabric Resource **Whelan Lace**

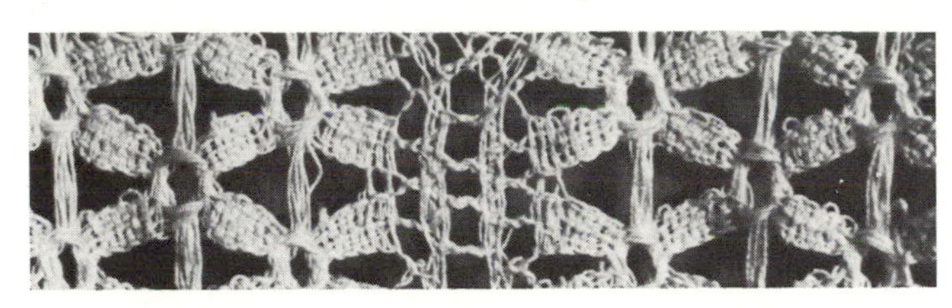

Cluny Lace

Description of Original Lace A coarse, thick, strong lace usually made with cotton or linen threads. Motifs are characterized by paddles, wheels or poinsetta flowers.
Original Lace-making Method bobbin
Origin Belgium; Germany; Italy
Origin of Sample unknown
Sample Courtesy of Design Laboratories, Fashion Institute of Technology

Cluny-type Designed Lace

Fiber Content 100% cotton
Fabric Structure Leavers machine
Width 40 inches (101.6 cm)
Weight medium-heavy
Hand firm; semi-soft
Fabric Resource **Associated Lace Corp.**

Crochet Lace

Description of Original Lace A variety of coarse, heavy laces usually made with cotton or linen threads. The loop-forming network of lace consists of medallions in a rose, shamrock or leaf motif. The designs are produced flat, raised or three dimensional. The network of loops connecting the medallions are edged with picots.
Original Lace-making Method crochet
Origin Ireland, c. 1820
Origin of Sample Belgium
Sample Courtesy of Design Laboratories, Fashion Institute of Technology

Crochet-type Designed Lace

Fiber Content 65% polyester/35% cotton
Fabric Structure Leavers machine
Width 50-54 inches (127-137.2 cm)
Weight medium-heavy
Hand semi-crisp
Fabric Resource **North American Lace Co.**

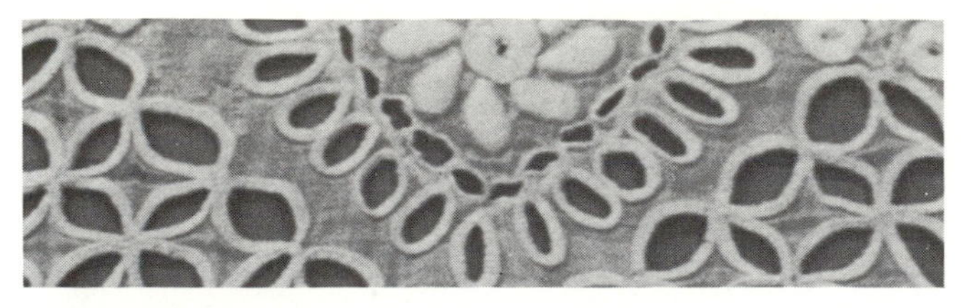

Cut Work Lace

Description of Original Lace Needlepoint cut work is a form of embroidery in which part of the fabric is cut away to produce a design with an openwork-lace effect. Buttonhole stitches are used to outline, finish and strengthen the worked design.
Original Lace-making Method needlepoint
Origin England, c. 1400
Origin of Sample United States, 19th century
Sample Courtesy of Design Laboratories, Fashion Institute of Technology

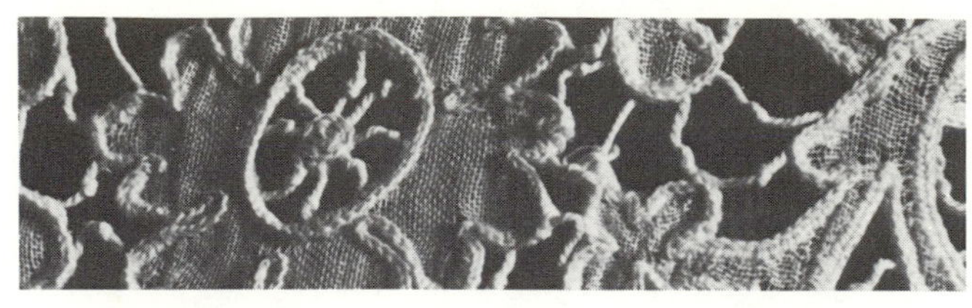

Cut Work/Openwork Lace

Description of Original Lace A cut work fabric utilizing the burnt-out method in which a pattern is applied to a woven ground or base fabric and then the nondesigned areas are removed to give an openwork effect (cut work embroidery). Stitching is made on the base fabric. The undesirable base fabric is later removed by heat or caustic chemicals leaving only the stitched outlined design areas. A cut work pattern is produced with motifs joined to one another.
Original Lace-making Method burnt-out method; machine embroidery on woven fabric
Origin Switzerland, c. 1860
Origin of Sample unknown
Sample Courtesy of Design Laboratories, Fashion Institute of Technology

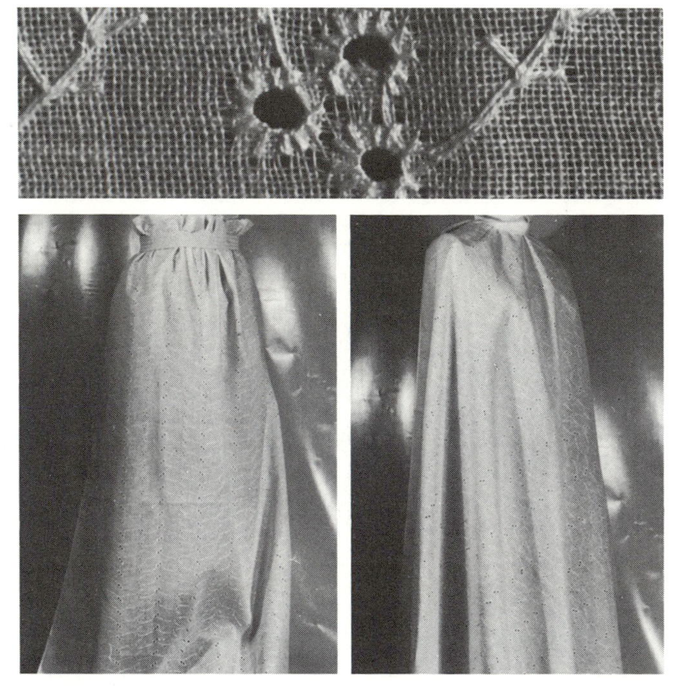

Cut Work-type Designed Lace/Eyelet-styled Openwork

Fiber Content 100% cotton
Fabric Structure Schiffli machine; openwork
Width 45 inches (114.3 cm)
Weight medium
Hand semi-crisp
Fabric Resource Applause Fabrics Ltd.

Burnt-out Method Openwork

Fiber Content 100% cotton
Fabric Structure Schiffli machine embroidery on woven base; printed and acid burnt-out method
Width 36 inches (91.4 cm)
Weight medium
Hand semi-soft
Sample Courtesy of Lita Konde

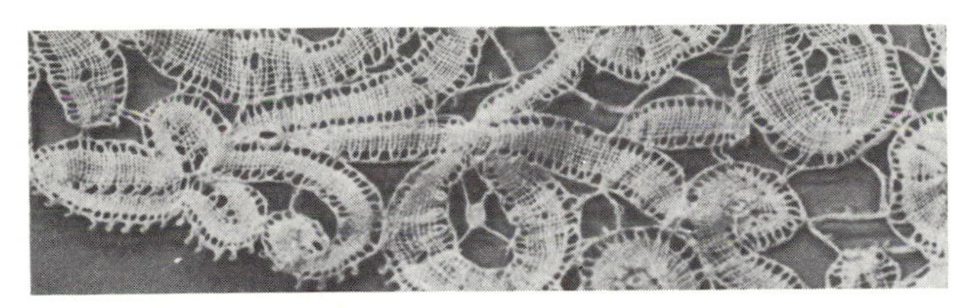

Duchesse Lace

Description of Original Lace A combination lace characterized by brides of bobbin ground with needle-point inserts. The tapelike needlepoint worked into floral sprays, flowers, or leaves is made separately. The detached motifs are joined by bobbin-made brides.
Original Lace-making Method mixed lace: needlepoint lace motifs joined by bobbin-made lace ground
Origin Flanders
Origin of Sample Belgium, 19th century
Sample Courtesy of Design Laboratories, Fashion Institute of Technology

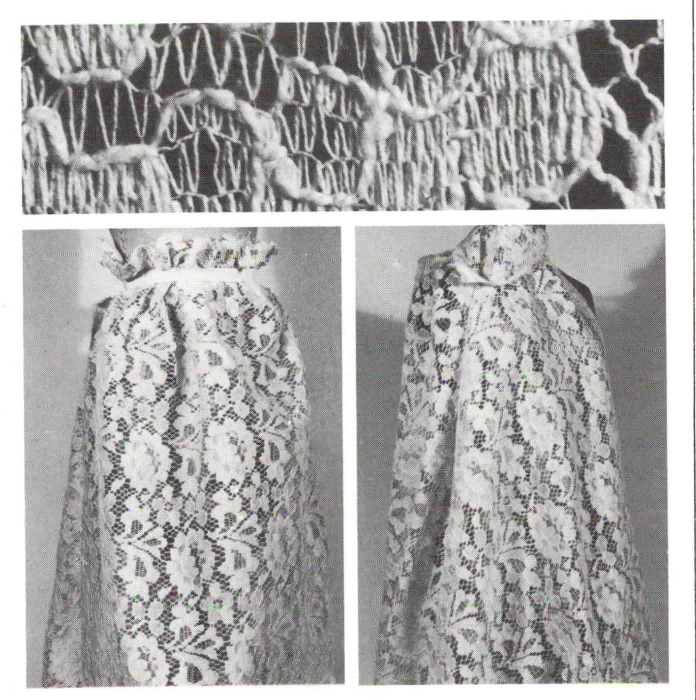

Duchesse-type Designed Lace

Fiber Content 100% cotton
Fabric Structure Leavers machine
Width 36 inches (91.4 cm)
Weight medium
Hand semi-crisp
Fabric Resource Associated Lace Corp.

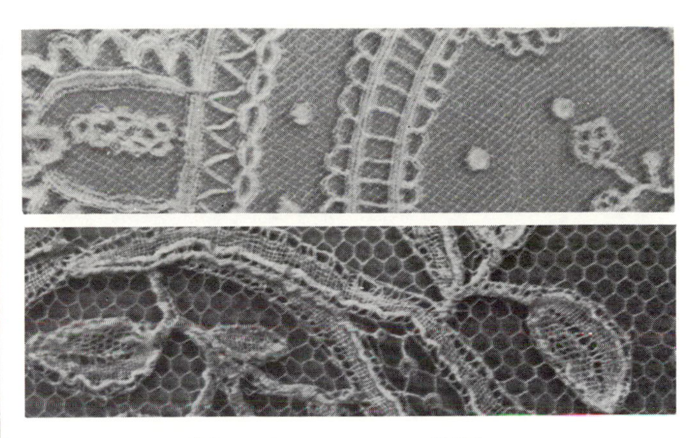

Honiton Bobbin Lace/Darned Netting

Description of Original Lace The application of an embroidered pattern or design on a machine-made net ground. The fine mesh/net ground may be worked with stitching or needle work by outlining the design with chain stitches and filling the spaces with darning or overcast stitches.
Original Lace-making Method mixed lace: hand embroidery on machine-made net ground
Origin England, during reign of Henry II
Origin of Sample France, 1890
Sample Courtesy of Design Laboratories, Fashion Institute of Technology

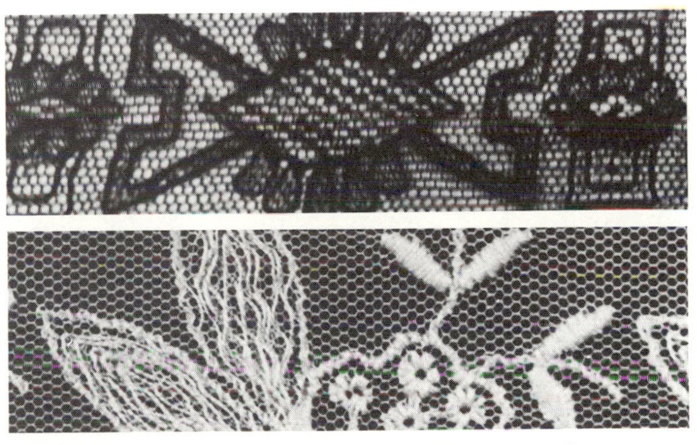

Embroidered/Appliqué-type Designed Lace

Fiber Content 100% nylon
Fabric Structure Schiffli machine
Width 16 inches (40.6 cm)
Weight fine to lightweight
Hand semi-crisp
Fabric Resource Associated Lace Corp.

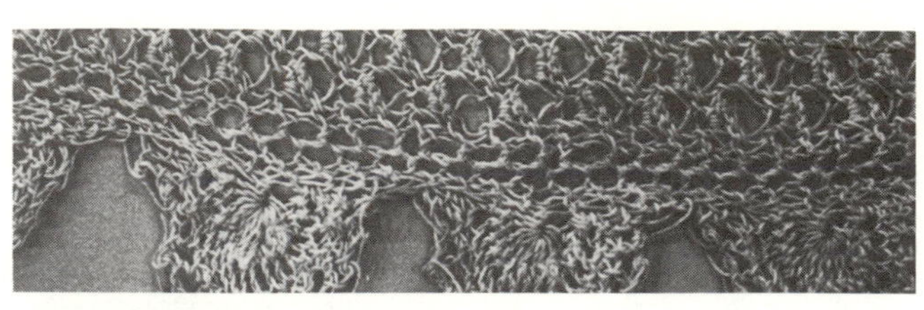

Filet/Filet Crochet Lace

Description of Original Lace A coarse, heavy crochet lace worked to resemble the result of filet lace or filet-darned lace. The pattern of the crocheted filet is formed by closely working certain areas of the lace to form closed-in spaces or squares producing a needle-worked darned lace design.
Original Lace-making Method crochet
Origin Italy
Origin of Sample Ireland, 19th century
Sample Courtesy of Design Laboratories, Fashion Institute of Technology

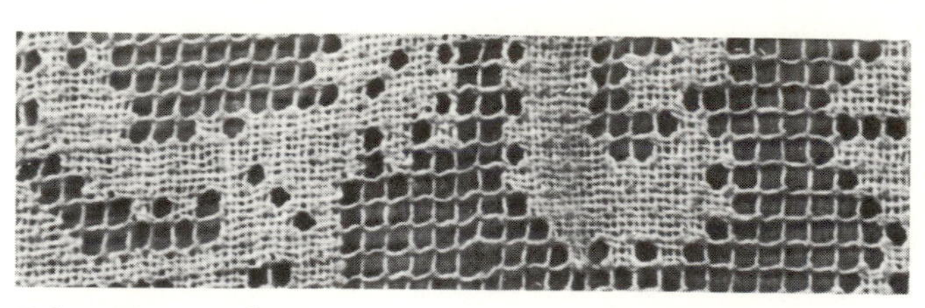

Filet/Darned Lace

Description of Original Lace A coarse, heavy, square mesh netting in which the pattern or design is embroidered or filled in by needlework creating a darned lace effect. The handmade filet netting has horizontal and vertical threads knotted at all points of intersection creating the square net. Only part of the knotted mesh foundation is filled with darning stitches.
Original Lace-making Method bobbin or needlepoint
Origin France
Origin of Sample Belgium
Sample Courtesy of Design Laboratories, Fashion Institute of Technology

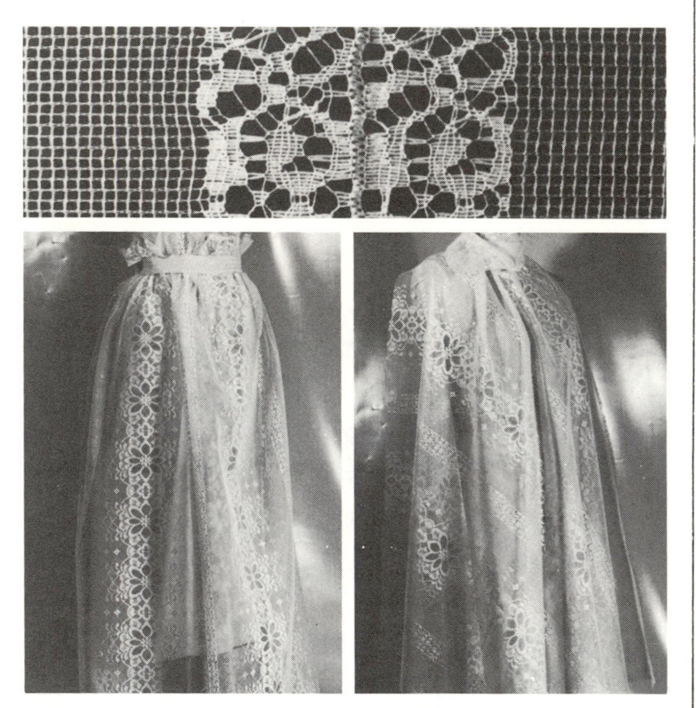

Filet-type Designed Lace/Crochet Styling

Fiber Content 100% polyester
Fabric Structure Raschel machine
Width 60 inches (152.4 cm)
Weight light
Hand stiff
Fabric Resource Native Textiles

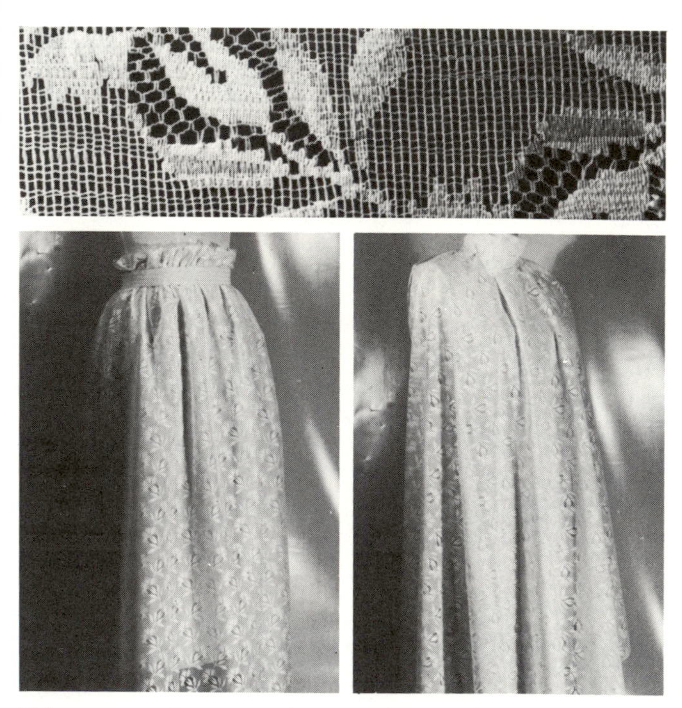

Filet-type Designed Lace/Darned Lace Styling

Fiber Content 100% nylon
Fabric Structure Raschel machine
Width 58 inches (147.3 cm)
Weight medium-light
Hand semi-crisp
Fabric Resource Native Textiles

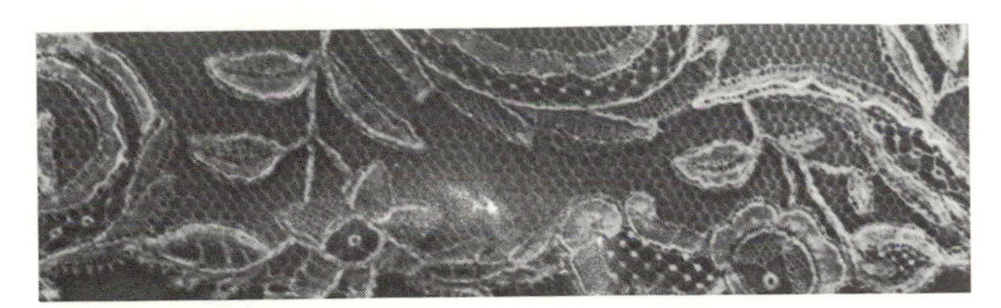

Honiton Lace

Description of Original Lace A lace characterized by fine, dainty motifs of flowers or leaf sprigs which are worked separately and then applied to a machine-made net ground by needlepoint or flat stitching.
Original Lace-making Method mixed lace: bobbin-made lace appliquéd to machine-made ground
Origin Flanders
Origin of Sample France
Sample Courtesy of Design Laboratories, Fashion Institute of Technology

Honiton-type Designed Lace

Fiber Content 100% nylon
Fabric Structure Raschel machine
Width 56 inches (142.2 cm)
Weight light
Hand soft
Fabric Resource Native Textiles

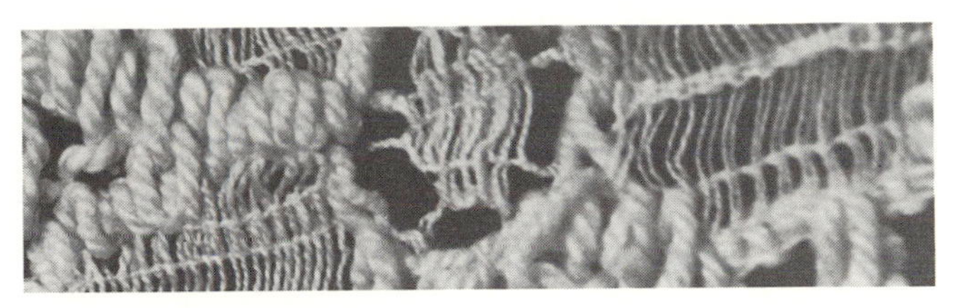

Knotted Lace

Description of Original Lace A lace made by knotting individual vertical threads to form a pattern. Knotting utilizes a combination of various knots. Single threads are allowed to hang loose when work is completed or are knotted to form motifs with a finished edge.
Original Lace-making Method knotting
Origin Italy, 15th century
Origin of Sample unknown
Sample Courtesy of Design Laboratories, Fashion Institute of Technology

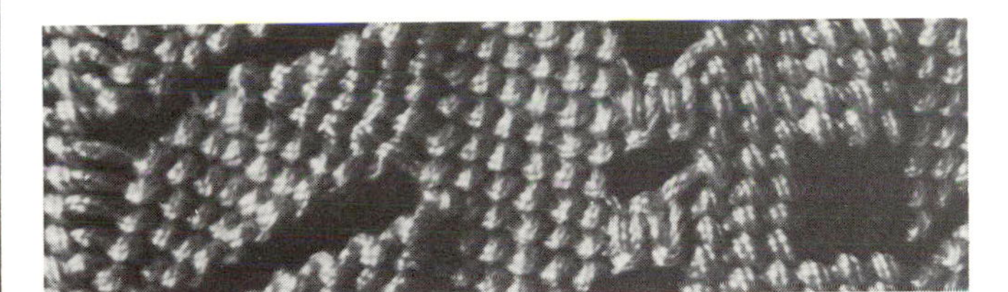

Macramé Lace

Fiber Content 100% jute
Fabric Structure hand knotted
Width determined by motif
Weight heavy
Hand coarse
Sample from Julie's Artisan Gallery (courtesy of author)

Machine/Nottingham Lace

Description of Original Lace A coarse heavy lace with intricate designs and motifs made with cotton or linen threads on a bobbinet lace-making machine. The gridwork effect of the lace with filled areas forming the design produces a slightly raised surface which resembles darned filet lace.
Original Lace-making Method bobbinet machine
Origin Nottingham, England
Origin of Sample unknown
Sample Courtesy of Design Laboratories, Fashion Institute of Technology

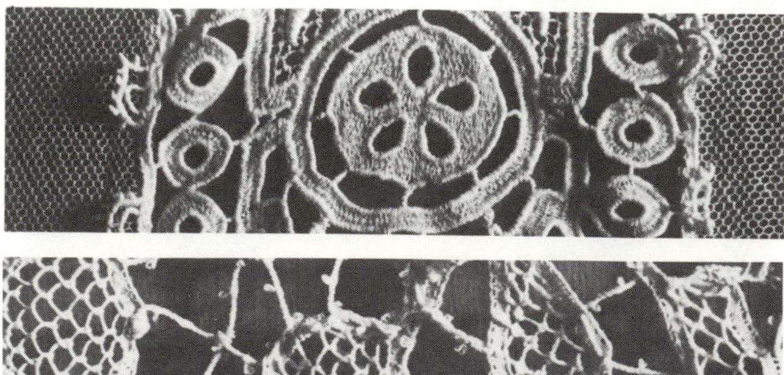

Plauen/Saxony Lace

Description of Original Lace A coarse machine-made lace utilizing two different fibers; one type for the embroidered design and another for the ground fabric. The design is applied by an embroidery machine to a ground of net or other fabric. When chemically treated, the ground fabric dissolves or is burnt-out leaving only the embroidered design which forms the lace. The coarse lace with overworked stitches forming the motifs, sprays or sprigs are joined one to the other. No ground or brides are used to connect the lace design.
Original Lace-making Method Saxony machine
Origin Switzerland, 1860
Origin of Sample unknown
Sample Courtesy of Design Laboratories, Fashion Institute of Technology

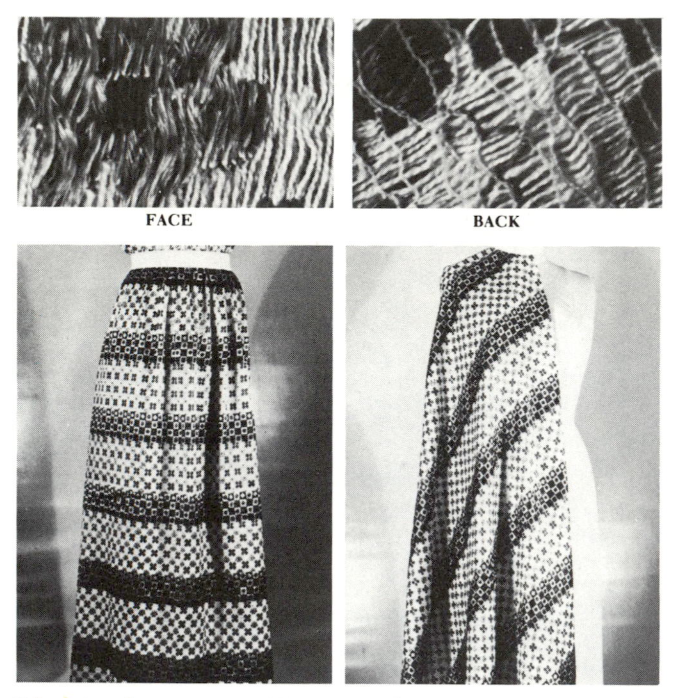

FACE BACK

Nottingham-type Designed Lace

Fiber Content 60% rayon/30% cotton/10% nylon
Fabric Structure bobbinet or Jacquard Raschel machine
Width 42 inches (106.7 cm)
Weight medium-heavy
Hand soft
Fabric Resource Whelan Lace

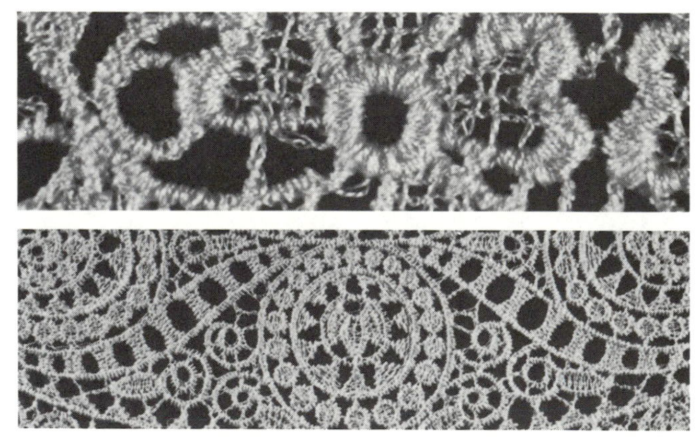

Plauen-type Designed/Burnt-out Lace

Fiber Content 100% cotton
Fabric Structure burnt-out method; Schiffli embroidery on chemically soluble ground fabric
Width all over design in 36-inch (91.4-cm) widths; component parts; strips
Weight medium to heavy
Hand soft to crisp depending on finish
Fabric Resource Associated Lace Corp.

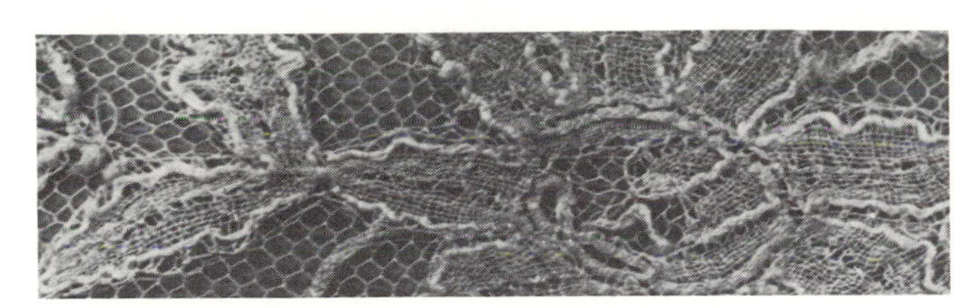

Point d'Angleterre

Description of Original Lace A combination lace characterized by a needlepoint design and bobbin-net ground. The needle lace is worked separately and joined to form a design. The ground areas, or brides, are worked in afterwards. The design joined to the net ground is outlined with plied threads forming a raised rib.
Original Lace-making Method mixed lace: needlepoint-lace motif on bobbin-made net ground
Origin Brussels
Origin of Sample Brussels
Sample Courtesy of Design Laboratories, Fashion Institute of Technology

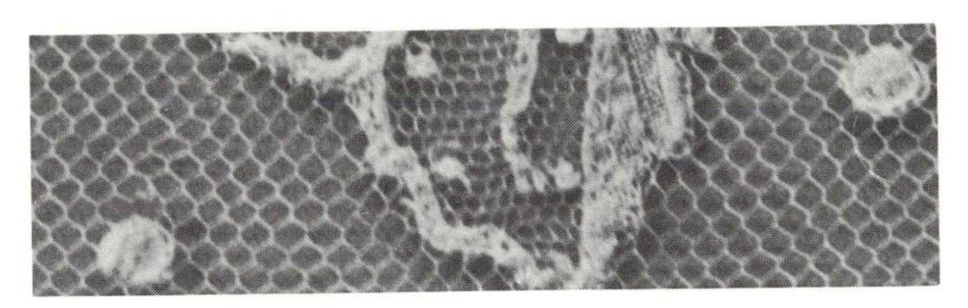

Normandy Lace

Description of Original Lace A simple dot, small oval or square embroidered on a bobbin-made net ground. Embroidery made by a Schiffli machine.
Original Lace-making Method mixed lace: machine embroidery on bobbin-made net ground
Origin France, 1884
Origin of Sample unknown
Sample Courtesy of Design Laboratories, Fashion Institute of Technology

Point d'Angleterre-type Designed Lace

Fiber Content 60% cotton/30% rayon/10% nylon
Fabric Structure Leavers machine
Width 45 inches (114.3 cm)
Weight medium
Hand crisp
Fabric Resource Whelan Lace

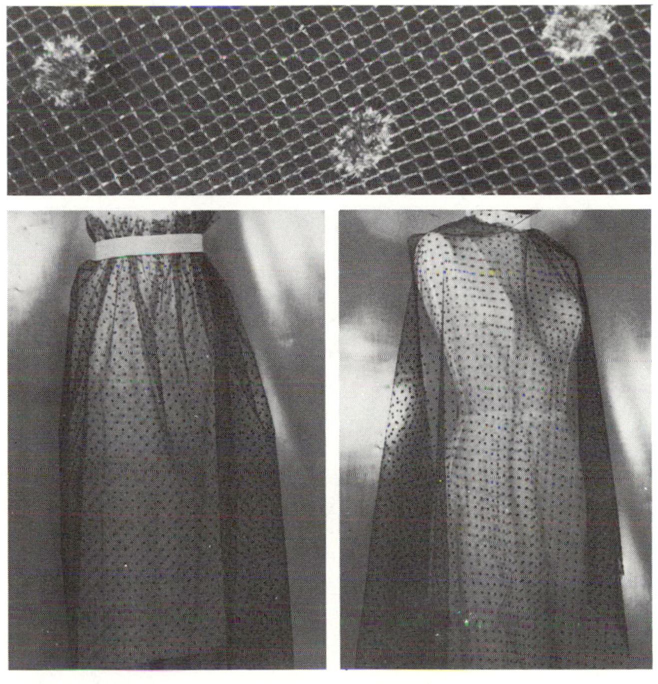

Point d'Esprit-type Designed Lace

Fiber Content 100% nylon
Fabric Structure Raschel machine net with adhesive-flocked dots
Width 54 inches (137.2 cm)
Weight light
Hand stiff
Fabric Resource Novik and Co. Inc.

Point de France

Description of Original Lace Flowers, baskets, birds or cherubs are worked into flat design motifs. The motifs are combined with or joined by hexagonal-shaped mesh. Picoted brides create a fond which connects the motifs. The defined motif is outlined with gimp to produce a dimensional effect.
Original Lace-making Method needlepoint
Origin France
Origin of Sample unknown
Sample Courtesy of Design Laboratories, Fashion Institute of Technology

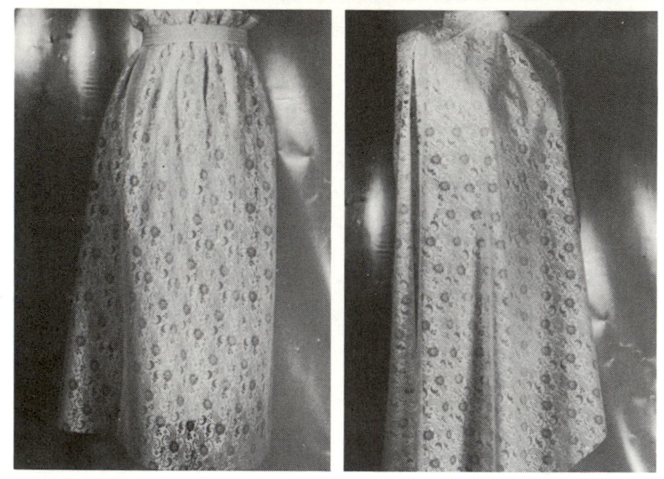

Point de France-type Designed Lace

Fiber Content 100% polyester
Fabric Structure Raschel machine
Width 56 inches (142.2 cm)
Weight medium-light
Hand semi-soft
Fabric Resource **Native Textiles**

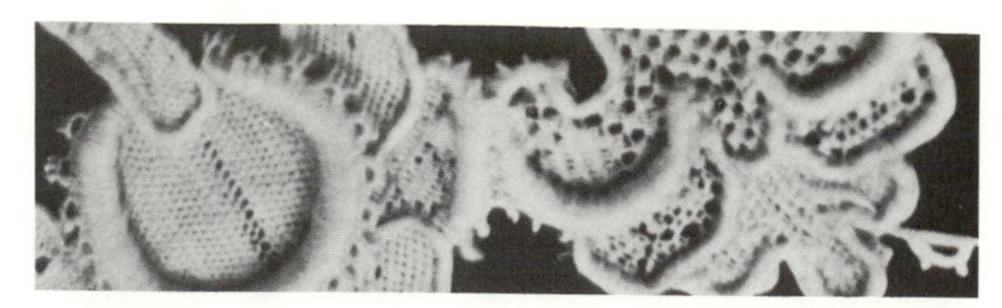

Point de Venice/Punto Avorio

Description of Original Lace A heavy, thick lace characterized by floral motifs designed in low relief. The built-up, raised-edge design area or motif is connected with irregularly placed bars of thread or picoted brides. The term Punto Avorio means a Venetian lace that is padded.
Original Lace-making Method needlepoint
Origin Venice, Italy
Origin of Sample unknown
Sample Courtesy of Design Laboratories, Fashion Institute of Technology

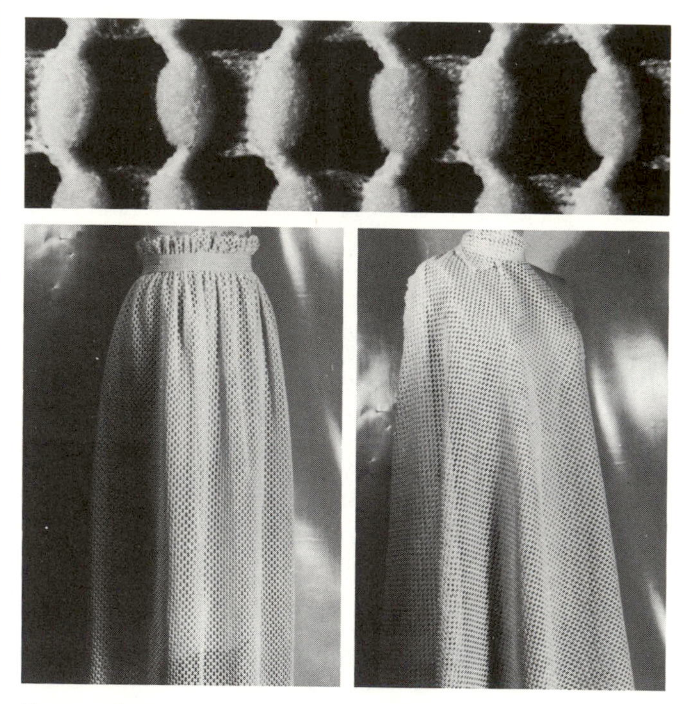

Point de Venice-type Designed Lace

Fiber Content 100% polyester
Fabric Structure Raschel machine
Width 52–54 inches (132.1–137.2 cm)
Weight medium
Hand soft; thick
Fabric Resource **Native Textiles**

Point Plat de Venice (bobbin lace)

Description of Original Lace A French Venetian lace *made with bobbins*. The flat designs of flowers and leaves, outlined with plied threads, are joined by *thread bars*. Plat is a term used to distinguish the flat treatment of Point de Venice or Venetian point from the raised type.
Original Lace-making Method bobbin
Origin France
Origin of Sample unknown
Sample Courtesy of Design Laboratories, Fashion Institute of Technology

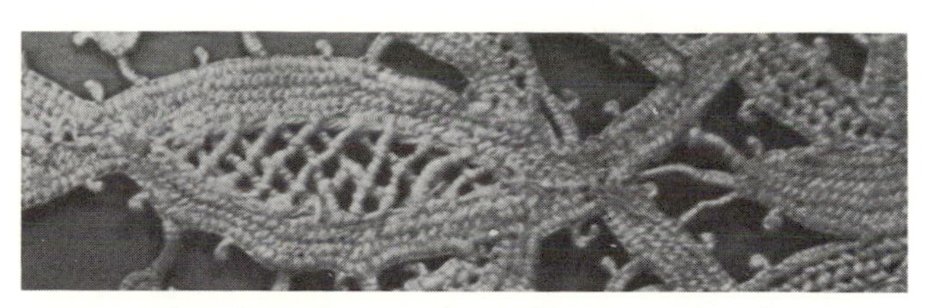

Point Plat de Venice (needlepoint lace)

Description of Original Lace An Italian Venetian lace made by needlepoint. The flowers and leaves forming the motif are made flat without relief, and are joined by *picoted brides*. The term point plat, meaning flat, is used to distinguish this lace from *gros point* meaning heavily raised.
Original Lace-making Method needlepoint
Origin Venice, Italy
Origin of Sample unknown
Sample Courtesy of Design Laboratories, Fashion Institute of Technology

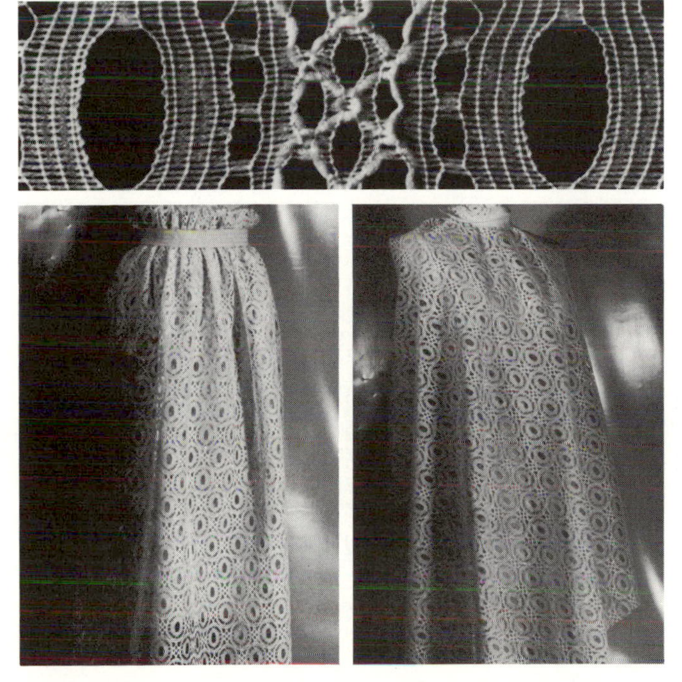

Point Plat de Venice-type Designed Lace (simulates bobbin lace)

Fiber Content 100% polyester
Fabric Structure Raschel machine
Width 60 inches (152.4 cm)
Weight medium-heavy
Hand stiff
Fabric Resource **Native Textiles**

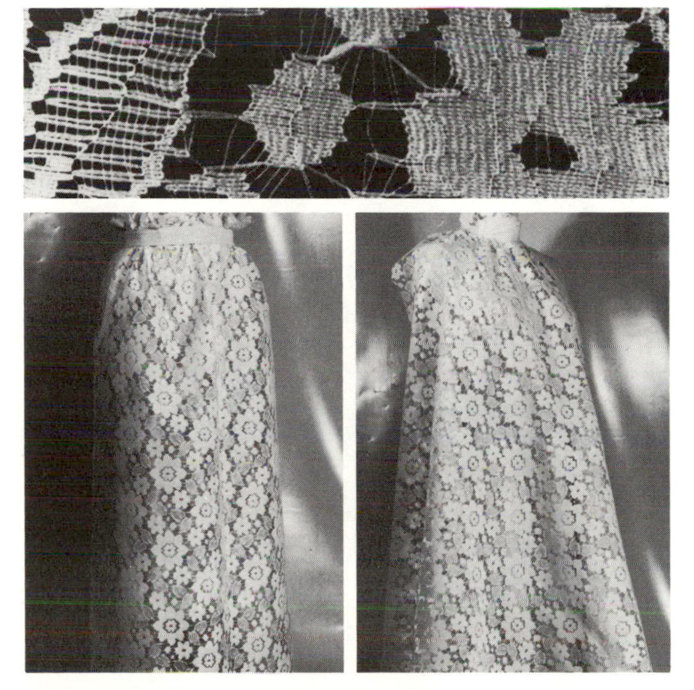

Point Plat de Venice-type Designed Lace (simulates needlepoint lace)

Fiber Content 100% polyester
Fabric Structure Raschel machine
Width 45 inches (114.3 cm)
Weight medium
Hand crisp
Fabric Resource **Native Textiles**

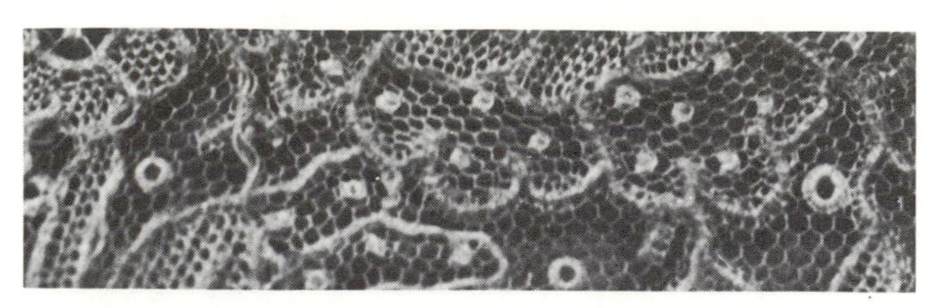

Rose Point (large motif)

Description of Original Lace A lace characterized by a large rose-like flower, leaf or scroll motif worked separately and arranged in a pattern. The designed pattern may be buttonhole-stitched onto a net ground or may be joined by brides. String cordonnet outlines the characteristic rose of the pattern and the superimposed petals.
Original Lace-making Method needlepoint
Origin Venice, Italy
Origin of Sample France, 18th century
Sample Courtesy of Design Laboratories, Fashion Institute of Technology

Rose Point-type Designed Lace (large motif)

Fiber Content 100% nylon
Fabric Structure Leavers machine
Width 50-54 inches (127-137.2 cm)
Weight medium-light
Hand semi-soft
Fabric Resource North American Lace Corp.

Rose Point (small motif)

Description of Original Lace A fine lace characterized by small rose-like flowers in raised work, connected by ornamental brides. The designed pattern may be buttonhole-stitched onto a net ground or may be joined by brides. String cordonnet outlines the characteristic rose of the pattern and the superimposed petals.
Original Lace-making Method needlepoint
Origin Venice, Italy
Origin of Sample France, 19th century
Sample Courtesy of Design Laboratories, Fashion Institute of Technology

Rose Point-type Designed Lace (small motif)

Fiber Content 100% nylon
Fabric Structure Raschel machine
Width 56 inches (142.2 cm)
Weight medium-light
Hand soft
Fabric Resource North American Lace Corp.

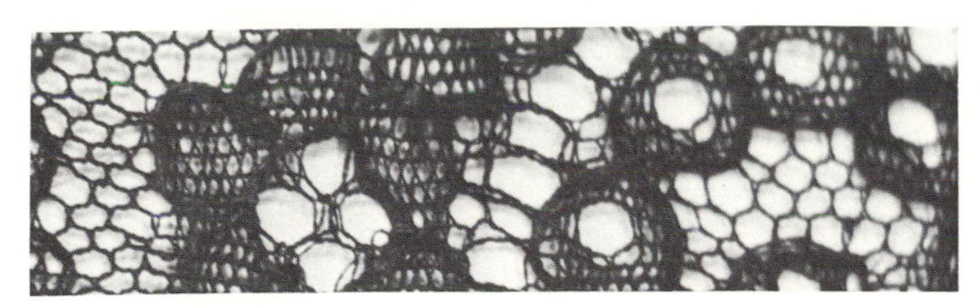

Point de Venice

Description of Original Lace A thin, filmy lace characterized by a flat surface and the shadowy appearance of the motif area. The effects of the coarsely worked areas and the fine net areas of shadow lace may be made in any design motif or character.
Original Lace-making Method needlepoint
Origin England; France; Germany
Origin of Sample Belgium, 19th century
Sample Courtesy of Design Laboratories, Fashion Institute of Technology

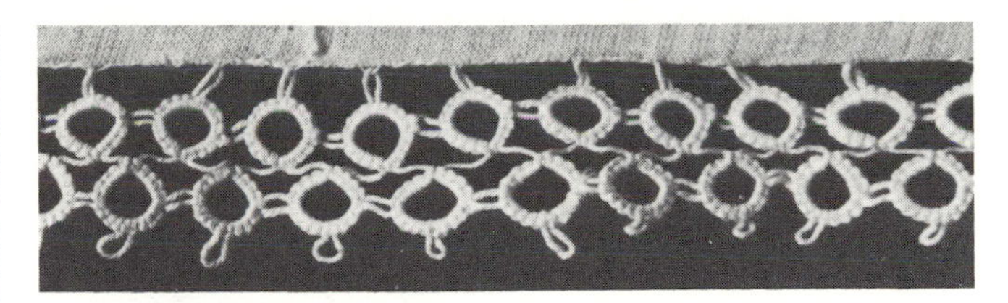

Tatting/Tatted Lace

Description of Original Lace A light, fine, fragile lace made with cotton or linen threads. A single thread wound in a shuttle is manipulated with hand and finger motions and worked into knots. The knotting procedure is formed into small loops which are developed into various designs, motifs or edgings.
Original Lace-making Method knotted; thread and shuttle worked with fingers
Origin Italy; Spain
Origin of Sample United States, 1875
Sample Courtesy of Design Laboratories, Fashion Institute of Technology

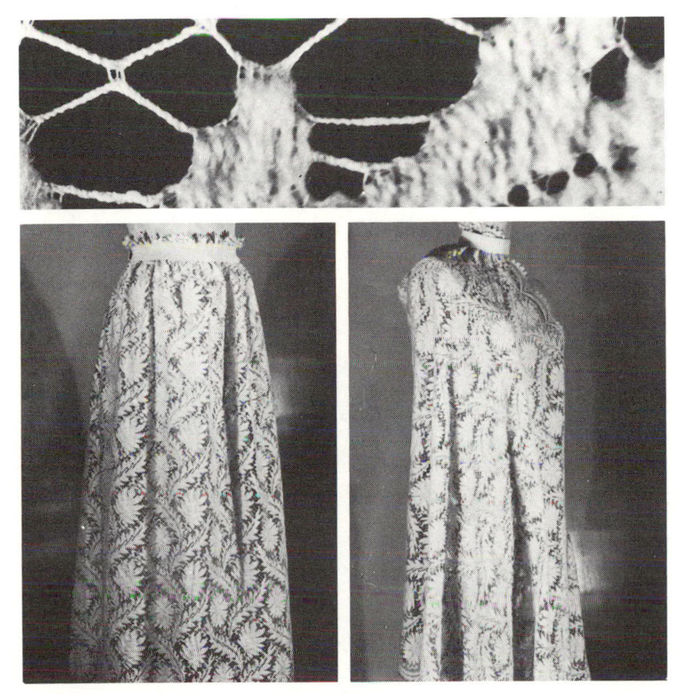

Shadow-type Designed Lace

Fiber Content 55% wool/45% cotton
Fabric Structure Leavers machine; 9 points
Width 30 inches (76.2 cm)
Weight medium-heavy
Hand soft
Fabric Resource **Whelan Lace**

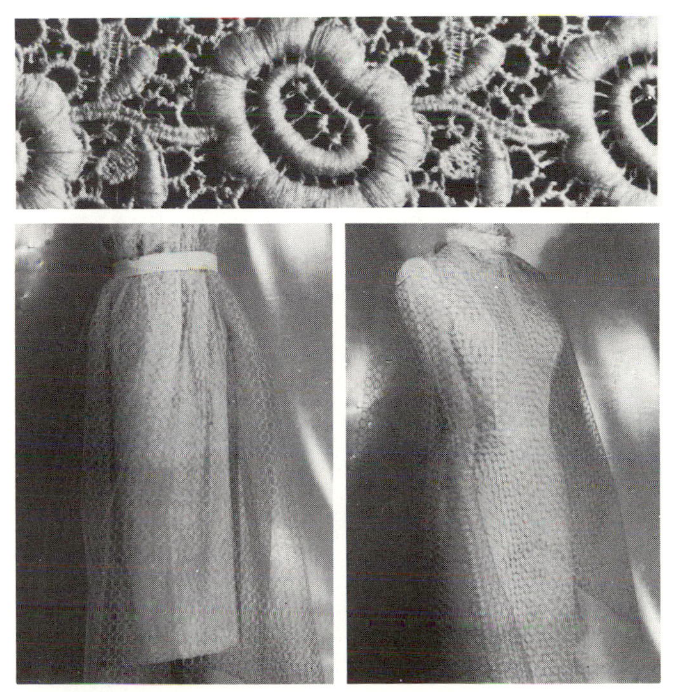

Tatting-type Designed Lace/Simulated Tatted Lace

Fiber Content nylon/acetate blend
Fabric Structure tricot machine
Width 36 inches (91.4 cm)
Weight light
Hand stiff
Fabric Resource **Native Textiles**

Val/Valenciennes Lace

Description of Original Lace A non-dimensional, fine, narrow lace worked in one piece in which the same threads form both the ground and the design. The trailing patterns of flowers, dots or sprigs are produced flat with the mesh ground. Val laces are produced with:

1. One scalloped edge (edgings);
2. Two straight edges (inserts);
3. Two picoted edges (banding);
4. Two scalloped edges (galloon).

Original Lace-making Method bobbin or needlepoint
Origin Venice, Italy
Origin of Sample Normandy, France, 19th century
Sample Courtesy of Design Laboratories, Fashion Institute of Technology

Val/Valenciennes-type Designed Lace

Fiber Content 100% nylon
Fabric Structure Raschel or Leavers machine
Width edgings from ⅛–4 inches (.32–10.16 cm)
Weight fine to lightweight
Hand soft to crisp depending on finish
Fabric Resources Native Textiles; North American Lace Corp.

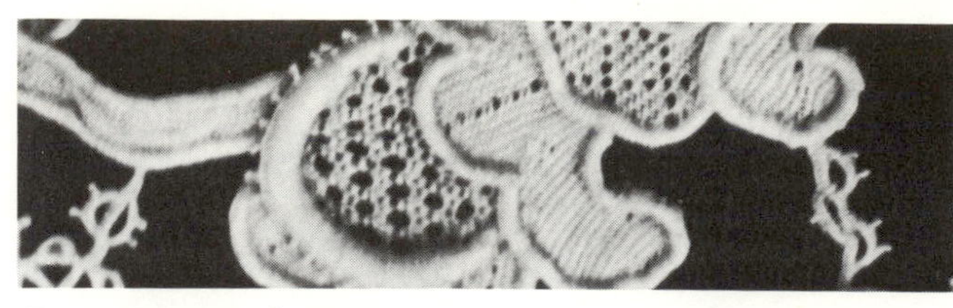

Gros Point de Venice

Description of Original Lace A thick, heavy needlepoint lace worked in high relief without a net ground. Gros point lace is characterized by large designs which are heavily worked with evenly spaced, close stitches producing a sculptured effect. The elaborate florals and scrolls, which form the solid area of the design, are connected by picoted brides. The term *gros point*, meaning heavily raised, is used to distinguish the lace from *point plat*, meaning flat.
Original Lace-making Method needlepoint
Origin Venice, Italy
Origin of Sample France, 16th century
Sample Courtesy of Design Laboratories, Fashion Institute of Technology

Venice-type/Gros Point-type Designed Lace

Fiber Content 100% cotton
Fabric Structure Schiffli machine; burnt-out method
Width determined by motif
Weight heavy
Hand coarse; firm
Fabric Resources Ambassador Lace and Embroidery Co. Inc.; Associated Lace Corp.

Performance Expectations

Based on fiber content, fabric structure, yarn type and finishes; subject to snagging and catching due to open areas of mesh or design; subject to tearing due to fragility

Recommended Construction Techniques & Aids

Hand Sewing Needles ball point, 5–10; beading, 10–13; cotton darners, 7–10; embroidery, 7–10; milliners, 7–10; sharps, 1–10
Machine Needles ball point, 9–11; round/set point, medium-fine 9–11
Threads silk, industrial size A; mercerized cotton, 50; cotton or poly-core waxed, 60 fine; cotton-polyester blend, extra fine; cotton-covered polyester core, all purpose; spun polyester, all purpose; nylon/Dacron, monocord A
Seams French; hairline; overedged; plain; safety-stitched; seams of lace; tissue-stitched
Seam Finishes net bound; Hong Kong bound; double-stitched; double-stitched and trimmed; double-ply or single-ply overcast; serging/single-ply overedged; untreated/plain; double-ply zigzag
Hems if edge is scalloped no hem finish is needed; bound/Hong Kong finish; net binding; double fold; horsehair; net faced; machine-stitched; overedged/turned flat; plain/turned flat
Pressing use pressing cloth; steam; safe temperature limit 325°F (164.1°C)
Care dry cleaning recommended; nylon laces used for lingerie or trimming may be laundered

13 ~ Net/Netting/Net-type Fabrics

Cable Net / Laundry Mesh
Fishnet
  Fine Hole Fishnet (soft finish)
  Small Hole Fishnet (crisp finish)
  Medium Hole Fishnet (stiff finish)
  Large Hole Fishnet (stretch properties)
Maline Net
Net
  Net
  Net with Yarn Embroidery
Point d'Esprit Net
Point d'Esprit Tulle
Tulle
Tulle Veiling / Illusion Veiling

A net fabric is characterized by a square, hexagonal or octagonal open-mesh construction. Natural, man-made or any combination fiber blend of thread, twine or yarn may be used to produce a net fabric.

Net or net-type fabrics may vary from:

- Fine to coarse
- Light to heavy weight
- Sheer and open to closed and heavy structure
- Soft to crisp hand
- Limp to stiff finish
- Regular- or irregular-sized openings
- Low to high point count construction

Quality and durability of net fabrics depend on:

- Fiber content
- Yarn type and construction
- Method of construction employed
- Number of point count
- Finishing processes applied
- End use

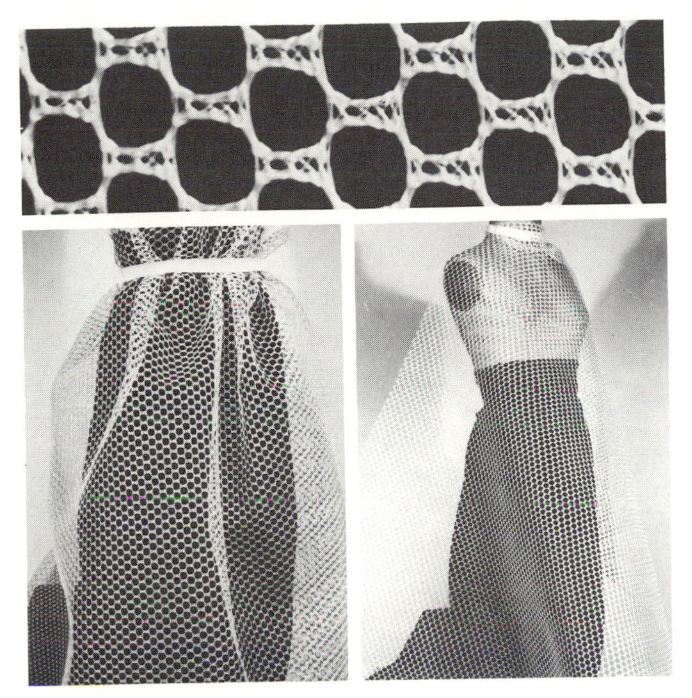

Cable Net/Laundry Mesh

Fiber Content 100% nylon
Yarn Type filament fiber
Yarn Construction 140-denier yarn; textured; slack twist
Fabric Structure Raschel knit
Finishing Processes flame retardant; heat-set stabilizing
Color Application solution dyed; yarn dyed
Width 20–60 inches (50.8–152.4 cm)
Weight medium
Hand soft to stiff depending on finish
Texture harsh; open-mesh knotted pattern; rough
Performance Expectations abrasion resistant; dimensional stability; durable for its type; resilient; high tensile strength
Drapability Qualities falls into wide flares with soft finish; falls into wide cones with stiff finish; fullness maintains crisp effect or soft fall depending on finish

Recommended Construction Aids & Techniques

Hand Needles betweens, 3; cotton darners, 1–5; yarn darners, 14–18; embroidery, 1–3; milliners, 3–5; sharps, 1–4
Machine Needles ball point, 11
Threads silk, industrial size A-B; poly-core waxed, 50; cotton-covered polyester core, all purpose; spun polyester, all purpose; nylon/Dacron, monocord A
Hems single-fold or double-fold bias binding; edge-stitched only; machine-stitched; double machine-stitched; overedged
Seams hairline; overedged; plain; safety-stitched; tissue-stitched; zigzagged
Seam Finishes net bound; double-ply bound; double-stitched; double-stitched and trimmed; serging/single-ply overedged; untreated/plain; double-ply zigzag
Pressing use pressing cloth; safe temperature limit 350°F (178.1°C)
Care launder
Fabric Resource **Kortex Associates Inc.**

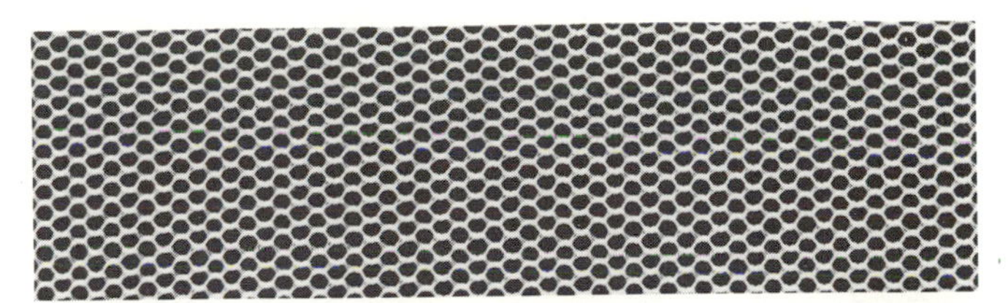

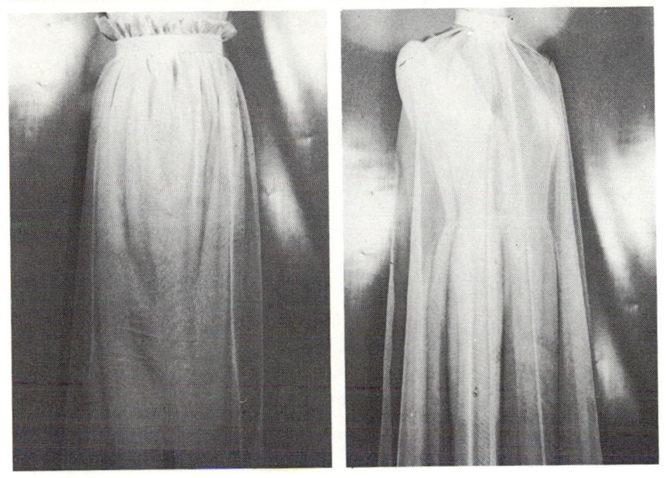

Fine Hole Fishnet (soft finish)

Fiber Content 100% nylon
Yarn Type filament fiber
Yarn Construction monofilament
Fabric Structure Raschel knit
Finishing Processes flame retardant; heat-set stabilizing; softening
Color Application solution dyed; yarn dyed
Width 45 inches (114.3 cm)
Weight light
Hand fine; sandy; soft
Texture flat; harsh; open mesh
Performance Expectations abrasion resistant; dimensional stability; durable for its type; resilient; high tensile strength
Drapability Qualities falls into soft flares; accommodates fullness by gathering; fullness retains soft graceful fall

Recommended Construction Aids & Techniques

Hand Needles betweens, 3; cotton darners, 1–5; yarn darners, 14–18; embroidery, 1–3; milliners, 3–5; sharps, 1–4
Machine Needles ball point, 9
Threads silk, industrial size A-B; poly-core waxed, 50; cotton-covered polyester core, all purpose; spun polyester, all purpose; nylon/Dacron, monocord A
Hems edge-stitched only; flat/plain; horsehair; machine-stitched; hand- or machine-rolled; wired
Seams French; hairline; overedged; plain; tissue-stitched
Seam Finishes net bound; double-ply bound; double-stitched; double-stitched and trimmed; serging/single-ply overedged; untreated/plain; double-ply zigzag
Pressing use pressing cloth; safe temperature limit 350°F (178.1°C)
Care launder; dry clean
Fabric Resource **Associated Lace Corp.**

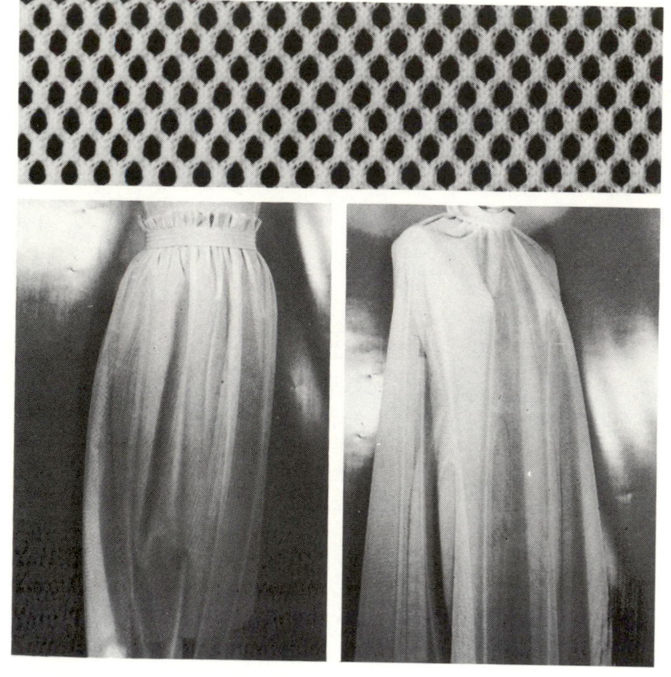

Small Hole Fishnet (crisp finish)

Fiber Content 100% nylon
Yarn Type filament fiber
Yarn Construction fine-denier yarn; textured; slack twist
Fabric Structure Raschel knit
Finishing Processes flame retardant; heat-set stabilizing; stiffening
Color Application solution dyed; yarn dyed
Width 72 inches (182.9 cm)
Weight medium
Hand crisp; harsh; rigid
Texture boardy; open mesh; rough
Performance Expectations abrasion resistant; dimensional stability; durable for its type; resilient; high tensile strength
Drapability Qualities falls into crisp flares; accommodates fullness by gathering; fullness maintains crisp effect

Recommended Construction Aids & Techniques

Hand Needles betweens, 3; cotton darners, 1–5; yarn darners, 14–18; embroidery, 1–3; milliners, 3–5; sharps, 1–4
Machine Needles ball point, 9–11
Threads silk, industrial size A-B; poly-core waxed, 50; cotton-covered polyester core, all purpose; spun polyester, all purpose; nylon/Dacron, monocord A
Hems edge-stitched only; flat/plain; machine-stitched; double machine-stitched; overedged
Seams hairline; overedged; plain; safety-stitched; tissue-stitched; zigzagged
Seam Finishes net bound; double-ply bound; double-stitched; double-stitched and trimmed; serging/single-ply overedged; untreated/plain; double-ply zigzag
Pressing use pressing cloth; safe temperature limit 350°F (178.1°C)
Care launder; dry clean
Fabric Resource Gehring Textiles

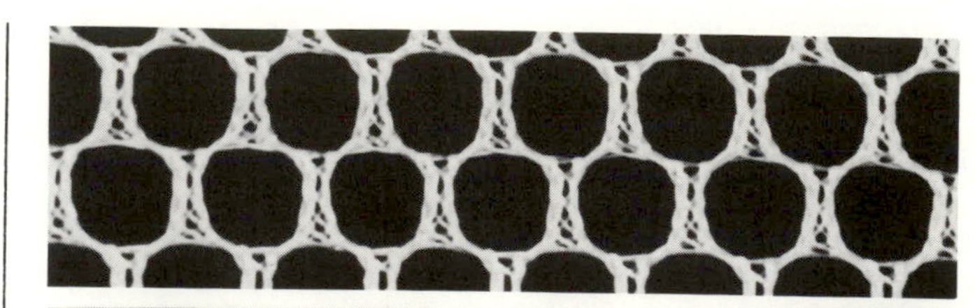

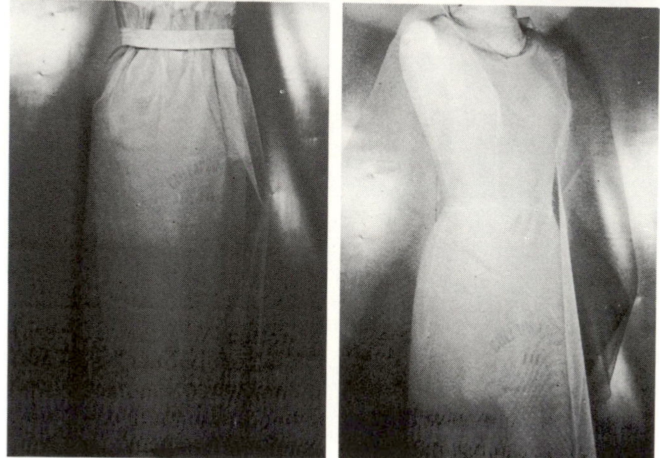

Medium Hole Fishnet (stiff finish)

Fiber Content 100% nylon
Yarn Type filament fiber
Yarn Construction fine-denier yarn; textured; slack twist
Fabric Structure Raschel knit
Finishing Processes flame retardant; heat-set stabilizing; stiffening
Color Application solution dyed; yarn dyed
Width 72 inches (182.9 cm)
Weight medium
Hand nonelastic; harsh; rigid; stiff
Texture[1] open mesh; rough; scratchy
Performance Expectations abrasion resistant; dimensional stability; durable for its type; resilient; high tensile strength
Drapability Qualities falls into wide cones; accommodates fullness by pleating, gathering; fullness maintains stiff bouffant effect; better utilized if used flat or darted

Recommended Construction Aids & Techniques

Hand Needles betweens, 3; cotton darners, 1–5; yarn darners, 14–18; embroidery, 1–3; milliners, 3–5; sharps, 1–4
Machine Needles ball point, 9–11
Threads silk, industrial size A-B; poly-core waxed, 50; cotton-covered polyester core, all purpose; spun polyester, all purpose; nylon/Dacron, monocord A
Hems single-fold or double-fold bias binding; edge-stitched only; machine-stitched; double machine-stitched; overedged
Seams hairline; overedged; plain; safety-stitched; tissue-stitched; zigzagged
Seam Finishes net bound; double-ply bound; double-stitched; double-stitched and trimmed; serging/single-ply overedged; untreated/plain; double-ply zigzag
Pressing use pressing cloth; safe temperature limit 350°F (178.1°C)
Care dry clean
Fabric Resources Gehring Textiles; Kortex Associates Inc.

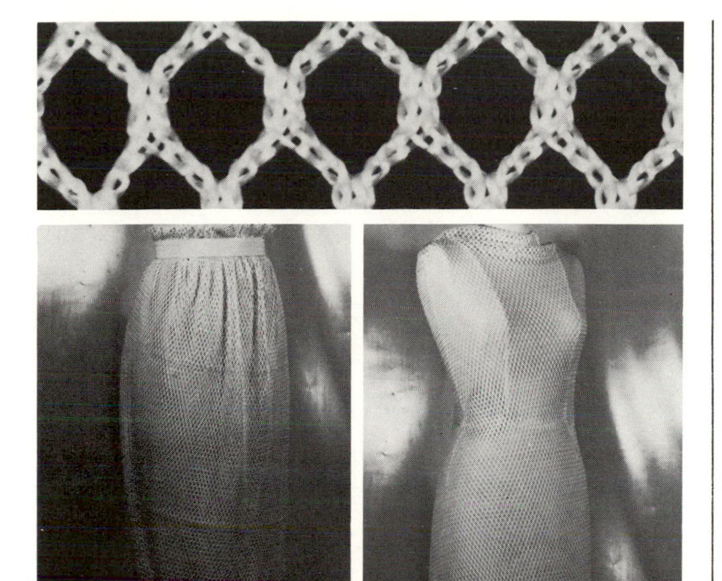

Large Hole Fishnet (stretch properties)

Fiber Content 100% nylon
Yarn Type filament fiber
Yarn Construction fine-denier yarn; textured; slack twist
Fabric Structure Raschel knit
Finishing Processes flame retardant; heat-set stabilizing; softening
Color Application solution dyed; yarn dyed
Width 45 inches (114.3 cm)
Weight medium-heavy
Hand pliable; soft; stretch in crosswise and lengthwise directions
Texture loose; large open mesh; rough
Performance Expectations abrasion resistant; dimensional stability; durable for its type; resilient; stretch properties; high tensile strength
Drapability Qualities falls into soft flares; clings to body contour; can be shaped or molded by stretching

Recommended Construction Aids & Techniques

Hand Needles betweens, 3; cotton darners, 1-5; yarn darners, 14-18; embroidery, 1-3; milliners, 3-5; sharps, 1-4
Machine Needles ball point, 14
Threads silk, industrial size A-B; poly-core waxed, 50; cotton-covered polyester core, all purpose; spun polyester, all purpose; nylon/Dacron, monocord A
Hems single-fold or double-fold bias binding; edge-stitched only; machine-stitched; double machine-stitched; overedged
Seams hairline; overedged; plain; safety-stitched; tissue-stitched; zigzagged
Seam Finishes net bound; double-ply bound; double-stitched; double-stitched and trimmed; serging/single-ply overedged; untreated/plain
Pressing use pressing cloth; safe temperature limit 350°F (178.1°C)
Care launder; dry clean
Fabric Resource Gehring Textiles

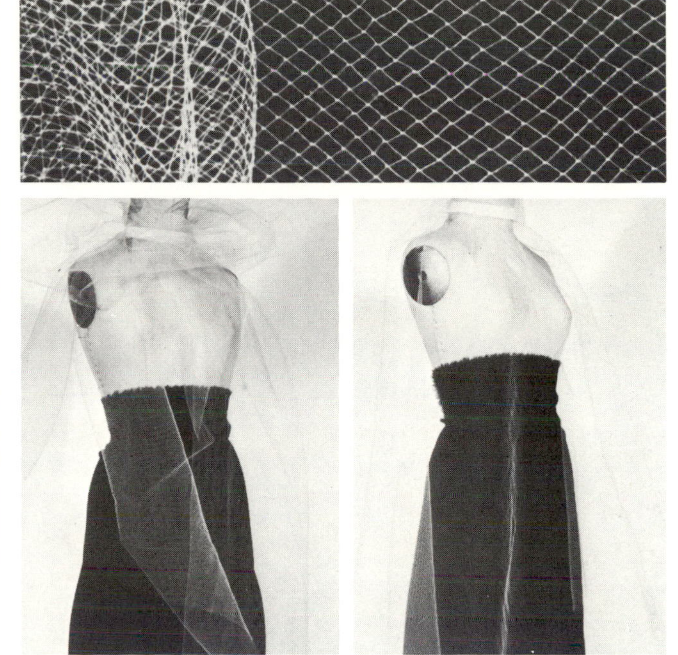

Maline Net

Fiber Content 100% nylon
Yarn Type filament fiber
Yarn Construction monofilament
Fabric Structure Raschel knit
Finishing Processes flame retardant; heat-set stabilizing; stiffening
Color Application solution dyed; yarn dyed
Width 27 inches (68.6 cm)
Weight fine to lightweight
Hand crisp; fine; fragile
Texture harsh; hexagonal-shaped open mesh
Performance Expectations subject to snagging and tearing due to fragile yarn and structure; subject to wrinkles, creases, folds
Drapability Qualities falls into stiff cones; accommodates fullness by pleating, gathering; fullness maintains crisp bouffant effect

Recommended Construction Aids & Techniques

Hand Needles betweens, 3; cotton darners, 5-10; embroidery, 3-7; milliners, 5-8; sharps, 1-4
Machine Needles ball point, 9
Threads silk, industrial size A; poly-core waxed, 60; cotton-covered polyester core, extra fine; spun polyester, fine; nylon/Dacron, monocord A
Hems plain/untreated raw edge
Seams French; hairline; overedged; plain; tissue-stitched
Seam Finishes net bound; self bound; double-stitched; double-stitched and trimmed; untreated/plain
Pressing use pressing cloth; safe temperature limit 350°F (178.1°C)
Care dry clean
Fabric Resource Novik and Co. Inc.

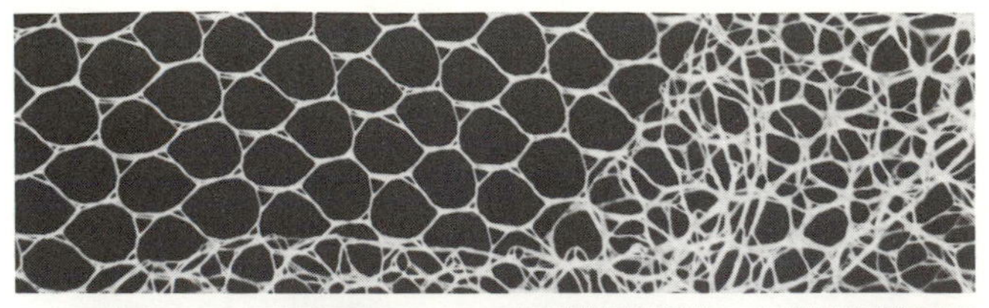

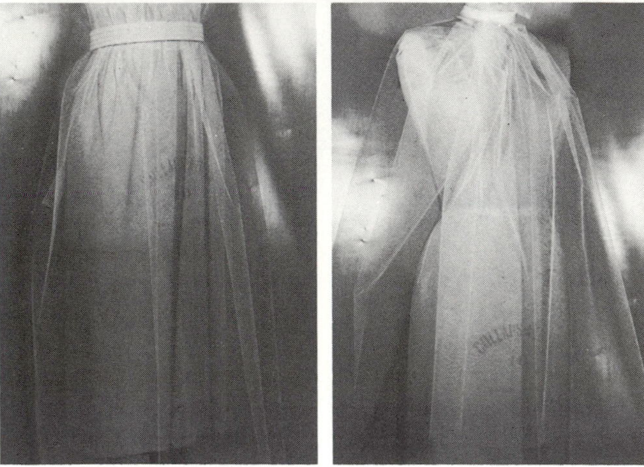

Net

Fiber Content 100% nylon
Yarn Type filament fiber
Yarn Construction 40-denier yarn; monofilament
Fabric Structure 2-bar Raschel knit
Finishing Processes flame retardant; heat-set stabilizing; stiffening
Color Application solution dyed; yarn dyed
Width 70-72 inches (177.8-182.9 cm)
Weight fine to lightweight; 20 points per inch
Hand nonelastic; fine; stiff
Texture harsh; open mesh
Performance Expectations abrasion resistant; dimensional stability; durable for its type; no elasticity
Drapability Qualities falls into crisp cones; accommodates fullness by pleating, gathering; fullness maintains crisp bouffant effect

Recommended Construction Aids & Techniques

Hand Needles betweens, 3; cotton darners, 5-10; embroidery, 3-7; milliners, 5-8; sharps, 1-4
Machine Needles ball point, 9
Threads silk, industrial size A; poly-core waxed, 60; cotton-covered polyester core, extra fine; spun polyester, fine; nylon/Dacron, monocord A
Hems plain/untreated raw edge
Seams French; hairline; overedged; plain; tissue-stitched
Seam Finishes net bound; self bound; double-stitched; double-stitched and trimmed; untreated/plain
Pressing use pressing cloth; safe temperature limit 350°F (178.1°C)
Care dry clean
Fabric Resources **Kortex Associates Inc.; Novik and Co. Inc.**

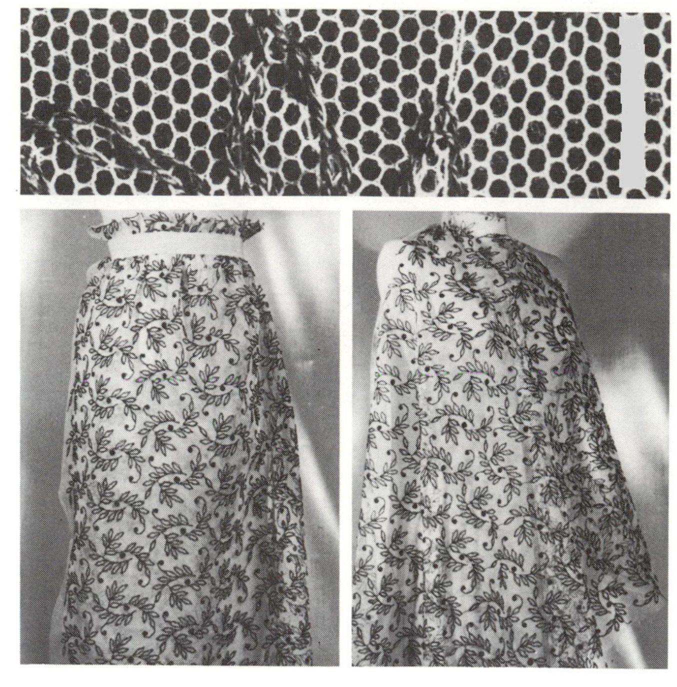

Net with Yarn Embroidery

Fiber Content 100% cotton
Yarn Type combed staple
Yarn Construction conventional; tight twist
Fabric Structure Schiffli embroidery on net structure base
Finishing Processes flame retardant; mercerizing; stiffening
Color Application base: piece dyed; embroidery yarn: skein or packaged dyed
Width 30 inches (76.2 cm)
Weight fine to lightweight
Hand crisp; delicate; fragile
Texture base: open mesh; embroidery yarn forms high and low raised pattern on surface
Performance Expectations air permeable; durable; elongation; high tensile strength; subject to wrinkling and creasing; raised embroidered design subject to snagging
Drapability Qualities falls into wide cones; accommodates fullness by gathering; fullness maintains crisp bouffant effect

Recommended Construction Aids & Techniques

Hand Needles beading, 10-13; betweens, 7; embroidery, 9-10; sharps, 9-10
Machine Needles ball point, 9
Threads silk, industrial size A; spun polyester, fine; nylon/Dacron, monocord A
Hems bound/Hong Kong finish; net binding; faced; horsehair; machine-stitched
Seams hairline; plain; tissue-stitched
Seam Finishes net bound; Hong Kong bound; double-stitched; double-stitched and trimmed; untreated/plain
Pressing use pressing cloth; safe temperature limit 400°F (206.1°C)
Care dry clean
Fabric Resource **Embroidery Council of America**

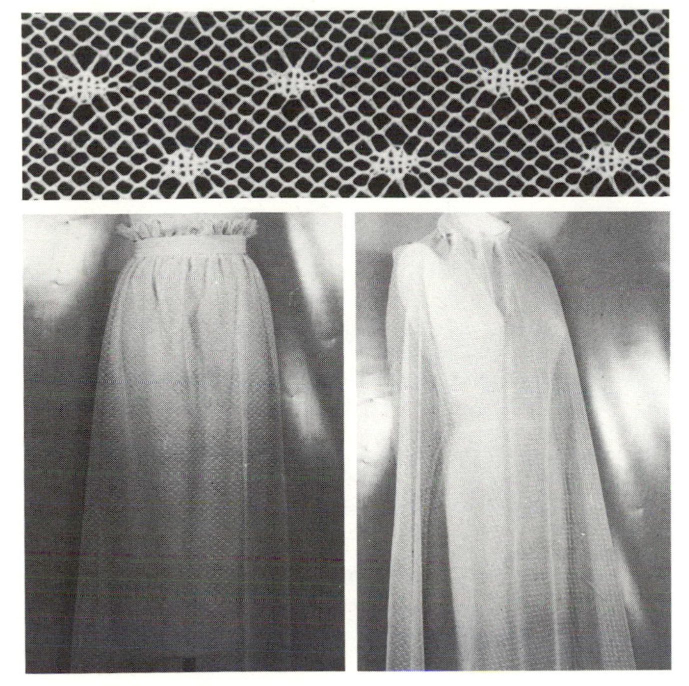

Point d'Esprit Net

Fiber Content 100% polyester
Yarn Type filament fiber
Yarn Construction fine-denier yarn; textured; low twist
Fabric Structure Raschel knit; pinching of yarns produces dot
Finishing Processes flame retardant; heat-set stabilizing
Color Application solution dyed; yarn dyed
Width 45 inches (114.3 cm)
Weight fine to lightweight
Hand crisp; delicate; rough
Texture open mesh with widely spaced small dots, ovals or squares
Performance Expectations abrasion resistant; dimensional stability; durable; elasticity; elongation; resilient; high tensile strength
Drapability Qualities falls into moderately crisp flares; accommodates fullness by gathering; fullness retains moderately soft fall

Recommended Construction Aids & Techniques

Hand Needles betweens, 3; cotton darners, 5–10; embroidery, 3–7; milliners, 5–8; sharps, 1–4
Machine Needles ball point, 9
Threads silk, industrial size A; poly-core waxed, 60; cotton-covered polyester core, extra fine; spun polyester, fine; nylon/Dacron, monocord A
Hems edge-stitched only; horsehair; machine-stitched; untreated raw edge; hand- or machine-rolled; wired
Seams hairline; overedged; plain; safety-stitched; tissue-stitched; zigzagged
Seam Finishes net bound; double-ply bound; double-stitched; double-stitched and trimmed; serging/single-ply overedged; untreated/plain; double-ply zigzag
Pressing use pressing cloth; safe temperature limit 350°F (178.1°C)
Care launder; dry clean
Fabric Resources **Associated Lace Corp.; Kortex Associates Inc.; Native Textiles**

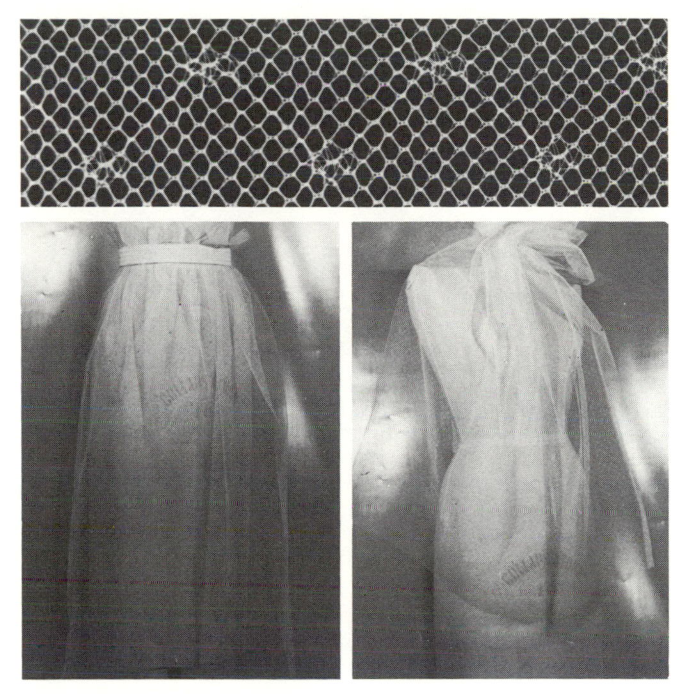

Point d'Esprit Tulle

Fiber Content 100% nylon
Yarn Type filament fiber
Yarn Construction 15-denier yarn; monofilament
Fabric Structure tricot knit; pinching of yarns produces dot
Finishing Processes flame retardant; heat-set stabilizing; stiffening
Color Application solution dyed; yarn dyed
Width 45 inches (114.3 cm)
Weight fine to lightweight
Hand crisp; delicate; rough
Texture open mesh with widely spaced small dots, ovals or squares
Performance Expectations abrasion resistant; dimensional stability; durable; elasticity; elongation; resilient; high tensile strength
Drapability Qualities falls into stiff wide cones; accommodates fullness by gathering; fullness maintains stiff bouffant effect

Recommended Construction Aids & Techniques

Hand Needles betweens, 3; cotton darners, 5–10; embroidery, 3–7; milliners, 5–8; sharps, 1–4
Machine Needles ball point, 9
Threads silk, industrial size A; poly-core waxed, 60; cotton-covered polyester core, extra fine; spun polyester, fine; nylon/Dacron, monocord A
Hems edge-stitched only; horsehair; machine-stitched; untreated raw edge; hand- or machine-rolled; wired
Seams hairline; overedged; plain; safety-stitched; tissue-stitched; zigzagged
Seam Finishes net bound; double-ply bound; double-stitched; double-stitched and trimmed; serging/single-ply overedged; untreated/plain; double-ply zigzag
Pressing use pressing cloth; safe temperature limit 350°F (178.1°C)
Care launder; dry clean
Fabric Resources **Kortex Associates Inc.; Native Textiles**

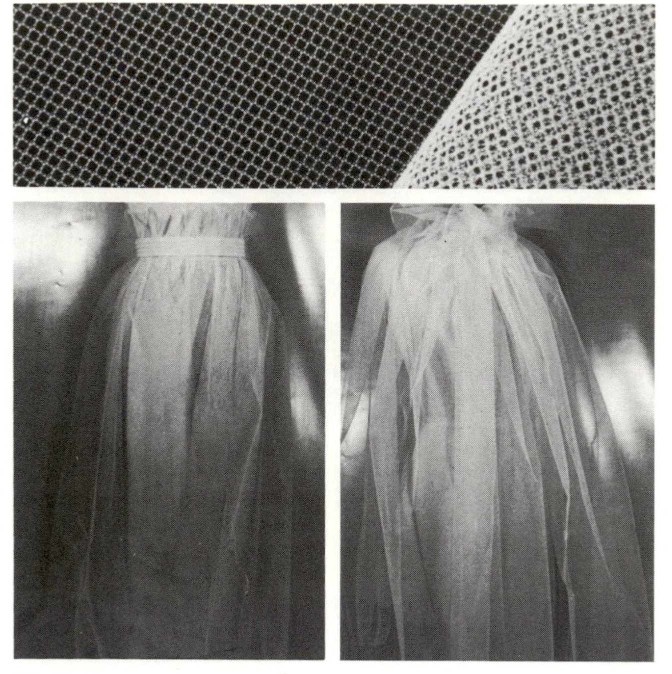

Tulle

Fiber Content 100% nylon
Yarn Type filament fiber
Yarn Construction 15-denier yarn; monofilament
Fabric Structure tricot knit
Finishing Processes flame retardant; heat-set stabilizing
Color Application solution dyed; yarn dyed
Width 52-54 inches (132.1-137.2 cm)
Weight fine to lightweight; 32 points per inch
Hand semi-crisp; fragile
Texture harsh; hexagonal-shaped open mesh
Performance Expectations abrasion resistant; dimensional stability; durable for its type; subject to snagging and tearing due to yarn and structure
Drapability Qualities falls into stiff flares; accommodates fullness by gathering; fullness maintains crisp effect

Recommended Construction Aids & Techniques

Hand Needles betweens, 3; cotton darners, 5-10; embroidery, 3-7; milliners, 5-8; sharps, 1-4
Machine Needles ball point, 6
Threads silk, industrial size A; poly-core waxed, 60; cotton-covered polyester core, extra fine; nylon/Dacron, monocord A
Hems untreated raw edge
Seams French; hairline; overedged; plain; tissue-stitched
Seam Finishes net bound; self bound; double-stitched; double-stitched and trimmed; untreated/plain
Pressing use pressing cloth; safe temperature limit 350°F (178.1°C)
Care dry clean
Fabric Resources **Gehring Textiles; Kortex Associates Inc.; Novik and Co. Inc.**

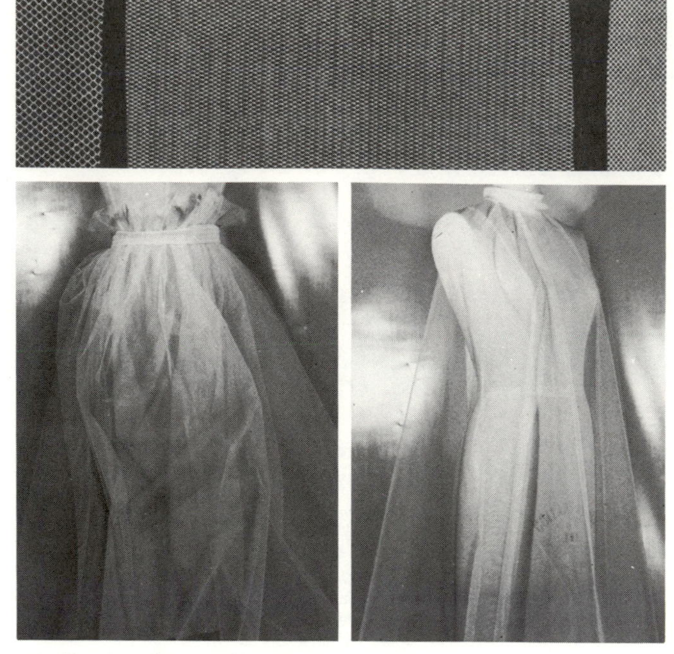

Tulle Veiling/Illusion Veiling

Fiber Content 100% nylon
Yarn Type filament fiber
Yarn Construction 15-denier yarn; monofilament
Fabric Structure tricot knit
Finishing Processes flame retardant; heat-set stabilizing
Color Application solution dyed; yarn dyed
Width 52-54 inches (132.1-137.2 cm)
Weight fine to lightweight; 48 points per inch
Hand semi-crisp; delicate; fragile
Texture harsh; hexagonal-shaped open mesh
Performance Expectations abrasion resistant; dimensional stability; durable for its type; subject to snagging and tearing due to fragile yarn and structure
Drapability Qualities falls into moderately crisp flares; accommodates fullness by gathering; fullness maintains moderately crisp effect

Recommended Construction Aids & Techniques

Hand Needles betweens, 3; cotton darners, 5-10; embroidery, 3-7; milliners, 5-8; sharps, 1-4
Machine Needles ball point, 9
Threads silk, industrial size A; poly-core waxed, 60; cotton-covered polyester core, extra fine; spun polyester, fine; nylon/Dacron, monocord A
Hems untreated raw edge
Seams French; hairline; overedged; plain; tissue-stitched
Seam Finishes net bound; self bound; double-stitched; double-stitched and trimmed; untreated/plain
Pressing use pressing cloth; safe temperature limit 350°F (178.1°C)
Care dry clean
Fabric Resources **Gehring Textiles; Kortex Associates Inc.; Novik and Co. Inc.**

14 ~ Pile Surface/Pile-type Fabrics

Corduroy
- Feathercord Corduroy
- Fine Wale / Pinwale Corduroy
- Mid / Medium / Regular Wale Corduroy
- Thick-set Corduroy
- Wide Wale Corduroy
- Broad Wale Corduroy
- Novelty Wale Corduroy

Fleece
- Nap-faced Fleece
- Baby Bunting Fleece
- Sweatshirt Fleece

Friezé (Frez) / Shadow Wale / Ribless Corduroy

Plush

Terry Cloth
- Double-faced Terry Cloth (loop face & back) / Turkish Toweling
- Double-faced Terry Cloth (loop face; velour back) (close up only)
- Jacquard Double-faced Terry Cloth (close up only)
- Sculptured Terry Cloth
- Terry Cloth Chenille (diagonal pattern)
- Terry Cloth Chenille (overall pattern)
- Stretch / Single-faced Terry Cloth

Tufted Cloth / Tufting
- Chenille Candlewick
- Chenille Velvet / Velcle®
- Chenille Yarn Cloth

Velour
- Velour
- Sealskin Velour

Velvet
- Velvet Brocade
- Cotton Velvet
- Crushed Velvet (shiny face & woven ground)
- Crushed Velvet (dull face & knit ground)
- Cut / Beaded Velvet
- Embossed / Sculptured Velvet
- Flocked / Simulated Velvet
- Lyons Velvet
- Lyons Velvet (dull face)
- Panne Velvet (knit ground)
- Panne Velvet (woven ground)
- Transparent / Chiffon Velvet
- Velvet Cord
- Velvet Pinstripe Cord

Velveteen

Simulated Fur Pile Cloth / High Pile Cloth
- Astrakhan / Persian Lamb Cloth
- Poodle / Poodle Fur Cloth
- Shag
- Sherpa / Sheared Lamb Cloth
- Teddy Bear
- Fox Cloth (close up only)
- Leopard Cloth (close up only)
- Lynx Cloth (close up only)
- Ocelot Cloth (close up only)
- Pieced Rabbit Cloth (close up only)
- Raccoon Cloth (close up only)
- Ranch Mink Cloth (close up only)
- Seal Cloth (close up only)
- Skunk Cloth (close up only)

Pile fabrics or pile surface fabrics are characterized by a raised or protruding surface from the base or ground of the fabric. The raised surface or pile is created by the use of additional warp or filling yarns during the construction of the fabric. This unit includes pile structure fabrics with:

- Looped surface on one or both sides
- Cut pile surface
- Crushed pile surface
- Patterned cut and uncut loops
- All over or patterned tufts
- High-and-low or thick-and-thin surface

Pile fabrics or pile surface fabrics may be constructed on a:

- Plain or twill weave ground
- Knitted- or malimo-based fabric
- Plain ground utilizing needling action

Pile fabrics or pile surface fabrics may be made with:

- Natural fiber blends
- Man-made fiber blends
- Natural and man-made fiber blends
- Mixed or combined yarns of different fiber origins
- Warp and filling yarns of different fiber origins
- Base and pile yarns of different fiber origins

The drapability, hand, performance, quality and durability of the individual pile fabrics depend on:

- Fiber content
- Yarn type
- Fabric structure of base or ground
- Type of weave interlacing
- Type and method of pile interlacing or tufting
- Thread count
- Density of pile
- Depth of pile
- Finishes applied to fibers, yarns or finished cloth

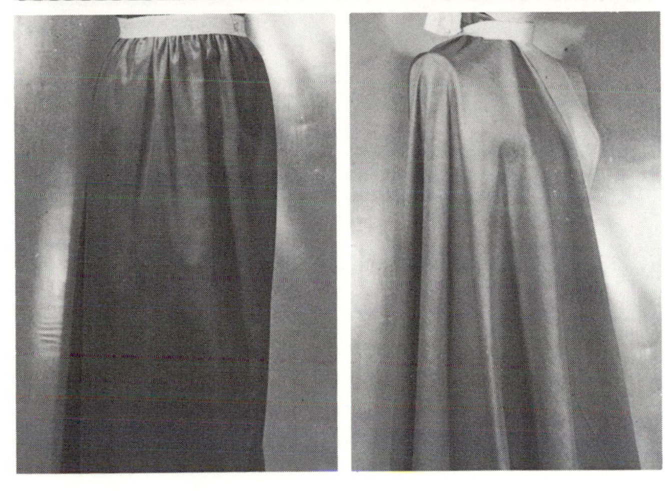

Feathercord Corduroy

Fiber Content 100% cotton
Yarn Type carded staple
Yarn Construction conventional
Fabric Structure filling pile face; plain weave back
Finishing Processes brushing; napping; singeing; stiffening
Color Application continuous and jig dyed
Width 46 inches (116.8 cm)
Weight medium
Hand firm; downy soft; thick
Texture uneven surface; 20 to 25 ribs per inch in lengthwise direction
Performance Expectations pile tends to abrade; absorbent; antistatic; durable; elongation; nonpilling; high tensile strength
Drapability Qualities falls into firm wide cones; fullness maintains crisp effect; retains shape of garment

Recommended Construction Aids & Techniques

Hand Needles betweens, 3–4; cotton darners, 6–8; embroidery, 7–8; milliners, 7–9; sharps, 4–6
Machine Needles round/set point, medium 11–14
Threads mercerized cotton, 50; six-cord cotton, 50; cotton- or poly-core waxed, 50; cotton-covered polyester core, all purpose
Hems book; single-fold bias binding; bound/Hong Kong finish; edge-stitched; machine-stitched; overedged; seam binding
Seams flat-felled; lapped; plain; safety-stitched; welt; double welt
Seam Finishes book; single-ply bound; Hong Kong bound; edge-stitched; pinked and stitched; serging/single-ply overedged; untreated/plain
Pressing steam or press; use needle board; safe temperature limit 400°F (206.1°C)
Care launder; dry clean
Fabric Resource Crompton-Richmond Inc.

Fine Wale/Pinwale Corduroy

Fiber Content 50% polyester/50% cotton
Yarn Type carded staple; filament staple
Yarn Construction conventional
Fabric Structure filling pile face; plain weave back
Finishing Processes brushing; napping; singeing; stiffening
Color Application continuous and jig dyed
Width 46 inches (116.8 cm)
Weight medium
Hand firm; soft; thick
Texture uneven surface; 16 to 23 ribs per inch in lengthwise direction
Performance Expectations pile subject to color crocking and edge abrasion; dimensional stability; durable; elongation; heat-set properties; high tensile strength
Drapability Qualities falls into soft wide cones; fullness maintains bouffant effect; retains shape of garment

Recommended Construction Aids & Techniques

Hand Needles betweens, 3–4; cotton darners, 6–8; embroidery, 7–8; milliners, 7–9; sharps, 4–6
Machine Needles round/set point, medium 14
Threads mercerized cotton, 50; six-cord cotton, 50; cotton- or poly-core waxed, 50; cotton-covered polyester core, all purpose
Hems book; single-fold bias binding; bound/Hong Kong finish; edge-stitched; machine-stitched; overedged; seam binding
Seams flat-felled; lapped; plain; safety-stitched; welt; double welt
Seam Finishes book; single-ply bound; Hong Kong bound; edge-stitched; pinked and stitched; serging/single-ply overedged; untreated/plain
Pressing steam or press; use needle board; safe temperature limit 325°F (164.1°C)
Care launder; dry clean
Fabric Resources Crompton-Richmond Inc.; J.P. Stevens & Co. Inc.

Mid/Medium/Regular Wale Corduroy

Fiber Content 100% cotton
Yarn Type carded staple
Yarn Construction conventional
Fabric Structure filling pile face; plain or twill weave back
Finishing Processes brushing; napping; singeing; stiffening
Color Application continuous and jig dyed
Width 46 inches (116.8 cm)
Weight medium-heavy
Hand firm; soft; thick
Texture uneven surface; 14 ribs per inch in lengthwise direction
Performance Expectations pile tends to abrade; absorbent; antistatic; durable; elongation; nonpilling; high tensile strength
Drapability Qualities falls into wide cones; fullness maintains lofty bouffant effect; retains shape of garment

Recommended Construction Aids & Techniques

Hand Needles cotton darners, 3–6; embroidery, 4–6; milliners, 3–7; sharps, 1–6
Machine Needles round/set point, medium 14
Threads mercerized cotton, heavy duty; six-cord cotton, 40; cotton- or poly-core waxed, 50; cotton-covered polyester core, all purpose
Hems book; single-fold and double-fold bias binding; bound/Hong Kong finish; double-stitched; faced; overedged; seam binding
Seams flat-felled; lapped; plain; safety-stitched; welt; double welt
Seam Finishes single-ply bound; Hong Kong bound; single-ply overcast; pinked; pinked and stitched; serging/single-ply overedged; untreated/plain
Pressing steam or press; use needle board; safe temperature limit 400°F (206.1°C)
Care launder; dry clean
Fabric Resource **Crompton-Richmond Inc.**

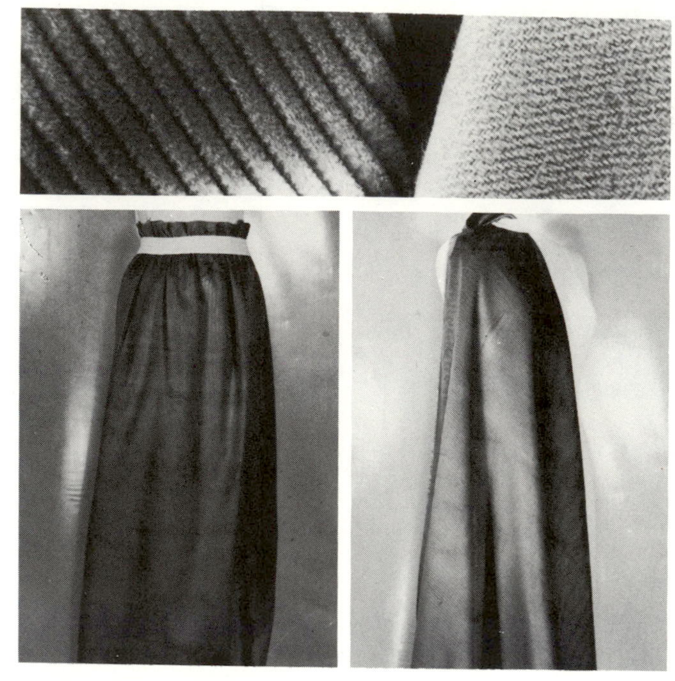

Thick-set Corduroy

Fiber Content 100% cotton
Yarn Type carded staple
Yarn Construction conventional
Fabric Structure filling pile face; twill weave back
Finishing Processes brushing; napping; singeing; stiffening
Color Application continuous and jig dyed
Width 46 inches (116.8 cm)
Weight medium-heavy
Hand firm; velvety soft; thick
Texture uneven surface; 11 ribs per inch in lengthwise direction
Performance Expectations pile tends to abrade; absorbent; antistatic; compressible; durable; elongation; nonpilling; high tensile strength
Drapability Qualities falls into crisp wide cones; fullness maintains crisp bouffant effect; retains shape of garment

Recommended Construction Aids & Techniques

Hand Needles cotton darners, 1–4; embroidery, 1–4; milliners, 3–5
Machine Needles round/set point, medium-coarse 14–16
Threads mercerized cotton, heavy duty; six-cord cotton, 60; cotton- or poly-core waxed, 40; cotton-covered polyester core, heavy duty
Hems double-fold or single-fold bias binding; bound/Hong Kong finish; double-stitched; faced; overedged; seam binding
Seams flat-felled; safety-stitched; plain; welt; double welt
Seam Finishes single-ply bound; Hong Kong bound; single-ply overcast; pinked; pinked and stitched; serging/single-ply overedged; untreated/plain
Pressing steam or press; use needle board; safe temperature limit 400°F (206.1°C)
Care launder; dry clean
Fabric Resource **Crompton-Richmond Inc.**

Wide Wale Corduroy

Fiber Content 100% cotton
Yarn Type carded staple
Yarn Construction conventional
Fabric Structure filling pile face; twill weave back
Finishing Processes brushing; napping; singeing; stiffening
Color Application continuous and jig dyed
Width 46 inches (116.8 cm)
Weight medium-heavy to heavy
Hand lofty; velvety soft; thick
Texture uneven surface; 6 to 10 ribs per inch in lengthwise direction
Performance Expectations pile tends to abrade; absorbent; antistatic; compressible; durable; elongation; nonpilling; resilient; high tensile strength
Drapability Qualities falls into soft wide flares; fullness maintains bouffant effect; retains shape of garment

Recommended Construction Aids & Techniques

Hand Needles cotton darners, 1–4; embroidery, 1–4; milliners, 3–5
Machine Needles round/set point, coarse 16–18
Threads mercerized cotton, heavy duty; six-cord cotton, 60; cotton- or poly-core waxed, 40; cotton-covered polyester core, heavy duty
Hems double-fold or single-fold bias binding; bound/Hong Kong finish; double-stitched; faced; overedged; seam binding
Seams flat-felled; plain; safety-stitched; welt; double welt
Seam Finishes single-ply bound; Hong Kong bound; single-ply overcast; pinked; pinked and stitched; serging/single-ply overedged; untreated/plain
Pressing steam or press; use needle board; safe temperature limit 400°F (206.1°C)
Care launder; dry clean
Fabric Resource Crompton-Richmond Inc.

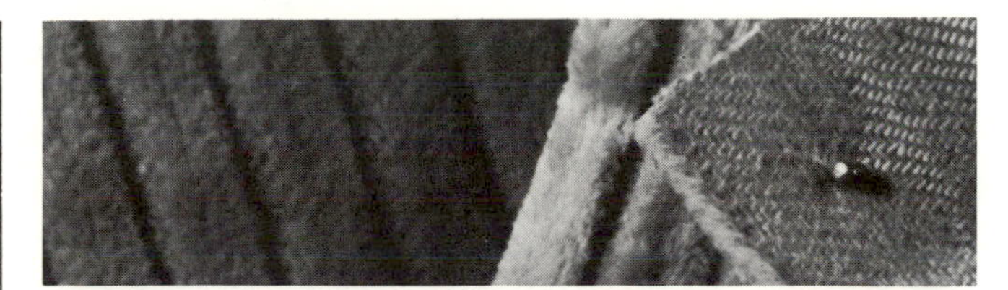

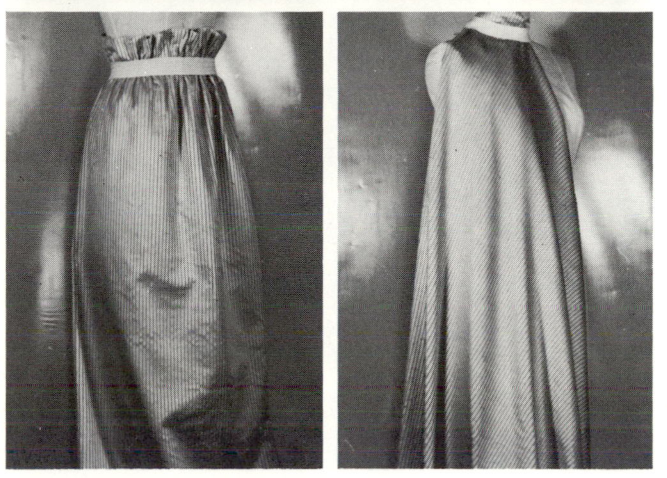

Broad Wale Corduroy

Fiber Content 100% cotton
Yarn Type carded staple
Yarn Construction conventional
Fabric Structure filling pile face; twill weave back
Finishing Processes brushing; napping; singeing; stiffening
Color Application continuous and jig dyed
Width 46 inches (116.8 cm)
Weight medium-heavy
Hand lofty; velvety soft; thick
Texture uneven surface; 3 to 5 ribs per inch in lengthwise direction
Performance Expectations pile tends to abrade; absorbent; antistatic; compressible; durable; elongation; nonpilling; resilient; high tensile strength
Drapability Qualities falls into soft wide flares; fullness maintains bouffant effect; retains shape of garment

Recommended Construction Aids & Techniques

Hand Needles cotton darners, 1–4; embroidery, 1–4; milliners, 3–5
Machine Needles round/set point, coarse 16–18
Threads mercerized cotton, heavy duty; six-cord cotton, 60; cotton- or poly-core waxed, 40; cotton-covered polyester core, heavy duty
Hems double-fold or single-fold bias binding; bound/Hong Kong finish; double-stitched; faced; overedged; seam binding
Seams plain; safety-stitched
Seam Finishes single-ply bound; Hong Kong bound; single-ply overcast; pinked and stitched; serging/single-ply overedged; untreated/plain
Pressing steam or press; use needle board; safe temperature limit 400°F (206.1°C)
Care launder; dry clean
Fabric Resource Crompton-Richmond Inc.

Novelty Wale Corduroy

Fiber Content 100% cotton
Yarn Type carded staple
Yarn Construction conventional
Fabric Structure filling pile face; twill weave back
Finishing Processes brushing; napping; singeing; stiffening
Color Application continuous and jig dyed
Width 46 inches (116.8 cm)
Weight medium-heavy
Hand lofty; velvety soft
Texture thick and thin wales; high and low rib formation; 3 to 10 ribs per inch in lengthwise direction
Performance Expectations pile tends to abrade; absorbent; antistatic; compressible; durable; elongation; nonpilling; high tensile strength
Drapability Qualities falls into soft wide flares; fullness maintains bouffant effect; retains shape of garment

Recommended Construction Aids & Techniques

Hand Needles cotton darners, 1–4; embroidery, 1–4; milliners, 3–5
Machine Needles round/set point, coarse 16–18
Threads mercerized cotton, heavy duty; six-cord cotton, 60; cotton- or poly-core waxed, 40; cotton-covered polyester core, heavy duty
Hems double-fold or single-fold bias binding; bound/Hong Kong finish; double-stitched; faced; overedged; seam binding
Seams plain; safety-stitched
Seam Finishes single-ply bound; Hong Kong bound; single-ply overcast; pinked and stitched; serging/single-ply overedged; untreated/plain
Pressing steam or press; use needle board; safe temperature limit 400°F (206.1°C)
Care launder; dry clean
Fabric Resource **Crompton-Richmond Inc.**

Nap-faced Fleece

Fiber Content 100% nylon
Yarn Type filament fiber
Yarn Construction bulked; textured
Fabric Structure pile face; jersey knit base
Finishing Processes brushing; napping; shearing; stabilizing
Color Application piece dyed
Width 54 inches (137.2 cm)
Weight medium
Hand lofty; downy soft; springy
Texture fuzzy; napped face; smooth back
Performance Expectations abrasion resistant; dimensional stability; durable; elasticity; elongation; resilient; high tensile strength; subject to pilling and static buildup
Drapability Qualities falls into soft flares; accommodates fullness by gathering; fullness maintains bouffant effect

Recommended Construction Aids & Techniques

Hand Needles cotton darners, 3–6; embroidery, 4–6; milliners, 3–7; sharps, 1–6
Machine Needles ball point, 11–14
Threads poly-core waxed, 70; cotton-polyester blend, extra fine; cotton-covered polyester core, extra fine; spun polyester, fine; nylon/Dacron, monocord A
Hems book; single-fold bias binding; bound/Hong Kong finish; double-stitched; edge-stitched; flat/plain; machine-stitched; overedged; seam binding
Seams flat-felled; lapped; overedged; plain; safety-stitched; welt; double welt
Seam Finishes single-ply bound; Hong Kong bound; serging/single-ply overedged; untreated/plain
Pressing *do not* steam; safe temperature limit 350°F (178.1°C)
Care launder; fluffs up in dryer and requires no ironing; dry clean
Fabric Resource **Gloversville Mills Inc.**

Baby Bunting Fleece

Fiber Content 100% acrylic
Yarn Type filament fiber; filament staple
Yarn Construction bulked; textured
Fabric Structure pile face and back; jersey knit base
Finishing Processes brushing; napping; shearing; stabilizing
Color Application piece dyed
Width 60 inches (154.9 cm)
Weight medium-heavy
Hand lofty; downy soft; thick
Texture fluffy, napped face and back
Performance Expectations compressible; dimensional stability; elasticity; elongation; heat-set and insulating properties; resilient; subject to pilling and static buildup
Drapability Qualities falls into soft wide flares; fullness maintains lofty bouffant effect; better utilized if fitted by seaming and eliminating excess fabric

Recommended Construction Aids & Techniques

Hand Needles cotton darners, 3-6; embroidery, 4-6; milliners, 3-7; sharps, 1-6
Machine Needles ball point, 14
Threads poly-core waxed, 70; cotton-polyester blend, extra fine; cotton-covered polyester core, extra fine; spun polyester, fine; nylon/Dacron, monocord A
Hems book; single-fold bias binding; bound/Hong Kong finish; double-stitched; edge-stitched; flat/plain; machine-stitched; overedged; seam binding
Seams flat-felled; lapped; overedged; plain; safety-stitched; welt; double welt
Seam Finishes single-ply bound; Hong Kong bound; serging/single-ply overedged; untreated/plain
Pressing *do not* steam; safe temperature limit 300°F (150.1°C)
Care launder; fluffs up in dryer and requires no ironing; dry clean
Fabric Resource **Gloversville Mills Inc.**

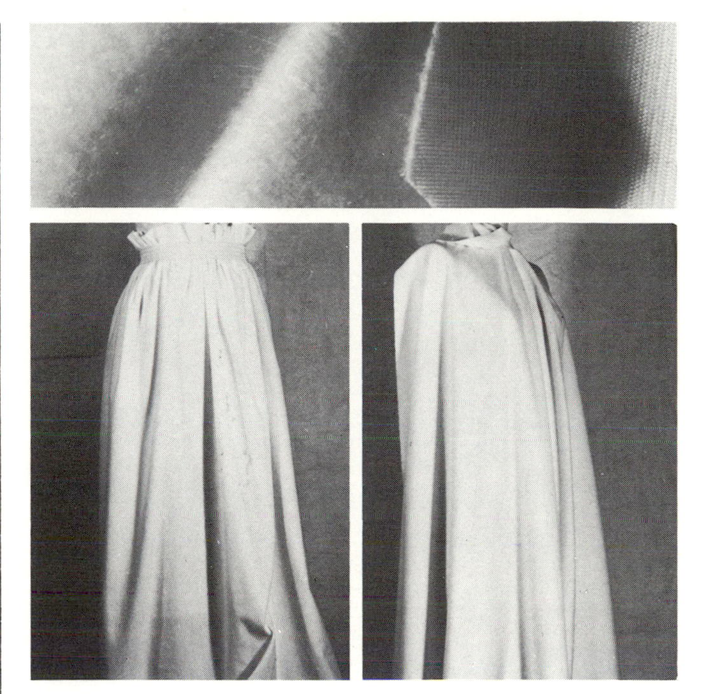

Sweatshirt Fleece

Fiber Content 100% nylon
Yarn Type filament fiber
Yarn Construction bulked; textured
Fabric Structure tricot knit face; pile back
Finishing Processes brushing; heat-set stabilizing; napping; shearing
Color Application piece dyed; yarn dyed
Width 70-72 inches (177.8-182.9 cm)
Weight medium-heavy
Hand lofty; downy soft; springy
Texture fuzzy; silky smooth face; napped back; pile surface is worn towards the body and smooth surface is used as face
Performance Expectations abrasion resistant; dimensional stability; durable; elasticity; elongation; resilient; high tensile strength; subject to pilling and static buildup
Drapability Qualities falls into moderate flares; fullness maintains bouffant effect; retains shape of garment

Recommended Construction Aids & Techniques

Hand Needles cotton darners, 3-6; embroidery, 4-6; milliners, 3-7; sharps, 1-6
Machine Needles ball point, 14
Threads poly-core, 50; cotton-polyester blend, all purpose; cotton-covered polyester core, all purpose; spun polyester, all purpose; nylon/Dacron, twist construction A
Hems double-fold or single-fold bias binding; bound/Hong Kong finish; double-stitched; flat/plain; overedged
Seams overedged; plain; safety-stitched; taped
Seam Finishes single-ply bound; Hong Kong bound; serging/single-ply overedged; untreated/plain
Pressing *do not* steam; safe temperature limit 350°F (178.1°C)
Care launder; fluffs up in dryer and requires no ironing; dry clean
Fabric Resource **Gloversville Mills Inc.**

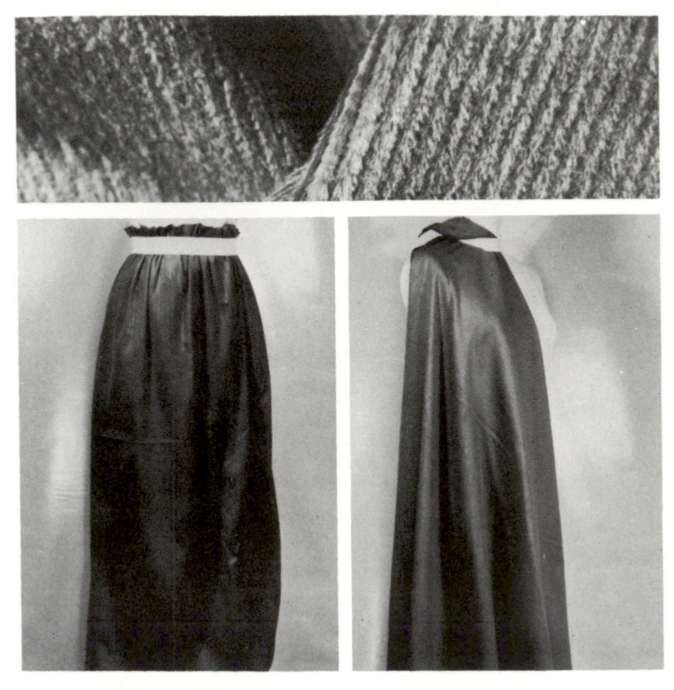

Friezé (Frez)/Shadow Wale/Ribless Corduroy

Fiber Content 100% cotton
Yarn Type carded staple
Yarn Construction conventional; spun
Fabric Structure warp pile face; twill weave back; constructed on wide flat wire loom
Finishing Processes brushing; napping; singeing; stiffening
Color Application continuous and jig dyed
Width 46 inches (116.8 cm)
Weight medium-light
Hand compact; lofty; velvety soft
Texture short pile face with no noticeable rib; smooth back
Performance Expectations pile tends to abrade; absorbent; antistatic; durable; elongation; nonpilling; high tensile strength
Drapability Qualities falls into moderately soft flares; accommodates fullness by pleating, gathering; fullness retains moderately soft fall

Recommended Construction Aids & Techniques

Hand Needles betweens, 3–4; cotton darners, 6–8; embroidery, 7–8; milliners, 7–9; sharps, 4–6
Machine Needles round/set point, medium 11–14
Threads mercerized cotton, 50; six-cord cotton, 50; cotton- or poly-core waxed, 50; cotton-covered polyester core, all purpose
Hems book; single-fold bias binding; bound/Hong Kong finish; edge-stitched; machine-stitched; overedged; seam binding
Seams flat-felled; lapped; plain; safety-stitched; welt; double welt
Seam Finishes book; single-ply bound; Hong Kong bound; edge-stitched; pinked; pinked and stitched; serging/single-ply overedged; untreated/plain
Pressing steam or press; use needle board; safe temperature limit 400°F (206.1°C)
Care launder; dry clean
Fabric Resource **Crompton-Richmond Inc.**

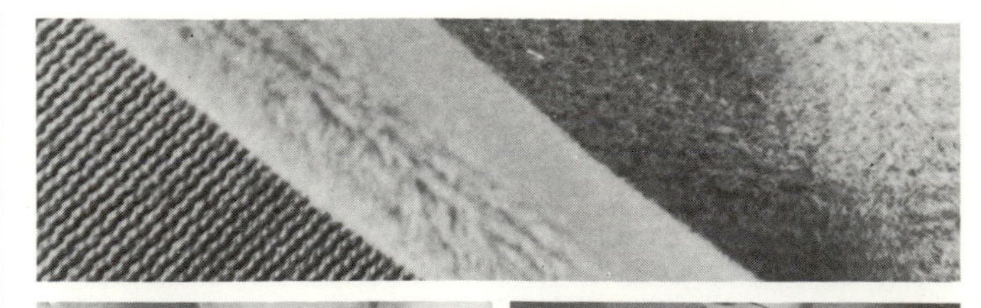

Plush

Fiber Content 70% acetate/30% nylon
Yarn Type filament fiber
Yarn Construction bulked; textured
Fabric Structure jersey knit with laid-in yarns
Finishing Processes brushing; delustering; shearing
Color Application piece dyed–union dyed
Width 60 inches (152.4 cm)
Weight medium-heavy
Hand velvety soft; spongy; wiry
Texture fuzzy; erect cut pile surface; deeper pile than velvet or velour fabric
Performance Expectations pile tends to abrade; dimensional stability; elongation; flexible; resilient; high tensile strength
Drapability Qualities falls into moderately soft flares; fullness maintains bouffant effect; retains shape of garment

Recommended Construction Aids & Techniques

Hand Needles cotton darners, 3–6; embroidery, 4–6; milliners, 3–7; sharps, 1–6
Machine Needles ball point, 14
Threads poly-core waxed, 70; cotton-polyester blend, extra fine; cotton-covered polyester core, extra fine; spun polyester, fine; nylon/Dacron, monocord A
Hems book; single-fold bias binding; bound/Hong Kong finish; double-stitched; machine-stitched; overedged
Seams flat-felled; overedged; plain; safety-stitched; welt; double welt
Seam Finishes single-ply bound; Hong Kong bound; single-ply overcast; serging/single-ply overedged; untreated/plain
Pressing steam; safe temperature limit 250°F (122.1°C)
Care launder; fluffs up in dryer and requires no ironing; dry clean
Fabric Resource **Gloversville Mills Inc.**

Double-faced Terry Cloth (loop face & back)/Turkish Toweling

Fiber Content 85% cotton/15% viscose rayon
Yarn Type filament staple; staple
Yarn Construction conventional
Fabric Structure uncut looped face and back; plain weave base
Finishing Processes brushing; mercerizing; stabilizing
Color Application piece dyed
Width 60 inches (152.4 cm)
Weight heavy
Hand heavy; rough; thick
Texture rough; loop-piled face and back; reversible
Performance Expectations absorbent; compressible; durable; elongation; resilient; high tensile strength; subject to snagging and pulling due to loop structure
Drapability Qualities falls into wide flares; fullness maintains bouffant effect; retains shape of garment

Recommended Construction Aids & Techniques

Hand Needles cotton darners, 1–4; embroidery, 1–4; milliners, 3–5
Machine Needles round/set point, medium 14
Threads mercerized cotton, heavy duty; six-cord cotton, 40; cotton- or poly-core waxed, 50; cotton-covered polyester core, regular
Hems double-fold or single-fold bias binding; bound/Hong Kong finish; double-stitched; faced; overedged; seam binding
Seams flat-felled; plain; safety-stitched; welt; double welt
Seam Finishes single-ply bound; Hong Kong bound; serging/single-ply overedged; untreated/plain
Pressing no pressing required
Care launder; fluffs up in dryer and requires no ironing
Fabric Resource Brunswick Associates Inc.

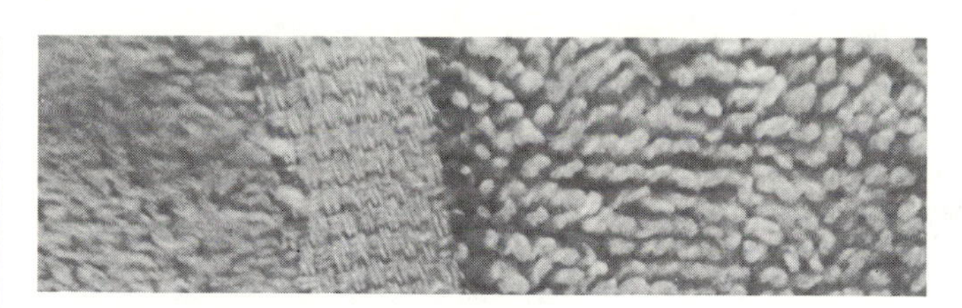

Double-faced Terry Cloth (loop face; velour back)

Fabric Resource Dan River Inc.

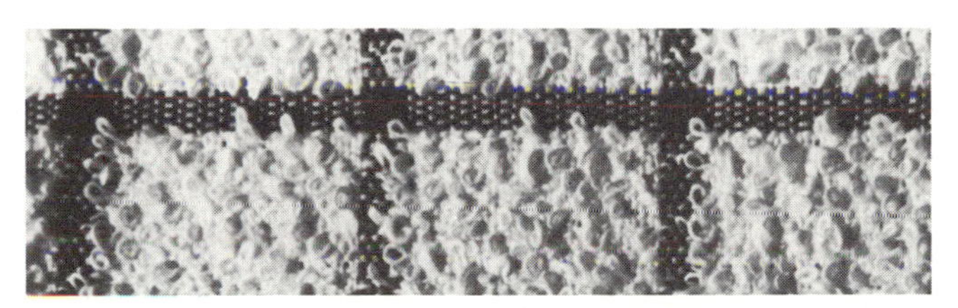

Jacquard Double-faced Terry Cloth

Fabric Resource Brunswick Associates Inc.

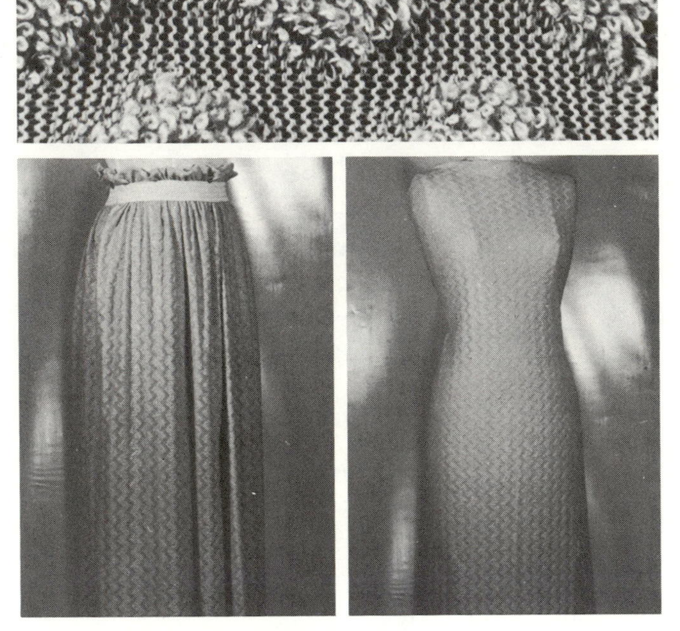

Sculptured Terry Cloth

Fiber Content 100% cotton
Yarn Type carded staple
Yarn Construction conventional; spun
Fabric Structure uncut looped pattern face; jersey knit base
Finishing Processes brushing; slack mercerization; softening
Color Application piece dyed
Width 72 inches (182.9 cm)
Weight medium-light
Hand lofty; pliable; soft
Texture uneven surface; high and low patterned looped face and back
Performance Expectations absorbent; compressible; durable; elasticity; elongation; resilient; high tensile strength; subject to snagging and pulling due to loop structure
Drapability Qualities falls into soft flares; fullness retains soft fall; can be shaped and molded by stretching

Recommended Construction Aids & Techniques

Hand Needles betweens, 3–4; cotton darners, 6–8; embroidery, 7–8; milliners, 7–9; sharps, 4–6
Machine Needles ball point, 14
Threads mercerized cotton, heavy duty; six-cord cotton, 40; cotton- or poly-core waxed, 50; cotton-covered polyester core, all purpose
Hems book; double-fold or single-fold bias binding; bound/Hong Kong finish; double-stitched; machine-stitched; overedged
Seams flat-felled; overedged; plain; safety-stitched; welt; double welt
Seam Finishes single-ply bound; Hong Kong bound; serging/single-ply overedged; untreated/plain
Pressing steam; safe temperature limit 400°F (206.1°C)
Care launder; fluffs up in dryer and requires no ironing; dry clean
Fabric Resource Stēvcoknit, Inc. (sample courtesy of College Town)

Terry Cloth Chenille (diagonal pattern)

Fiber Content 85% cotton (face)/15% nylon (base)
Yarn Type carded staple; filament fiber
Yarn Construction conventional (face); textured (back)
Fabric Structure uncut pile face; jersey knit base
Finishing Processes brushing; slack mercerization; stabilizing
Color Application piece dyed–union dyed
Width 72 inches (182.9 cm)
Weight medium-heavy
Hand lofty; rough; supple
Texture uneven surface; uncut loop surface forms diagonal-patterned base visible between rows
Performance Expectations absorbent; compressible; durable; elasticity; elongation; resilient; high tensile strength; subject to snagging and pulling due to loop structure
Drapability Qualities falls into moderately soft flares; fullness retains moderately soft fall; can be shaped and molded by stretching

Recommended Construction Aids & Techniques

Hand Needles cotton darners, 3–6; embroidery, 4–6; milliners, 3–7; sharps, 1–6
Machine Needles ball point, 14
Threads mercerized cotton, heavy duty; six-cord cotton, 40; cotton- or poly-core waxed, 50; cotton-covered polyester core, all purpose
Hems book; Hong Kong bound; double-fold or single-fold bias binding; bound/Hong Kong finish; double-stitched; machine-stitched; overedged
Seams flat-felled; overedged; plain; safety-stitched; welt; double welt
Seam Finishes single-ply bound; Hong Kong bound; serging/single-ply overedged; untreated/plain
Pressing steam; safe temperature limit 350°F (178.1°C)
Care launder; fluffs up in dryer and requires no ironing;
Fabric Resource Gloversville Mills Inc.

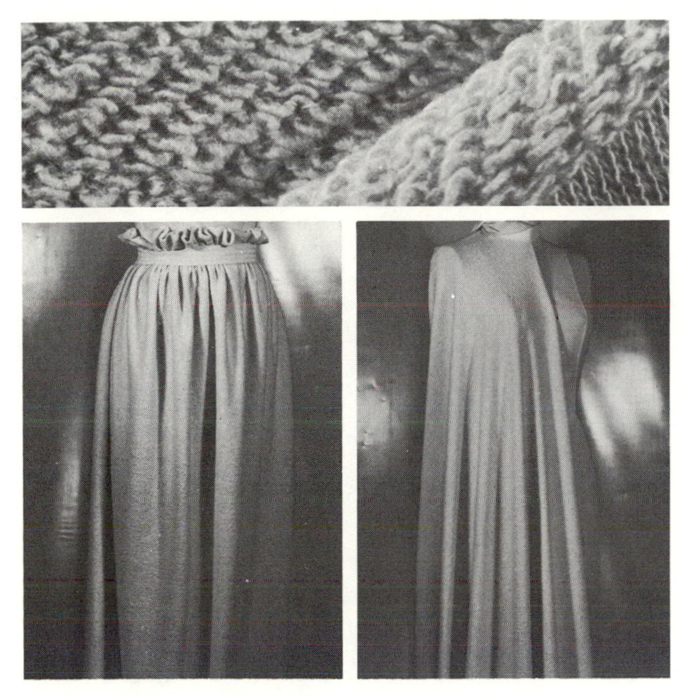

Terry Cloth Chenille (overall pattern)

Fiber Content 100% cotton
Yarn Type carded staple
Yarn Construction conventional; thick-and-thin
Fabric Structure thick laid-in yarns form loop pattern on jersey knit base
Finishing Processes brushing; mercerizing; stabilizing
Color Application piece dyed
Width 72 inches (182.9 cm)
Weight heavy
Hand semi-soft; spongy; thick
Texture coarse; overall loop-patterned face; flat back
Performance Expectations absorbent; compressible; durable; elasticity; elongation; resilient; high tensile strength; subject to snagging and pulling due to loop structure
Drapability Qualities falls into moderately soft flares; fullness maintains bouffant effect; retains shape of garment

Recommended Construction Aids & Techniques

Hand Needles cotton darners, 1-4; embroidery, 1-4; milliners, 3-5
Machine Needles ball point, 14
Threads mercerized cotton, heavy duty; six-cord cotton, 40; cotton- or poly-core waxed, 50; cotton-covered polyester core, all purpose
Hems book; double-fold or single-fold bias binding; bound/Hong Kong finish; double-stitched; machine-stitched; overedged
Seams flat-felled; overedged; plain; safety-stitched; welt; double welt
Seam Finishes single-ply bound; Hong Kong bound; serging/single-ply overedged; untreated/plain
Pressing steam; safe temperature limit 400°F (206.1°C)
Care launder; fluffs up in dryer and requires no ironing; dry clean
Fabric Resource **Gloversville Mills Inc.**

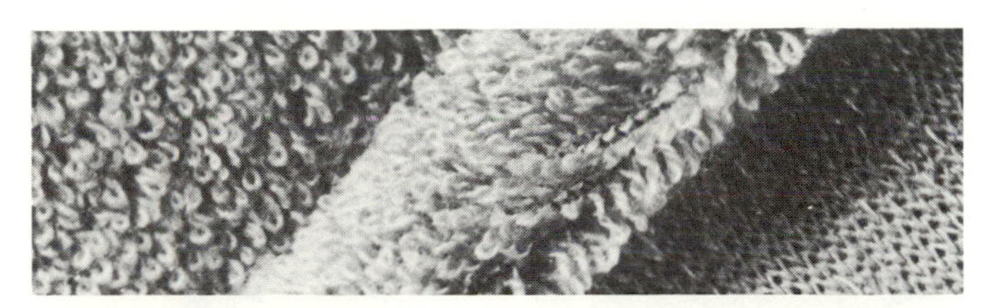

Stretch/Single-faced Terry Cloth

Fiber Content 70% cotton (face)/30% polyester (back)
Yarn Type carded staple; filament staple
Yarn Construction conventional (face); textured (back)
Fabric Structure uncut pile face; jersey knit base
Finishing Processes brushing; slack mercerization; stabilizing
Color Application piece dyed-union dyed
Width 72 inches (182.9 cm)
Weight medium-heavy
Hand pliable; rough; spongy; stretch in crosswise direction
Texture widely spaced, high loop pile face; flat back
Performance Expectations compressible; dimensional stability; durable; elasticity; elongation; resilient; high tensile strength; subject to snagging and pulling due to loop structure
Drapability Qualities falls into moderately soft flares; fullness retains moderately soft fall; can be shaped or molded by stretching

Recommended Construction Aids & Techniques

Hand Needles cotton darners, 3-6; embroidery, 4-6; milliners, 3-7; sharps, 1-6
Machine Needles ball point, 14
Threads poly-core waxed, 70; cotton-polyester blend, extra fine; cotton-covered polyester core, extra fine; spun polyester, fine; nylon/Dacron, monocord A
Hems book; single-fold bias binding; bound/Hong Kong finish; double-stitched; machine-stitched; overedged
Seams flat-felled; overedged; plain; safety-stitched; welt; double welt
Seam Finishes single-ply bound; Hong Kong bound; serging/single-ply overedged; untreated/plain
Pressing steam; safe temperature limit 325°F (164.1°C)
Care launder; fluffs up in dryer and requires no ironing
Fabric Resource **Gloversville Mills Inc.**

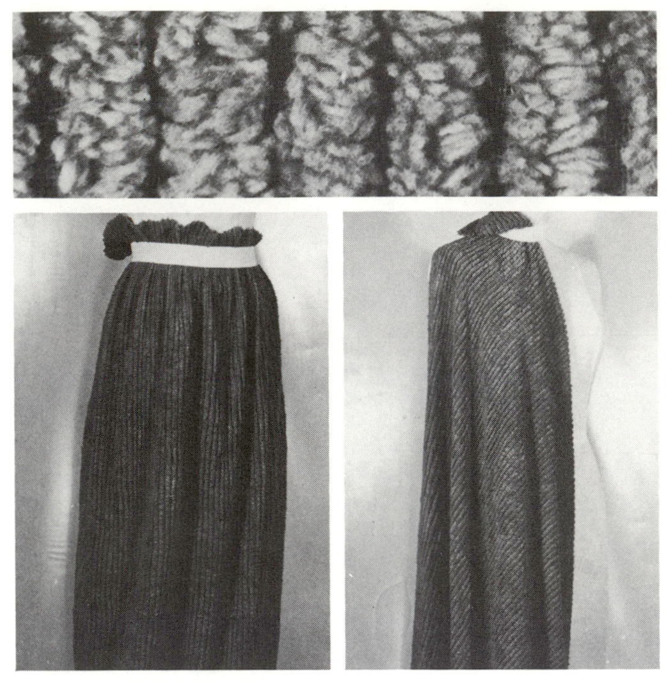

Chenille Candlewick

Fiber Content 100% cotton
Yarn Type carded staple
Yarn Construction conventional; thick slack twist (tufts); thin high twist (base)
Fabric Structure tufted face; plain weave base
Finishing Processes brushing; napping; shearing; stabilizing
Color Application piece dyed
Width 47–48 inches (119.4–121.9 cm)
Weight heavy
Hand bulky; heavy; thick
Texture uneven surface; raised plush cut cord form lengthwise, crosswise or patterned ridges on face
Performance Expectations absorbent; durable; elongation; insulating properties; resilient; tufts tend to pull out with wear
Drapability Qualities falls into soft wide flares; fullness retains soft fall; retains shape of garment

Recommended Construction Aids & Techniques

Hand Needles cotton darners, 1–4; embroidery, 1–4; milliners, 3–5
Machine Needles round/set point, coarse 16–18
Threads mercerized cotton, heavy duty; six-cord cotton, 60; cotton- or poly-core waxed, 40; cotton-covered polyester core, heavy duty
Hems double-fold or single-fold bias binding; bound/Hong Kong finish; double-stitched; machine-stitched; overedged
Seams plain; safety-stitched
Seam Finishes single-ply bound; Hong Kong bound; serging/single-ply overedged; untreated/plain
Pressing steam; safe temperature limit 400°F (206.1°C)
Care launder; fluffs up in dryer and requires no ironing; dry clean
Fabric Resource Brookhaven Textiles Inc. (sample courtesy of College Town)

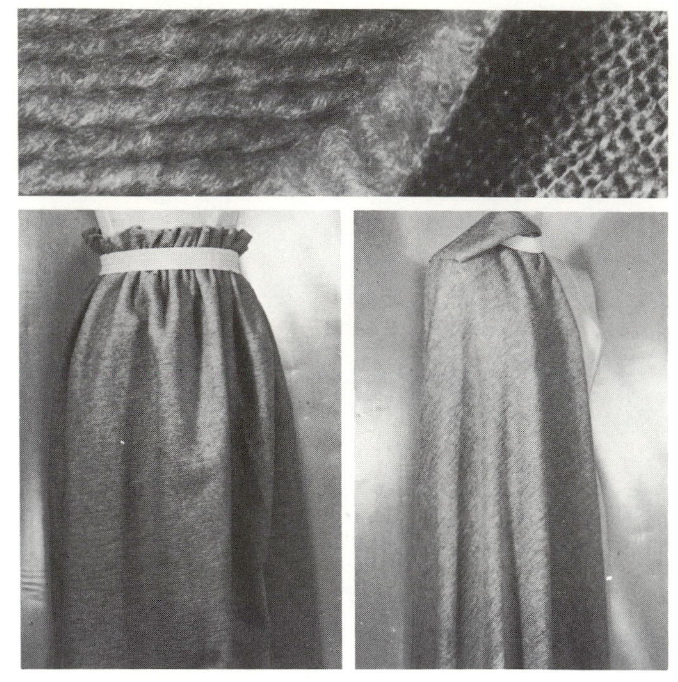

Chenille Velvet/Velcle®

Fiber Content acrylic/modacrylic/polyester blend
Yarn Type filament fiber
Yarn Construction bulked; textured
Fabric Structure looped pile face; jersey knit base
Finishing Processes brushing; fulling; napping; shearing
Color Application piece dyed–union dyed
Width 68 inches (172.7 cm)
Weight medium
Hand bulky; lofty; downy soft
Texture bulky without added weight; fuzzy; napped face and back
Performance Expectations compressible; dimensional stability; durable; elongation; resilient; high tensile strength
Drapability Qualities falls into moderately soft flares; retains shape or silhouette of garment; better utilized if fitted by seaming and eliminating excess fabric

Recommended Construction Aids & Techniques

Hand Needles cotton darners, 3–6; embroidery, 4–6; milliners, 3–7; sharps, 1–6
Machine Needles ball point, 14
Threads poly-core waxed, 50; cotton-polyester blend, all purpose; cotton-covered polyester core, all purpose; spun polyester, all purpose; nylon/Dacron, twist A
Hems double-fold or single-fold bias binding; bound/Hong Kong finish; double-stitched; machine-stitched; overedged
Seams overedged; plain; safety-stitched
Seam Finishes single-ply bound; Hong Kong bound; serging/single-ply overedged; untreated/plain
Pressing *do not* steam; use needle board; safe temperature limit 225°F (108.1°C)
Care dry clean
Fabric Resource Furtex

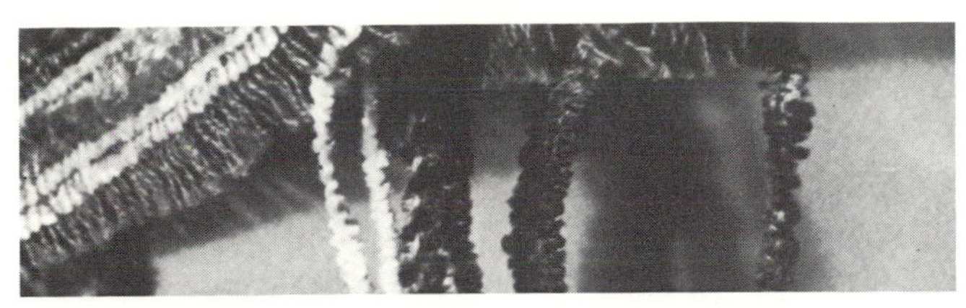

Chenille Yarn Cloth

Fiber Content natural and man-made fibers
Yarn Type carded staple; filament fiber; filament staple
Yarn Construction conventional warp yarns; novelty chenille filling yarn
Fabric Structure woven or knit; pile formed by yarn
Finishing Processes brushing; stabilizing
Color Application yarn dyed
Width 45 inches (114.3 cm)
Weight heavy
Hand firm; thick; velvety soft
Texture fuzzy cut pile; high and low areas; same face and back
Performance Expectations depends on fiber content of yarn or cloth; subject to yarn and seam slippage due to yarn and structure; subject to bagging due to structure
Drapability Qualities falls into wide cones; fullness maintains bouffant effect; better utilized if fitted by seaming and eliminating excess fabric

Recommended Construction Aids & Techniques

Hand Needles chenilles, 18; cotton darners, 1-3; embroidery, 1-5; milliners, 3
Machine Needles woven goods: round/set point, coarse 18; knit goods: ball point, 16
Threads poly-core waxed, 50; cotton-polyester blend, heavy duty; cotton-covered polyester core, heavy duty; spun polyester, heavy duty
Hems double-fold or single-fold bias binding; bound/Hong Kong finish; machine-stitched; overedged
Seams overedged; plain; safety-stitched; taped
Seam Finishes single-ply bound; Hong Kong bound; double-stitched; serging/single-ply overedged; untreated/plain
Pressing steam; use needle board; safe temperature limits depend on fiber content
Care dry clean
Fabric Resource **Lawrence Textile Co. Inc.**

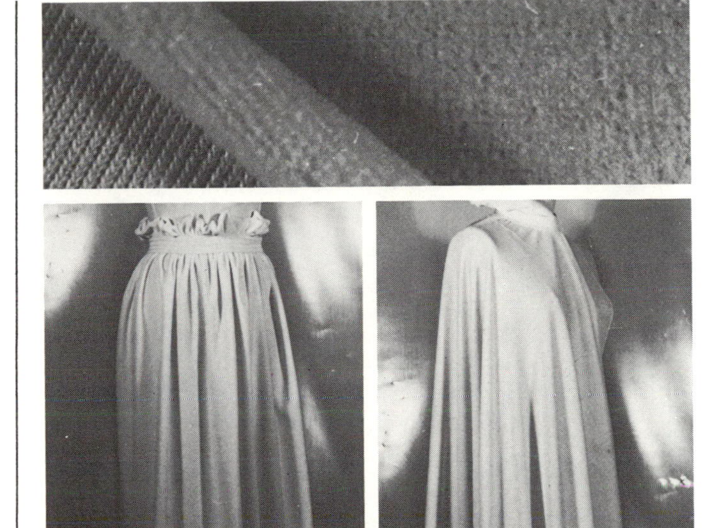

Velour

Fiber Content 100% cotton
Yarn Type combed staple
Yarn Construction conventional
Fabric Structure cut pile face; jersey knit base
Finishing Processes brushing; napping; shearing; stabilizing
Color Application continuous and jig dyed
Width 62 inches (157.5 cm)
Weight medium
Hand pliable; downy soft; spongy; stretch in crosswise direction
Texture cut pile pressed flat in one direction; velvety soft
Performance Expectations absorbent; compressible; durable; elasticity; elongation; resilient; high tensile strength
Drapability Qualities falls into moderately soft flares; fullness maintains bouffant effect; can be shaped and molded by stretching

Recommended Construction Aids & Techniques

Hand Needles betweens, 3-4; cotton darners, 6-8; embroidery, 7-8; milliners, 7-9; sharps, 4-6
Machine Needles ball point, 11-14
Threads mercerized cotton, heavy duty; six-cord cotton, 40; cotton- or poly-core waxed, 50; cotton-covered polyester core, all purpose
Hems book; single-fold bias binding; bound/Hong Kong finish; double-stitched; machine-stitched; overedged
Seams overedged; plain; safety-stitched
Seam Finishes book; single-ply bound; Hong Kong bound; serging/single-ply overedged; untreated/plain
Pressing steam; safe temperature limit 400°F (206.1°C)
Care launder; fluffs up in dryer and requires no ironing; dry clean
Fabric Resources **Loomskill/Gallery Screen Print; Dan River Inc.**

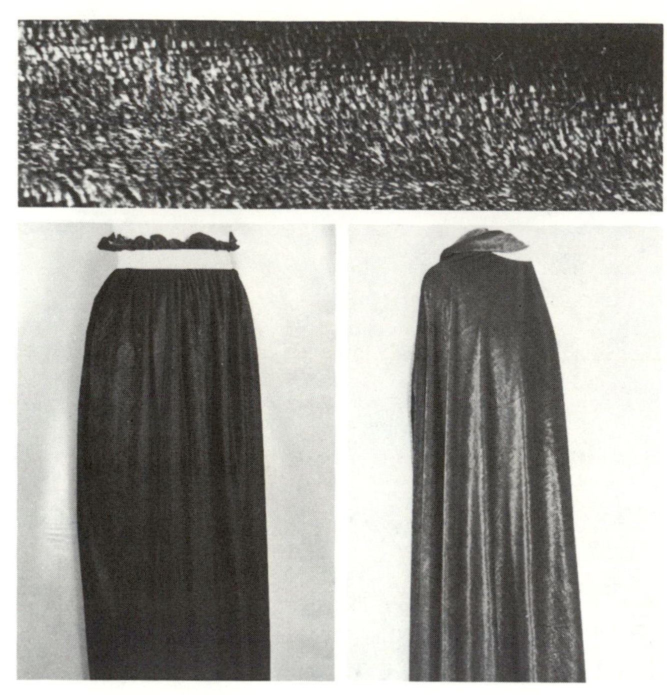

Sealskin Velour

Fiber Content 70% acetate/30% nylon
Yarn Type filament fiber
Yarn Construction conventional; slack twist
Fabric Structure pile face; jersey knit base
Finishing Processes brushing; roller pressed in one direction; shearing
Color Application piece dyed–union dyed; yarn dyed
Width 60 inches (152.4 cm)
Weight medium-heavy
Hand pliable; soft; stretch in crosswise direction; supple
Texture short dense pile pressed flat in one direction on face; flat back
Performance Expectations pile tends to abrade and flatten with wear; edges subject to curling; dimensional stability; elongation; flexible
Drapability Qualities falls into soft languid flares; can be shaped and molded by stretching; fullness retains soft fall

Recommended Construction Aids & Techniques

Hand Needles betweens, 3–4; cotton darners, 6–8; embroidery, 7–8; milliners, 7–9; sharps, 4–6
Machine Needles ball point, 11–14
Threads poly-core waxed, 50; cotton-polyester blend, all purpose; cotton-covered polyester core, all purpose; spun polyester, all purpose; nylon/Dacron, twist A
Hems book; single-fold bias binding; bound/Hong Kong finish; double-stitched; flat/plain; machine-stitched; overedged
Seams overedged; plain; safety-stitched
Seam Finishes single-ply bound; Hong Kong bound; serging/single-ply overedged; untreated/plain
Pressing steam; use needle board; safe temperature limit 250°F (122.1°C)
Care dry clean
Fabric Resource Gloversville Mills Inc.

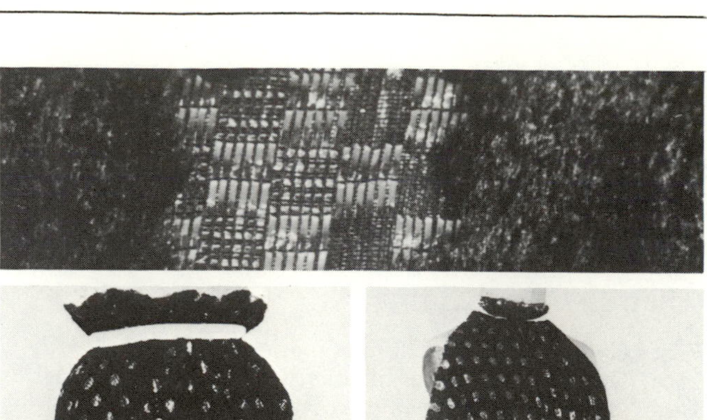

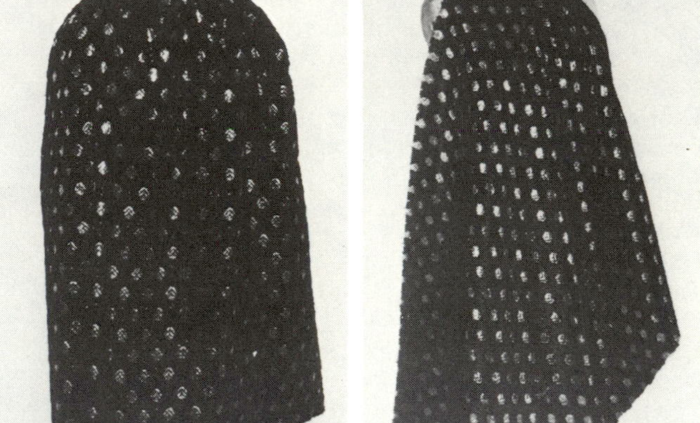

Velvet Brocade

Fiber Content silk/viscose rayon/metallic
Yarn Type extruded film; natural and man-made filament fibers
Yarn Construction conventional; cut film; textured
Fabric Structure Jacquard weave pile
Finishing Processes crush and spot resistant; shearing; stabilizing
Color Application piece dyed–union dyed; yarn dyed
Width 36 inches (91.4 cm)
Weight medium-heavy
Hand soft; springy
Texture uneven surface; high, raised pile and recessed flat woven areas on face; flat back
Performance Expectations pile tends to abrade and flatten with wear; satin areas subject to snagging; elongation; good tensile strength
Drapability Qualities falls into wide flares; fullness maintains lofty effect; retains shape of garment

Recommended Construction Aids & Techniques

Hand Needles cotton darners, 3–6; embroidery, 4–6; milliners, 3–7; sharps, 1–6
Machine Needles round/set point, medium-fine 9–11
Threads silk, industrial size A; poly-core waxed, 70; cotton-polyester blend, extra fine; cotton-covered polyester core, extra fine; spun polyester, fine; nylon/Dacron, monocord A
Hems double-fold or single-fold bias binding; bound/Hong Kong finish; edge-stitched; machine-stitched; overedged; seam binding
Seams plain; safety-stitched
Seam Finishes single-ply bound; Hong Kong bound; single-ply overcast; pinked; pinked and stitched; serging/single-ply overedged; untreated/plain
Pressing steam; use needle board; safe temperature limit 300°F (150.1°C)
Care dry clean
Fabric Resource Sormani Co. Inc./Diochon

Cotton Velvet

Fiber Content 100% cotton
Yarn Type combed staple
Yarn Construction conventional; high twist
Fabric Structure warp pile face (wire method); plain weave back
Finishing Processes brushing; crush and spot resistant; shearing
Color Application piece dyed–beck dyed/reel dyed
Width 39–40 inches (99.1–101.6 cm)
Weight heavy
Hand lofty; velvety soft; spongy
Texture dense; thick pile face; flat back
Performance Expectations pile tends to abrade; absorbent; durable; elongation; high tensile strength
Drapability Qualities falls into soft wide cones; fullness maintains lofty effect; retains shape of garment

Recommended Construction Aids & Techniques

Hand Needles cotton darners, 3–6; embroidery, 4–6; milliners, 3–7; sharps, 1–6
Machine Needles round/set point, medium 11–14
Threads mercerized cotton, heavy duty; six-cord cotton, 40; cotton- or poly-core waxed, 50; cotton-covered polyester core, all purpose
Hems book; double-fold or single-fold bias binding; bound/Hong Kong finish; double-stitched; faced; machine-stitched; overedged; seam binding
Seams flat-felled; plain; safety-stitched; welt; double welt
Seam Finishes single-ply bound; Hong Kong bound; single-ply overcast; pinked; pinked and stitched; serging/single-ply overedged; untreated/plain
Pressing steam; use needle board; safe temperature limit 400°F (206.1°C)
Care machine wash warm and tumble dry, remove promptly; dry clean
Fabric Resources **American Silk Mills Corp.; J. B. Martin Co. Inc.**

Crushed Velvet (shiny face & woven ground)

Fiber Content 100% rayon
Yarn Type filament fiber
Yarn Construction conventional
Fabric Structure warp pile face; plain weave base
Finishing Processes brushing; crush resistant; shearing; steamed and roller pressed for crushed effect
Color Application piece dyed–beck dyed/reel dyed
Width 45 inches (114.3 cm)
Weight medium
Hand lofty; soft; springy
Texture raised and flattened areas; light-reflecting effect on face; smooth back
Performance Expectations pile tends to abrade and flatten with wear; elongation; good tensile strength
Drapability Qualities falls into soft wide flares; fullness retains soft fall; retains shape of garment

Recommended Construction Aids & Techniques

Hand Needles cotton darners, 3–6; embroidery, 4–6; milliners, 3–7; sharps, 1–6
Machine Needles round/set point, medium 11–14
Threads poly-core waxed, 50; cotton-polyester blend, all purpose; cotton-covered polyester core, all purpose; spun polyester, all purpose; nylon/Dacron, twist A
Hems book; single-fold bias binding; bound/Hong Kong finish; edge-stitched; machine-stitched; overedged; seam binding
Seams plain; safety-stitched; horsehair tape for zipper insertion
Seam Finishes single-ply bound; Hong Kong bound; single-ply overcast; pinked; pinked and stitched; serging/single-ply overedged; untreated/plain
Pressing steam; use needle board; safe temperature limit 350°F (178.1°C)
Care dry clean
Fabric Resource **Deluxe Velvet Co.**

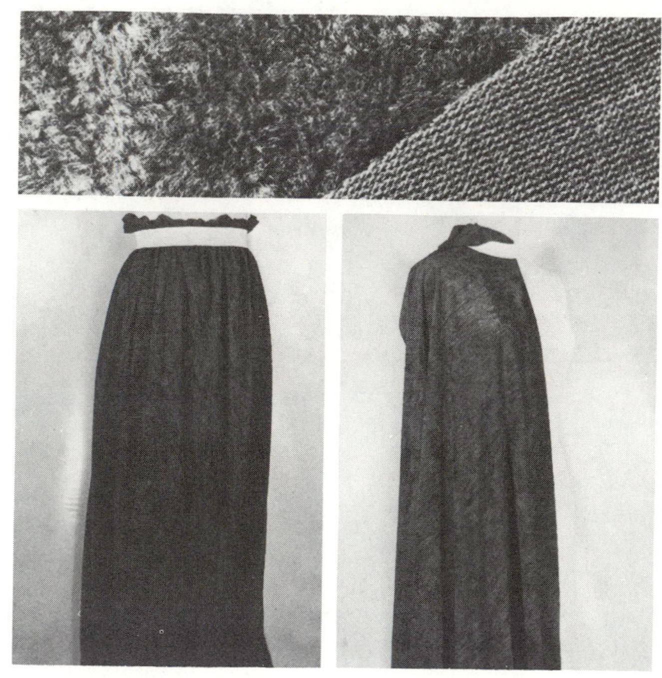

Crushed Velvet (dull face & knit ground)

Fiber Content 70% dull acetate/30% dull nylon
Yarn Type filament fiber
Yarn Construction textured
Fabric Structure pile insertion double knit
Finishing Processes brushing; delustering; shearing; steamed and roller pressed for crushed effect
Color Application piece dyed–union dyed; yarn dyed
Width 68–72 inches (172.7–182.9 cm)
Weight medium-heavy
Hand furry; soft
Texture high and low areas; unevenly cut pile face; flat back
Performance Expectations elasticity; elongation; flexible; mildew and moth resistant; resilient
Drapability Qualities falls into moderately soft flares; accommodates fullness by gathering; fullness retains soft fall

Recommended Construction Aids & Techniques

Hand Needles cotton darners, 3–6; embroidery, 4–6; milliners, 3–7; sharps, 1–6
Machine Needles ball point, 11–14
Threads poly-core waxed, 50; cotton-polyester blend, all purpose; cotton-covered polyester core, all purpose; spun polyester, all purpose; nylon/Dacron, twist A
Hems book; single-fold bias binding; bound/Hong Kong finish; double-stitched; machine-stitched; overedged
Seams plain; overedged; safety-stitched
Seam Finishes single-ply bound; Hong Kong bound; serging/single-ply overedged; untreated/plain
Pressing steam; use needle board; safe temperature limit 250°F (122.1°C)
Care launder; dry clean
Fabric Resource **Gloversville Mills Inc.**

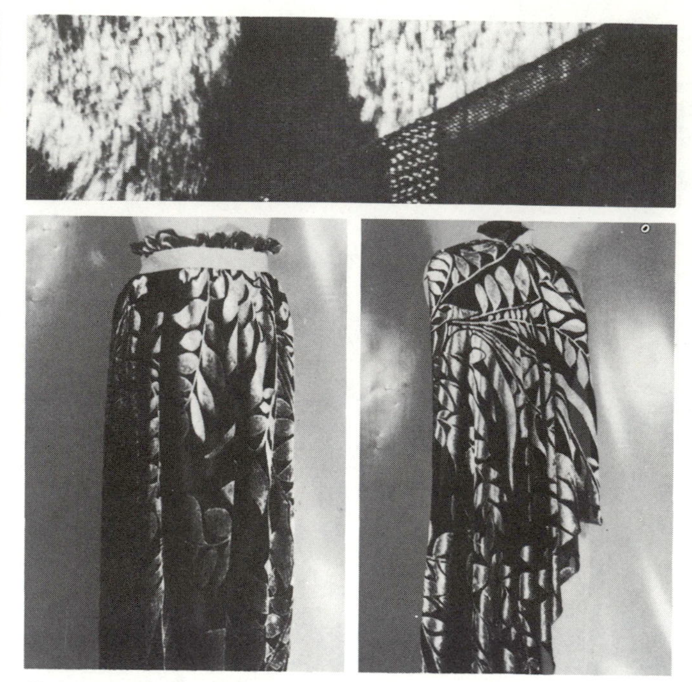

Cut/Beaded Velvet

Fiber Content 65% viscose rayon/35% silk
Yarn Type filament fiber
Yarn Construction bulked; conventional; high twist; textured
Fabric Structure pile; Jacquard loom
Finishing Processes brushing; cutting; shearing; stabilizing
Color Application piece dyed–cross dyed; yarn dyed
Width 46–47 inches (116.8–119.4 cm)
Weight medium-light
Hand fragile; limp; soft
Texture velvet brocade pattern on sheer ground
Performance Expectations pile tends to abrade and pull out with wear; dimensional stability; flexible
Drapability Qualities falls into soft languid flares; accommodates fullness by pleating, gathering, elasticized shirring; fullness retains soft graceful fall

Recommended Construction Aids & Techniques

Hand Needles beading, 10–13; betweens, 6–7; cotton darners, 9–10; embroidery, 10; milliners, 10; sharps, 7–10
Machine Needles round/set point, fine 9
Threads silk, industrial size A; poly-core waxed, 70; cotton-polyester blend, extra fine; cotton-covered polyester core, extra fine; spun polyester, fine; nylon/Dacron, monocord A
Hems bound/Hong Kong finish; edge-stitched; horsehair; machine-stitched; hand- or machine-rolled
Seams hairline; plain; safety-stitched
Seam Finishes single-ply or double-ply bound; Hong Kong bound; double-stitched and trimmed; single-ply overcast; pinked; pinked and stitched; serging/single-ply overedged; untreated/plain
Pressing steam; use needle board; safe temperature limit 300°F (150.1°C)
Care dry clean
Fabric Resources **American Silk Mills Corp./Bucol; Sormani Co. Inc./Diochon**

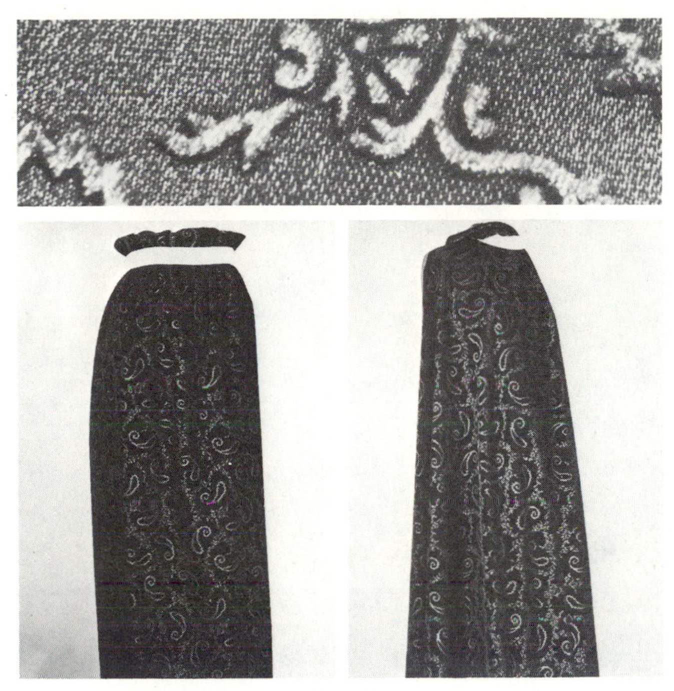

Embossed/Sculptured Velvet

Fiber Content 100% rayon
Yarn Type filament fiber
Yarn Construction textured
Fabric Structure warp pile
Finishing Processes brushing; heat-set embossing; shearing
Color Application piece dyed; yarn dyed
Width 39–40 inches (99.1–101.6 cm)
Weight medium
Hand velvety soft; springy
Texture high and low pile areas form pattern; uneven surface on face; flat back
Performance Expectations pile tends to abrade and flatten with wear; elongation; good tensile strength
Drapability Qualities falls into moderately soft flares; fullness maintains lofty effect; retains shape of garment

Recommended Construction Aids & Techniques

Hand Needles betweens, 3–4; cotton darners, 6–8; embroidery, 7–8; milliners, 7–9; sharps, 4–6
Machine Needles round/set point, medium 11–14
Threads poly-core waxed, 50; cotton-polyester blend, all purpose; cotton-covered polyester core, all purpose; spun polyester, all purpose; nylon/Dacron, twist A
Hems book; single-fold bias binding; bound/Hong Kong finish; edge-stitched; machine-stitched; overedged; seam binding
Seams flat-felled; plain; safety-stitched; welt; double welt; horsehair tape for zipper insertion
Seam Finishes single-ply bound; Hong Kong bound; single-ply overcast; pinked; pinked and stitched; serging/single-ply overedged; untreated/plain
Pressing steam; use needle board; safe temperature limit 350°F (178.1°C)
Care dry clean only
Fabric Resources **Deluxe Velvet Co.; J. B. Martin Co. Inc.**

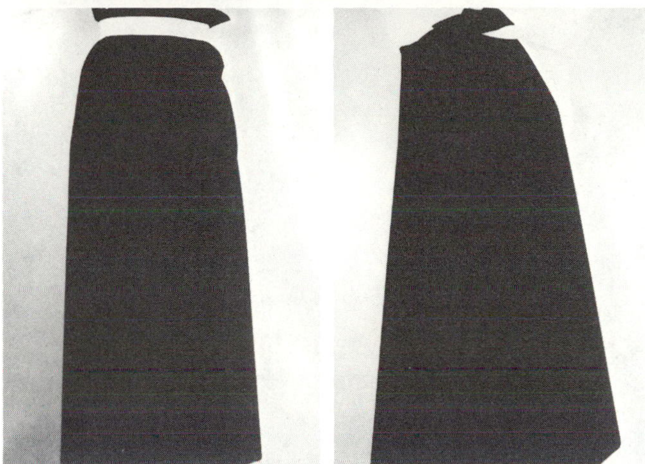

Flocked/Simulated Velvet

Fiber Content 100% nylon (face); cotton/polyester blend (back)
Yarn Type face: filament staple, flock fibers; back: staple, filament fiber
Yarn Construction cut staple for flocked face; conventional (base)
Fabric Structure flocked face; plain weave back
Finishing Processes brushing; electrocoating flocking; mercerizing
Color Application piece dyed–union dyed
Width 44–45 inches (111.8–114.3 cm)
Weight heavy
Hand soft face; firm; papery
Texture fuzzy, close pile face; flat back
Performance Expectations subject to color crocking and flex abrasion; dimensional stability; durable
Drapability Qualities falls into moderately firm wide cones; fullness maintains crisp effect; retains shape of garment

Recommended Construction Aids & Techniques

Hand Needles betweens, 3–4; cotton darners, 6–8; embroidery, 7–8; milliners, 7–9; sharps, 4–6
Machine Needles leather point, 13–16
Threads poly-core waxed, 50; cotton-polyester blend, heavy duty; cotton-covered polyester core, heavy duty; spun polyester, heavy duty
Hems book; bonded; single-fold bias binding; bound/Hong Kong finish; double-stitched; edge-stitched; flat/plain; machine-stitched; overedged; seam binding
Seams plain; safety-stitched
Seam Finishes single-ply bound; Hong Kong bound; single-ply overcast; pinked; pinked and stitched; serging/single-ply overedged; untreated/plain
Pressing steam; use needle board; safe temperature limit 325°F (164.1°C)
Care dry clean only
Fabric Resource **Pervel Inc.**

Lyons Velvet

Fiber Content 65% cotton/35% rayon
Yarn Type combed staple; filament fiber
Yarn Construction conventional; textured
Fabric Structure warp pile face; plain weave back
Finishing Processes brushing; crush and spot resistant; shearing
Color Application piece dyed–union dyed; yarn dyed
Width 39–40 inches (99.1–101.6 cm)
Weight medium-heavy
Hand semi-crisp; lofty; springy
Texture short, dense, erect pile face; flat back
Performance Expectations pile tends to abrade; absorbent; elongation; good tensile strength
Drapability Qualities falls into firm wide cones; fullness maintains lofty bouffant effect; retains shape of garment

Recommended Construction Aids & Techniques

Hand Needles betweens, 3–4; cotton darners, 6–8; embroidery, 7–8; milliners, 7–9; sharps, 4–6
Machine Needles round/set point, medium 11–14
Threads poly-core waxed, 50; cotton-polyester blend, all purpose; cotton-covered polyester core, all purpose; spun polyester, all purpose; nylon/Dacron, twist A
Hems book; single-fold bias binding; bound/Hong Kong finish; edge-stitched; machine-stitched; overedged; seam binding
Seams flat-felled; plain; safety-stitched; welt; double welt; horsehair tape for zipper insertion
Seam Finishes single-ply bound; Hong Kong bound; single-ply overcast; pinked; pinked and stitched; serging/single-ply overedged; untreated/plain
Pressing steam; use needle board; safe temperature limit 350°F (178.1°C)
Care machine wash warm and tumble dry, remove promptly; dry clean
Fabric Resource **J. B. Martin Co. Inc.**

Lyons Velvet (dull face)

Fiber Content 100% rayon
Yarn Type filament fiber
Yarn Construction delustered yarn; textured
Fabric Structure warp pile face; plain weave back
Finishing Processes delustering; crush and spot resistant
Color Application piece dyed–beck dyed/reel dyed
Width 39–40 inches (99.1–101.6 cm)
Weight medium
Hand lofty; springy; wiry
Texture dull face; short, dense erect pile face; flat back
Performance Expectations pile tends to abrade; absorbent; elongation; good tensile strength
Drapability Qualities falls into firm wide cones; fullness maintains lofty bouffant effect; retains shape of garment

Recommended Construction Aids & Techniques

Hand Needles betweens, 3–4; cotton darners, 6–8; embroidery, 7–8; milliners, 7–9; sharps, 4–6
Machine Needles round/set point, medium 11–14
Threads poly-core waxed, 50; cotton-polyester blend, all purpose; cotton-covered polyester core, all purpose; spun polyester, all purpose; nylon/Dacron, twist A
Hems book; single-fold bias binding; bound/Hong Kong finish; edge-stitched; machine-stitched; overedged; seam binding
Seams flat-felled; plain; safety-stitched; welt; double welt; horsehair tape for zipper insertion
Seam Finishes single-ply bound; Hong Kong bound; single-ply overcast; pinked; pinked and stitched; serging/single-ply overedged; untreated/plain
Pressing steam; use needle board; safe temperature limit 350°F (178.1°C)
Care dry clean only
Fabric Resources **Deluxe Velvet Co.; J. B. Martin Co. Inc.**

Panne Velvet (knit ground)

Fiber Content 70% acetate/30% nylon
Yarn Type filament fiber
Yarn Construction conventional
Fabric Structure warp pile face; jersey knit base
Finishing Processes brushing; shearing; steamed and roller pressed in one direction
Color Application piece dyed–union dyed; yarn dyed
Width 68–72 inches (172.7–182.9 cm)
Weight medium-heavy
Hand soft; stretch in crosswise direction
Texture pile pressed flat in one direction; satiny lustrous face; flat dull back
Performance Expectations subject to abrasion; edges subject to curling; dimensional stability; elongation; flexible; stretch properties
Drapability Qualities falls into soft flares; can be shaped or molded by stretching; falls close to body contour

Recommended Construction Aids & Techniques

Hand Needles betweens, 3–4; cotton darners, 6–8; embroidery, 7–8; milliners, 7–9; sharps, 4–6
Machine Needles ball point, 11–14
Threads poly-core waxed, 50; cotton-polyester blend, all purpose; cotton-covered polyester core, all purpose; spun polyester, all purpose; nylon/Dacron, twist A
Hems book; single-fold bias binding; bound/Hong Kong finish; double-stitched; flat/plain; machine-stitched; overedged
Seams overedged; plain; safety-stitched
Seam Finishes single-ply bound; Hong Kong bound; serging/single-ply overedged; untreated/plain
Pressing steam; use needle board; safe temperature limit 250°F (122.1°C)
Care machine wash warm and tumble dry, remove promptly; dry clean
Fabric Resource **Gloversville Mills Inc.**

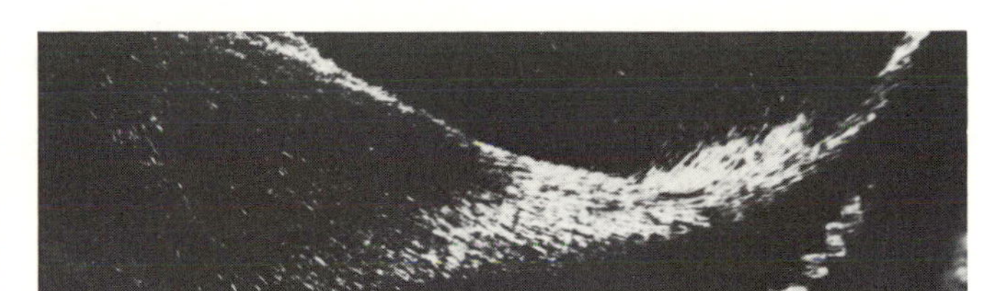

Panne Velvet (woven ground)

Fiber Content 100% rayon
Yarn Type filament fiber
Yarn Construction textured
Fabric Structure warp pile
Finishing Processes Sylmer® silicone crush and spot resistant
Color Application piece dyed; yarn dyed
Width 40 inches (101.6 cm)
Weight medium
Hand limp; soft; supple
Texture pile pressed flat in one direction; lustrous face; flat dull back
Performance Expectations pile tends to abrade; absorbent; elongation; good tensile strength
Drapability Qualities falls into soft flares; fullness retains soft fall; retains shape of garment

Recommended Construction Aids & Techniques

Hand Needles betweens, 3–4; cotton darners, 6–8; embroidery, 7–8; milliners, 7–9; sharps, 4–6
Machine Needles round/set point, medium 11–14
Threads poly-core waxed, 50; cotton-polyester blend, all purpose; cotton-covered polyester core, all purpose; spun polyester, all purpose; nylon/Dacron, twist A
Hems book; single-fold bias binding; bound/Hong Kong finish; double-stitched; edge-stitched; machine-stitched; overedged; seam binding
Seams flat-felled; plain; safety-stitched; welt; double welt
Seam Finishes single-ply bound; Hong Kong bound; serging/single-ply overedged; untreated/plain
Pressing steam; use needle board; safe temperature limit 350°F (178.1°C)
Care dry clean only
Fabric Resource **American Silk Mills Corp.**

Transparent/Chiffon Velvet

Fiber Content 100% rayon
Yarn Type filament fiber
Yarn Construction conventional
Fabric Structure warp pile face; plain weave back
Finishing Processes brushing; crush and spot resistant; shearing
Color Application piece dyed-beck dyed/reel dyed
Width 39-40 inches (99.1-101.6 cm)
Weight medium-light
Hand lofty; soft; thin
Texture short, dense erect pile face; satiny smooth back; light-admitting thinness
Performance Expectations pile tends to crush and flatten with wear; elongation; good tensile strength
Drapability Qualities falls into soft wide cones; fullness maintains lofty effect; retains shape of garment

Recommended Construction Aids & Techniques

Hand Needles betweens, 3-4; cotton darners, 6-8; embroidery, 7-8; milliners, 7-9; sharps, 4-6
Machine Needles round/set point, medium 11
Threads poly-core waxed, 70; cotton-polyester blend, fine; cotton-covered polyester core, fine; spun polyester, fine; nylon/Dacron, monocord A
Hems book; single-fold bias binding; bound/Hong Kong finish; edge-stitched; machine-stitched; overedged; seam binding
Seams flat-felled; plain; safety-stitched; welt; double welt; horsehair tape for zipper insertion
Seam Finishes single-ply bound; Hong Kong bound; single-ply overcast; pinked; pinked and stitched; serging/single-ply overedged; untreated/plain
Pressing steam; use needle board; safe temperature limit 350°F (178.1°C)
Care dry clean only
Fabric Resources **Deluxe Velvet Co.; J. B. Martin Co. Inc.**

Velvet Cord

Fiber Content 100% cotton
Yarn Type combed staple
Yarn Construction conventional
Fabric Structure warp pile
Finishing Processes brushing; crush and spot resistant; cutting; napping; shearing
Color Application continuous and jig dyed
Width 39-40 inches (99.1-101.6 cm)
Weight heavy
Hand compact; firm; soft
Texture uneven surface; high and low wide rib in lengthwise direction on face; flat back
Performance Expectations pile tends to abrade; absorbent; antistatic; durable; elongation; nonpilling; high tensile strength
Drapability Qualities falls into firm wide flares; fullness maintains lofty bouffant effect; retains shape of garment

Recommended Construction Aids & Techniques

Hand Needles cotton darners, 1-4; embroidery, 1-4; milliners, 3-5
Machine Needles round/set point, medium-coarse 14-16
Threads mercerized cotton, heavy duty; six-cord cotton, 60; cotton- or poly-core waxed, 40; cotton-covered polyester core, heavy duty
Hems double-fold or single-fold bias binding; bound/Hong Kong finish; faced; machine-stitched; overedged; seam binding
Seams plain; safety-stitched
Seam Finishes single-ply bound; Hong Kong bound; single-ply overcast; pinked; pinked and stitched; serging/single-ply overedged; untreated/plain
Pressing steam; use needle board; safe temperature limit 400°F (206.1°C)
Care machine wash warm and tumble dry, remove promptly; dry clean
Fabric Resource **J. B. Martin Co. Inc.**

Velvet Pinstripe Cord

Fiber Content 100% cotton
Yarn Type combed staple
Yarn Construction conventional
Fabric Structure warp pile
Finishing Processes Sylmer® silicone crush and spot resistant
Color Application continuous and jig dyed
Width 39-40 inches (99.1-101.6 cm)
Weight heavy
Hand compact; firm; soft
Texture uneven surface; high and low narrow rib in lengthwise direction on face; flat back
Performance Expectations pile tends to abrade; absorbent; antistatic; durable; elongation; nonpilling; high tensile strength
Drapability Qualities falls into moderately firm wide flares; fullness maintains lofty effect; retains shape of garment

Recommended Construction Aids & Techniques

Hand Needles cotton darners, 1-4; embroidery, 1-4; milliners, 3-5
Machine Needles round/set point, medium-coarse 14-16
Threads mercerized cotton, heavy duty; six-cord cotton, 60; cotton- or poly-core waxed, 40; cotton-covered polyester core, heavy duty
Hems book; double-fold or single-fold bias binding; bound/Hong Kong finish; faced; machine-stitched; overedged; seam binding
Seams plain; safety-stitched
Seam Finishes single-ply bound; Hong Kong bound; single-ply overcast; pinked; pinked and stitched; serging/single-ply overedged; untreated/plain
Pressing steam; use needle board; safe temperature limit 400°F (206.1°C)
Care machine wash warm and tumble dry, remove promptly; dry clean
Fabric Resource **American Silk Mills Corp.**

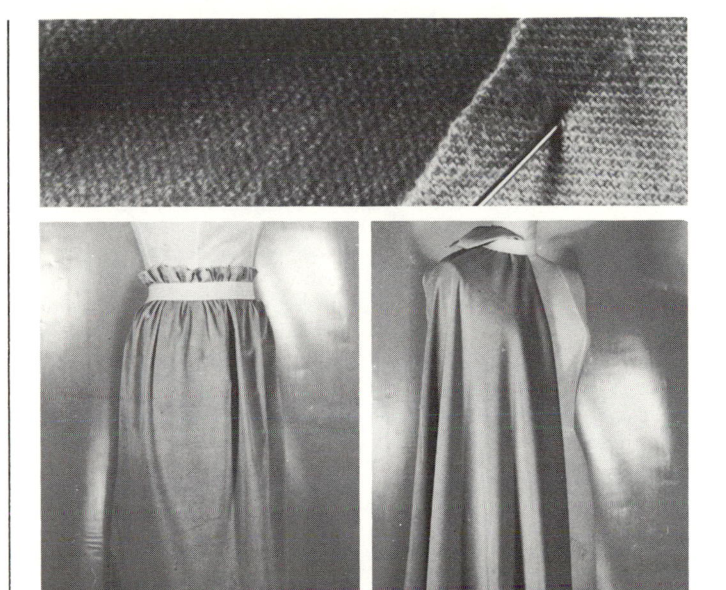

Velveteen

Fiber Content 100% cotton
Yarn Type combed staple
Yarn Construction conventional; fine
Fabric Structure filling pile face (wire method); twill weave back
Finishing Processes cross brushing; crush resistant; pressing; shearing
Color Application continuous and jig dyed
Width 44-45 inches (111.8-114.3 cm)
Weight medium-heavy
Hand compact; firm; soft
Texture short, closely set even pile on face; pile slopes slightly; flat back
Performance Expectations pile tends to abrade; absorbent; antistatic; durable; elongation; nonpilling; high tensile strength
Drapability Qualities falls into moderately soft flares; fullness maintains lofty effect; retains shape of garment

Recommended Construction Aids & Techniques

Hand Needles cotton darners, 3-6; embroidery, 4-6; milliners, 3-7; sharps, 1-6
Machine Needles round/set point, medium 11-14
Threads mercerized cotton, heavy duty; six-cord cotton, 40; cotton- or poly-core waxed, 50; cotton-covered polyester core, all purpose
Hems book; single-fold bias binding; bound/Hong Kong finish; edge-stitched; machine-stitched; overedged; seam binding
Seams flat-felled; plain; safety-stitched; welt; double welt
Seam Finishes single-ply bound; Hong Kong bound; single-ply overcast; pinked; pinked and stitched; serging/single-ply overedged; untreated/plain
Pressing steam; use needle board; safe temperature limit 400°F (206.1°C)
Care machine wash warm and tumble dry, remove promptly; dry clean
Fabric Resource **Crompton-Richmond Inc.**

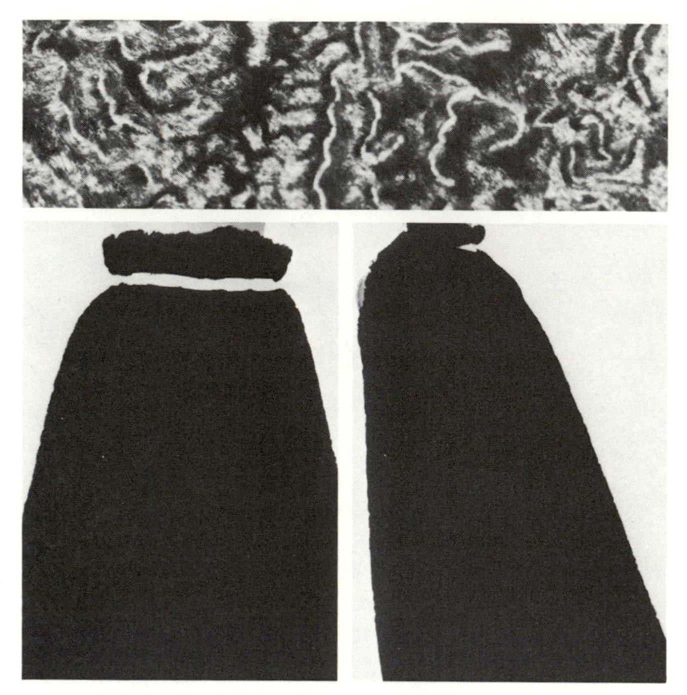

Astrakhan/Persian Lamb Cloth

Fiber Content 100% acrylic (face); 100% polyester (back)
Yarn Type filament fiber
Yarn Construction textured; loose twist; thick heat-set curled yarn (face); thin yarn (base)
Fabric Structure high pile face; jersey knit base
Finishing Processes face: crush resistant, randomly sheared, swirl brushed; back: polypropylene-resin bonded
Color Application piece dyed-union dyed; solution dyed; yarn dyed
Width 58-60 inches (147.3-172.7 cm)
Weight heavy
Hand soft; thick; woolly
Texture high and low curled face simulates pelt of karakul lamb; flat back
Performance Expectations subject to abrasion; compressible; dimensional stability; insulating properties; resilient
Drapability Qualities falls into wide cones; retains shape or silhouette of garment; better utilized if fitted by seaming and eliminating excess fabric

Recommended Construction Aids & Techniques

Hand Needles chenilles, 18-22; cotton darners, 1-3; yarn darners, 14-18; embroidery, 1-3; tapestry, 18-22
Machine Needles ball point, 16
Threads silk, industrial size B-C-D; poly-core waxed, 40 heavy duty; cotton-covered polyester core, heavy duty; spun polyester, heavy duty; nylon/Dacron, twist A
Hems bound/Hong Kong finish; faced; flat/plain; overedged
Seams overedged; plain; safety-stitched; stayed; taped
Seam Finishes Hong Kong bound; serging/single-ply overedged; untreated/plain; single-ply zigzag; clip or shear pile from seam allowance as close to backing as possible
Pressing steam; use needle board; safe temperature limit 300°F (150.1°C)
Care dry clean
Fabric Resource Malda Fabrics Inc. (sample courtesy of College Town)

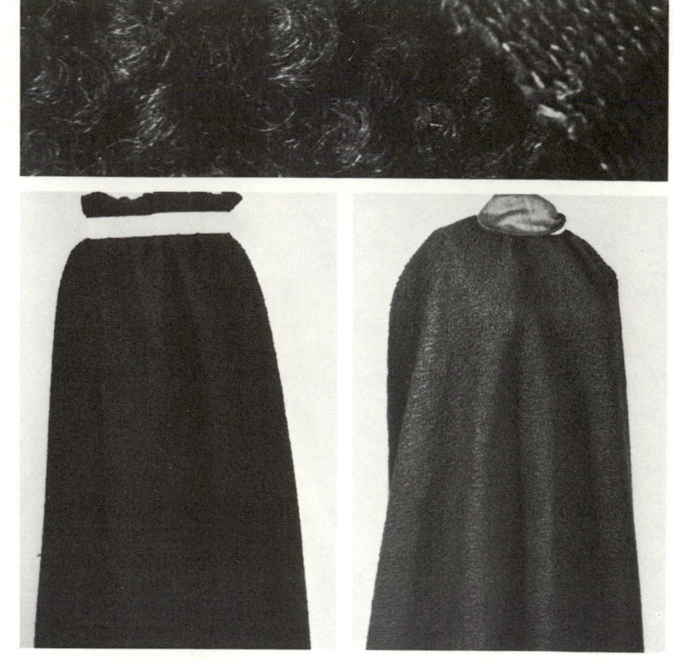

Poodle/Poodle Fur Cloth

Fiber Content 100% polyester
Yarn Type filament fiber
Yarn Construction looped and bouclé novelty yarns; textured
Fabric Structure looped pile face; jersey knit base
Finishing Processes brushing; crush resistant; napping
Color Application solution dyed; piece dyed; yarn dyed
Width 58-60 inches (147.3-152.4 cm)
Weight medium-heavy
Hand lofty; silky; springy
Texture uneven looped face simulates tightly curled hair of poodle; flat back
Performance Expectations looped pile tends to abrade; compressible; dimensional stability; durable; elongation; resilient; high tensile strength; subject to snagging due to looped yarn
Drapability Qualities falls into soft wide cones; fullness maintains lofty bouffant effect; retains shape of garment

Recommended Construction Aids & Techniques

Hand Needles chenilles, 18-22; cotton darners, 1-3; yarn darners, 14-18; embroidery, 1-3; tapestry, 18-22
Machine Needles ball point, 14
Threads silk, industrial size B-C; poly-core waxed, 50; cotton-covered polyester core, all purpose; spun polyester, all purpose; nylon/Dacron, twist A
Hems bound/Hong Kong finish; double-stitched; faced; flat/plain; overedged
Seams overedged; plain; safety-stitched; stayed; taped
Seam Finishes Hong Kong bound; serging/single-ply overedged; untreated/plain; single-ply zigzag; clip or shear pile from seam allowance as close to backing as possible
Pressing steam; use needle board; safe temperature limit 350°F (178.1°C)
Care dry clean
Fabric Resource Boverman Fabrics (sample courtesy of Fabric Room, FIT)

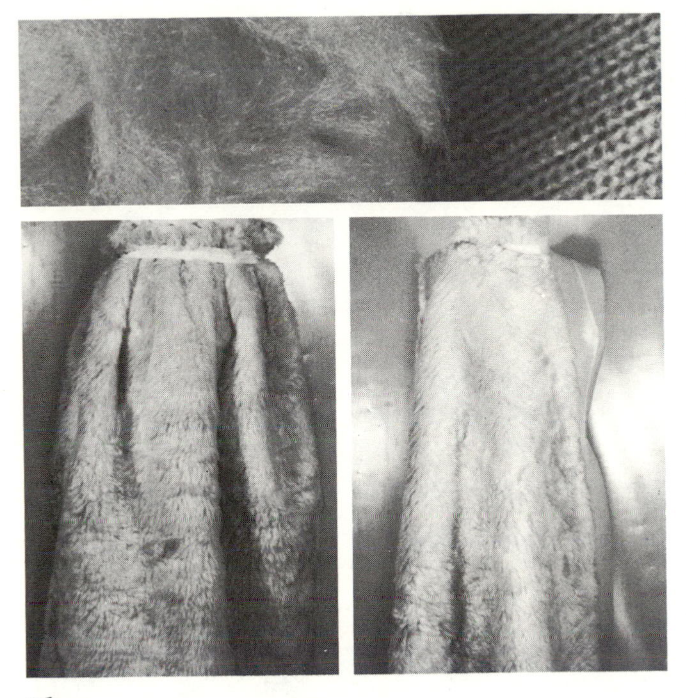

Shag

Fiber Content 100% acrylic (face)
Yarn Type filament fiber; filament staple
Yarn Construction textured
Fabric Structure pile face; jersey knit base
Finishing Processes face: brushing, crush resistant, napping; back: polypropylene-resin bonded
Color Application solution dyed; piece dyed; yarn dyed
Width 72 inches (182.9 cm)
Weight heavy
Hand soft; thick; woolly
Texture furry, high, thick, fuzzy pile face; flat back
Performance Expectations pile tends to abrade and pill; compressible; dimensional stability; insulating properties; resilient
Drapability Qualities falls into wide cones; retains shape or silhouette of garment; better utilized if fitted by seaming and eliminating excess fabric

Recommended Construction Aids & Techniques

Hand Needles chenilles, 18–22; cotton darners, 1–3; yarn darners, 14–18; embroidery, 1–3; tapestry, 18–22
Machine Needles ball point, 16
Threads silk, industrial size B-C-D; poly-core waxed, 40 heavy duty; cotton-covered polyester core, heavy duty; spun polyester, heavy duty; nylon/Dacron, twist A
Hems bound/Hong Kong finish; faced; flat/plain; overedged
Seams overedged; plain; safety-stitched; stayed
Seam Finishes Hong Kong bound; serging/single-ply overedged; untreated/plain; single-ply zigzag; clip or shear pile from seam allowance as close to backing as possible
Pressing steam; use needle board; safe temperature limit 300°F (150.1°C)
Care dry clean
Fabric Resource Furtex Inc.

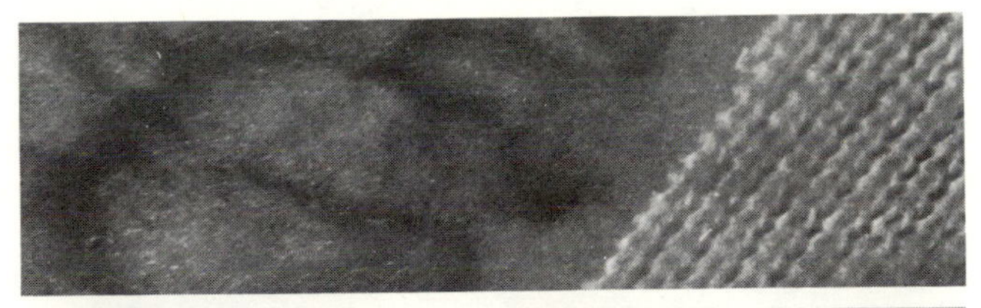

Sherpa/Sheared Lamb Cloth

Fiber Content 100% polyester
Yarn Type filament fiber; filament staple
Yarn Construction textured; slack twist
Fabric Structure high pile face; jersey knit base
Finishing Processes face: swirl and cross brushed, crush resistant, shearing; back: polypropylene-resin bonded
Color Application solution dyed; piece dyed; yarn dyed
Width 72 inches (182.9 cm)
Weight heavy
Hand soft; spongy; thick
Texture high and low woolly patterned pile on face simulates sheared lamb; flat back
Performance Expectations subject to color crocking and edge abrasion; compressible; dimensional stability; durable; elongation; resilient
Drapability Qualities falls into wide cones; retains shape or silhouette of garment; better utilized if fitted by seaming and eliminating excess fabric

Recommended Construction Aids & Techniques

Hand Needles chenilles, 18–22; cotton darners, 1–3; yarn darners, 14–18; embroidery, 1–3; tapestry, 18–22
Machine Needles ball point, 16
Threads silk, industrial size B-C-D; poly-core waxed, 40 heavy duty; cotton-covered polyester core, heavy duty; spun polyester, heavy duty; nylon/Dacron, twist A
Hems bound/Hong Kong finish; faced; flat/plain; overedged
Seams overedged; plain; safety-stitched; stayed
Seam Finishes Hong Kong bound; serging/single-ply overedged; untreated/plain; single-ply zigzag; clip or shear pile from seam allowance as close to backing as possible
Pressing steam; use needle board; safe temperature limit 350°F (178.1°C)
Care dry clean
Fabric Resources Furtex Inc.; Gloversville Mills Inc.

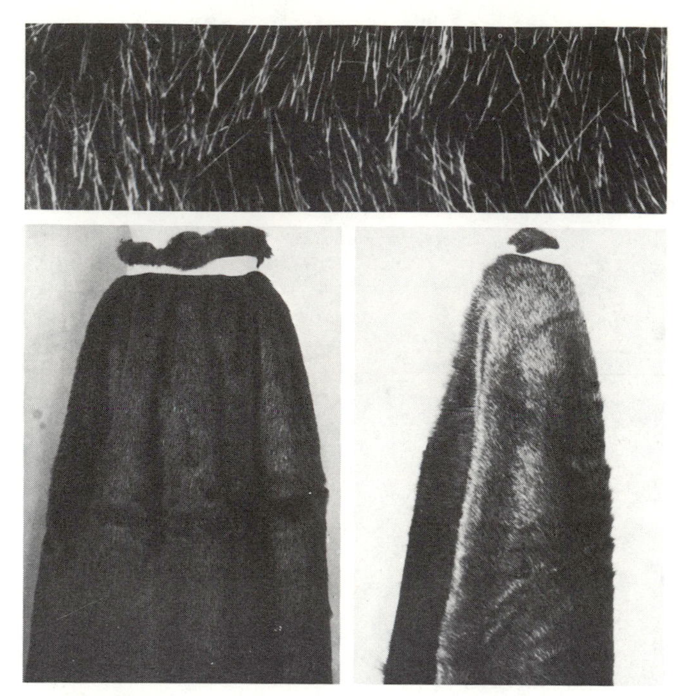

Teddy Bear

Fiber Content 75% acrylic/25% modacrylic
Yarn Type filament fiber; filament staple
Yarn Construction textured
Fabric Structure high pile face; jersey knit base
Finishing Processes face: brushing, crush resistant, napping; back: polypropylene-resin bonded
Color Application solution dyed; piece dyed–cross dyed; yarn dyed
Width 72 inches (182.9 cm)
Weight heavy
Hand soft; thick; woolly
Texture furry, high, frizzy pile on face simulates bear fur; flat back
Performance Expectations pile tends to abrade; compressible; dimensional stability; insulating properties; resilient
Drapability Qualities falls into wide cones; retains shape or silhouette of garment; better utilized if fitted by seaming and eliminating excess fabric

Recommended Construction Aids & Techniques

Hand Needles chenilles, 18–22; cotton darners, 1–3; yarn darners, 14–18; embroidery, 1–3; tapestry, 18–22
Machine Needles ball point, 16
Threads silk, industrial size B-C-D; poly-core waxed, 40 heavy duty; cotton-covered polyester core, heavy duty; spun polyester, heavy duty; nylon/Dacron, twist A
Hems bound/Hong Kong finish; faced; flat/plain; overedged
Seams overedged; plain; safety-stitched; stayed
Seam Finishes Hong Kong bound; serging/single-ply overedged; untreated/plain; single-ply zigzag; clip or shear pile from seam allowance as close to backing as possible
Pressing steam; use needle board; safe temperature limit 225°F (108.1°C)
Care dry clean
Fabric Resource **Furtex Inc.**

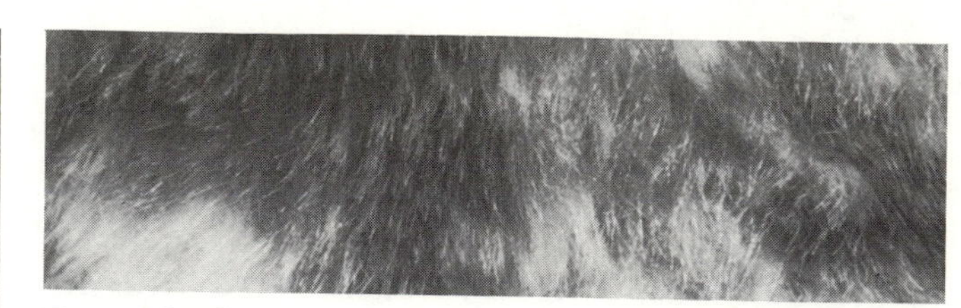

Fox Cloth

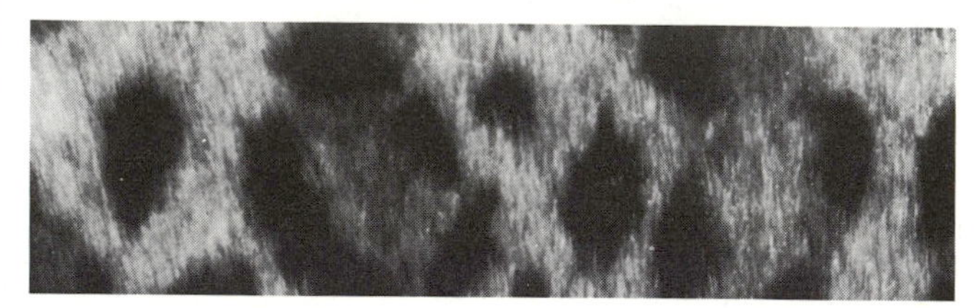

Leopard Cloth

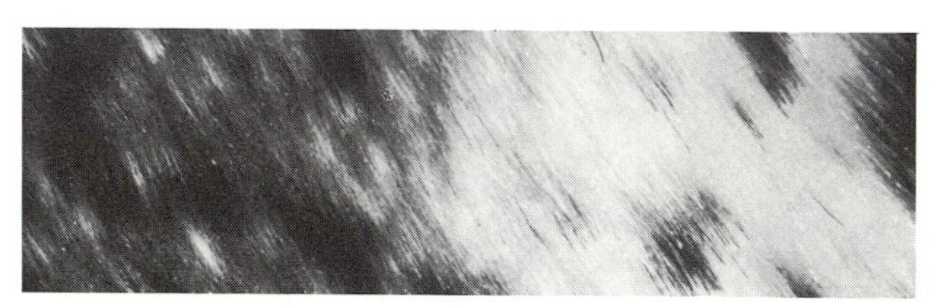

Lynx Cloth

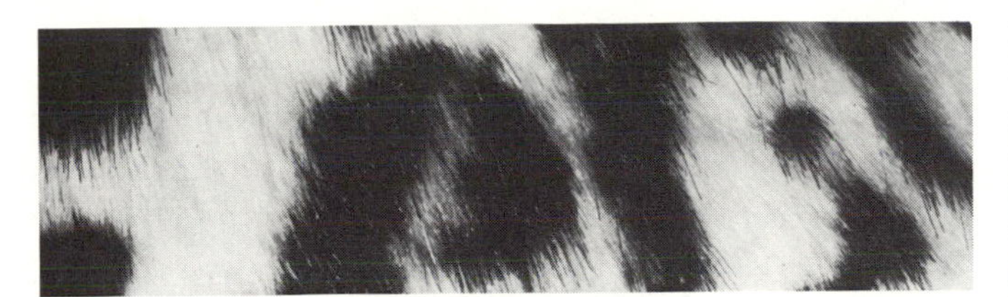

Ocelot Cloth

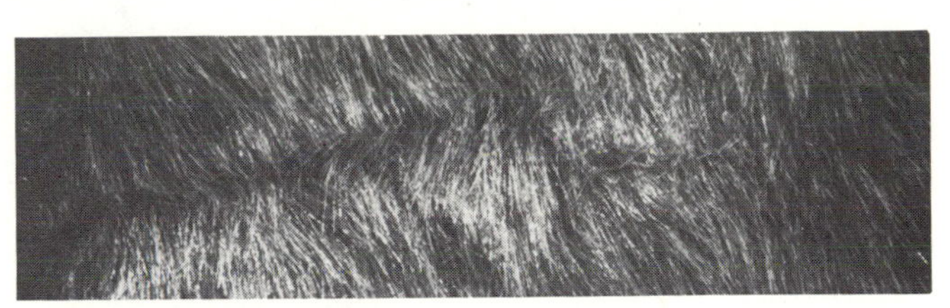

Ranch Mink Cloth

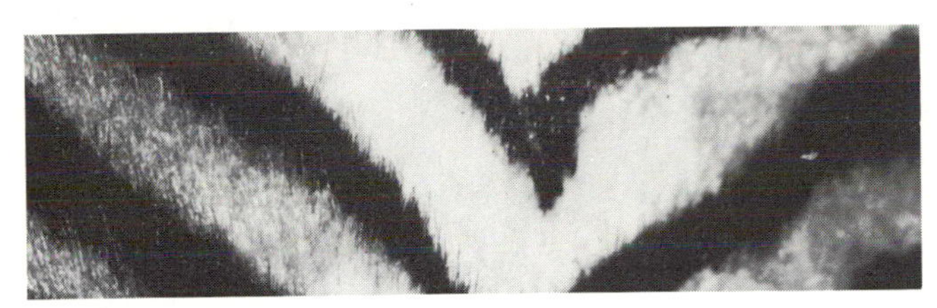

Pieced Rabbit Cloth

Seal Cloth

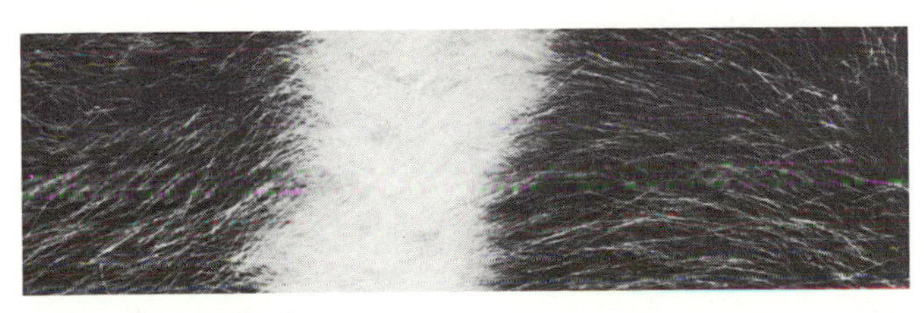

Raccoon Cloth

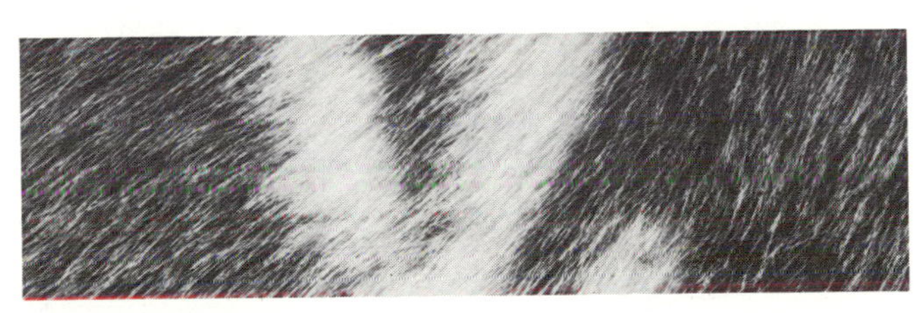

Skunk Cloth

15 ~ Stretch/Stretch-type Fabrics

Fiber Stretch
- Lycra® (180–220% fiber stretch)
- Lycra® (140–175% fiber stretch)
- Lycra® (110–140% fiber stretch)

Yarn Stretch
- Canvas (yarn stretch)
- Chino (yarn stretch)
- Denim (yarn stretch)
- Gabardine (yarn stretch)
- Poplin (yarn stretch)
- Satin (yarn stretch)
- Twill (yarn stretch)
- Whipcord (yarn stretch)

Fabric Structure & Yarn Stretch
- Lightweight Knit (structure & yarn stretch)
- Medium-weight Knit (structure & yarn stretch)
- Heavyweight Knit (structure & yarn stretch)
- Stretch Terry Cloth (structure & yarn stretch)
- Stretch Lace (structure & yarn stretch)

Finish Stretch
- Crimped Cloth (finish stretch)
- Crinkled Double Cloth (finish stretch)

A stretch fabric is characterized by its "snap-back" or stretch and recovery properties which provide give during movement of the body.

Stretch fabrics may be made of:

- Natural or man-made rubber fibers
- Uncovered or bare Spandex or Lycra yarns
- Single-covered or core-spun elastomeric yarns
- Twist-texturing or false-twist type stretch yarns
- Heat-set, long-lasting stretch-type yarns
- Mechanically or chemically finished on finished goods
- Woven or knitted structure

Core-spun elastomeric yarns are produced by an elastomeric fiber core covered with other fibers. The yarns maintain the hand and appearance of the covering yarn or yarns. Stretch and recovery qualities of the stretch fabric containing core-spun elastomeric yarns can be controlled. Durability and snap-back properties are high.

Twist-texturing yarns are produced by a process applied to man-made thermoplastic filament fibers. The individual yarns are twisted or crimped, heat-set in the textured condition, then untwisted prior to weaving. Twist-texturing yarns in stretch fabrics extend when stretched and recoil or recover elastic properties in the relaxed state. Stretch and recovery properties are built in. Durability and snap-back properties are considered good.

Mechanical or chemical stretch finish applied to finished fabric. Extra crimp is imparted to the finished fabric *chemically, mechanically,* or by using both methods, providing the finished fabrics with stretch properties. Recovery or snap-back properties are slow. Durability of stretch qualities is poor.

Characteristics of stretch fabrics include:

- Give or action in the lengthwise, crosswise or both directions
- Restraining power depending on yarn and structure
- Suppleness and flexibility at the shoulder, elbow, hip, knee and seat areas
- Greater comfort in active sportswear garments

Stretch fabric appearance and hand depend on:

- Fiber content
- Type of yarns
- Methods of imparting stretch
- Weaving or knitting process of fabric structure
- Type of finishing processes

Lycra® (180–220% fiber stretch)

Fiber Content 24% spandex/76% nylon
Yarn Type elastomeric fiber; filament fiber
Yarn Construction core: elastic yarn; outer layer: 70-denier filament yarn
Fabric Structure knit
Finishing Processes calendering; flame retardant; stabilizing
Color Application piece dyed–union dyed; solution dyed
Width 90–92 inches (228.6–233.7 cm)
Weight light
Hand fine; semi-crisp; stretchy
Texture semi-sheer; silky; smooth
Performance Expectations elongation and recovery with 180–220% stretch; power stretch properties; resilient; thermoplastic qualities
Drapability Qualities falls into soft flares; fullness retains soft fall; can be molded or shaped by stretching

Recommended Construction Aids & Techniques

Hand Needles betweens, 3–5; cotton darners, 8–10; embroidery, 8–10; milliners, 9–10; sharps, 8–10
Machine Needles ball point, 9
Threads poly-core waxed, 60 fine; spun polyester, fine; nylon/Dacron, monocord A
Hems double-stitched; edge-stitched; flat/plain; machine-stitched; overedged
Seams hairline; overedged; plain; safety-stitched; taped; tissue-stitched
Seam Finishes double-ply bound; double-stitched; serging/single-ply overedged; untreated/plain; double-ply zigzag
Pressing steam; use pressing cloth; safe temperature limit 300°F (150.1°C)
Care launder; remove from dryer promptly
Fabric Resource **Liberty Fabrics of New York**

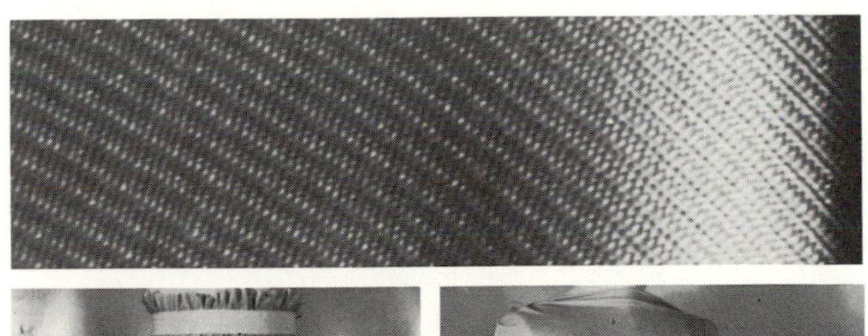

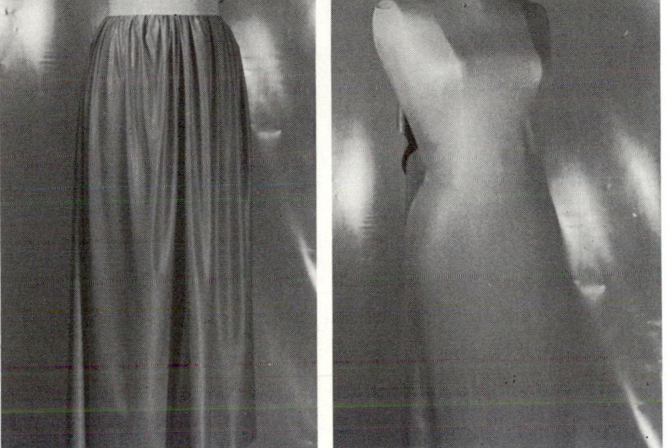

Lycra® (140–175% fiber stretch)

Fiber Content 39% spandex/61% nylon
Yarn Type elastomeric fiber; filament fiber
Yarn Construction core: extruded elastic; outer layer: spun yarn
Fabric Structure knit
Finishing Processes calendering; flame retardant; stabilizing
Color Application piece dyed–union dyed; solution dyed
Width 60–62 inches (152.4–157.5 cm)
Weight medium-light
Hand rubbery; slippery; stretchy
Texture flat; silky; smooth
Performance Expectations elongation and recovery with 140–175% stretch; power stretch properties; resistant to sun, oil, salt water, cleaning agents; subject to damage by chlorine; resilient; thermoplastic qualities
Drapability Qualities falls into moderately soft flares; can be molded or shaped by stretching; can be molded by heat setting

Recommended Construction Aids & Techniques

Hand Needles ball point, 5–10
Machine Needles ball point, 9–11
Threads cotton-polyester blend, all purpose; cotton-covered polyester core, all purpose; nylon/Dacron, monocord A
Hems double-stitched; flat/plain; machine-stitched; overedged
Seams overedged; plain; safety-stitched; taped; tissue-stitched; zigzagged
Seam Finishes double-stitched; serging/single-ply overedged; untreated/plain; single-ply zigzag
Pressing steam; use pressing cloth; safe temperature limit 300°F (150.1°C)
Care launder; remove from dryer promptly
Fabric Resource **Liberty Fabrics of New York**

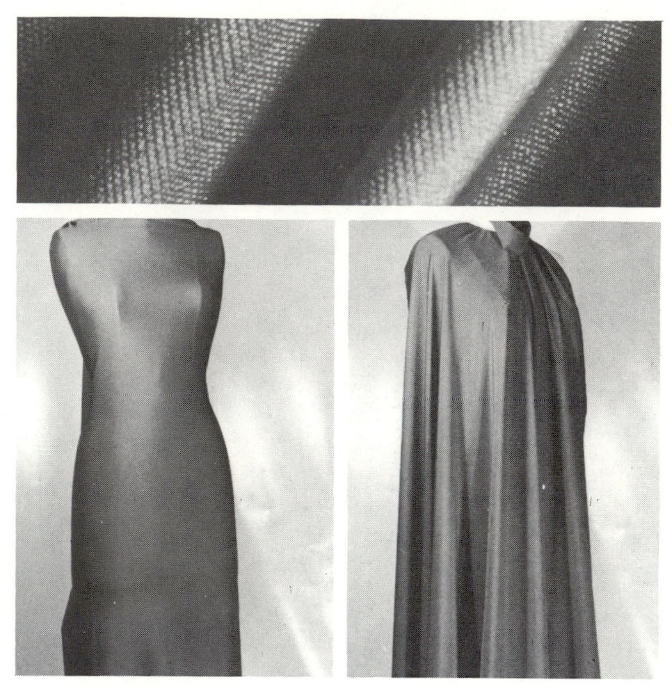

Lycra® (110–140% fiber stretch)

Fiber Content 15% spandex/85% nylon
Yarn Type elastomeric fiber; filament fiber
Yarn Construction core: extruded elastic yarn; outer layer: spun yarn
Fabric Structure knit
Finishing Processes calendering; flame retardant; stabilizing
Color Application piece dyed–union dyed; solution dyed; yarn dyed
Width 68–70 inches (172.7–177.8 cm)
Weight medium
Hand rubbery; slippery; stretchy
Texture flat; silky
Performance Expectations elongation and recovery with 110–140% stretch; power stretch properties; resistant to sun, oil, salt water, cleaning agents; subject to damage by chlorine; resilient; thermoplastic qualities; wrinkle resistant
Drapability Qualities falls into moderately soft flares; can be molded or shaped by stretching; can be molded by heat setting

Recommended Construction Aids & Techniques

Hand Needles ball point, 5–10
Machine Needles ball point, 9–11
Threads cotton-polyester blend, all purpose; cotton-covered polyester core, all purpose; nylon/Dacron, monocord A
Hems double-stitched; flat/plain; overedged
Seams overedged; plain; safety-stitched; taped; tissue-stitched; zigzagged
Seam Finishes double-stitched; serging/single-ply overedged; untreated/plain; single-ply zigzag
Pressing steam; use pressing cloth; safe temperature limit 300°F (150.1°C)
Care launder; remove from dryer promptly
Fabric Resource **Liberty Fabrics of New York**

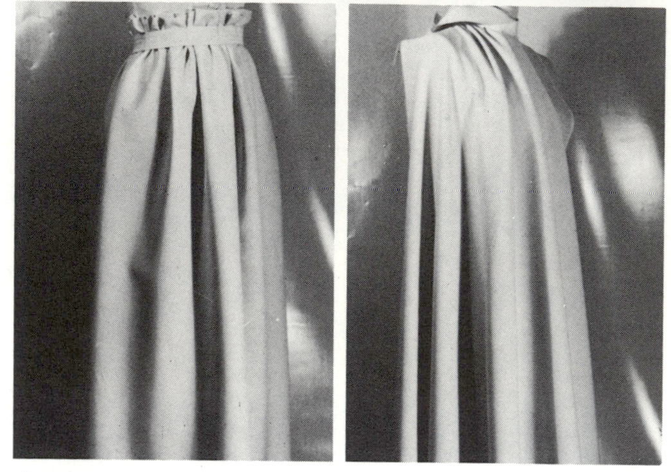

Canvas (yarn stretch)

Fiber Content 100% polyester
Yarn Type filament fiber
Yarn Construction heat-set textured stretch yarn
Fabric Structure plain weave
Finishing Processes compressive shrinkage; mercerizing; singeing
Color Application piece dyed
Width 60–62 inches (152.4–157.5 cm)
Weight medium-heavy
Hand crisp; firm; worsted hard feel; controlled stretch in crosswise direction
Texture coarse; grainy; harsh
Performance Expectations subject to color crocking and edge abrasion; dimensional stability; durable; heat-set properties; resilient; high tensile strength; wrinkle resistant
Drapability Qualities falls into firm wide cones; retains shape or silhouette of garment; better utilized if fitted by seaming and eliminating excess fabric

Recommended Construction Aids & Techniques

Hand Needles cotton darners, 1–4; embroidery, 1–4; milliners, 3–8; sharps, 1–4
Machine Needles round/set point, medium-coarse 14–16
Threads poly-core waxed, heavy duty; spun polyester, heavy duty; nylon/Dacron, monocord A
Hems double-fold or single-fold bias binding; bound/Hong Kong finish; overedged
Seams flat-felled; lapped; plain; safety-stitched; welt; double welt
Seam Finishes single-ply bound; single-ply overcast; pinked; pinked and stitched; serging/single-ply overedged; untreated/plain; single-ply zigzag
Pressing steam; safe temperature limit 325°F (164.1°C)
Care launder; remove from dryer promptly
Fabric Resource **Bloomsburg Mills**

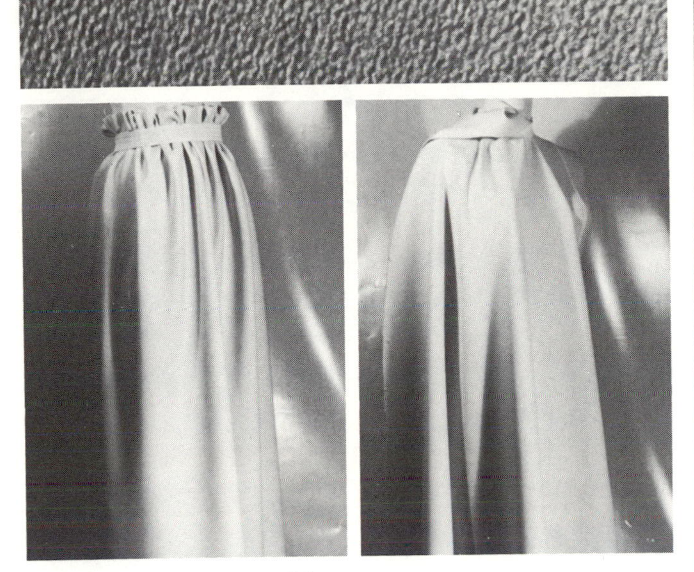

Chino (yarn stretch)

Fiber Content 100% Dacron polyester
Yarn Type filament fiber
Yarn Construction heat-set textured stretch yarn
Fabric Structure twill weave
Finishing Processes mercerizing; napping; shearing; heat-set stabilizing
Color Application piece dyed
Width 58–60 inches (147.3–152.4 cm)
Weight medium-heavy
Hand semi-crisp; firm; hard; controlled stretch in crosswise direction
Texture slightly napped face
Performance Expectations subject to color crocking and edge abrasion; dimensional stability; durable; heat-set properties; resilient; high tensile strength; wrinkle resistant
Drapability Qualities falls into wide cones; retains shape of garment; better utilized if fitted by seaming and eliminating excess fabric

Recommended Construction Aids & Techniques

Hand Needles cotton darners, 1–4; embroidery, 1–4; milliners, 3–8; sharps, 1–4
Machine Needles round/set point, medium-coarse 14–16
Threads poly-core waxed, heavy duty; spun polyester, heavy duty; nylon/Dacron, monocord A
Hems double-fold or single-fold bias binding; bound/Hong Kong finish; faced; overedged
Seams flat-felled; lapped; plain; safety-stitched; welt; double welt
Seam Finishes single-ply bound; single-ply overcast; pinked; pinked and stitched; serging/single-ply overedged; untreated/plain; single-ply zigzag
Pressing steam; safe temperature limit 325°F (164.1°C)
Care launder; remove from dryer promptly
Fabric Resource **Bloomsburg Mills**

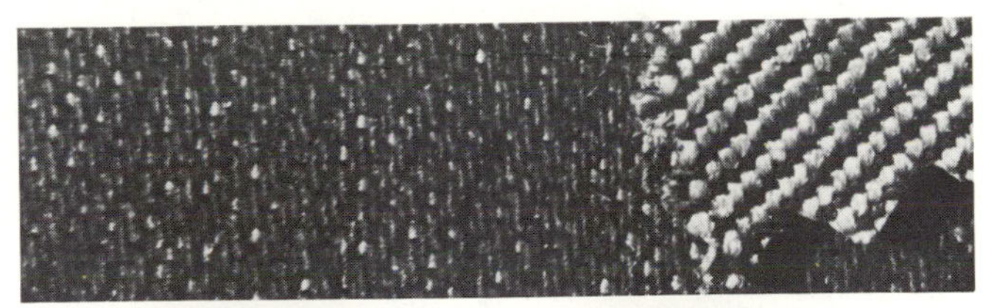

Denim (yarn stretch)

Fiber Content 63% cotton/37% Dacron polyester
Yarn Type combed staple; filament fiber
Yarn Construction heat-set textured stretch yarn
Fabric Structure twill weave
Finishing Processes mercerizing; napping; pre-shrunk; heat-set stabilizing
Color Application piece dyed–cross dyed
Width 48 inches (121.9 cm)
Weight heavy
Hand compact; crisp; firm; controlled stretch in crosswise direction
Texture coarse; diagonal ridge on face
Performance Expectations subject to color crocking and edge abrasion; dimensional stability; durable; heat-set properties; resilient; high tensile strength; wrinkle resistant
Drapability Qualities falls into firm wide cones; retains shape or silhouette of garment; better utilized if fitted by seaming and eliminating excess fabric

Recommended Construction Aids & Techniques

Hand Needles cotton darners, 1–4; embroidery, 1–4; milliners, 3–8; sharps, 1–4
Machine Needles round/set point, medium-coarse 14–16
Threads poly-core waxed, heavy duty; spun polyester, heavy duty; nylon/Dacron, monocord A
Hems double-fold or single-fold bias binding; bound/Hong Kong finish; overedged
Seams flat-felled; lapped; plain; safety-stitched; welt; double welt
Seam Finishes single-ply bound; single-ply overcast; pinked; pinked and stitched; serging/single-ply overedged; untreated/plain; single-ply zigzag
Pressing steam; safe temperature limit 325°F (164.1°C)
Care launder
Fabric Resource **Cone Mills Corporation**

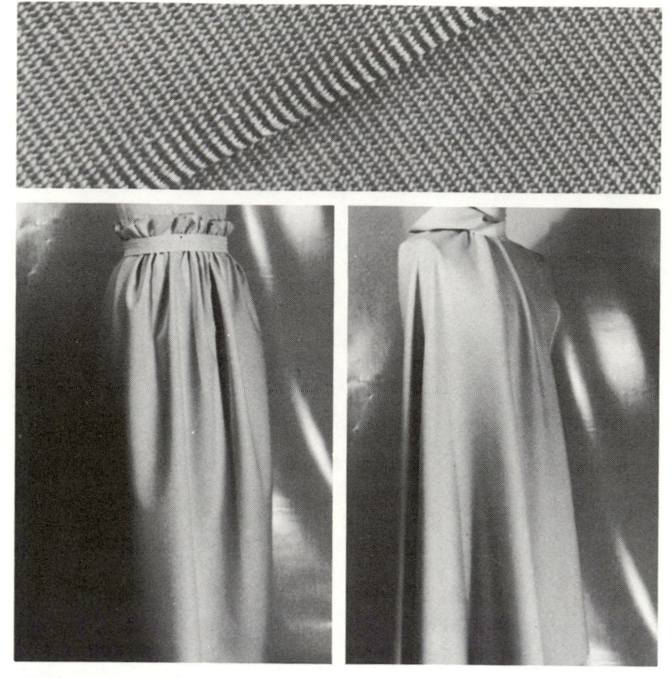

Gabardine (yarn stretch)

Fiber Content 100% polyester
Yarn Type filament fiber
Yarn Construction heat-set textured, two-way stretch yarn
Fabric Structure twill weave
Finishing Processes compressive shrinkage; heat-set stabilizing and nonsnagging
Color Application piece dyed
Width 60 inches (152.4 cm)
Weight medium-heavy
Hand crisp; firm; worsted hard feel; controlled stretch in crosswise and lengthwise directions
Texture rough; diagonal ridge on face
Performance Expectations subject to color crocking and edge abrasion; dimensional stability; durable; heat-set properties; resilient; high tensile strength; wrinkle resistant
Drapability Qualities falls into wide cones; retains shape of garment; better utilized if fitted by seaming and eliminating excess fabric

Recommended Construction Aids & Techniques

Hand Needles cotton darners, 1–4; embroidery, 1–4; milliners, 3–8; sharps, 1–4
Machine Needles round/set point, medium-coarse 14–16
Threads poly-core waxed, heavy duty; spun polyester, heavy duty; nylon/Dacron, monocord A
Hems double-fold or single-fold bias binding; bound/ Hong Kong finish; overedged
Seams flat-felled; lapped; plain; safety-stitched; welt; double welt
Seam Finishes single-ply bound; single-ply overcast; pinked; pinked and stitched; serging/single-ply overedged; untreated/plain; single-ply zigzag
Pressing steam; safe temperature limit 325°F (164.1°C)
Care launder; remove from dryer promptly
Fabric Resources Bloomsburg Mills; Klopman Mills

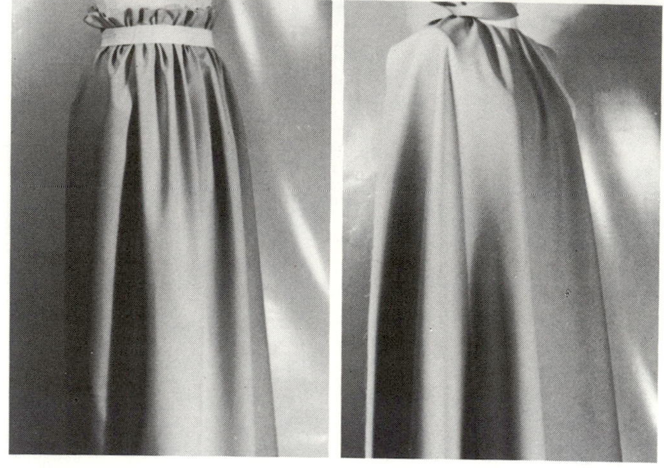

Poplin (yarn stretch)

Fiber Content 100% polyester
Yarn Type filament fiber
Yarn Construction heat-set textured stretch yarn
Fabric Structure plain weave; thicker filling yarns form rib in crosswise direction
Finishing Processes compressive shrinkage; singeing
Color Application piece dyed; solution dyed; yarn dyed
Width 58–60 inches (147.3–152.4 cm)
Weight medium
Hand crisp; firm; worsted hard feel; controlled stretch in crosswise direction
Texture harsh; pronounced crosswise rib
Performance Expectations subject to color crocking and edge abrasion; dimensional stability; durable; heat-set properties; resilient; high tensile strength; wrinkle resistant
Drapability Qualities falls into wide cones; retains shape of garment; better utilized if fitted by seaming and eliminating excess fabric

Recommended Construction Aids & Techniques

Hand Needles cotton darners, 4–8; embroidery, 4–8; milliners, 6–8; sharps, 4–8
Machine Needles round/set point, medium 14
Threads poly-core waxed, 50; spun polyester, all purpose; nylon/Dacron, monocord A
Hems double-fold or single-fold bias binding; bound/ Hong Kong finish; overedged
Seams plain; safety-stitched; welt; double welt
Seam Finishes book; single-ply bound; edge-stitched; pinked; pinked and stitched; serging/single-ply overedged; untreated/plain; single-ply zigzag
Pressing steam; safe temperature limit 325°F (164.1°C)
Care launder; remove from dryer promptly
Fabric Resource Bloomsburg Mills

Satin (yarn stretch)

Fiber Content 57% triacetate/43% Helenca® nylon
Yarn Type filament fiber
Yarn Construction heat-set crimped and textured stretch yarn
Fabric Structure satin weave
Finishing Processes calendering; stabilizing
Color Application piece dyed–union dyed
Width 56 inches (142.2 cm)
Weight medium-heavy
Hand crisp; firm; silky; controlled stretch in crosswise and lengthwise directions
Texture smooth face; grainy back
Performance Expectations subject to abrasion due to weave; dimensional stability; elongation and recovery; heat-set properties
Drapability Qualities falls into crisp wide flares; retains shape or silhouette of garment; better utilized if fitted by seaming and eliminating excess fabric

Recommended Construction Aids & Techniques

Hand Needles cotton darners, 4–8; embroidery, 4–8; milliners, 6–8; sharps, 4–8
Machine Needles round/set point, medium 11–14
Threads poly-core waxed, 50; spun polyester, all purpose; nylon/Dacron, monocord A
Hems double-fold or single-fold bias binding; bound/Hong Kong finish; overedged
Seams plain; safety-stitched; welt; double welt
Seam Finishes single-ply bound; single-ply overcast; pinked; pinked and stitched; serging/single-ply overedged; untreated/plain; single-ply zigzag
Pressing steam; safe temperature limit 350°F (178.1°C)
Care dry clean
Fabric Resource **American Silk Mills Corp./Bucol**

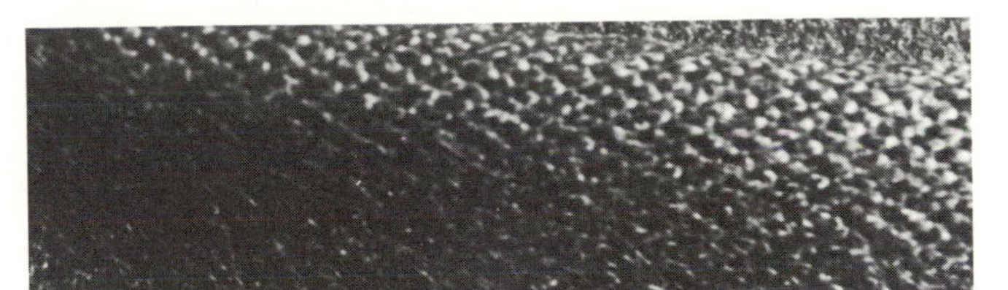

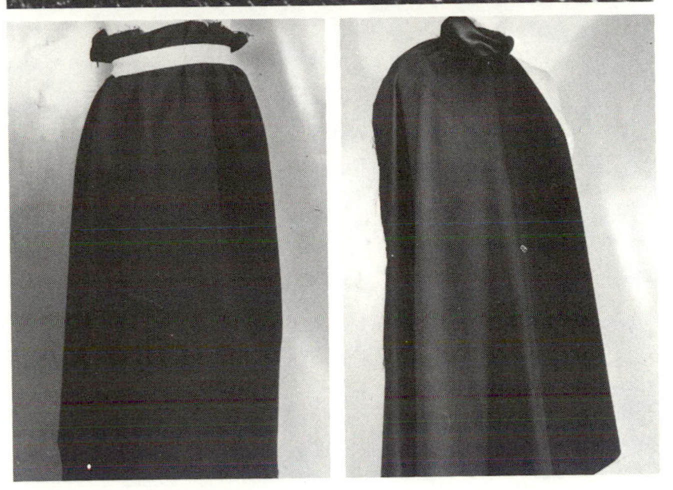

Twill (yarn stretch)

Fiber Content nylon/Helenca® nylon
Yarn Type filament fiber
Yarn Construction heat-set crimped and textured stretch yarn
Fabric Structure broken twill weave
Finishing Processes brushing; napping; shearing; stabilizing
Color Application piece dyed–union dyed
Width 45 inches (114.3 cm)
Weight heavy
Hand compact; lofty; rubbery; controlled stretch in crosswise and lengthwise directions
Texture napped face; diagonal ridge on back
Performance Expectations bags at knee, elbow, seat areas; durable; resilient; 2% nonrecovery stretch; subject to static buildup
Drapability Qualities falls into wide cones; retains shape or silhouette of garment; better utilized if fitted by seaming and eliminating excess fabric

Recommended Construction Aids & Techniques

Hand Needles cotton darners, 1–4; embroidery, 1–4; milliners, 3–8; sharps, 1–4
Machine Needles round/set point, medium-coarse 14–16
Threads poly-core waxed, heavy duty; spun polyester, heavy duty; nylon/Dacron, monocord A
Hems double-fold or single-fold bias binding; bound/Hong Kong finish; overedged
Seams plain; safety-stitched; welt; double welt
Seam Finishes single-ply bound; single-ply overcast; pinked; pinked and stitched; serging/single-ply overedged; untreated/plain; single-ply zigzag
Pressing steam; safe temperature limit 350°F (178.1°C)
Care launder; dry clean
Fabric Resource **Lawrence Textile Co. Inc.**

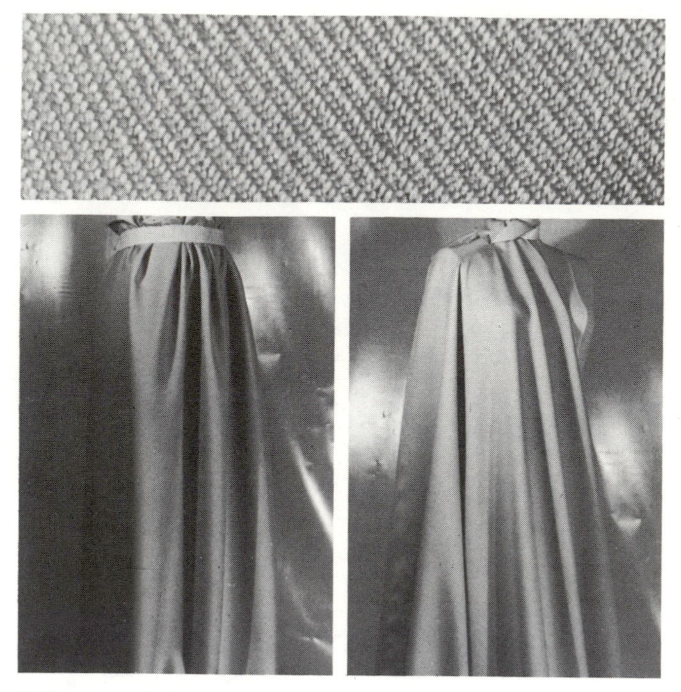

Whipcord (yarn stretch)

Fiber Content 100% Dacron® polyester
Yarn Type filament fiber
Yarn Construction heat-set textured, two-way stretch yarn
Fabric Structure left-hand twill weave
Finishing Processes delustering; heat-set stabilizing
Color Application piece dyed; solution dyed; yarn dyed
Width 59-60 inches (149.9-152.4 cm)
Weight medium-light
Hand crisp; firm; worsted hard feel; controlled stretch in crosswise and lengthwise directions
Texture rough; pronounced left-hand diagonal ridge on face and back
Performance Expectations subject to color crocking and edge abrasion; dimensional stability; durable; heat-set properties; resilient; 2% nonrecovery stretch; high tensile strength
Drapability Qualities falls into moderately crisp flares; fullness maintains crisp effect; retains shape of garment

Recommended Construction Aids & Techniques

Hand Needles cotton darners, 4-8; embroidery, 4-8; milliners, 6-8; sharps, 4-8
Machine Needles round/set point, medium 11-14
Threads poly-core waxed, 50; spun polyester, all purpose; nylon/Dacron, monocord A
Hems book; single-fold bias binding; bound/Hong Kong finish; edge-stitched; machine-stitched; overedged
Seams flat-felled; lapped; plain; safety-stitched; welt; double welt
Seam Finishes book; single-ply bound; edge-stitched; pinked; pinked and stitched; serging/single-ply overedged; untreated/plain; single-ply zigzag
Pressing steam; safe temperature limit 325°F (164.1°C)
Care launder; remove from dryer promptly
Fabric Resource Klopman Mills

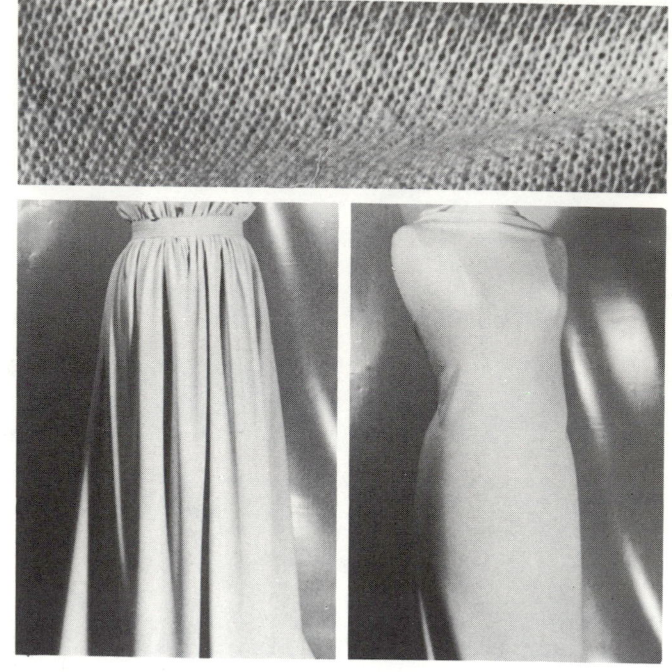

Lightweight Knit (structure & yarn stretch)

Fiber Content 85% polyester/15% cotton
Yarn Type combed staple; filament fiber; filament staple
Yarn Construction heat-set, back-twisted stretch yarn
Fabric Structure double knit
Finishing Processes brushing; napping; shearing; stabilizing
Color Application piece dyed; solution dyed; yarn dyed
Width 60-62 inches (152.4-157.5 cm)
Weight medium-light
Hand silky; stretch in crosswise and lengthwise directions; supple
Texture flat; smooth
Performance Expectations durable; resilient; action stretch properties; shape retention; high tensile strength
Drapability Qualities falls into soft flares; can be shaped or molded by stretching; retains shape of garment

Recommended Construction Aids & Techniques

Hand Needles ball point, 5-10
Machine Needles ball point, 11-14
Threads cotton-polyester blend, all purpose; cotton-covered polyester core, all purpose; nylon/Dacron, monocord A
Hems double-stitched; flat/plain; overedged
Seams overedged; plain; safety-stitched; taped; zigzagged
Seam Finishes serging/single-ply overedged; untreated/plain; single-ply zigzag
Pressing steam; safe temperature limit 325°F (164.1°C)
Care launder; dry clean
Fabric Resource Russell Corporation

Medium-weight Knit (structure & yarn stretch)

Fiber Content 100% nylon
Yarn Type filament staple
Yarn Construction heat-set, back-twisted stretch yarn
Fabric Structure double knit
Finishing Processes napping; shearing; stabilizing
Color Application piece dyed; solution dyed; yarn dyed
Width 54–60 inches (137.2–152.4 cm)
Weight medium
Hand coarse; pliable; stretch in crosswise and lengthwise directions
Texture flat face; grainy back
Performance Expectations abrasion resistant; dimensional stability; durable; elasticity; elongation; resilient; action stretch properties; high tensile strength; subject to static buildup
Drapability Qualities falls into wide flares; can be moderately shaped by stretching; retains shape of garment

Recommended Construction Aids & Techniques

Hand Needles ball point, 5–10
Machine Needles ball point, 14
Threads poly-core waxed, heavy duty; spun polyester, heavy duty; nylon/Dacron, monocord A
Hems double-stitched; flat/plain; overedged
Seams overedged; plain; safety-stitched; taped; zigzagged
Seam Finishes serging/single-ply overedged; untreated/plain; single-ply zigzag
Pressing steam; safe temperature limit 350°F (178.1°C)
Care launder; dry clean
Fabric Resource Russell Corporation

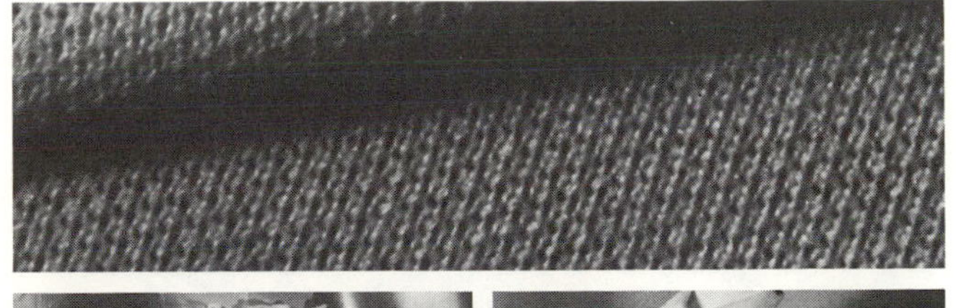

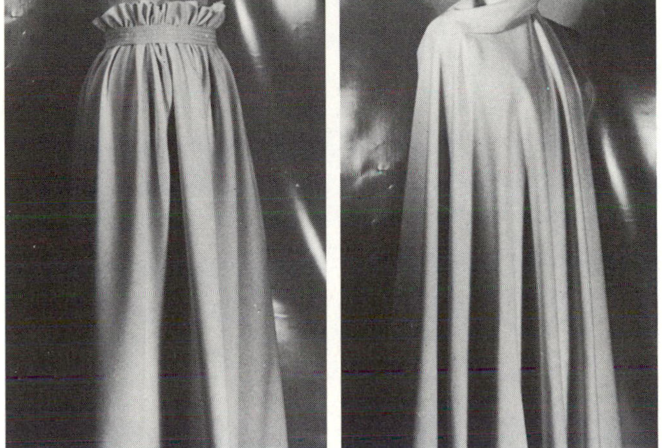

Heavyweight Knit (structure & yarn stretch)

Fiber Content 65% Kodel® polyester (face)/ 35% cotton (back)
Yarn Type combed staple; filament staple
Yarn Construction heat-set, back-twisted stretch yarn
Fabric Structure double knit
Finishing Processes napping; shearing; stabilizing
Color Application piece dyed; solution dyed; yarn dyed
Width 60–62 inches (152.4–157.5 cm)
Weight medium-heavy
Hand pliable; semi-soft; stretch in crosswise and lengthwise directions; thick
Texture flat face; grainy back
Performance Expectations cotton-back provides absorbency properties; dimensional stability; durable; elongation; action stretch properties; high tensile strength
Drapability Qualities falls into soft wide flares; can be moderately shaped by stretching; retains shape of garment

Recommended Construction Aids & Techniques

Hand Needles ball point, 5–10
Machine Needles ball point, 14–16
Threads cotton-polyester blend, heavy duty; cotton-covered polyester core, heavy duty
Hems double-stitched; flat/plain; overedged
Seams overedged; plain; safety-stitched; taped; tissue-stitched; zigzagged
Seam Finishes serging/single-ply overedged; untreated/plain; single-ply zigzag
Pressing steam; safe temperature limit 325°F (164.1°C)
Care launder; dry clean
Fabric Resource Russell Corporation

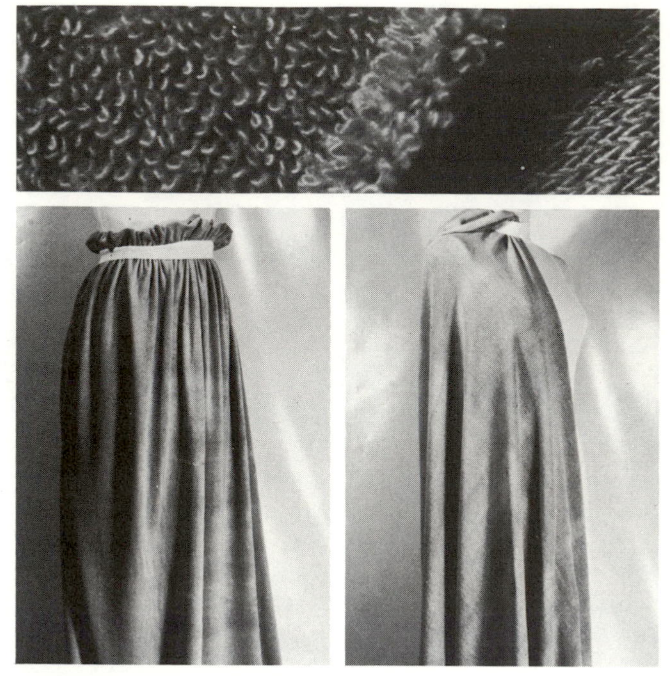

Stretch Terry Cloth (structure & yarn stretch)

Fiber Content 100% cotton (face) (85% of fabric); 100% nylon (back) (15% of fabric)
Yarn Type combed staple; filament fiber
Yarn Construction slack mercerized yarns (face); heat-set textured stretch yarns (back)
Fabric Structure looped pile face; knit base
Finishing Processes brushing; stabilizing
Color Application piece dyed–union dyed; yarn dyed
Width 60 inches (152.4 cm)
Weight heavy
Hand pliable; soft; spongy; stretch in crosswise and lengthwise directions
Texture short dense looped face; flat back
Performance Expectations absorbent; durable; elongation and recovery; resilient; stretch properties; high tensile strength; subject to snagging and pulling due to loop structure; wrinkle resistant
Drapability Qualities falls into soft flares; can be shaped or molded by stretching; fullness retains soft fall

Recommended Construction Aids & Techniques

Hand Needles ball point, 5–10
Machine Needles ball point, 14–16
Threads cotton-polyester blend, all purpose; cotton-covered polyester core, all purpose
Hems double-stitched; flat/plain; overedged
Seams overedged; plain; safety-stitched; taped; zigzagged
Seam Finishes serging/single-ply overedged; untreated/plain; single-ply zigzag
Pressing steam; safe temperature limit 350°F (178.1°C)
Care launder
Fabric Resource **Gloversville Mills Inc.**

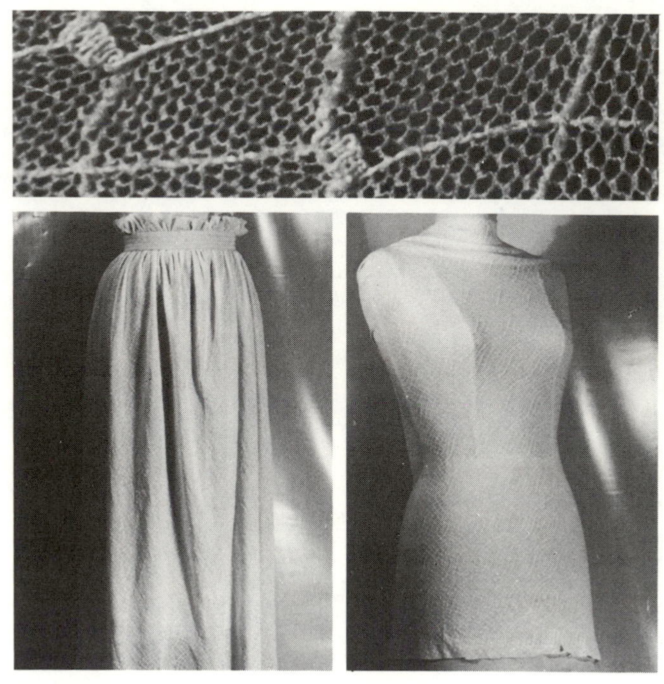

Stretch Lace (structure & yarn stretch)

Fiber Content 100% Helenca® nylon
Yarn Type filament fiber
Yarn Construction heat-set crimped stretch yarns
Fabric Structure purl knit variation; Jacquard knit
Finishing Processes brushing; stabilizing
Color Application piece dyed; solution dyed; yarn dyed
Width 63 inches (160 cm)
Weight light
Hand delicate; semi-soft; stretch in crosswise and lengthwise directions
Texture crepey; open structure
Performance Expectations abrasion resistant; elasticity; elongation and recovery; resilient; two-way stretch properties; high tensile strength; subject to pilling and static buildup; subject to snagging due to structure
Drapability Qualities falls into moderately soft flares; can be shaped or molded by stretching; can be molded by heat setting

Recommended Construction Aids & Techniques

Hand Needles ball point, 5–10
Machine Needles ball point, 9
Threads poly-core waxed, 60 fine; spun polyester, fine; nylon/Dacron, monocord A
Hems double-stitched; flat/plain; machine-stitched; overedged
Seams overedged; plain; safety-stitched; taped; tissue-stitched; zigzagged
Seam Finishes single-ply and double-ply bound; double-stitched and overcast; double-stitched and trimmed; serging/single-ply overedged; untreated/plain; double-ply zigzag
Pressing steam; use pressing cloth; safe temperature limit 350°F (178.1°C)
Care launder
Fabric Resource **Native Lace**

Crimped Cloth (finish stretch)

Fiber Content 50% polyester/50% cotton
Yarn Type filament fiber; staple
Yarn Construction conventional; textured
Fabric Structure gauze weave
Finishing Processes mercerizing; singeing; chemical and mechanical slack mercerizing
Color Application piece dyed–union dyed
Width 36 inches (91.4 cm)
Weight medium
Hand coarse; loose; soft; stretch in crosswise direction
Texture uneven surface created by deep crinkled effect in lengthwise direction on face and back
Performance Expectations crimped stretch *is not* permanent; elongation; high tensile strength; subject to snagging due to openweave
Drapability Qualities falls into moderately soft wide flares; accommodates fullness by shirring; fullness retains moderately soft fall

Recommended Construction Aids & Techniques

Hand Needles cotton darners, 1–4; embroidery, 1–4; milliners, 3–8; sharps, 1–4
Machine Needles round/set point, medium-coarse 14–16
Threads cotton-polyester blend, heavy duty; cotton-covered polyester core, heavy duty
Hems double-fold or single-fold bias binding; bound/Hong Kong finish; faced; overedged; seam binding
Seams plain; safety-stitched; taped; welt; double welt
Seam Finishes single-ply bound; single-ply overcast; pinked; pinked and stitched; serging/single-ply overedged; untreated/plain; single-ply zigzag
Pressing steam; safe temperature limit 325°F (164.1°C)
Care launder
Fabric Resource **Printsiples and Company B**

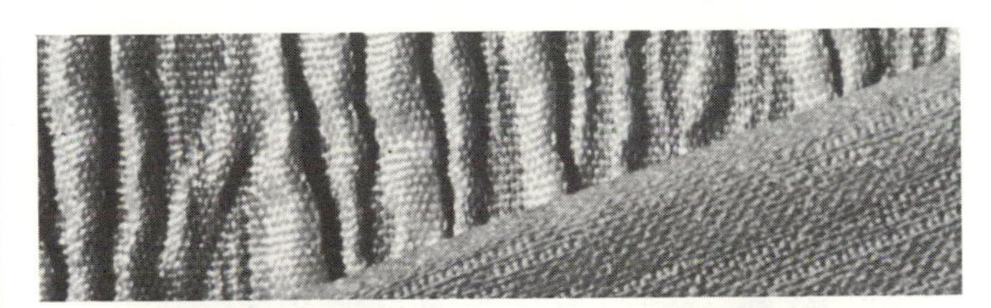

Crinkled Double Cloth (finish stretch)

Fiber Content 100% polyester
Yarn Type filament fiber
Yarn Construction textured
Fabric Structure double cloth weave
Finishing Processes calendering; heat-set crinkle (face)
Color Application piece dyed
Width 36 inches (91.4 cm)
Weight heavy
Hand crisp; rubbery; thick
Texture uneven surface created by crinkled effect in lengthwise direction on face; flat back
Performance Expectations subject to color crocking and edge abrasion; dimensional stability; durable; elongation; crimped stretch *is* permanent; controlled stretch properties
Drapability Qualities falls into wide cones; retains shape or silhouette of garment; better utilized if fitted by seaming and eliminating excess fabric

Recommended Construction Aids & Techniques

Hand Needles cotton darners, 4–8; embroidery, 4–8; milliners, 6–8; sharps, 4–8
Machine Needles round/set point, medium-coarse 14–16
Threads poly-core waxed, heavy duty; spun polyester, heavy duty; nylon/Dacron, monocord A
Hems double-fold or single-fold bias binding; bound/Hong Kong finish; faced; overedged
Seams plain; safety-stitched; welt; double welt
Seam Finishes single-ply bound; single-ply overcast; pinked; pinked and stitched; serging/single-ply overedged; untreated/plain; single-ply zigzag
Pressing steam; safe temperature limit 325°F (164.1°C)
Care dry clean
Fabric Resource **American Silk Mills Corp./Bucol**

16 ~ Metallic/Metallic-type Fabrics

Ciré / Metallic-coated Cloth
Embroidered Cloth
  Sheer Embroidered / Metallic Embroidered Cloth
Knit
  Knit Jersey (weft-inserted metallic yarn)
  Tricot Knit (laid-in metallic yarn)
Lamé
  Lamé Gauze
  Lamé Jersey
  Woven Lamé (close up only)
Metallic Chantilly Lace
Metallic Jacquard Cloth
Metallic Net
Metallic Pile
Metallic Point d'Esprit Lace
Sparkle or Glitter on Cloth
  Adhesive-bonded Sparkle or Glitter on Opaque Cloth
  Adhesive-bonded Sparkle or Glitter on Sheer Cloth
  Metallic Sparkle- or Glitter-printed Design

Metallic is a generic term for a manufactured fiber composed of a metal compound of aluminum, silver, gold or stainless steel. Metallic fabric is a term that refers to any fabric that has a gold, silver, tinsel or other metal thread interspersed throughout the design.

Metallic fibers or yarns may be composed of:

- Metal
- Plastic coated with metal
- A core completely covered with metal
- Metal coated with plastic of Mylar® polyester film, acetate film
- Color added to adhesive during bonding process

Metallic fabrics may be made:

- In any type of fabric structure
- With all metallic yarns
- By combining metallic yarns with other natural or man-made fiber yarns
- With blended fiber yarns
- Embossed

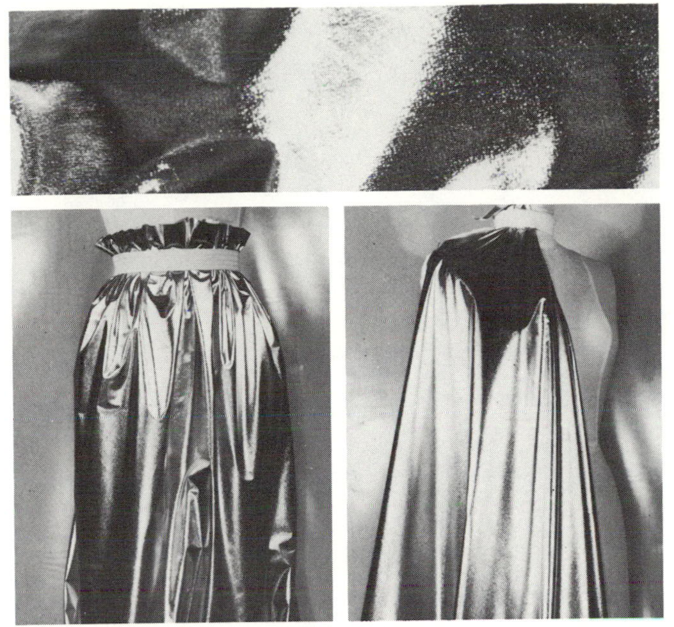

Ciré/Metallic-coated Cloth

Fiber Content Mylar® metallic-coated (face); 100% nylon (back)
Yarn Type filament fiber
Yarn Construction textured
Fabric Structure metallic-coated face; tricot knit back
Finishing Processes ciré calendering; metallic-coated or plated; water repellent
Color Application pigment coating (face); piece dyed (base)
Width 36 inches (91.4 cm)
Weight medium
Hand clammy; slippery; soft
Texture light-reflecting, high-sheen face; dull back; plastic; satiny smooth
Performance Expectations high lustrous finish tends to crock and abrade with wear; dimensional stability; durable; elongation; heat-set properties; resilient; high tensile strength; water repellent due to finish
Drapability Qualities falls into crisp flares; fullness maintains crisp effect; retains shape of garment

Recommended Construction Aids & Techniques

Hand Needles beading, 10–13; betweens, 4–6; cotton darners, 9–10; embroidery, 6–8; milliners, 7–9; sharps, 5–8
Machine Needles ball point, 9
Threads silk, industrial size A; poly-core waxed, all purpose; cotton-polyester blend, all purpose; cotton-covered polyester core, all purpose; spun polyester, all purpose; nylon/Dacron, monocord medium
Hems book; bonded; bound/Hong Kong finish; edge-stitched; flat/plain; pinked flat; stitched and pinked flat; machine-stitched; hand- or machine-rolled; wired
Seams flat-felled; French; lapped; overedged; plain; safety-stitched; taped; tissue-stitched; welt; double welt; zigzagged
Seam Finishes book; single-ply bound; Hong Kong bound; edge-stitched; pinked and stitched; serging/single-ply overedged; untreated/plain; single-ply zigzag
Pressing steam; use pressing cloth; safe temperature limit 350°F (178.1°C)
Care launder; dry clean
Fabric Resource **Walden Textiles/Barry M. Richards Inc.**

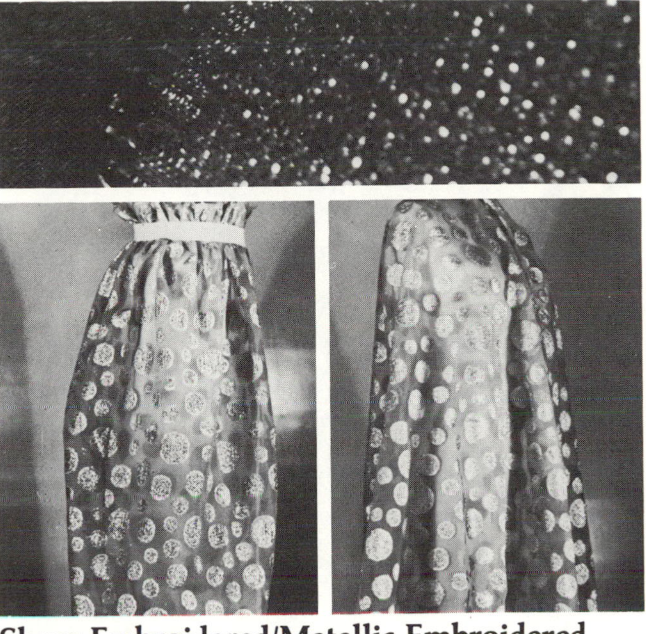

Sheer Embroidered/Metallic Embroidered Cloth

Fiber Content 100% nylon (base); Mylar® (design)
Yarn Type base: extruded film, filament fiber; embroidery: extruded metallic film fiber
Yarn Construction monofilament (embroidery); textured (base)
Fabric Structure Schiffli embroidery on plain weave base
Finishing Processes calendering; heat-set stabilizing; stiffening
Color Application piece dyed
Width 45 inches (114.3 cm)
Weight fine to lightweight
Hand crisp; delicate; sheer
Texture metallic fringe forms raised pattern design
Performance Expectations dimensional stability; durable; elasticity; elongation; heat-set properties; resilient; high tensile strength; embroidery tends to snag and abrade with wear
Drapability Qualities falls into crisp cones; accommodates fullness by pleating, gathering; fullness maintains crisp bouffant effect

Recommended Construction Aids & Techniques

Hand Needles beading, 10–13; betweens, 6–7; embroidery, 9–10; milliners, 10; sharps, 8–10
Machine Needles round/set point, fine 9
Threads silk, industrial size A; poly-core waxed, 70; cotton-polyester blend, extra fine; cotton-covered polyester core, extra fine; spun polyester, fine; nylon/Dacron, monocord fine
Hems double fold; bound/Hong Kong finish; edge-stitched; horsehair; machine-stitched; hand- or machine-rolled; wired
Seams French; hairline; plain; tissue-stitched
Seam Finishes self bound; Hong Kong bound; double-stitched; double-stitched and overcast; double-stitched and trimmed; serging/single-ply overedged; untreated/plain; double-ply zigzag
Pressing steam; use pressing cloth; safe temperature limit 350°F (178.1°C)
Care dry clean
Fabric Resource **Silverman Textile Co. (sample courtesy of Lita Konde)**

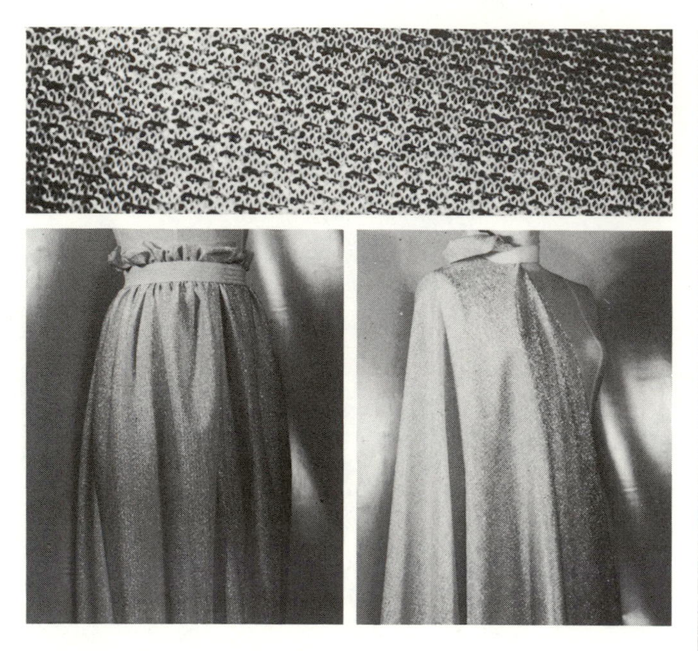

Knit Jersey (weft-inserted metallic yarn)

Fiber Content 100% nylon (base); Mylar® inserted yarn
Yarn Type extruded metallic film; filament fiber
Yarn Construction monofilament laid-in yarn; textured (base)
Fabric Structure jersey knit with laid-in yarn
Finishing Processes calendering; heat-set stabilizing; stretching
Color Application piece dyed; solution dyed; yarn dyed
Width 60 inches (152.4 cm)
Weight light
Hand limp; rough; supple
Texture metallic yarn raised on face; sandy; scratchy
Performance Expectations abrasion resistant; durable; elasticity; elongation; heat-set properties; resilient; high tensile strength; inserted yarn subject to snagging
Drapability Qualities falls into moderately soft flares; accommodates fullness by gathering, elasticized shirring; fullness retains moderately soft fall

Recommended Construction Aids & Techniques

Hand Needles ball point, 6-10; beading, 10-13; betweens, 4-6; cotton darners, 9-10; embroidery, 6-8; milliners, 7-9; sharps, 5-8
Machine Needles ball point, 9-11
Threads silk, industrial size A; poly-core waxed, all purpose; cotton-polyester blend, all purpose; cotton-covered polyester core, all purpose; spun polyester, all purpose; nylon/Dacron, monocord medium
Hems book; bonded; bound/Hong Kong finish; double-stitched; edge-stitched; faced; pinked; pinked and stitched flat; machine-stitched; overedged
Seams flat-felled; overedged; plain; safety-stitched; taped; welt; double welt; zigzagged
Seam Finishes single-ply bound; Hong Kong bound; double-stitched; serging/single-ply overedged; untreated/plain; single-ply zigzag
Pressing steam; use pressing cloth; safe temperature limit 350°F (178.1°C)
Care dry clean
Fabric Resource **Applause Fabrics Ltd.**

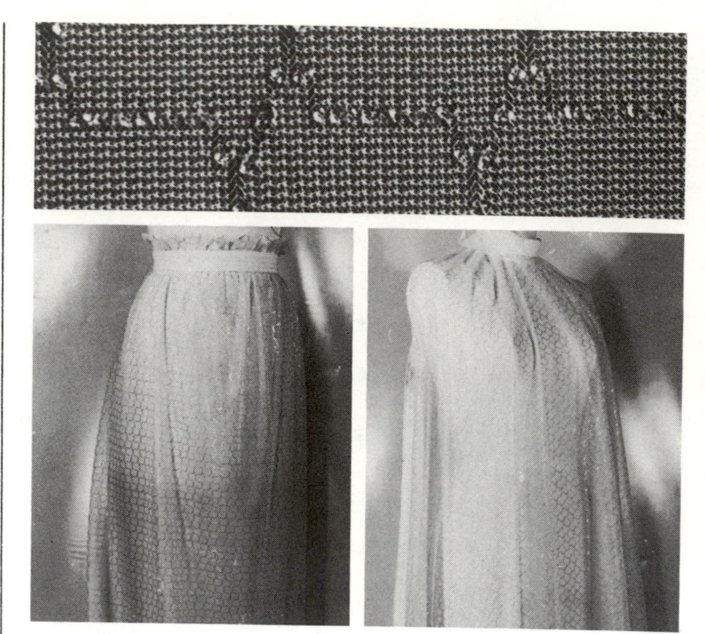

Tricot Knit (laid-in metallic yarn)

Fiber Content 100% nylon (base); Mylar® laid-in yarn
Yarn Type extruded metallic film; filament fiber
Yarn Construction monofilament laid-in yarn; textured (base)
Fabric Structure tricot knit with laid-in yarn
Finishing Processes calendering; heat-set stabilizing; stretching
Color Application piece dyed; solution dyed; yarn dyed
Width 46 inches (116.8 cm)
Weight fine to lightweight
Hand delicate; soft; springy
Texture harsh, raised metallic design; sheer
Performance Expectations abrasion resistant; durable; elasticity; elongation; heat-set properties; resilient; high tensile strength; inserted yarn subject to snagging
Drapability Qualities falls into moderately soft flares; accommodates fullness by gathering; fullness retains moderately soft fall

Recommended Construction Aids & Techniques

Hand Needles ball point, 6-10; beading, 10-13; betweens, 6-7; embroidery, 9-10; milliners, 10; sharps, 8-10
Machine Needles ball point, 9
Threads silk, industrial size A; poly-core waxed, 70; cotton-polyester blend, extra fine; cotton-covered polyester core, extra fine; spun polyester, fine; nylon/Dacron, monocord fine
Hems net binding; double fold; double-stitched; edge-stitched; horsehair; machine-stitched; hand- or machine-rolled; wired
Seams French; hairline; overedged; plain; safety-stitched; taped; tissue-stitched; zigzagged
Seam Finishes net bound; self bound; Hong Kong bound; double-stitched; double-stitched and overcast; double-stitched and trimmed; serging/single-ply overedged; untreated/plain; double-ply zigzag
Pressing steam; use pressing cloth; safe temperature limit 350°F (178.1°C)
Care dry clean
Fabric Resource **Native Textiles**

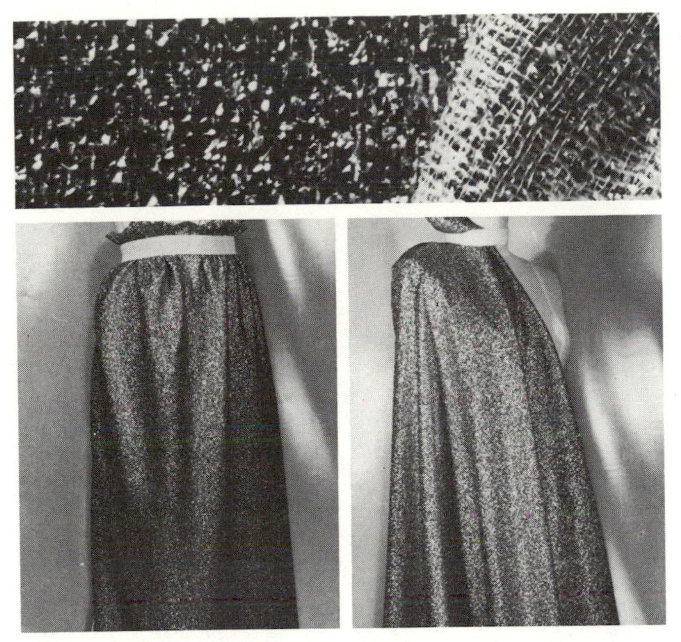

Lamé Gauze

Fiber Content 60% rayon; 40% Lurex® yarn
Yarn Type extruded metallic film; filament fiber
Yarn Construction monofilament metallic film (face); textured (base)
Fabric Structure satin weave face; gauze weave back
Finishing Processes brushing; heat-set stabilizing
Color Application solution dyed; yarn dyed
Width 44 inches (111.8 cm)
Weight medium-light
Hand lofty; rough; supple
Texture metallic yarn raised on face; sandy; scratchy
Performance Expectations subject to abrasion, shrinkage, bagging; elongation; nonpilling; good tensile strength; *face* subject to snagging due to weave
Drapability Qualities falls into moderately soft flares; accommodates fullness by pleating; gathering, elasticized shirring; fullness retains moderately soft fall

Recommended Construction Aids & Techniques

Hand Needles beading, 10–13; betweens, 6–7; embroidery, 9–10; milliners, 10; sharps, 8–10
Machine Needles round/set point, medium-fine 9–11
Threads silk, industrial size A; poly-core waxed, all purpose; cotton-polyester blend, all purpose; cotton-covered polyester core, all purpose; spun polyester, all purpose; nylon/Dacron, monocord medium
Hems book; bonded; bound/Hong Kong finish; edge-stitched; faced; pinked; pinked and stitched; machine-stitched; overedged
Seams French; hairline; overedged; plain; safety-stitched; taped; tissue-stitched; zigzagged
Seam Finishes book; single-ply bound; Hong Kong bound; edge-stitched; pinked and stitched; serging/single-ply overedged; untreated/plain; single-ply zigzag
Pressing steam; use pressing cloth; safe temperature limit 350°F (178.1°C)
Care dry clean
Fabric Resource Lawrence Textile Co. Inc.

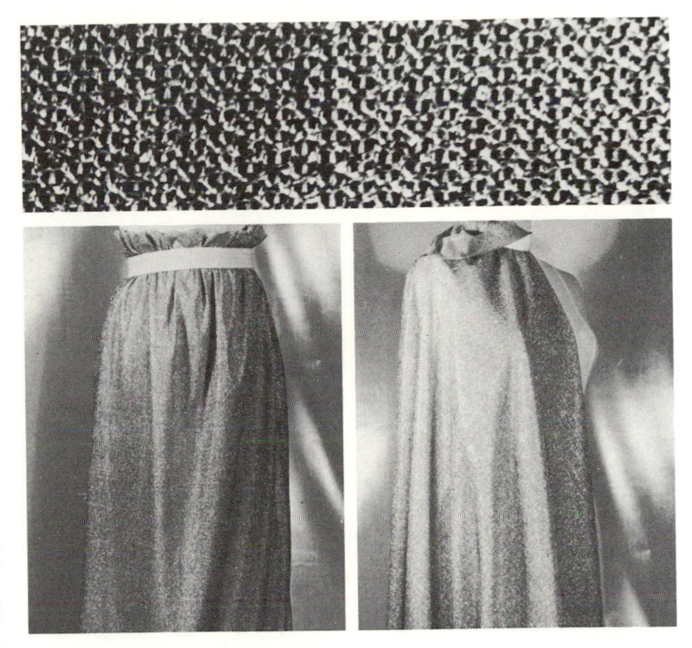

Lamé Jersey

Fiber Content 40% acetate/20% cotton/20% rayon/13% Mylar®/7% ribbon
Yarn Type carded staple; extruded metallic film; filament fiber
Yarn Construction monofilament; textured
Fabric Structure jersey knit
Finishing Processes heat-set stabilizing
Color Application solution dyed; yarn dyed
Width 45 inches (114.3 cm)
Weight light
Hand light; rough; supple
Texture coarse; sandy
Performance Expectations dimensional stability; elongation; heat-set properties; nonpilling; subject to flex abrasion and snagging
Drapability Qualities falls into soft flares; accommodates fullness by gathering, elasticized shirring; fullness retains soft fall

Recommended Construction Aids & Techniques

Hand Needles ball point, 6–10; beading, 10–13; betweens, 4–6; cotton darners, 9–10; embroidery, 6–8; milliners, 7–9; sharps, 5–8
Machine Needles ball point, 9
Threads silk, industrial size A; poly-core waxed, all purpose; cotton-polyester blend, all purpose; cotton-covered polyester core, all purpose; spun polyester, all purpose; nylon/Dacron, monocord medium
Hems book; bonded; bound/Hong Kong finish; edge-stitched; pinked; pinked and stitched; machine-stitched; overedged
Seams flat-felled; overedged; plain; safety-stitched; tissue-stitched; taped; welt; double welt; zigzagged
Seam Finishes single-ply bound; Hong Kong bound; double-stitched; serging/single-ply overedged; untreated/plain; single-ply zigzag
Pressing steam; use pressing cloth; safe temperature limit 250°F (122.1°C)
Care dry clean
Fabric Resource Lawrence Textile Co. Inc.

Woven Lamé

Fiber Content 53% copper metal/47% silk
Yarn Type extruded metallic film; natural filament fiber
Yarn Construction filament fiber; monofilament metal film
Fabric Structure plain weave
Finishing Processes decating; stabilizing
Color Application yarn dyed
Width 36 inches (91.4 cm)
Weight light to medium-light
Hand crisp; rigid; thin
Texture flat; smooth; metallic yarn creates high light-reflecting qualities and two-tone effect
Performance Expectations abrasion resistant; dimensional stability; durable; flexible
Drapability Qualities falls into moderately stiff cones; accommodates fullness by gathering, pleating; fullness maintains crisp effect

Recommended Construction Aids & Techniques

Hand Needles beading, 10–13; betweens, 4–6; cotton darners, 9–10; embroidery, 6–8; milliners, 7–9; sharps, 5–8
Machine Needles round/set point, fine 9
Threads silk, industrial size A; poly-core waxed, all purpose; cotton-polyester blend, all purpose; cotton-covered polyester core, all purpose; spun polyester, all purpose; nylon/Dacron, monocord medium
Hems book; bonded; bound/Hong Kong finish; edge-stitched; faced; pinked; pinked and stitched; machine-stitched; overedged; seam binding
Seams flat-felled; French; lapped; plain; safety-stitched; welt; double welt
Seam Finishes book; single-ply bound; Hong Kong bound; edge-stitched; pinked and stitched; serging/single-ply overedged; untreated/plain; single-ply zigzag
Pressing steam; use pressing cloth; safe temperature limit 250°F (122.1°C)
Care dry clean
Fabric Resource **American Silk Mills Corp./Bucol**

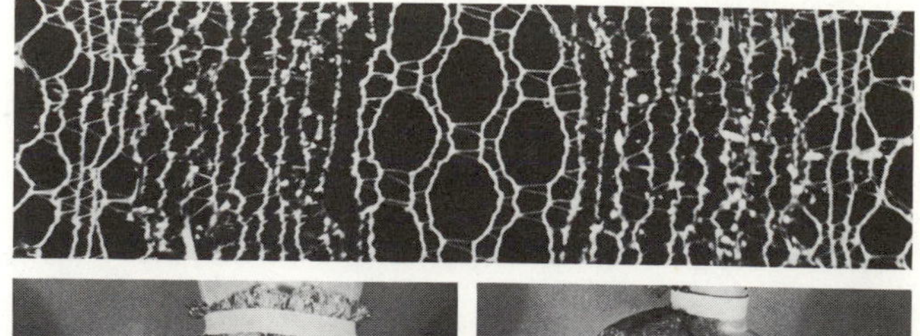

Metallic Chantilly Lace

Fiber Content 20% nylon/80% Mylar®
Yarn Type extruded metallic film; filament fiber
Yarn Construction monofilament; monofilament metallic film
Fabric Structure Leavers machine-made lace
Finishing Processes heat-set stabilizing
Color Application solution dyed; yarn dyed
Width 56 inches (142.2 cm)
Weight fine to lightweight
Hand crisp; delicate; thin
Texture open-mesh structure; rough, scratchy; sheer
Performance Expectations abrasion resistant; durable; elasticity; elongation; heat-set properties; resilient; high tensile strength; subject to snagging due to structure
Drapability Qualities falls into moderately soft flares; accommodates fullness by gathering; fullness retains moderately soft fall

Recommended Construction Aids & Techniques

Hand Needles beading, 10–13; betweens, 6–7; embroidery, 9–10; milliners, 10; sharps, 8–10
Machine Needles round/set point, fine 9
Threads silk, industrial size A; poly-core waxed, 70; cotton-polyester blend, extra fine; cotton-covered polyester core, extra fine; spun polyester, fine; nylon/Dacron, monocord fine
Hems net binding; double fold; edge-stitched; flat/plain; horsehair; machine-stitched; hand- or machine-rolled; wired; if scalloped no hem finish is necessary
Seams French; hairline; lace; plain; tissue-stitched
Seam Finishes net bound; self bound; Hong Kong bound; double-stitched; double-stitched and overcast; double-stitched and trimmed; serging/single-ply overedged; untreated/plain; double-ply zigzag
Pressing steam; use pressing cloth; safe temperature limit 350°F (178.1°C)
Care dry clean
Fabric Resource **Whelan Lace**

Metallic Jacquard Cloth

Fiber Content acetate; Mylar®
Yarn Type extruded metallic film; filament fiber
Yarn Construction monofilament; textured
Fabric Structure Jacquard weave
Finishing Processes calendering; heat-set stabilizing
Color Application yarn dyed
Width 45 inches (114.3 cm)
Weight medium
Hand rough; sandy; supple
Texture light-reflecting qualities; papery; scratchy
Performance Expectations subject to abrasion; dimensional stability; elongation; flexible; heat-set properties; nonpilling
Drapability Qualities falls into moderately soft flares; accommodates fullness by gathering, elasticized shirring; fullness retains moderately soft fall

Recommended Construction Aids & Techniques

Hand Needles beading, 10–13; betweens, 4–6; cotton darners, 9–10; embroidery, 6–8; milliners, 7–9; sharps, 5–8
Machine Needles round/set point, fine 9
Threads silk, industrial size A; poly-core waxed, all purpose; cotton-polyester blend, all purpose; cotton-covered polyester core, all purpose; spun polyester, all purpose; nylon/Dacron, monocord medium
Hems book; bonded; bound/Hong Kong finish; edge-stitched; faced; pinked; pinked and stitched; machine-stitched; overedged; seam binding
Seams flat-felled; French; lapped; plain; safety-stitched; welt; double welt
Seam Finishes book; single-ply bound; Hong Kong bound; edge-stitched; pinked and stitched; serging/single-ply overedged; untreated/plain; single-ply zigzag
Pressing steam; use pressing cloth; safe temperature limit 250°F (122.1°C)
Care dry clean
Fabric Resource **Lawrence Textile Co. Inc.**

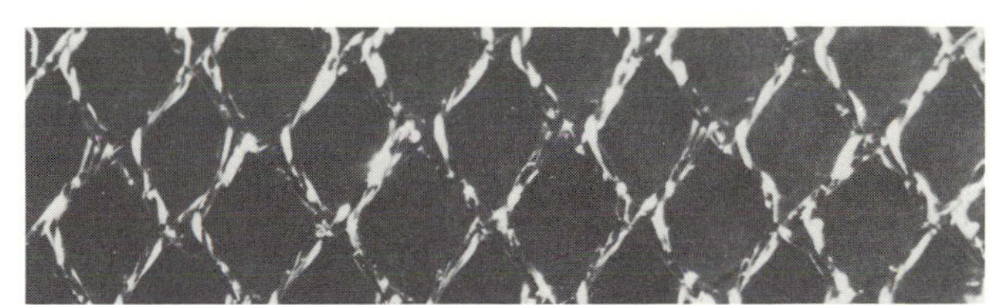

Metallic Net

Fiber Content 100% Mylar®
Yarn Type extruded metallic film
Yarn Construction monofilament
Fabric Structure Raschel knit
Finishing Processes heat-set stabilizing
Color Application solution dyed; yarn dyed
Width 36 inches (91.4 cm)
Weight medium-light
Hand delicate; scratchy; stiff
Texture harsh; open-mesh structure
Performance Expectations abrasion resistant; dimensional stability; durable; flexible; resistant to moths, mildew, sunlight, and chemicals; high tensile strength
Drapability Qualities falls into wide cones; accommodates fullness by gathering; fullness maintains stiff bouffant effect

Recommended Construction Aids & Techniques

Hand Needles betweens, 3; cotton darners, 7–8; embroidery, 4–7; milliners, 5–7; sharps, 3–5
Machine Needles ball point, 9–11
Threads silk, industrial size A-B-C; poly-core waxed, 50; cotton-polyester blend, all purpose; cotton-covered polyester core, heavy duty; spun polyester, heavy duty; nylon/Dacron, monocord medium
Hems net binding; bound/Hong Kong finish; double fold; double-stitched; edge-stitched; horsehair; machine-stitched; untreated raw edge
Seams French; hairline; overedged; plain; safety-stitched; taped; tissue-stitched; zigzagged
Seam Finishes net bound; self bound; Hong Kong bound; double-stitched; double-stitched and trimmed; serging/single-ply overedged; untreated/plain; double-ply zigzag
Pressing steam; use pressing cloth; safe temperature limit 300°F (150.1°C)
Care dry clean
Fabric Resource **Kortex Associates Inc.**

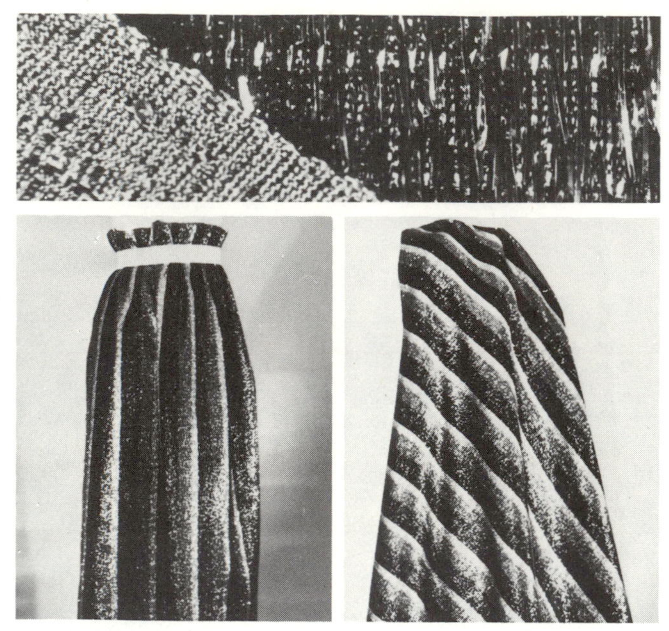

Metallic Pile

Fiber Content nylon/Mylar®
Yarn Type extruded metallic film (pile); filament fiber (base)
Yarn Construction monofilament metallic film (face); textured (base)
Fabric Structure pile (face); plain weave (base)
Finishing Processes brushing; heat-set stabilizing; shearing
Color Application solution dyed; yarn dyed
Width 40 inches (101.6 cm)
Weight medium
Hand rough; supple
Texture pile formed by raised metallic yarns; scratchy
Performance Expectations dimensional stability; durable; elasticity; elongation; heat-set properties; resilient; high tensile strength; metallic pile tends to abrade and dislodge with wear and cleaning
Drapability Qualities falls into soft wide cones; accommodates fullness by gathering; fullness retains moderately soft fall

Recommended Construction Aids & Techniques

Hand Needles beading, 10–13; betweens, 4–6; cotton darners, 9–10; embroidery, 6–8; milliners, 7–9; sharps, 5–8
Machine Needles round/set point, medium-fine 9–11
Threads silk, industrial size A-B-C; poly-core waxed, 50; cotton-polyester blend, all purpose; cotton-covered polyester core, heavy duty; spun polyester, heavy duty; nylon/Dacron, monocord medium
Hems bonded; double-fold and single-fold bias binding; bound/Hong Kong finish; double-stitched; faced; pinked; pinked and stitched; machine-stitched; overedged
Seams flat-felled; French; lapped; plain; safety-stitched; tissue-stitched; welt; double-welt
Seam Finishes single-ply and double-ply bound; Hong Kong bound; double-stitched; edge-stitched; pinked and stitched; serging/single-ply overedged; untreated/plain; single-ply zigzag
Pressing steam; use pressing cloth; safe temperature limit 350°F (178.1°C)
Care dry clean
Fabric Resource **Silverman Textile Co. (sample courtesy of Lita Konde)**

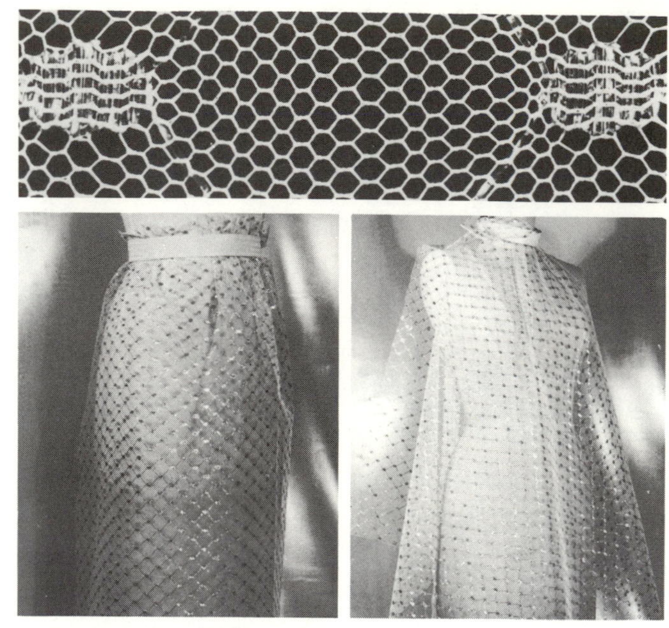

Metallic Point d'Esprit Lace

Fiber Content 100% nylon (base); Mylar® raised pattern
Yarn Type extruded metallic film; filament fiber
Yarn Construction monofilament; monofilament metallic film
Fabric Structure Raschel machine-made lace
Finishing Processes heat-set stabilizing
Color Application solution dyed; yarn dyed
Width 72 inches (182.9 cm)
Weight fine to lightweight
Hand delicate; rigid; rough; stiff
Texture open-mesh structure with raised pattern; design may be small or large; slightly or highly raised
Performance Expectations dimensional stability; elongation; heat-set properties; resilient; high tensile strength; open-mesh structure subject to tearing and snagging
Drapability Qualities falls into wide cones; accommodates fullness by gathering; fullness maintains stiff bouffant effect

Recommended Construction Aids & Techniques

Hand Needles beading, 10–13; betweens, 6–7; embroidery, 9–10; milliners, 10; sharps, 8–10
Machine Needles ball point, 9
Threads silk, industrial size A; poly-core waxed, 70; cotton-polyester blend, extra fine; cotton-covered polyester core, extra fine; spun polyester, fine; nylon/Dacron, monocord fine
Hems net binding; double fold; edge-stitched; horsehair; machine-stitched; hand- or machine-rolled; wired; if scalloped no hem finish is necessary
Seams French; hairline; plain; tissue-stitched
Seam Finishes net bound; self bound; Hong Kong bound; double-stitched; double-stitched and overcast; double-stitched and trimmed; ; serging/single-ply overedged; untreated/plain; double-ply zigzag
Pressing steam; use pressing cloth; safe temperature limit 350°F (178.1°C)
Care dry clean
Fabric Resource **Kortex Associates Inc.**

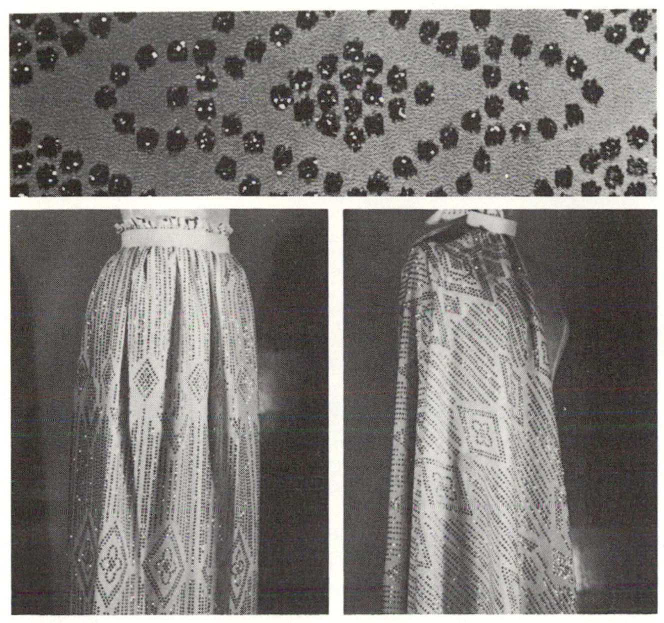

Adhesive-bonded Sparkle or Glitter on Opaque Cloth

Fiber Content 100% polyester (base); Mylar® glitter
Yarn Type extruded cut-up metallic film; filament fiber
Yarn Construction creped; high twist
Fabric Structure plain weave base; paste-on glitter
Finishing Processes crepeing; heat-set stabilizing
Color Application piece dyed; adhesive printing
Width 45 inches (114.3 cm)
Weight medium-light
Hand crepey; supple
Texture raised design areas rough and scratchy
Performance Expectations dimensional stability; elongation; resilient; high tensile strength; glitter tends to abrade and dislodge with wear and cleaning
Drapability Qualities falls into soft flares; accommodates fullness by pleating, gathering, elasticized shirring; fullness retains soft fall

Recommended Construction Aids & Techniques

Hand Needles beading, 10–13; betweens, 4–6; cotton darners, 9–10; embroidery, 6–8; milliners, 7–9; sharps, 5–8
Machine Needles round/set point, medium-fine 9–11
Threads silk, industrial size A; cotton-polyester blend, all purpose; poly-core waxed, all purpose; cotton-covered polyester core, all purpose; spun polyester, all purpose; nylon/Dacron, monocord medium
Hems book; bound/Hong Kong finish; faced; edge-stitched; pinked; pinked and stitched; machine-stitched; overedged
Seams flat-felled; French; lapped; plain; safety-stitched; tissue-stitched; welt; double welt
Seam Finishes book; Hong Kong bound; single-ply bound; edge-stitched; pinked and stitched; serging/single-ply overedged; untreated/plain; single-ply zigzag
Pressing steam; use pressing cloth; safe temperature limit 325°F (164.1°C)
Care dry clean
Fabric Resource **Silverman Textile Co. (sample courtesy of Lita Konde)**

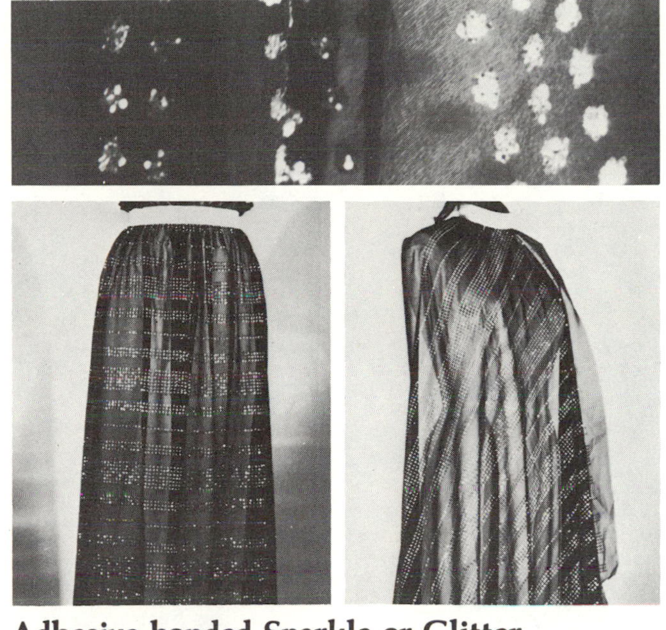

Adhesive-bonded Sparkle or Glitter on Sheer Cloth

Fiber Content 100% polyester (base); Mylar® glitter
Yarn Type extruded cut-up metallic film; filament fiber
Yarn Construction fine; textured
Fabric Structure plain weave base; paste-on glitter
Finishing Processes heat-set stabilizing
Color Application piece dyed; adhesive printing
Width 45 inches (114.3 cm)
Weight fine to lightweight
Hand delicate; soft; supple
Texture raised design areas rough and scratchy ; sheer base
Performance Expectations dimensional stability; elongation; resilient; high tensile strength; glitter tends to abrade and dislodge with wear and cleaning
Drapability Qualities falls into soft languid flares; accommodates fullness by gathering; fullness retains soft graceful fall

Recommended Construction Aids & Techniques

Hand Needles beading, 10–13; betweens, 6–7; embroidery, 9–10; milliners, 10; sharps, 8–10
Machine Needles round/set point, fine 9
Threads silk, industrial size A; poly-core waxed, 70; cotton-polyester blend, extra fine; cotton-covered polyester core, extra fine; spun polyester, fine; nylon/Dacron, monocord fine
Hems bound/Hong Kong finish; double-fold; edge-stitched; horsehair; machine-stitched; hand- or machine-rolled; wired
Seams French; mock French; hairline; plain; tissue-stitched
Seam Finishes self bound; Hong Kong bound; double-stitched; double-stitched and overcast; double-stitched and trimmed; serging/single-ply overedged; untreated/plain; double-ply zigzag
Pressing steam; use pressing cloth; safe temperature limit 325°F (164.1°C)
Care dry clean
Fabric Resource **Silverman Textile Co. (sample courtesy of Lita Konde)**

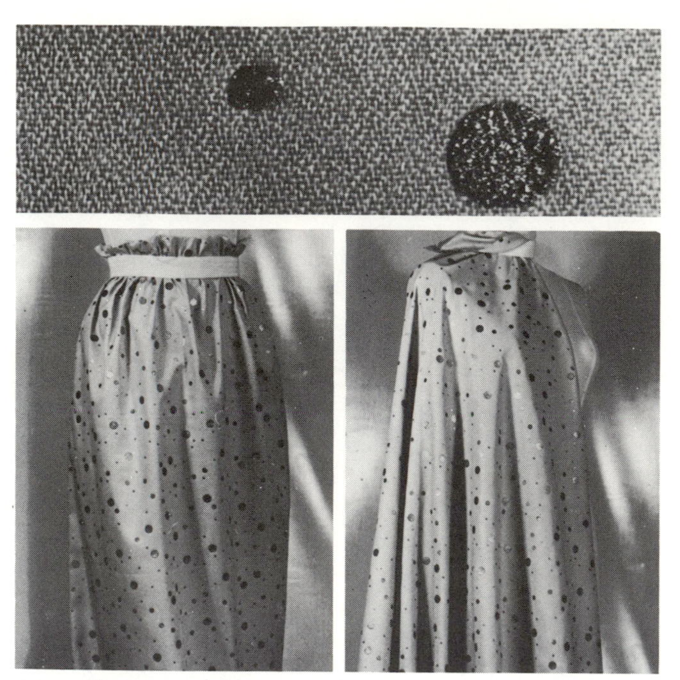

Metallic Sparkle- or Glitter-Printed Design

Fiber Content 100% acetate
Yarn Type filament fiber
Yarn Construction creped; textured
Fabric Structure tightly woven plain weave
Finishing Processes crepeing; heat-set stabilizing
Color Application piece dyed; pigment glitter adhesive printing
Width 45 inches (114.3 cm)
Weight medium
Hand coarse; crepey; supple
Texture raised design areas rough and scratchy
Performance Expectations dimensional stability; elongation; flexible; heat-set properties; mildew and moth resistant; glitter tends to abrade and dislodge with wear and cleaning
Drapability Qualities falls into moderately wide cones; accommodates fullness by gathering; fullness maintains crisp effect

Recommended Construction Aids & Techniques

Hand Needles betweens, 3; cotton darners, 7–8; embroidery, 4–7; milliners, 5–7; sharps, 3–5
Machine Needles round/set point, medium-fine 9–11
Threads silk, industrial size A; poly-core waxed, all purpose; cotton-polyester blend, all purpose; cotton-covered polyester core, all purpose; spun polyester, all purpose; nylon/Dacron, monocord medium
Hems bonded; double-fold or single-fold bias binding; bound/Hong Kong finish; faced; pinked; pinked and stitched; machine-stitched; overedged
Seams flat-felled; plain; safety-stitched; welt; double welt
Seam Finishes book; single-ply bound; Hong Kong bound; edge-stitched; pinked; pinked and stitched; serging/single-ply overedged; untreated/plain; single-ply zigzag
Pressing steam; use pressing cloth; safe temperature limit 250°F (122.1°C)
Care dry clean
Fabric Resource **Iselin Jefferson Division, Dan River Inc. (sample courtesy of Jonan Enterprises Inc.)**

17 ~ Multicomponent/Layered Fabrics

Bonded Fabrics
- Adhesive-bonded Fabric
- Laminated-bonded Fabric

Film-faced Bonded Fabrics
- Simulated Leather (lightweight)
- Simulated Leather (heavyweight)
- Simulated Leather (textured)

Flock-faced Bonded Fabrics
- Simulated Suede
- Simulated Suede (patterned)
- Simulated Suede (crushed)
- Simulated Double-faced Suede

Water-repellent Film-faced Bonded Fabrics
- Rubber Slicker (solid)
- Rubber Slicker (perforated)
- Vinyl Film to Nonwoven Back
- Vinyl / Clear Film

Water-repellent Resin-coated Fabrics
- Resin-coated Duck
- Resin-coated Gabardine
- Resin-coated Muslin Sheeting
- Resin-coated Poplin
- Resin-coated Rip-stop Nylon / Parachute Silk
- Resin-coated Taffeta

Quilted Fabrics
- Double-faced Quilted Fabric
- Gauze-backed Quilted Fabric
- Knit-backed Quilted Fabric

Multicomponent or layered is a term applied to fabrics which are produced with two or more layers of cloth. The multilayers of cloth may be stitched, bonded or sealed together to complete the structure.

A multicomponent fabric may utilize a face, filling and back fabric of any type of fabric structure with compatible fiber content and care factors.

Multicomponent or layered fabrics in this unit include:

- Bonded fabrics
- Coated fabrics
- Laminated fabrics
- Quilted fabrics

Adhesive-bonded Fabric

Fiber Content 100% wool (face); 100% acetate (back)
Yarn Type carded staple; filament fiber
Yarn Construction bulked; textured
Fabric Structure multicomponent structure: twill-weave face adhesive-bonded to tricot-knit back
Finishing Processes heat-set stabilizing; tentering
Color Application each fabric piece dyed separately prior to bonding
Width 60 inches (152.4 cm)
Weight heavy
Hand lofty; spongy; woolly
Texture coarse; napped
Performance Expectations elongation; insulating properties; resilient; subject to snagging, bagging, pilling, abrasion; subject to uneven shrinkage and separation during cleaning
Drapability Qualities falls into soft wide cones; fullness maintains bouffant effect; retains shape of garment

Recommended Construction Aids & Techniques

Hand Needles cotton darners, 1–4; embroidery, 1–3; milliners, 3–4; sharps, 1–3
Machine Needles round/set point, coarse 16–18
Threads six-cord cotton, 30; poly-core waxed, heavy duty; cotton-polyester blend, heavy duty; cotton-covered polyester core, heavy duty; spun polyester, heavy duty; nylon/Dacron, twist A
Hems double-fold and single-fold bias binding; bound/Hong Kong finish; faced; overedged; seam binding
Seams plain; safety-stitched
Seam Finishes single-ply bound; Hong Kong bound; single-ply overcast; pinked; pinked and stitched; serging/single-ply overedged; untreated/plain; single-ply zigzag
Pressing steam; safe temperature limit 250°F (122.1°C)
Care dry clean only
Fabric Resource **Lawrence Textile Co. Inc.**

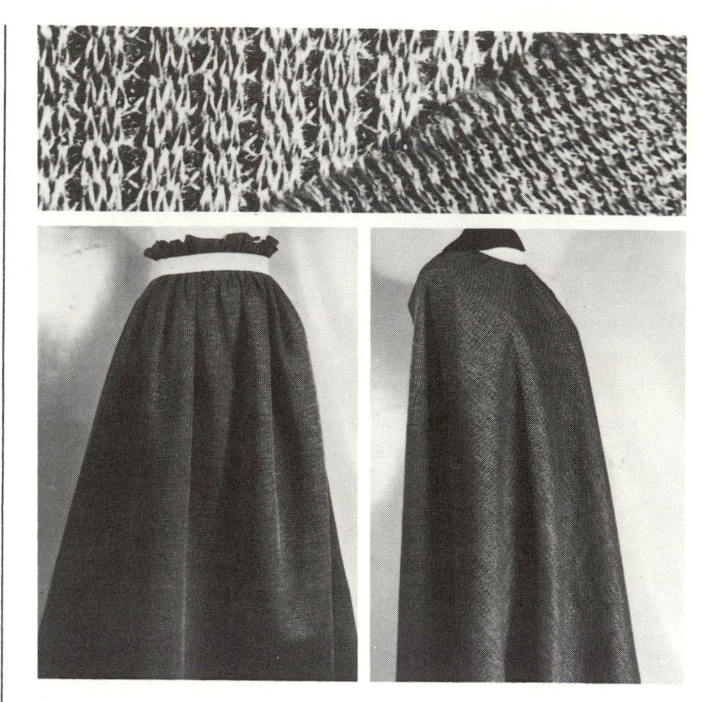

Laminated-bonded Fabric

Fiber Content 100% acrylic (face); 100% acetate (back); urethane (filling)
Yarn Type filament fiber
Yarn Construction bulked; textured
Fabric Structure multicomponent structure: rib-knit face laminated-bonded to tricot-knit back
Finishing Processes heat-set stabilizing; tentering
Color Application each fabric piece dyed separately prior to laminating
Width 60 inches (152.4 cm)
Weight medium-heavy
Hand lofty; spongy; thick
Texture uneven surface; permanently set vertical rib
Performance Expectations compressible; dimensional stability; elongation; elasticity; resilient; subject to separation during cleaning
Drapability Qualities falls into wide cones; fullness maintains lofty bouffant effect; retains shape or silhouette of garment

Recommended Construction Aids & Techniques

Hand Needles cotton darners, 1–4; embroidery, 1–3; milliners, 3–4; sharps, 1–3
Machine Needles ball point, 14
Threads six-cord cotton, 30; poly-core waxed, heavy duty; cotton-polyester blend, heavy duty; cotton-covered polyester core, heavy duty; spun polyester, heavy duty; nylon/Dacron, twist A
Hems double-fold and single-fold bias binding; bound/Hong Kong finish; faced; overedged; seam binding
Seams plain; safety-stitched
Seam Finishes single-ply bound; Hong Kong bound; single-ply overcast; pinked; pinked and stitched; serging/single-ply overedged; untreated/plain; single-ply zigzag
Pressing steam; safe temperature limit 250°F (122.1°C)
Care dry clean only
Fabric Resource **Lawrence Textile Co. Inc.**

Simulated Leather (lightweight)

Fiber Content 100% vinyl (face); acetate-nylon blend (back)
Yarn Type extruded vinyl (face); filament fiber (back)
Yarn Construction solid film (face); textured (back)
Fabric Structure multicomponent structure: vinyl-coated face bonded to tricot-knit back
Finishing Processes face: calendering, softening; back: brushing, stabilizing, napping
Color Application solution dyed (face); piece dyed (back)
Width 54–56 inches (137.2–142.2 cm)
Weight medium-light
Hand rigid; soft; solid
Texture leathery; smooth
Performance Expectations high bursting strength; durable; backing subject to abrasion; water-repellent properties (face)
Drapability Qualities falls into soft wide flares; retains shape of garment; better utilized if fitted by seaming and removing excess fabric

Recommended Construction Aids & Techniques

Hand Needles cotton darners, 7–9; embroidery, 1–3; milliners, 3–4; sharps, 1–3; glovers; sailmakers
Machine Needles leather point, 13–16
Threads mercerized cotton, heavy duty; six-cord cotton, 40; poly-core waxed, regular; cotton-polyester blend, heavy duty; cotton-covered polyester core, all purpose; spun polyester, heavy duty; nylon/Dacron, twist A
Hems bonded; bound/Hong Kong finish; double-stitched; flat/plain; glued
Seams butted; lapped; lapped with raw edges; plain; seam of skins; tissue-stitched; top-stitched; double top-stitched; welt
Seam Finishes single-ply bound; Hong Kong bound; glued; serging/single-ply overedged; untreated/plain
Pressing *do not* press
Care *do not* dry clean; machine wash in warm water, gentle cycle; wash surface with soap and water; air cycle of dryer; *no heat*
Fabric Resources **Concord Fabrics Inc.; Pervel Inc.**

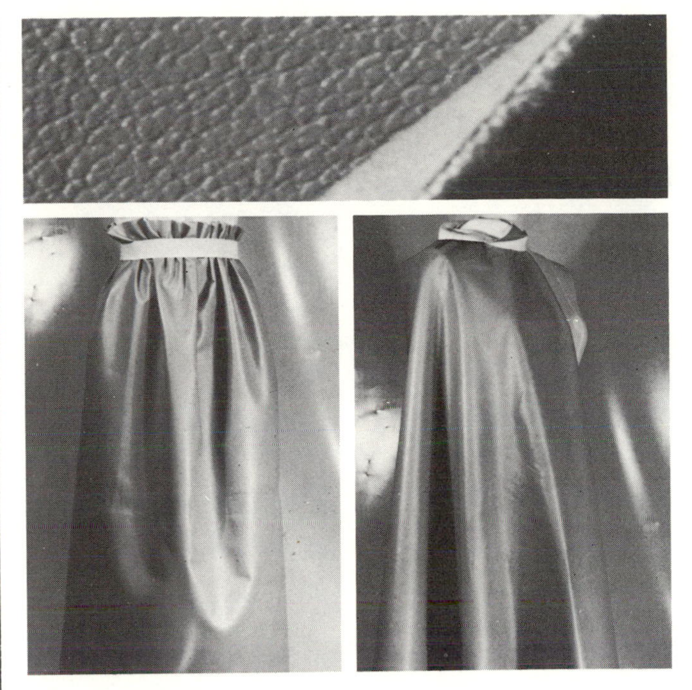

Simulated Leather (heavyweight)

Fiber Content 100% vinyl (face); 100% cotton (back)
Yarn Type extruded film (face); staple (back)
Yarn Construction solid film (face); conventional (back)
Fabric Structure multicomponent structure: vinyl-coated face bonded to woven-flannel back
Finishing Processes face: calendering, softening; back: brushing, napping, shearing
Color Application solution dyed (face); piece dyed (back)
Width 54–56 inches (137.2–142.2 cm)
Weight medium-heavy
Hand rigid; soft; solid
Texture leathery; smooth
Performance Expectations cotton back provides absorbency; high bursting strength; durable; insulating properties; water-repellent properties (face)
Drapability Qualities falls into wide cones; retains shape of garment; better utilized if fitted by seaming and removing excess fabric

Recommended Construction Aids & Techniques

Hand Needles cotton darners, 7–9; embroidery, 1–3; milliners, 3–4; sharps, 1–3; glovers; sailmakers
Machine Needles leather point, 13–16
Threads six-cord cotton, 30; poly-core waxed, heavy duty; cotton-polyester blend, heavy duty; cotton-covered polyester core, heavy duty; spun polyester, heavy duty; nylon/Dacron, twist A
Hems bonded; bound/Hong Kong finish; double-stitched; flat/plain; glued
Seams butted; lapped; lapped with raw edges; plain; seam of skins; tissue-stitched; top-stitched; double top-stitched; welt
Seam Finishes single-ply bound; Hong Kong bound; glued; serging/single-ply overedged; untreated/plain
Pressing *do not* press
Care *do not* dry clean; machine wash in warm water, gentle cycle; wash surface with soap and water; air cycle of dryer; *no heat*
Fabric Resource **Pervel Inc.**

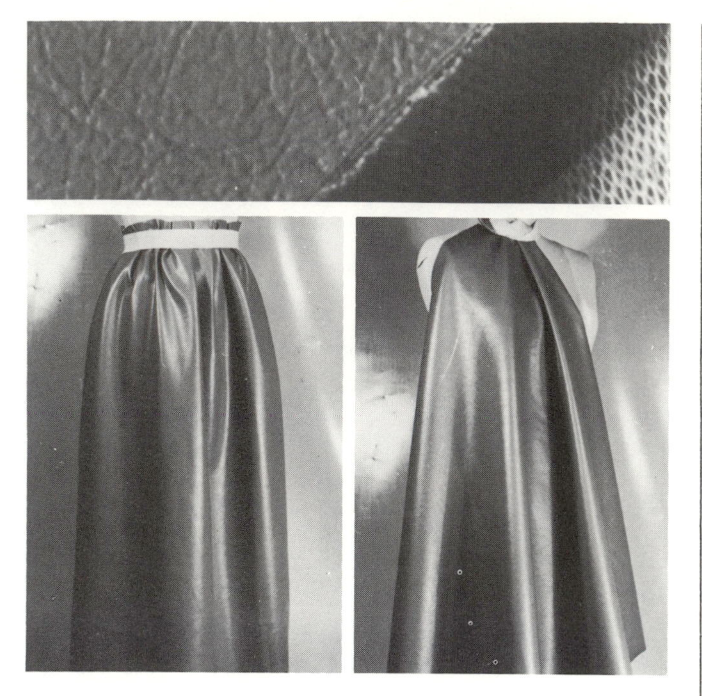

Simulated Leather (textured)

Fiber Content 100% vinyl (face); 100% cotton (back)
Yarn Type extruded film (face); combed staple (back)
Yarn Construction solid film (face); conventional (back)
Fabric Structure multicomponent structure: vinyl-coated face bonded to tricot-knit back
Finishing Processes face: textured by heat embossing; back: mercerizing, singeing, stabilizing
Color Application solution dyed (face); piece dyed (back)
Width 54-56 inches (137.2-142.2 cm)
Weight medium
Hand rigid; soft; solid
Texture grainy textured surface; leathery
Performance Expectations cotton back provides absorbency; high bursting strength; durable; insulating properties; water-repellent properties (face)
Drapability Qualities falls into wide cones; retains shape of garment; better utilized if fitted by seaming and removing excess fabric

Recommended Construction Aids & Techniques

Hand Needles cotton darners, 7-9; embroidery, 1-3; milliners, 3-4; sharps, 1-3; glovers; sailmakers
Machine Needles leather point, 13-16
Threads mercerized cotton, heavy duty; six-cord cotton, 40; poly-core waxed, all purpose; cotton-polyester blend, heavy duty; cotton-covered polyester core, all purpose; spun polyester, heavy duty; nylon/Dacron, twist A
Hems bonded; bound/Hong Kong finish; double-stitched; flat/plain; glued
Seams butted; lapped; lapped with raw edges; plain; seam of skins; tissue-stitched; top-stitched; double top-stitched; welt
Seam Finishes single-ply bound; Hong Kong bound; glued; serging/single-ply overedged; untreated/plain
Pressing *do not* press
Care *do not* dry clean; machine wash in warm water, gentle cycle; wash surface with soap and water; air cycle of dryer; *no heat*
Fabric Resource **Pervel Inc.**

Simulated Suede

Fiber Content 100% rayon (face); 100% cotton (back)
Yarn Type filament fiber (face); staple (back)
Yarn Construction cut staple (face); conventional (back)
Fabric Structure compound structure: flocked face; twill weave back
Finishing Processes face: electrostatic flocking; back: heat-set stabilizing, shearing
Color Application piece dyed
Width 53-54 inches (134.6-137.2 cm)
Weight medium
Hand compact; downy soft; thick
Texture flat sueded effect; velvety
Performance Expectations face subject to edge abrasion and color crocking; antistatic; nonpilling; high tensile strength
Drapability Qualities falls into soft wide flares; fullness maintains bouffant effect; retains shape of garment

Recommended Construction Aids & Techniques

Hand Needles cotton darners, 7-9; embroidery, 1-3; milliners, 3-4; sharps, 1-3
Machine Needles round/set point, medium 14
Threads mercerized cotton, heavy duty; six-cord cotton, 40; poly-core waxed, all purpose; cotton-polyester blend, heavy duty; cotton-covered polyester core, all purpose; spun polyester, heavy duty; nylon/Dacron, twist A
Hems bonded; bound/Hong Kong finish; double-stitched; flat/plain; glued
Seams butted; lapped; lapped with raw edges; plain; seam of skins; tissue-stitched; top-stitched; double top-stitched; welt
Seam Finishes single-ply bound; Hong Kong bound; glued; serging/single-ply overedged; untreated/plain
Pressing steam; safe temperature limit 350°F (178.1°C)
Care dry clean only
Fabric Resource **Pervel Inc.**

Simulated Suede (patterned)

Fiber Content 100% urethane (face); 100% cotton (back)
Yarn Type filament staple; staple
Yarn Construction cut staple (face); conventional (back)
Fabric Structure compound structure: flocked (face); twill weave (back)
Finishing Processes face: heat-set embossing, electrostatic flocking; back: mercerizing, stabilizing
Color Application resist printing (face); piece dyed (back)
Width 52-54 inches (132.1-137.2 cm)
Weight medium
Hand nonelastic, semi-soft; solid
Texture flat napped effect; velvety
Performance Expectations face subject to edge abrasion and color crocking; antistatic; nonpilling; high tensile strength
Drapability Qualities falls into soft wide flares; fullness maintains bouffant effect; retains shape of garment

Recommended Construction Aids & Techniques

Hand Needles cotton darners, 7-9; embroidery, 1-3; milliners, 3-4; sharps, 1-3
Machine Needles round/set point, medium 14
Threads mercerized cotton, heavy duty; six-cord cotton, 40; poly-core waxed, all purpose; cotton-polyester blend, heavy duty; cotton-covered polyester core, all purpose; spun polyester, heavy duty; nylon/Dacron, twist A
Hems bonded; bound/Hong Kong finish; double-stitched; flat/plain; glued
Seams butted; lapped; lapped with raw edges; plain; seam of skins; tissue-stitched; top-stitched; double top-stitched; welt
Seam Finishes single-ply bound; Hong Kong bound; glued; serging/single-ply overedged; untreated/plain
Pressing steam; use pressing cloth; safe temperature limit 275°F (136.1°C)
Care dry clean; machine wash in cold water; medium cycle of dryer
Fabric Resource **Pervel Inc.**

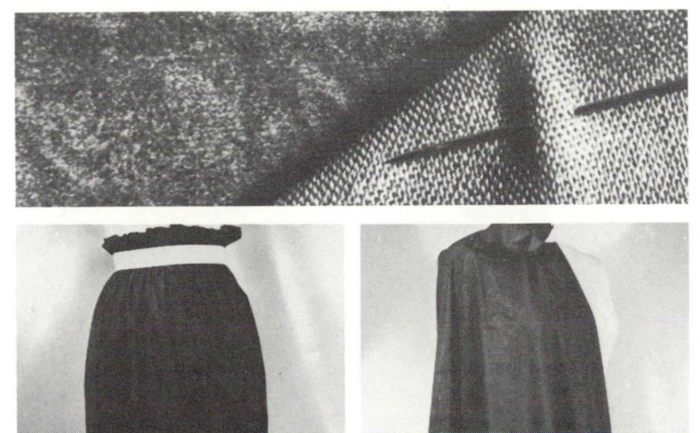

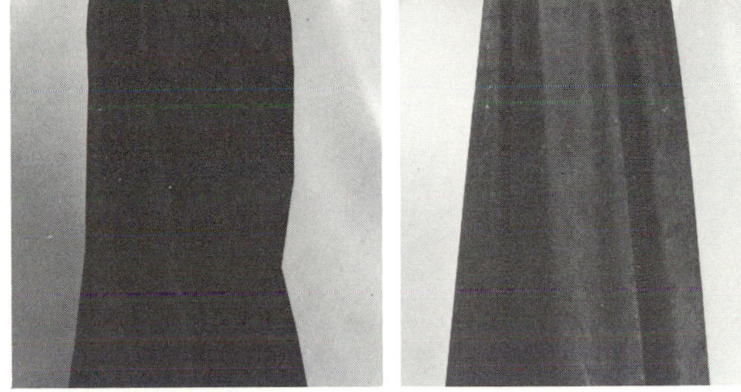

Simulated Suede (crushed)

Fiber Content 90% rayon/10% cotton (face); 100% cotton (back)
Yarn Type filament staple; carded staple
Yarn Construction cut staple (face); conventional (back)
Fabric Structure compound structure: flocked face; twill weave back
Finishing Processes face: embossing, electrostatic flocking; back: brushing, shearing
Color Application piece dyed
Width 52-54 inches (132.1-137.2 cm)
Weight medium
Hand soft; solid; supple
Texture flat napped effect; velvety
Performance Expectations face subject to edge abrasion and color crocking; antistatic; nonpilling; high tensile strength
Drapability Qualities falls into soft wide flares; fullness maintains bouffant effect; retains shape of garment

Recommended Construction Aids & Techniques

Hand Needles cotton darners, 7-9; embroidery, 1-3; milliners, 3-4; sharps, 1-3
Machine Needles round/set point, medium 14
Threads mercerized cotton, heavy duty; six-cord cotton, 40; poly-core waxed, all purpose; cotton-polyester blend, heavy duty; cotton-covered polyester core, all purpose; spun polyester, heavy duty; nylon/Dacron, twist A
Hems bonded; bound/Hong Kong finish; double-stitched; flat/plain; glued
Seams butted; lapped; lapped with raw edges; plain; seam of skins; tissue-stitched; top-stitched; double top-stitched; welt
Seam Finishes single-ply bound; Hong Kong bound; glued; serging/single-ply overedged; untreated/plain
Pressing steam; safe temperature limit 350°F (178.1°C)
Care dry clean; machine wash, gentle cycle; medium cycle of dryer
Fabric Resources **M. Lowenstein & Sons Inc.; Pervel Inc.**

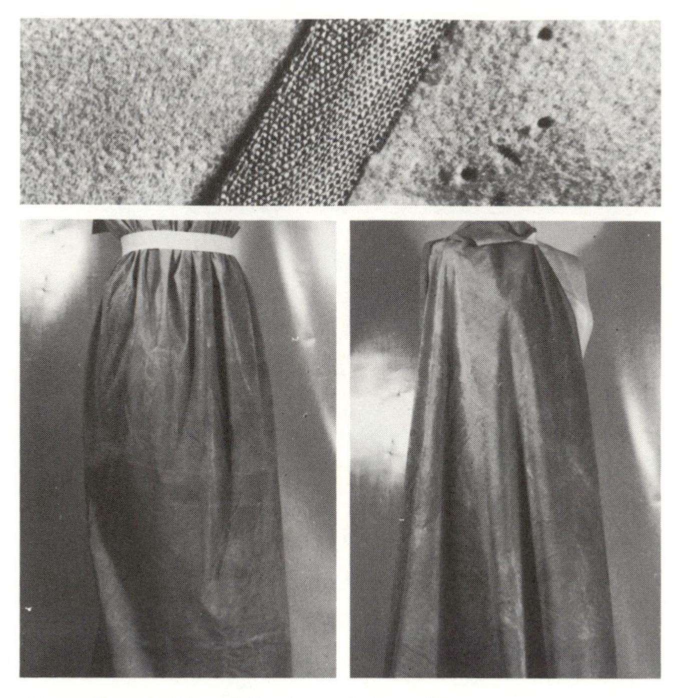

Simulated Double-faced Suede

Fiber Content 100% rayon (face and back); 100% cotton (filling)
Yarn Type filament staple flock; staple fiber
Yarn Construction cut staple (face and back); conventional (filling)
Fabric Structure compound structure: flocked face and back; plain weave filling
Finishing Processes electrostatic flocking; water and stain repellent
Color Application piece dyed
Width 46 inches (116.8 cm)
Weight heavy
Hand nonelastic; soft; thick
Texture sueded face and back; reversible; velvety
Performance Expectations face and back subject to abrasion and color crocking; antistatic; insulating properties; nonpilling; resilient; high tensile strength
Drapability Qualities falls into crisp wide cones; retains shape or silhouette of garment; better utilized if fitted by seaming and eliminating excess fabric

Recommended Construction Aids & Techniques

Hand Needles cotton darners, 7–9; embroidery, 1–3; milliners, 3–4; sharps, 1–3
Machine Needles round/set point, coarse 16–18
Threads six-cord cotton, 30; poly-core waxed, heavy duty; cotton-polyester blend, heavy duty; cotton-covered polyester core, heavy duty; spun polyester, heavy duty; nylon/Dacron, twist A
Hems bonded; bound/Hong Kong finish; double-stitched; flat/plain; glued
Seams butted; lapped with raw edges; plain; seam of skins; tissue-stitched; top-stitched; double top-stitched; welt
Seam Finishes single-ply bound; Hong Kong bound; glued; serging/single-ply overedged; untreated/plain
Pressing steam; safe temperature limit 350°F (178.1°C)
Care dry clean only
Fabric Resource Springs Mills Inc.

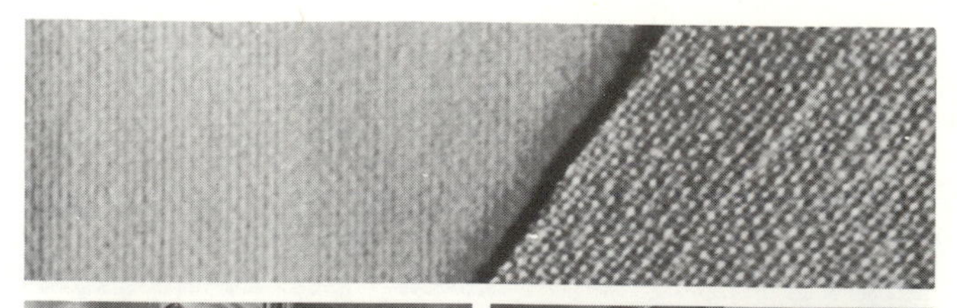

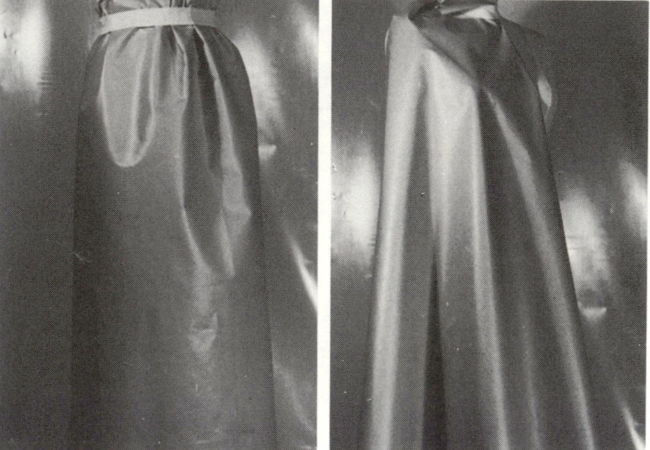

Rubber Slicker (solid)

Fiber Content neoprene rubber (face); 100% cotton (back)
Yarn Type extruded film; staple
Yarn Construction solid film (face); conventional (back)
Fabric Structure multicomponent structure: rubber film face bonded to plain weave back
Finishing Processes heat-set stabilizing
Color Application printed; solution dyed (face); grey goods (back)
Width 43 inches (109.2 cm)
Weight heavy
Hand clammy; rubbery
Texture flat; smooth
Performance Expectations subject to flex abrasion; non air permeable; resilient; subject to stiffening in cold temperatures; water-repellent properties (face)
Drapability Qualities falls into crisp wide cones; retains shape or silhouette of garment; better utilized if fitted by seaming and eliminating excess fabric

Recommended Construction Aids & Techniques

Hand Needles glovers, sailmakers
Machine Needles leather point, 10
Threads six-cord cotton, 30; poly-core waxed, heavy duty; cotton-polyester blend, heavy duty; cotton-covered polyester core, heavy duty; spun polyester, heavy duty; nylon/Dacron, twist A
Hems bonded; bound/Hong Kong finish; double-stitched; flat/plain; glued
Seams butted; lapped with raw edges; plain; seam of skins; tissue-stitched; top-stitched; double top-stitched; welt
Seam Finishes single-ply bound; Hong Kong bound; glued; serging/single-ply overedged; untreated/plain
Pressing *do not* press
Care sponge wipe to clean
Fabric Resource Waldon Textiles/Barry M. Richards Inc.

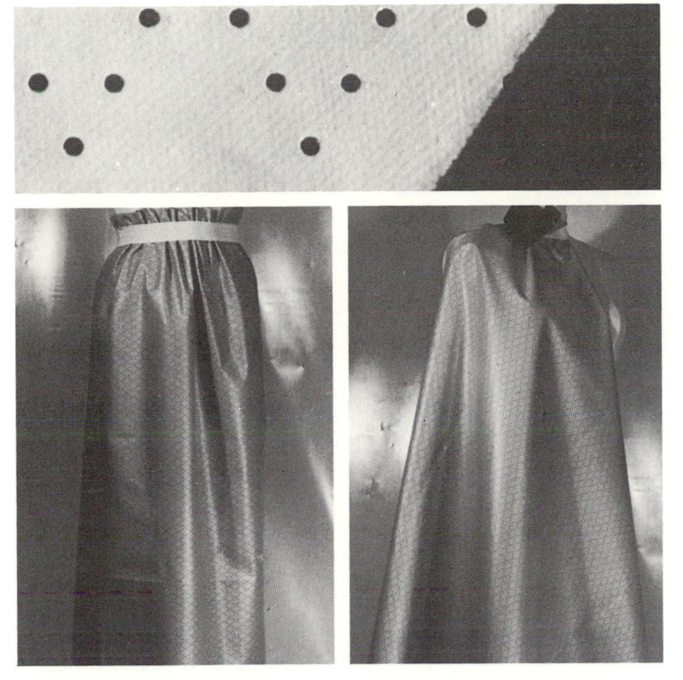

Rubber Slicker (perforated)

Fiber Content neoprene rubber (face); 100% cotton (back)
Yarn Type extruded film; staple
Yarn Construction solid film (face); conventional (back)
Fabric Structure multicomponent structure: rubber film face bonded to plain weave back
Finishing Processes heat-set stabilizing; perforating
Color Application printed; solution dyed (face); grey goods (back)
Width 43 inches (109.2 cm)
Weight heavy
Hand clammy; rubbery
Texture flat; perforated design
Performance Expectations perforations eliminate water-repellent properties, but allow for air permeability; resilient; subject to flex abrasion; subject to stiffening in cold temperatures
Drapability Qualities falls into crisp wide cones; retains shape or silhouette of garment; better utilized if fitted by seaming and eliminating excess fabric

Recommended Construction Aids & Techniques

Hand Needles glovers; sailmakers
Machine Needles leather point, 10
Threads six-cord cotton, 30; poly-core waxed, heavy duty; cotton-polyester blend, heavy duty; cotton-covered polyester core, heavy duty; spun polyester, heavy duty; nylon/Dacron, twist A
Hems bonded; bound/Hong Kong finish; double-stitched; flat/plain; glued
Seams butted; lapped with raw edges; plain; seam of skins; tissue-stitched; top-stitched; double top-stitched; welt
Seam Finishes single-ply bound; Hong Kong bound; glued; serging/single-ply overedged; untreated/plain
Pressing *do not* press
Care sponge wipe to clean
Fabric Resource Waldon Textiles/Barry M. Richards Inc.

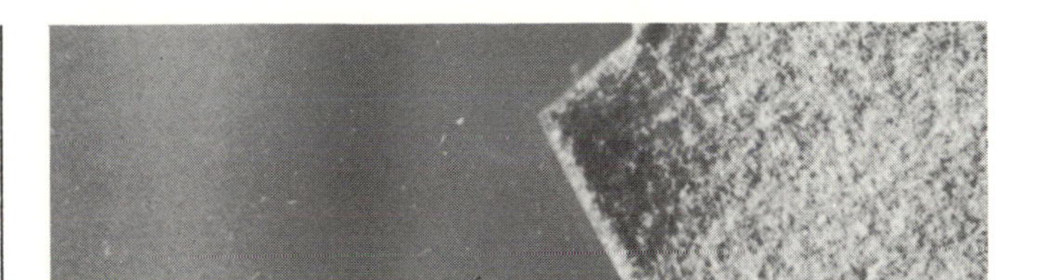

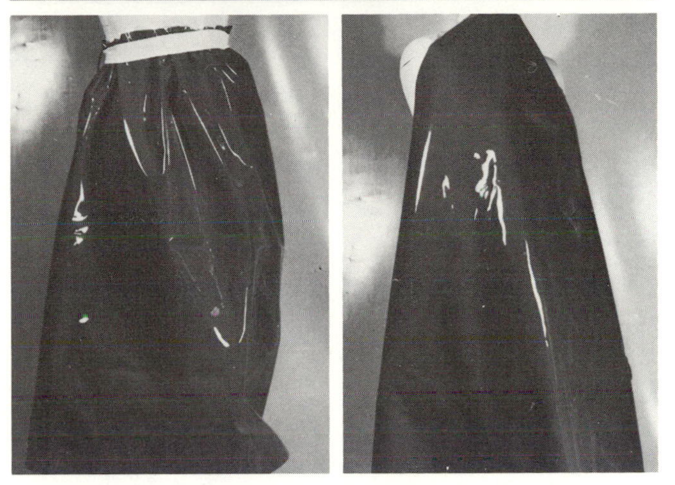

Vinyl Film to Nonwoven Back

Fiber Content 100% vinyl (face); 100% rayon (back)
Yarn Type extruded film; filament staple
Yarn Construction solid film (face); cut staple (back)
Fabric Structure multicomponent structure: vinyl face bonded to nonwoven back
Finishing Processes heat-set stabilizing
Color Application solution dyed
Width 54 inches (137.2 cm)
Weight medium-heavy
Hand plastic; rigid; stiff
Texture opaque; slippery smooth
Performance Expectations subject to edge abrasion; non air permeable; subject to stiffening and cracking in cold temperatures; water-repellent properties
Drapability Qualities falls into stiff wide cones; retains shape or silhouette of garment; better utilized if fitted by seaming and eliminating excess fabric

Recommended Construction Aids & Techniques

Hand Needles glovers; sailmakers
Machine Needles leather point, 10
Threads mercerized cotton, heavy duty; six-cord cotton, 40; poly-core waxed, all purpose; cotton-polyester blend, heavy duty; cotton-covered polyester core, all purpose; spun polyester, heavy duty; nylon/Dacron, twist A
Hems bonded; bound/Hong Kong finish; double-stitched; flat/plain; glued
Seams butted; lapped with raw edges; plain; seam of skins; tissue-stitched; top-stitched; double top-stitched; welt
Seam Finishes single-ply bound; Hong Kong bound; serging/single-ply overedged; untreated/plain
Pressing *do not* press
Care sponge wipe to clean
Fabric Resource Lawrence Textile Co. Inc.

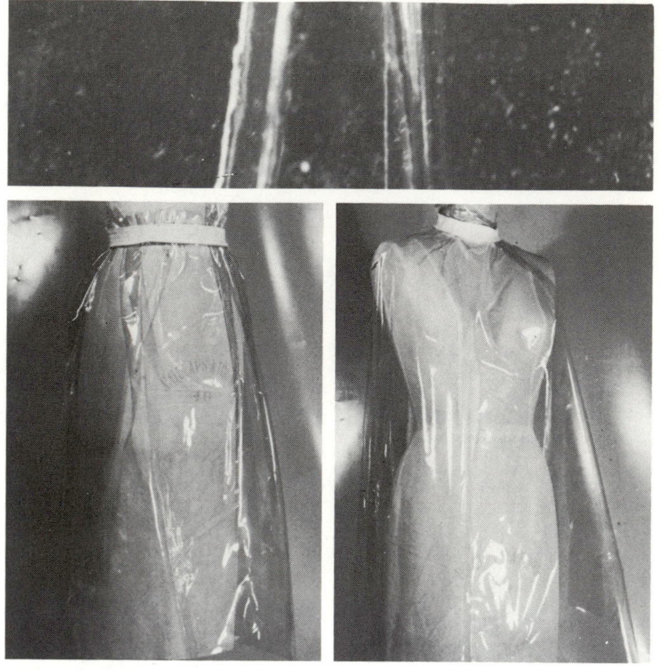

Vinyl/Clear Film

Fiber Content 100% vinyl
Yarn Type extruded film
Yarn Construction solid film
Fabric Structure extruded; poured
Finishing Processes stabilizing
Color Application solution dyed
Width 54 inches (137.2 cm)
Weight[1] medium
Hand clammy; plastic; rigid
Texture clear/transparent; slippery smooth
Performance Expectations subject to edge abrasion; non air permeable; subject to stiffening and cracking in cold temperatures; water-repellent properties
Drapability Qualities falls into stiff wide cones; retains shape or silhouette of garment; better utilized if fitted by seaming and eliminating excess fabric

Recommended Construction Aids & Techniques

Hand Needles glovers; sailmakers
Machine Needles leather point, 10
Threads mercerized cotton, heavy duty; six-cord cotton, 40; poly-core waxed, regular; cotton-polyester blend, heavy duty; cotton-covered polyester core, all purpose; spun polyester, heavy duty; nylon/Dacron, twist A
Hems double-stitched; electrostatic stitching; machine-stitched; untreated/plain
Seams butted; fused; lapped with raw edges; plain; seam of skins; electrostatic stitching; tissue-stitched; top-stitched; double top-stitched
Seam Finishes untreated/plain
Pressing *do not* press
Care sponge wipe to clean
Fabric Resources Lawrence Textile Co. Inc.; Waldon Textiles/Barry M. Richards Inc.

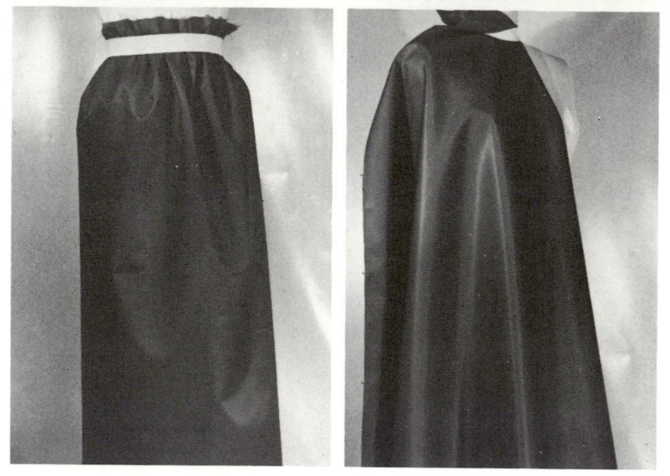

Resin-coated Duck

Fiber Content 60% cotton/40% nylon
Yarn Type combed staple; filament fiber
Yarn Construction conventional; textured
Fabric Structure closely woven plain weave
Finishing Processes calendering; stabilizing; water-repellent resin-coated and cured
Color Application piece dyed
Width 44–45 inches (111.8–114.3 cm)
Weight medium
Hand crisp; papery; rigid
Texture boardy; harsh
Performance Expectations durable; elongation; heat-set properties; high tensile strength; water-repellent properties
Drapability Qualities falls into crisp wide flares; fullness maintains crisp bouffant effect; retains shape or silhouette of garment

Recommended Construction Aids & Techniques

Hand Needles betweens, 3–5; cotton darners, 4–7; embroidery, 4–6; milliners, 5–6; sharps, 4–6
Machine Needles round/set point, medium-coarse 14–16
Threads mercerized cotton, 50; six-cord cotton, 50; poly-core waxed, fine; cotton-polyester blend, all purpose; cotton-covered polyester core, all purpose; spun polyester, all purpose; nylon/Dacron, twist A
Hems book; single-fold or double-fold bias binding; bound/Hong Kong finish; edge-stitched; machine-stitched; overedged; seam binding
Seams flat-felled; lapped; plain; safety-stitched; welt; double welt
Seam Finishes book; single-ply bound; Hong Kong bound; edge-stitched; serging/single-ply overedged; untreated/plain; single-ply zigzag
Pressing steam; safe temperature limit 350°F (178.1°C)
Care dry clean
Fabric Resource Waldon Textiles/Barry M. Richards Inc.

Resin-coated Gabardine

Fiber Content 65% polyester/35% cotton
Yarn Type filament fiber; staple
Yarn Construction conventional; textured
Fabric Structure steep right-hand twill weave
Finishing Processes calendering; stabilizing; Zepel® water/stain repeller
Color Application piece dyed–union dyed
Width 59–60 inches (144.9–152.4 cm)
Weight medium-heavy
Hand harsh; scroopy; stiff
Texture grainy; rough; diagonal line of twill weave visible on face
Performance Expectations subject to color crocking and edge abrasion; dimensional stability; durable; elongation; heat-set properties; high tensile strength; water-repellent properties
Drapability Qualities falls into wide cones; retains shape or silhouette of garment; better utilized if fitted by seaming and eliminating excess fabric

Recommended Construction Aids & Techniques

Hand Needles cotton darners, 7–9; embroidery, 1–3; milliners, 3–4; sharps, 1–3
Machine Needles round/set point, coarse 16–18
Threads mercerized cotton, heavy duty; six-cord cotton, 40; poly-core waxed, regular; cotton-polyester blend, heavy duty; cotton-covered polyester core, all purpose; spun polyester, heavy duty; nylon/Dacron, twist A
Hems single-fold or double-fold bias binding; bound/Hong Kong finish; faced; overedged; seam binding
Seams plain; safety-stitched
Seam Finishes single-ply bound; Hong Kong bound; single-ply overcast; pinked; pinked and stitched; serging/single-ply overedged; untreated/plain; single-ply zigzag
Pressing steam; safe temperature limit 325°F (164.1°C)
Care dry clean
Fabric Resource **Reeves Brothers Inc.**

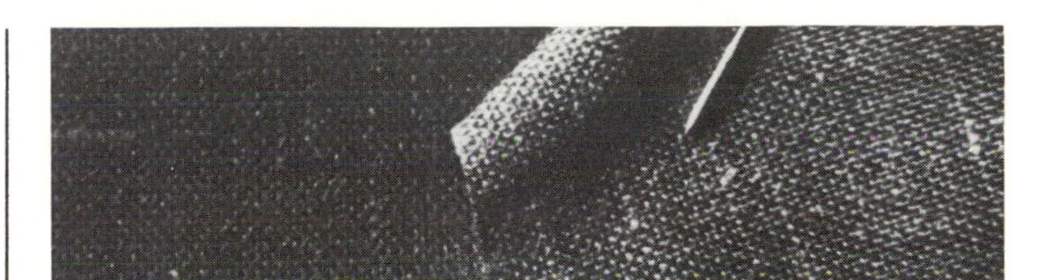

Resin-coated Muslin Sheeting

Fiber Content 100% cotton
Yarn Type carded staple
Yarn Construction conventional
Fabric Structure plain weave
Finishing Processes calendering; glazing; water-repellent resin-coated and cured
Color Application piece dyed
Width 44–45 inches (111.8–114.3 cm)
Weight medium-light
Hand clammy; rigid; semi-soft
Texture flat; smooth
Performance Expectations durable; elongation; high tensile strength; water-repellent properties
Drapability Qualities falls into wide cones; fullness maintains crisp effect; retains shape of garment

Recommended Construction Aids & Techniques

Hand Needles betweens, 6–7; cotton darners, 8–10; embroidery, 7–8; milliners, 7–8; sharps, 7–8
Machine Needles round/set point, medium 11–14
Threads mercerized cotton, 50; six-cord cotton, 50; poly-core waxed, fine; cotton-polyester blend, all purpose; cotton-covered polyester core, all purpose
Hems book; single-fold bias binding; bound/Hong Kong finish; edge-stitched; machine-stitched; overedged; seam binding
Seams flat-felled; lapped; plain; safety-stitched; welt; double welt
Seam Finishes self bound; Hong Kong bound; double-stitched; double-stitched and trimmed; edge-stitched; serging/single-ply overedged; untreated/plain; double-ply zigzag
Pressing steam; safe temperature limit 325°F (164.1°C)
Care dry clean
Fabric Resource **Waldon Textiles/Barry M. Richards Inc.**

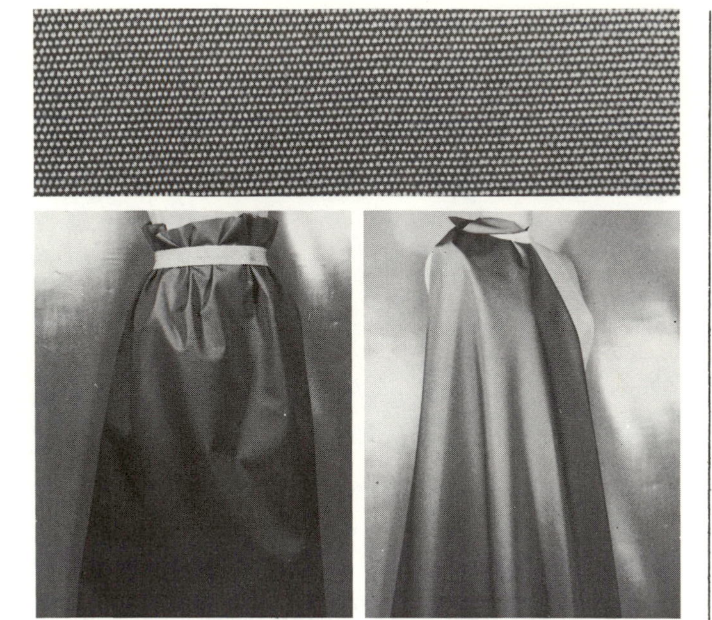

Resin-coated Poplin

Fiber Content 75% polyester/25% cotton
Yarn Type combed staple; filament fiber
Yarn Construction conventional; textured
Fabric Structure plain weave; thicker yarns in filling produce rib in crosswise direction
Finishing Processes calendering; water-repellent resin-coated and cured
Color Application piece dyed–union dyed
Width 45–46 inches (114.3–116.8 cm)
Weight medium-heavy
Hand harsh; papery; stiff
Texture pronounced crosswise rib; rough
Performance Expectations durable; elongation; heat-set properties; high tensile strength; water-repellent properties
Drapability Qualities falls into stiff wide cones; retains shape or silhouette of garment; better utilized if fitted by seaming and eliminating excess fabric

Recommended Construction Aids & Techniques

Hand Needles cotton darners, 7–9; embroidery, 1–3; milliners, 3–4; sharps, 1–3
Machine Needles round/set point, medium-coarse 14–16
Threads mercerized cotton, heavy duty; six-cord cotton, 40; poly-core waxed, all purpose; cotton-polyester blend, heavy duty; cotton-covered polyester core, all purpose; spun polyester, heavy duty; nylon/Dacron, twist A
Hems book; single-fold bias binding; bound/Hong Kong finish; edge-stitched; machine-stitched; overedged; seam binding
Seams flat-felled; lapped; plain; safety-stitched; welt; double welt
Seam Finishes book; single-ply bound; Hong Kong bound; edge-stitched; serging/single-ply overedged; untreated/plain; single-ply zigzag
Pressing steam; safe temperature limit 325°F (164.1°C)
Care dry clean
Fabric Resource **Klopman Mills**

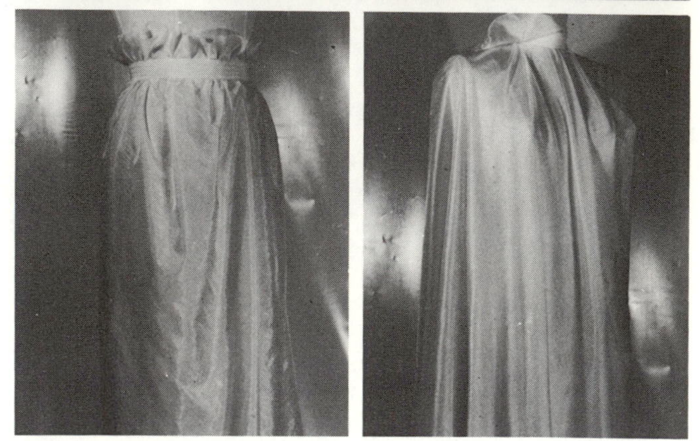

Resin-coated Rip-stop Nylon/Parachute Silk

Fiber Content 100% nylon
Yarn Type filament fiber
Yarn Construction textured
Fabric Structure plain weave variation; cords in lengthwise and crosswise directions produce box effect
Finishing Processes calendering; glazing; water-repellent resin-coated and cured
Color Application piece dyed; solution dyed; yarn dyed
Width 48 inches (121.9 cm)
Weight fine to lightweight
Hand semi-crisp; fine; scroopy
Texture flat; satiny smooth
Performance Expectations abrasion resistant; non air permeable; dimensional stability; durable; elasticity; elongation; resilient; subject to static buildup; high tensile strength; water-repellent properties; wind resistant; lengthwise and crosswise cords prevent fabric from ripping
Drapability Qualities falls into moderately crisp flares; accommodates fullness by gathering, elasticized shirring; fullness maintains moderately crisp effect

Recommended Construction Aids & Techniques

Hand Needles beading, 10–13; betweens, 6–7; embroidery, 9–10; milliners, 9–10; sharps, 9–10
Machine Needles round/set point, fine 9
Threads silk, industrial size A; poly-core waxed, extra fine; cotton-polyester blend, extra fine; spun polyester, fine; nylon/Dacron, monocord A
Hems book; bound/Hong Kong finish; machine-stitched; overedged
Seams flat-felled; French; hairline; lapped; safety-stitched; tissue-stitched
Seam Finishes self bound; Hong Kong bound; double-stitched; double-stitched and trimmed; edge-stitched; serging/single-ply overedged; double-ply zigzag
Pressing steam; safe temperature limit 350°F (178.1°C)
Care launder
Fabric Resource **Waldon Textiles/Barry M. Richards Inc.**

Resin-coated Taffeta

Fiber Content 100% nylon
Yarn Type filament fiber
Yarn Construction fine; textured
Fabric Structure plain weave
Finishing Processes Alumini-Zed® water-repellent coating; calendering
Color Application piece dyed
Width 44–45 inches (111.8–114.3 cm)
Weight light
Hand nonelastic; rubbery; thin
Texture flat; satiny smooth
Performance Expectations durable; elongation; heat-set properties; high tensile strength; water and stain repellent
Drapability Qualities falls into crisp flares; fullness maintains crisp effect; retains shape of garment

Recommended Construction Aids & Techniques

Hand Needles beading, 10–13; betweens, 6–7; embroidery, 9–10; milliners, 9–10; sharps, 9–10
Machine Needles round/set point, fine 9
Threads silk, industrial size A; poly-core waxed, extra fine; cotton-polyester blend, extra fine; spun polyester, fine; nylon/Dacron, monocord A
Hems book; bound/Hong Kong finish; machine-stitched; overedged
Seams flat-felled; French; hairline; lapped; safety-stitched; tissue-stitched; welt
Seam Finishes self bound; Hong Kong bound; double-stitched; double-stitched and trimmed; edge-stitched; serging/single-ply overedged; untreated/plain; double-ply zigzag
Pressing steam; use pressing cloth; safe temperature limit 200°F (94.1°C)
Care dry clean
Fabric Resource **Waldon Textiles/Barry M. Richards Inc.**

Double-faced Quilting

Fiber Content 100% nylon (face and back); 100% polyester (filling)
Yarn Type filament fiber; filament staple
Yarn Construction textured, fine yarn (face and back); textured, bulked batting (filling)
Fabric Structure multicomponent structure: 4 layers of fabric stitched together
Quilting Information woven (face and back); nonwoven (backing); bulked batting (filling)
Color Application fabrics are piece dyed separately prior to quilting
Width 45 inches (114.3 cm)
Weight medium
Hand bulky; lofty; soft
Texture high and low puffed effect; patterned stitching
Performance Expectations bulky but no added weight; dimensional stability; durable; elasticity; elongation; insulating properties; resilient; subject to static buildup
Drapability Qualities falls into wide cones; retains shape or silhouette of garment; better utilized if fitted by seaming and eliminating excess fabric

Recommended Construction Aids & Techniques

Hand Needles cotton darners, 1–4; embroidery, 1–3; milliners, 3–4; sharps, 1–3
Machine Needles round/set point, medium 11–14
Threads six-cord cotton, 30; poly-core waxed, heavy duty; cotton-polyester blend, heavy duty; cotton-covered polyester core, heavy duty; spun polyester, heavy duty; nylon/Dacron, twist A
Hems single-fold or double-fold bias binding; bound/Hong Kong finish; faced; overedged; seam binding
Seams plain; safety-stitched
Seam Finishes single-ply bound; Hong Kong bound; single-ply overcast; pinked; pinked and stitched; serging/single-ply overedged; untreated/plain; single-ply zigzag
Pressing steam; safe temperature limit 350°F (178.1°C)
Care dry clean
Fabric Resources **Lawrence Textile Co. Inc.; Waldon Textiles/Barry M. Richards Inc.**

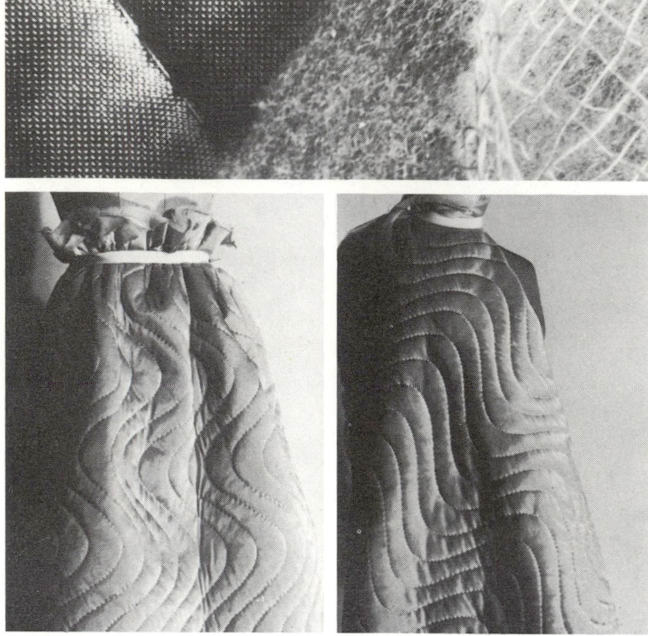

Gauze-backed Quilting

Fiber Content 100% acetate (face); 100% cotton (back); 100% polyester (filling)
Yarn Type filament fiber (face); staple (back); filament staple (filling)
Yarn Construction textured (face); conventional (back); cut staple (filling)
Fabric Structure multicomponent structure: 3 layers of fabric stitched together
Quilting Information tricot knit (face); gauze (backing); nonwoven batting (filling)
Color Application fabrics are piece dyed separately prior to quilting
Width 45 inches (114.3 cm)
Weight medium-light
Hand bulky; soft; spongy
Texture high and low puffed effect; patterned stitching
Performance Expectations subject to abrasion; bulky but no added weight; dimensional stability; insulating properties
Drapability Qualities falls into wide cones; retains shape or silhouette of garment; better utilized if fitted by seaming and eliminating excess fabric

Recommended Construction Aids & Techniques

Hand Needles cotton darners, 1–4; embroidery, 1–3; milliners, 3–4; sharps, 1–3
Machine Needles ball point, 11–14
Threads six-cord cotton, 30; poly-core waxed, heavy duty; cotton-polyester blend, heavy duty; cotton-covered polyester core, heavy duty; spun polyester, heavy duty; nylon/Dacron, twist A
Hems single-fold or double-fold bias binding; bound/Hong Kong finish; faced; overedged; seam binding
Seams plain; safety-stitched
Seam Finishes single-ply bound; Hong Kong bound; single-ply overcast; pinked; pinked and stitched; serging/single-ply overedged; untreated/plain; single-ply zigzag
Pressing steam; safe temperature limit 250°F (122.1°C)
Care launder; dry clean
Fabric Resource Springs Mills Inc.

Knit-backed Quilting

Fiber Content 100% cotton (face); 100% polyester (back); 100% polyester (filling)
Yarn Type staple; filament fiber; filament staple
Yarn Construction conventional (face); textured (back); cut staple (filling)
Fabric Structure multicomponent structure: 3 layers of fabric stitched together
Quilting Information woven (face); tricot knit (back); nonwoven batting (filling)
Color Application fabrics are piece dyed separately prior to quilting
Width 42–43 inches (106.7–109.2 cm)
Weight medium
Hand bulky; lofty; soft
Texture high and low puffed effect; patterned stitching
Performance Expectations bulky but no added weight; dimensional stability; elongation; durable; insulating properties; high tensile strength
Drapability Qualities falls into wide cones; retains shape or silhouette of garment; better utilized if fitted by seaming and eliminating excess fabric

Recommended Construction Aids & Techniques

Hand Needles cotton darners, 1–4; embroidery, 1–3; milliners, 3–4; sharps, 1–3
Machine Needles round/set point, medium 14
Threads six-cord cotton, 30; poly-core waxed, heavy duty; cotton-polyester blend, heavy duty; cotton-covered polyester core, heavy duty; spun polyester, heavy duty; nylon/Dacron, twist A
Hems single-fold or double-fold bias binding; bound/Hong Kong finish; faced; overedged; seam binding
Seams plain; safety-stitched
Seam Finishes single-ply bound; Hong Kong bound; single-ply overcast; pinked; pinked and stitched; serging/single-ply overedged; untreated/plain; single-ply zigzag
Pressing steam; safe temperature limit 325°F (164.1°C)
Care launder; dry clean
Fabric Resource Springs Mills Inc.

Notes

1—Cotton/Cotton-type Fabrics

1. Variations in weight, hand and texture are due to the fineness of yarns, thread count and finishes used in the construction of the fabric.
2. Printcloth may be a broadcloth, muslin, percale or sheeting that has been printed.
3. True calico is not available today. Muslin, cambric and other plain weave fabrics are printed with "calico print."

2—Linen/Linen-type Fabrics

1. Variations in weight and opacity are due to differences in the fineness of yarn and thread count used in the construction of the fabric.
2. Homespun denotes hand-loomed quality or texture. Variations of homespun range from close to open weave and from finely spun to heavy yarns.

4—Silk/Silk-type Fabrics

1. The fabric draped on the form is a heavyweight, textured figured designed damask made of thick-and-thin yarns. Other silk damask fabrics are made of lightweight, fine yarns with a soft drape and fall.

6—Satin/Satin-type Fabrics

1. Weights range from light- to medium-heavy. Weights will affect types of needles, threads, hems, seams and seam finishes used in construction of garments.

8—Wool/Wool-type Fabrics

1. Most unclassified twill weave woolen fabrics are referred to as cassimere.
2. Flannels vary in closeness or fineness of weave, degree of napping and weight.
3. Some kasha cloth may contain specialty hair fibers. Natural color of hair fiber produces streaked, mottled appearance.
4. Originally a thick woolen homespun tweed handwoven by Irish peasants. Donegal tweed refers to a fabric woven with colorful, thick slubbed yarns which are prominent on the surface.
5. Unfinished worsteds may be named after their place of origin, e.g., English or Norfolk worsted.

11—Knit/Knit-type Fabrics

1. Byrnes Cloth® is the trade designation for fabrics used for thermal underwear.
2. The term sweater knit applies to fabric knitted with a rib band at the lower edge. The rib band is part of the knitting procedure.

13—Net/Netting/Net-type Fabrics

1. Available with soft finish and in a variety of mesh sizes.

16—Metallic/Metallic-type Fabrics

1. Points or openings may be larger or smaller. Available with a soft or crisp finish for a soft or crisp hand or drape.

17—Multicomponent/Layered Fabrics

1. Resin-coated Rip-stop Nylon is available:
   - A. without a water-repellent glaze;
   - B. in different denier yarns for thinner or thicker fabric;
   - C. in different finishes for soft or crisp hand or drape.
2. Weight of vinyl/clear film is measured from fine and thin to heavy and thick. Different weights (gauges) will affect types of needles, threads, hems, seams, and seam finishes used in construction of garments.

Fabric Resources

Alba-Waldensian Knits
350 Fifth Avenue
New York, New York 10001

Ambassador Lace and Embroidery Co. Inc.
4219 Palisade Avenue
Union City, New Jersey 07087

American Silk Mills Corp.
111 West 40th Street
New York, New York 10018

Ameritex Division
United Merchants and Manufacturers, Inc.
1407 Broadway
New York, New York 10018

Amical Fabrics
1040 Avenue of the Americas
New York, New York 10018

Anglo Fabrics Co. Inc.
1407 Broadway
New York, New York 10018

Applause Fabrics Ltd.
119 West 40 Street
New York, New York 10018

Associated Lace Corp.
108 West 39th Street
New York, New York 10018

Auburn Fabrics Inc.
215 West 40th Street
New York, New York 10018

Bloomsburg Mills
1430 Broadway
New York, New York 10036

Brunswick Associates Inc.
c/o AERFAB Corp.
1290 Avenue of the Americas
New York, New York 10019

Burlington Industries, Inc.
1345 Avenue of the Americas
New York, New York 10019

Burlington Menswear
Division of Burlington Industries, Inc.
1345 Avenue of the Americas
New York, New York 10019

Cameo Fabrics Inc.
20 West 36th Street
New York, New York 10018

Carleton Woolen Mills Inc.
1412 Broadway
New York, New York 10018

Cinderella Knitting Mills
Division of Reeves Brothers, Inc.
1271 Avenue of the Americas
New York, New York 10020

Concord Fabrics Inc.
1411 Broadway
New York, New York 10018

Cone Mills Corporation
1440 Broadway
New York, New York 10018

Continental Felt Division
Daniel H. Price Inc.
15 East 48th Street
New York, New York 10017

Crompton-Richmond Inc.
1071 Avenue of the Americas
New York, New York 10018

Deluxe Velvet Co.
29 West 36th Street
New York, New York 10018

Drexler Associates Inc.
1290 Avenue of the Americas
New York, New York 10018

Earl Glo/Erlanger Blumgart
and Co. Inc.
1450 Broadway
New York, New York 10018

Embroidery Council of America
c/o McGrath/Power Association, Inc.
500 Fifth Avenue
New York, New York 10036

Furtex
450 Seventh Avenue
New York, New York 10001

Gehring Textiles
200 Madison Avenue
New York, New York 10016

Gloversville Mills Inc.
1460 Broadway
New York, New York 10036

Hamilton Adams Imports Ltd.
104 West 40th Street
New York, New York 10036

Hargro Fabrics Inc.
1040 Avenue of the Americas
New York, New York 10018

Kabat Textile Corp.
215 West 40th Street
New York, New York 10018

Klopman Mills
Division of Burlington Industries, Inc.
Rockleigh, New Jersey 07647

Kortex Associates Inc.
39 West 28th Street
New York, New York 10001

Lace of France
989 Avenue of the Americas
New York, New York 10018

Lafitte Inc.
151 West 40th Street
New York, New York 10036

Jack Larsen Incorporated
232 East 59 Street
New York, New York 10022

Lawrence Textile Co. Inc.
1412 Broadway
New York, New York 10018

Liberty Fabrics of New York, Inc.
Two Park Avenue
New York, New York 10016

Loomskill/Gallery Screen Print
1412 Broadway
New York, New York 10018

Loom Tex Corp
469 Seventh Avenue
New York, New York 10018

M. Lowenstein & Sons, Inc.
111 West 40th Street
New York, New York 10036

Marcus Brothers Textile Inc.
1460 Broadway
New York, New York 10036

J.B. Martin Co. Inc.
1290 Avenue of the Americas
New York, New York 10036

Milan Textile Machines Inc.
492 Grand Boulevard
Westbury, New York 11590

Moiré Corporation
20-21 Wagaraw Road
Fair Lawn, New Jersey 07410

Native Textiles
1185 Avenue of the Americas
New York, New York 10036

North American Lace Co. Inc.
512 31st Street
Union City, New Jersey 07087

Novik and Co. Inc.
41 West 38th Street
New York, New York 10018

Pervel
180 Madison Avenue
New York, New York 10016

Printsiples and Company B
1450 Broadway
New York, New York 10018

Private Collections Fabrics Ltd.
1412 Broadway
New York, New York 10018

Reeves Brothers, Inc.
1271 Avenue of the Americas
New York, New York 10020

Dan River Inc.
111 West 40th Street
New York, New York 10018

Russell Corporation
1114 Avenue of the Americas
New York, New York 10036

Sequin International Corp.
108 West 39th Street
New York, New York 10018

Milton Sherlip Inc.
108 West 30th Street
New York, New York 10018

Sormani Co. Inc./Taroni
108 West 39th Street
New York, New York 10018

Springs Mills Inc.
1430 Broadway
New York, New York 10018

Stēvcoknit, Inc.
1450 Broadway
New York, New York 10018

J.P. Stevens & Co., Inc.
Wool and Cotton Divisions
Stevens Tower
1185 Avenue of the Americas
New York, New York 10036

Successful Creations/Shartex Inc.
273 Livingston Street
Northvale, New Jersey 07647

Truemark Discount Fabrics Inc.
261 Seventh Avenue
New York, New York 10001

Twintella Fabrics
1460 Broadway
New York, New York 10036

Universal Knitting Machines Corp.
3080 Atlantic Avenue
Brooklyn, New York 11208

Walden Textiles/Barry M. Richards Inc.
566 Seventh Avenue
New York, New York 10018

Whelan Lace
58 West 40th Street
New York, New York 10018

White Rose Fabrics
Division of A.E. Nathan Inc.
108 West 39th Street
New York, New York 10018

Arthur Zeiler Woolens Inc.
205 West 39th Street
New York, New York 10018

Fiber Resources

Allied Chemical Corporation
Fibers Division
Friendship Center Park Office
Suite 108
Greensboro, North Carolina 27409

American Cyanamid Company
Wayne, New Jersey 07470

American Enka Company
Marketing Technical Department
Enka, North Carolina 28728

Avtex Fibers Inc.
1185 Avenue of the Americas
New York, New York 10036

Belgian Linen Association
280 Madison Avenue
New York, New York 10016

Celanese Corporation
Celanese Fibers Marketing Company
1211 Avenue of the Americas
New York, New York 10036

Cotton Inc.
1370 Avenue of the Americas
New York, New York 10019

Courtaulds North America Inc.
104 West 40th Street
New York, New York 10018

Dow Badische Company
Williamsburg, Virginia 23185

Eastman Chemicals Products Inc.
Kingsport, Tennessee 37662

E.I. du Pont de Nemours & Company
Textile Fibers Department
Technical Service Section
Wilmington, Delaware 19898

Hercules Incorporated
Wilmington, Delaware 19899

Hoechst Fibers Division
American Hoechst Corporation
1515 Broadway
New York, New York 10036

International Silk Association
c/o Rudolph Desco & Co.
580 Sylvan Avenue
Englewood Cliffs, New Jersey 07632

Mohair Council of America
183 Madison Avenue
New York, New York 10016

Monsanto Textiles Company
1114 Avenue of the Americas
New York, New York 10036

Wool Bureau, Inc.
360 Lexington Avenue
New York, New York 10017

Bibliography

Books

Advances in False-twist Texturing Processes. Manchester, England: Shirley Institute, 1974.

Alth, Max and Simon Alth. *The Stain Removal Handbook.* New York: Hawthorn Books, Inc., 1977.

American Association of Textile Chemists and Colorists. *AATCC Handbook on Bonded and Laminated Fabrics.* Research Triangle Park, N.C.: American Association of Textile Chemists and Colorists.

American Fabrics Encyclopedia of Textiles. 2nd ed. Englewood Cliffs, N.J.: Prentice-Hall, Inc., 1972.

American Home Economics Association. *Textile Handbook.* 4th ed. Washington, D.C.: American Home Economics Association, 1970.

The Art of Sewing (series). New York: TIME-LIFE Books, 1976.

Basic Tailoring
Boutique Attire
The Classical Technique
Creative Design
The Custom Look
Delicate Wear
The Personal Touch
Separates That Travel
Short Cuts to Elegance
The Sporting Scene
Traditional Favorites

Bath, Virginia Churchill. *Lace.* Indiana: Regnery/Gateway, Inc., 1974.

Beech, W.F. *Fiber-reactive Dyes.* New York: SAF International, 1970.

Bhatnagar, V.M., editor. *Advances in Fire Retardant Textiles.* Westport, Conn.: Technomic Publishing Co., Inc., 1975.

Birrell, V.L. *The Textile Arts: A Handbook of Fabric Structure & Design Processes.* New York: Harper & Row, 1959.

Blackman, A.G. *Manual of Standard Fabric Defects in the Textile Industry.* Graniteville, S.C.: Graniteville Co., 1975.

Boyle, Michael. *Textile Dyes, Finishes, and Auxiliaries.* New York: Garland Publishing, Inc., 1977.

Brockman, Helen. *The Theory of Fashion Design.* New York: John Wiley & Sons Inc., 1967.

Carter, Mary E. *Essential Fiber Chemistry.* New York: Marcel Dekker, Inc., 1971.

Casper, M.S. *Nonwoven Textiles.* Park Ridge, N.J.: Noyes Data Corporation, 1975.

Celanese Fibers Marketing Co. *Fabric Performace Standards for Trademark Licensing.* Celanese Corp., 1972.

Chambers, Bernice G. *Color and Design.* Englewood Cliffs, N.J.: Prentice-Hall, Inc., 1951.

Clark, George L. *The Encyclopedia of Microscopy.* New York: Reinhold Book Co., 1961.

Clarke, W. *An Introduction to Textile Printing.* 4th ed. New York: Halsted Press, 1974.

Clifford, C.R. *Lace Dictionary.* New York: Clifford & Lawton Pub., 1913.

Corbman, Dr. Bernard P. *Textiles: Fiber to Fabric.* New York: McGraw-Hill Inc., 1975.

Dunbar, John Telfer. *The Official Tartan Map.* London, England: Elm Tree Books, Hamish Hamilton Ltd., 1976.

Emery, Irene. *The Primary Structures of Fabrics.* Washington, D.C.: Textile Museum, 1966.

Feldman, Annette. *Handmade Lace & Pattern.* New York: Harper & Row, 1975.

Dictionary of Textile Terms, A. 12th ed. New York: Dan River Inc., 1976.

Fuhrmann, Brigita. *Bobbin Lace.* New York: Watson-Guptill Publications, 1976.

Goswami, B.B. *Textile Yarns.* New York: John Wiley & Sons Inc., 1977.

Greenwood, K. *Weaving: Control of Fabric Structure.* Watford, England: Merrow, 1975.

Hathorne, B.L. *Woven Stretch and Textured Fabrics.* New York: John Wiley & Sons Inc., 1975.

Hall, A.J. *The Standard Handbook of Textiles.* 8th ed. New York: John Wiley & Sons Inc., 1975.

———. *Textile Finishing.* 3rd ed. New York: American Elsevier Press Inc., 1966.

Hearle, J.W.S. *Structural Mechanics of Fibers, Yarns and Fabrics.* New York: John Wiley & Sons Inc., 1969.

Henshaw, D.E. *Self-twist Yarn.* Plainfield, N.J.: Textile Book Service, 1971.

Hilado, Carlos J. *Handbook of Flammability Regulations.* Westport, Conn.: Technomic Publishing Co., Inc., 1975.

———. *Flammability of Fabrics.* Westport, Conn.: Technomic Publishing Co., 1974.

Holker, J.R. *Bonded Fabrics.* Watford, England: Merrow, 1975.

Hollen, Norma and Jane Saddler. *Textiles.* 4th ed. New York: Macmillan Inc., 1973.

Huelson, T.L. *Lace & Bobbin: A History and Collectors Guide.* Cranbury, N.J.: A.S. Barnes & Co. Inc., 1973.

Identification of Textile Materials. 7th ed. Manchester, England: The Textile Institute, 1975.

International Nonwovens & Disposables Association. *Guide to Nonwoven Fabrics.* New York: International Nonwovens & Disposables Association, 1978.

Johanson, Sally. *Traditional Lace Making*. New York: Van Nostrand Reinhold Company, 1974.

Johnson, Thomas. *Tricot Fashion Design*. New York: McGraw-Hill Inc., 1946.

Jones, Mary Erwen. *Romance of Lace*. London, England: Spring Books, n.d.

Joseph, Marjory L. *Essentials of Textiles*. New York: Holt, Rinehart & Winston, 1976.

———. *Introductory Textile Science*. New York: Holt, Rinehart & Winston, 1977.

Jourdain, M. *Old Lace: Handbook for Collectors*. B.T. Batsford, 1908.

Kleeberg, Irene Cummings, editor. *The Butterick Fabric Handbook*. New York: Butterick Publishing Co., 1975.

Koch, P.A. *Microscopic and Chemical Testing of Textiles*. Plainfield, N.J.: Textile Book Service, 1963.

Kushel, Lillian. *Fashion Textiles and Laboratory Workbook*. 2nd ed. New York: Taylor Career Programs, 1971.

Larsen, Jack Lenor and Jeanne Weeks. *Fabrics for Interiors*. Litton Education Publishing, Inc., 1975.

Lewis, Virginia Stolpe. *Comparative Clothing Construction Techniques*. Minneapolis, Minn.: Burgess Publishing Co., 1976.

Ley, Sandra. *America's Sewing Book*. New York: Charles Scribner's Sons, 1972.

Linton, George E. *Applied Basic Textiles*. 2nd ed. (revised). Plainfield, N.J.: Textile Book Service, 1966.

———. *The Modern Textile and Apparel Dictionary*. 4th ed. Plainfield, N.J.: Textile Book Service, 1973.

———. *Natural and Man-made Textile Fibers*. 2nd ed. (revised). Plainfield, N.J.: Textile Book Service, 1973.

Lyle, Dorothy S. *Focus on Fabrics*. Silver Spring, Md.: International Fabricare Institute, 1964.

———. *Modern Textiles*. New York: John Wiley & Sons Inc., 1976.

———. *Performance of Textiles*. New York: John Wiley & Sons Inc., 1977.

Lynch, Mary. *Sewing Made Easy*. Garden City, N.Y.: Garden City Books, 1952.

Magic Seamstress Co. *Magic Seamstress*. American Fashion Institute, 1970.

Manchester Silk. H.T. Gaddum & Co. Limited, 1961.

Manly, Robert H. *Durable Press Treatments of Fabrics*. Park Ridge, N.J.: Noyes Data Corporation, 1976.

Margolis, Adele P. *Fashion Sewing for Everyone*. Garden City, N.Y.: Doubleday & Co., Inc., 1974.

———. *How to Make Clothes that Fit and Flatter*. Garden City, N.Y.: Doubleday & Co., Inc., 1969.

Marsh, J.T. *An Introduction to Textile Finishing*. 2nd ed. Plainfield, N.J.: Textile Book Service, 1966.

Moncrieff, R.W. *Man-made Fibers*. 6th ed. New York: John Wiley & Sons Inc., 1975.

Moon, Alma Chestnut. *How to Clean Everything*. New York: Simon & Schuster, 1968.

Moore, H. *Lace Book*. New York: Frederick A. Stokes Co., 1904.

Morton, W.E. and J.W.S. Hearle. *Physical Properties of Textile Fibers*. 2nd ed. New York: John Wiley & Sons Inc., 1975.

Moyer, Earl D. *Principles of Double Knitting*. Brooklyn, N.Y.: Montrose Supply & Equipment Co., 1972.

Musheno, Elizabeth, editor. *The Vogue Sewing Book*. New York: Vogue Pattern Co., 1975.

Paling, D.F. *Warp Knitting Technology*. Manchester, England: Columbine Press, 1972.

Pankowski, Edith and Dallas Pankowski. *Basic Textiles*. New York: Macmillan Inc., 1972.

Perry, Patricia, editor, *Butterick Sewing Book: Ready Set Sew*. New York: Butterick Publishing Co., 1971.

Pfannschmidt, Ernst Erich. *Twentieth Century Lace*. New York: Charles Scribner's Sons, 1975.

Picken, Mary B., editor. *The Fashion Dictionary: Fabric, Sewing and Apparel as Expressed in the Language of Fashion*. New York: Funk and Wagnalls, Inc., 1972.

Piller, B. *Bulked Yarns*. Plainfield, N.J.: Textile Book Service, 1973.

Pizzuto, Joseph H. *Fabric Science*. 4th ed. revised by Arthur Price, Allen C. Cohen. New York: Fairchild Publications, 1980.

Pond, Gabrielle. *An Introduction to Lace*. New York: Charles Scribner's Sons, 1973.

Powys, Marion. *Lace and Lace Making*. Boston, Mass.: Charles T. Branford Co., 1959.

Press, Jack J. *Man-made Textile Encyclopedia*. New York: John Wiley & Sons Inc., 1959.

Reichman, Charles. *Advanced Knitting Principles*. New York: National Knitted Outerwear Association, 1964.

———. *Double Knit Fabric Manual*. New York: National Knitted Outerwear Association, 1961.

———. *Electronics in Knitting*. New York: National Knitted Outerwear Association, 1972.

———. *Guide to Manufacturer of Sweaters, Knit Shirts and Swimwear*. New York: National Knitted Outerwear Association, 1963.

———. *Knitted Fabric Primer*. New York: National Knitted Outerwear Association, 1967.

———. *Knitted Fabric Technology*. New York: National Knitted Outerwear Association, 1974.

———. *Knitting Encyclopedia.* New York: National Knitted Outerwear Association, 1972.

———. *Principles of Knitting Outerwear Fabrics and Garments.* New York: National Knitted Outerwear Association, 1961.

Reisfeld, Aaron. *Control of Defects in Raschel Fabrics.* New York: National Knitted Outerwear Association, 1955.

———. *Fundamentals of Raschel Knitting.* New York: National Knitted Outerwear Association, 1958.

———. *Warp Knit Engineering.* New York: National Knitted Outerwear Association, 1966.

Rotenstein, Charles. *Lace Manufacture on Raschel Machines.* New York: National Knitted Outerwear Association, 1958.

Schneider, Coleman. *Embroidery: Schiffli and Multihead.* New Jersey: C. Schneider, 1978.

———. *Machine-made Embroidery.* New Jersey: Schneider International Corporation, 1968.

Schwebke, Phyllis W. and Margaret B. Krohn. *How to Sew Leather, Suede, Fur.* The Bruce Publishing Co., 1970.

Scottish Clans & Their Tartans, The. 37th ed. Edinburgh: W. & A.K. Johnston & G.W. Bacon Limited, 1954.

Scottish Tartans, The. Edinburgh: W. & A.K. Johnston & G.W. Bacon Limited, n.d.

Seydel, Paul V. and James R. Hunt. *Textile Warp Sizing.* Atlanta, Ga.; Long & Clopton, 1972.

Smirfitt, J.A. *An Introduction to Weft Knitting.* Watford, England: Merrow, 1975.

Storey, Joyce. *Van Nostrand Reinhold Manual of Textile Printing.* New York: Van Nostrand Reinhold Company, 1974.

Stout, Evelyn E. *Introduction to Textiles.* 3rd ed. New York: John Wiley & Sons Inc., 1976.

Textile Institute. *Identification of Textile Materials.* Manchester, England: Textile Institute, 1975.

———. *Textile Terms and Definitions.* Manchester, England: Textile Institute, 1976.

Textile Research Institute. *Symposium on Transfer Printing.* Princeton, N.J.: Textile Research Institute, 1976.

Thomas, D.G.B. *An Introduction to Warp Knitting.* Watford, England: Merrow, 1971.

Von Henneberg, F.A. *Art & Craft of Old Lace.* 4 Weyhe, 794 Lexington Avenue, New York

Waidle, Patricia. *Victorian Lace.* London, England: Herbert Jenkins, 1968.

Warch, Constance. *Illustrated Guide to Sewing.* Plycon Press, 1975.

Weber, Klaus Peter. *An Introduction to the Stitch Formations in Warp Knitting.* West Germany: Employees Association Karl Mayer E.V., 6053 Obertshausen.

Williams, Alex. *Flame Resistant Fabrics.* Park Ridge, New Jersey: Noyes Data Corporation, 1974.

Wingate, Dr. Isabel B. *Fairchild's Dictionary of Textiles.* 6th ed. New York: Fairchild Publications, 1979.

———. *Textile Fabrics and Their Selection.* 7th ed. Englewood Cliffs, N.J.: Prentice-Hall, Inc., 1976.

Wingo, Caroline E. *The Clothes You Buy and Make.* New York: McGraw-Hill Inc., 1953.

Zielinski, S.A. *Encyclopedia of Hand Weaving.* New York: Funk & Wagnalls, Inc., 1959.

Znamerouwski, Nell. *Step by Step Weaving.* New York: Golden Press, 1967.

Pamphlets & Bulletins

Allied Chemical Corporation. *Caprolon Nylon.* A biography of Allied Chemical's nylon fiber, 1966.

———. *The Performance Fibers.*

American Cynamid Corp. *A Report on the Rocketing Sweater-Shirt Market.*

American Enka Co. *Encron Polyester Yarn.* General information and properties. Bulletin #PFP 1B. March 1971.

———. *Rayon Yarn: Enka Rayon Filament Yarn.* General information and properties. 1972.

———. *Enka Rayon Staple.* General information and properties. Bulletin #RSP 2A. March 1976.

American Fabrics. *American Fabrics.* 1954.

———. *Rayon & Acetate.* 1969.

American Sheep Producers Council. *Glossary of Wool Fabric Terms.* Educational Pamphlet 6.

American Textile Manufacturing Institute. *All about Textiles.* January 1978.

———. *How Textiles Are Made.*

———. *101 Textile Terms.*

———. *Save Energy with Textiles.*

Avtex Fibers, Inc. *Vinyon HH Staple.* Technical Service Bulletin. S28R.

———. *Vinyon Staple.* Technical Service Bulletin S28R-2.

———. *Crimped Fiber & Rayon.* Technical Service Bulletin S53.

Belgian Linen Association. *Belgian Linen.*

Burlington Industries, Inc. *Textile Fibers & Their Properties.* 1965.

Burlington Menswear. *Pyramid Cloth.*

Coats & Clark, Inc. *Lingerie: Sewing on Tricot.* 1970. (*Stitch in Time,* Vol. 40, No. 1-T)

Cooper Union Museum, New York. *Textile Arts.* January 1964. (The Greenleaf Collection)

Courtaulds North America, Inc. *From Fiber to Fashion.*

Crestlan Cyanamid Corp. *Crestlan 67-A & 67-H.*

———. *Crestlan Type 61-L.*

———. *Important Difference between Sweatshirts.*

———. *Laminated Sweat Jackets.*

Crompton-Richmond Co. *A Pile Fabric Primer.*

E.I. du Pont de Nemours & Company. *Dupont Textile Products.*

———. *Facts about Fabric.*

———. *If You're Thinking Sweaters, Craft Yarns or Other Knits.*

———. *Information about Nandel (Acrylic Yarn).*

———. Technical Information Bulletins: Fibers, Processing, Dyeing, Finishing.

———. *Qiana® Nylon.*

Eastman Kodak Company. *Estron® Acetate Yarn: Specifications & Properties.* Textile Fibers Division. Publication #A-201B. June 1974.

———. *Kodel® Polyester Filament Used as Filling in Spun Warp.* Textile Fibers Division. Publication #K-199A. January 1975.

———. *Processing Kodel® Polyester Staple on the Cotton System.* Textile Fibers Division. Publication #K-103B-1. April 1976.

———. *Processing Kodel® Polyester Staple on the Cotton System.* Textile Fibers Division. Publication #K-103C-1x. December 1976.

———. *Properties of Verel Modacrylic Fibers.* Textile Fibers Division. Publication #V-355B. October 1975.

———. *Textured Kodel® Polyester Yarns in Woven Fabrics.* Textile Fibers Division. Publication #K-212. October 1974.

FMC Corp. *Avril® Rayon: The Dependable Fiber.* 1968.

Hamilton Adams Imports Ltd. *Moygashel® Linens.*

Herculon Inc. *Olefin Fiber: A New Look at Herculon.*

Hoechst Fibers Inc. *Dyeing and Finishing of Hoechst Polyester.*

———. *How to Sew and Knit.*

International Fabric Institute. *International Fair Claims Guide for Consumer Textile Products.* 1973.

Man-Made Fiber Producers Association, Inc. *Man-made Fibers Fact Book.* 1978.

———. *Facts on Man-made Fibers.* No. 3. Spring 1976.

———. *Facts on Man-made Fibers.* No. 4. Summer 1977.

———. *Guide to Man-made Fibers.* 1977.

J.B. Martin & Co., New York. *How to Sew and Care for Velvet.*

Mayer, Christa C. *Three Centuries of Bobbin Lace.* New York: Cooper Union Museum, August 1966.

———. *Two Centuries of Needle Lace.* New York: Cooper Union Museum, February 1965.

Mohair Council, New York. *Mohair: Production and Marketing in the United States.*

Monsanto Company. *Draw-textured Yarn Technology.*

———. *Fashion Hang Up.*

———. *Tech Talk.* August 1975.

———. *Tech Talk.* 1979.

———. *Technical Publications on Yarns and Fibers.*

———. *Yarns & Fibers Catalog.* 1977.

Dan River Inc. *Dan River: A Brief History.*

———. *Dan River Textile Process.*

The Singer Company. *How to Sew Fashion Knits.* 1972.

J.P. Stevens & Co. Inc. *How to Sew with Elastic.* 1976.

Wool Bureau Inc. *Knitwear.* (The Wool Library, Vol. 4.)

———. *Men's Wear.* (The Wool Library, Vol. 3.)

———. *Uniform & Career Apparel.* (The Wool Library, Vol. 5.)

———. *Women's Fashion.* (The Wool Library, Vol. 2.)

———. *Wool Fiber to Fabric.* (The Wool Library, Vol. 1.)

Textile Periodicals

American Drycleaner
American Trade Magazines, Inc.
500 North Dearborn Street
Chicago, Illinois 60601
Monthly. Covers all areas of interest to cleaners, including management and sales.

American Dyestuff Reporter
SAF International, Inc.
44 East 23 Street
New York, New York 10010
Monthly. Covers wet-processing operations in textile mills, such as dyeing, printing and other chemistry-related areas.

American Fabrics Magazine
Doric Publishing Company, Inc.
24 East 38 Street
New York, New York 10016
Quarterly. Reviews trends and new developments in fabrics. Special articles on special topics and problems. Includes numerous fabric swatches of the latest fabric designs.

American Textiles—Knitter/Apparel Edition
Clark Publishing Company
106 East Stone Avenue
P. O. Box 88
Greenville, South Carolina 29602
Monthly. For manufacturers of hosiery and knitted wear. Emphasis on new developments and financial conditions.

America's Textiles Reporter/Bulletin
Clark Publishing Company
106 East Stone Avenue
P. O. Box 88
Greenville, South Carolina 29602
Monthly. For managers in the textile industry. Includes textile business/financial and manufacturing sections.

Clemson University Review of Industrial Management and Textile Science
Clemson University
College of Industrial Management and Textile Science
Clemson, South Carolina 29631
Biannually. Lectures, papers, seminars and addresses in the fields of management and textile science.

Daily News Record
Fairchild Publications, Inc.
7 East 12 Street
New York, New York 10003
Daily. Contains news about the textile industry, including much current information for textile manufacturers.

Industrial Fabric Products Review
Canvas Products Association International
600 Endicott Building
St. Paul, Minnesota 55101
Thirteen times a year. Published for the industrial canvas fabric industry.

Journal of Home Economics
Official Publication of the American Home Economics Association
2010 Massachusetts Avenue, N.W.
Washington, D.C. 20036
Monthly. Includes articles in the field of textiles and clothing which are primarily consumer oriented.

Knitting Times
Official Publication of the National Knitted Outerwear Association
51 Madison Avenue
New York, New York 10010
Weekly. Covers business conditions, technical developments, trends and forecasts in knitted fabrics and knitted apparel.

Modern Knitting Management
Rayon Publishing Corporation
303 Fifth Avenue
New York, New York 10016
Monthly. Articles deal primarily with management aspects of the knitting industry.

Modern Textiles
Rayon Publishing Corporation
303 Fifth Avenue
New York, New York 10016
Monthly. Covers marketing and technical activities in the man-made textile industry. Includes a special section on carpets.

Nonwovens Industry
Rodman Publications Inc.
P. O. Box 555
26 Lake Street
Ramsey, New Jersey 07446
Monthly. Features articles on manufacturing processes, distribution and use applications of nonwoven textile products.

Textile Chemist and Colorist
Journal of the American Association of Textile Chemists and Colorists P. O. Box 12215 Research Triangle Park, North Carolina 27709
Monthly. Deals primarily with the chemistry of textiles and color. Very technically oriented.

Textile Industries
W.R.C. Smith Publishing Company
1760 Peachtree Road, N.W.
Atlanta, Georgia 30309
Monthly. Aimed at management in the textile mill.

Textile Marketing Letter
Clemson University
Clemson, South Carolina 29631
Monthly. Covers items of current interest to persons involved in sales and marketing of textiles.

Textile Organon
Textile Economics Bureau, Inc.
489 Fifth Avenue
New York, New York 10017
Monthly. Market data for both the natural and man-made fiber industry.

Textile Technology Digest
Institute of Textile Technology
Charlottesville, Virginia 22902
Monthly. Provides abstract coverage of current periodicals, books and patents in all areas of the textile industry. World-wide in scope, with abstracts of foreign language material translated into English.

Textile World
McGraw-Hill Publications
1175 Peachtree Street, N.E.
Atlanta, Georgia 30309
Monthly. Covers technical developments in the textile industry.

Textracts
J. B. Goldberg
225 East 46 Street
New York, New York 10017
Monthly. Provides brief abstracts (citing original sources) of articles dealing with the technical and business news of fibers, yarns, fabrics, dyes and finishes. Domestic and foreign coverage.

Textile Trade & Professional Associations

American Apparel Manufacturers Association
1611 North Kent Street
Arlington, Virginia 22209

American Association for Textile Technology
1040 Avenue of the Americas
New York, New York 10036

American Association of Textile Chemists and Colorists
Box 12215
Research Triangle Park, North Carolina 27709

American National Standards Institute
1430 Broadway
New York, New York 10018

American Sheep Producers Council
200 Clayton Street
Denver, Colorado 80206

American Society for Testing and Materials
1916 Race Street
Philadelphia, Pennsylvania 19103

American Textile Machinery Association
1730 M Street, N.W.
Washington, D.C. 20036

American Textile Manufacturers Institute, Inc.
1101 Connecticut Avenue, N.W.
Washington, D.C. 20036

American Yarn Spinners Association, Inc.
601 West Franklin Avenue
Gastonia, North Carolina 28052

Belgian Linen Association
280 Madison Avenue
New York, New York 10016

Carpet & Rug Institute
310 South Holliday Avenue
Dalton, Georgia 30720

Color Association of the United States, Inc.
200 Madison Avenue
New York, New York 10016

Cotton Inc.
1370 Avenue of the Americas
New York, New York 10019

International Fabricare Institute
P. O. Box 940, Joliet, Illinois 60434
12251 Tech Road, Silver Spring, Maryland 20904

International Nonwovens & Disposables Association
10 East 40 Street
New York, New York 10017

International Silk Association
c/o Rudolph Desco & Co.
580 Sylvan Avenue
Englewood Cliffs, New Jersey 07632

Laundry & Cleaners Allied Trades Association
543 Valley Road
Upper Montclair, New Jersey 07043

Man-Made Fiber Producers Association, Inc.
1150 Seventeenth Street, N.W.
Washington, D.C. 20036

Mohair Council of America
183 Madison Avenue
New York, New York 10016

National Knitted Outerwear Association
51 Madison Avenue
New York, New York 10010

National Knitwear Manufacturers Association
350 Fifth Avenue
New York, New York 10001

National Retail Merchants Association
100 West 31 Street
New York, New York 10001

Neighborhood Cleaners Association
116 East 27 Street
New York, New York 10016

Textile Distributors Association
1040 Avenue of the Americas
New York, New York 10018

United States Department of Agriculture
Southern Regional Research Center
1100 Robert E. Lee Boulevard
New Orleans, Louisiana 70179

Wool Bureau, Inc.
360 Lexingon Avenue
New York, New York 10017

Fabric Index